MATHEMATICAL ASPECTS OF WEYL QUANTIZATION AND PHASE

MATHEMATICAL ASPECTS OF WEYL QUANTIZATION AND PHASE

D.A.Dubin

The Open University, UK

M.A.Hennings

Sidney Sussex College,
University of Cambridge, UK

T.B.Smith

The Open University, UK

World Scientific

Singapore • New Jersey • London • Hong Kong

Published by

World Scientific Publishing Co. Pte. Ltd.

P O Box 128, Farrer Road, Singapore 912805

USA office: Suite 1B, 1060 Main Street, River Edge, NJ 07661

UK office: 57 Shelton Street, Covent Garden, London WC2H 9HE

British Library Cataloguing-in-Publication Data
A catalogue record for this book is available from the British Library.

MATHEMATICAL ASPECTS OF WEYL QUANTIZATION AND PHASE

ISBN 981-02-3919-X

Printed in Singapore.

for
Diana, Lynne and Susie

CONTENTS

Part I

Fundamentals

Chapter 1

Background

THEN THE LORD ANSWERED JOB OUT OF THE WHIRLWIND AND SAID: *Who is this that darkeneth counsel by words without knowledge? Gird up now thy loins like a man: for I will demand of thee and answer thou me: where wast thou when I laid the foundations of the earth? Declare if thou hast understanding. Who hath laid the measure thereof, if thou knowest? Or who hath stretched the line upon it? Whereupon are the foundations thereof fastened? Or who laid the cornerstone thereof: when the morning stars sang together; and all the sons of GOD shouted for joy?*

– Job, *38, vv 1 - 7.*

One of the most important technological developments of this century has undoubtedly been the laser. It was the developments in microwave generation during World War II that paved the way for the creation first of masers and then of lasers. If we date the start of the age of lasers at 1960, to use a round number, over the last forty years we can chart the development of the laser from being an arcane curiosity to becoming a laboratory tool and now a ubiquitous component of our electronic age. To the physicist, the laser was first interesting because it displayed such a novel state of the electromagnetic field. It then graduated to becoming a laboratory tool with a wide variety of uses, including the creation of even more exotic states of radiation [42, 116, 101, 148, 153, 155, 164, 178].

What is special about laser radiation is not only its potential to produce a high energy density, but also its *coherence in phase*, which is achieved because a laser creates a *macroscopic quantum state*. Such states are exceedingly interesting, and characterize superconductivity, superfluidity and Bose-Einstein condensation. A theory for any such phenomenon requires a model of how ordinary quantum interactions combine to bring about the required collective effect. For laser radiation the question is how a pumped source of ordinary incoherent radiation is transformed into an output of coherent radiation. General principles tell us that a necessary (but not sufficient) condition for such a transformation is that the radiation be a subsystem driven by a "reservoir" with infinitely many degrees of freedom. Another important point is the fact that the continued input of energy, which drives and sustains the state, implies that such a state is one far from thermal equilibrium.

The question then arises as to which quantum mechanical observables describe this coherence property, as opposed to other important properties of the state (such as intensity). Our reading of the literature has led us to the conclusion that most physicists approach this problem by first considering the polar decomposition of the lowering operator of a Bose oscillator,

$$A_r = E_r N_r^{1/2}, \tag{1.1}$$

as was proposed by Dirac in 1927 [47]. Here the label r is a mode number, corresponding to the Fourier decomposition of the vector potential. By analogy with classical electromagnetism, E_r might be regarded as the operator phase factor for, were we to assume it unitary, we could write it in the form

$$E_r \;=\; e^{-i\theta_r}, \tag{1.2}$$

defining thereby a self-adjoint phase operator θ_r for each mode. The canonical commutation relations (see Chapter 5) would then tell us that

$$N_r\theta_s - \theta_s N_r \;=\; -i\,\delta_{rs}, \tag{1.3}$$

so that θ_r would be an operator canonically conjugate to N_r.

Restricting ourselves to one mode (which is no real loss of generality since the modes are independent) θ would then be the quantum phase operator, and would satisfy a number-phase uncertainty relation

$$\Delta N\,\Delta\theta \;\geqslant\; 1 \tag{1.4}$$

in all pertinent states. Heitler [105] shows how this relation can be used to find bounds on the accuracy with which a light beam can be used to determine the position of a particle. Unfortunately, none of the above argument is valid, since the operator E is not unitary and so the canonical self-adjoint operator θ simply does not exist. Indeed, it can be shown that no sensible self-adjoint operator exists which is canonically conjugate to N.

The proof of this fact is an operator-theoretic result [154]; the physics is in the application. The only input is that the number operator has a basis of eigenvectors corresponding to the nondegenerate spectrum $\{0, 1, 2, \dots\}$. That being so, the Theorem and proof can be given an abstract form.

Proposition 1.1 *(The No-Go Theorem) Let $\mathcal{H}$ be a separable Hilbert space and $(e_n)_{n \geqslant 0}$ an orthonormal basis for $\mathcal{H}$. Let Q_n be the orthogonal projection operator onto the subspace spanned by e_n and let N be the unbounded self-adjoint operator defined by*

$$N = \sum_{n \geqslant 0} n Q_n. \tag{1.5}$$

Let $\mathcal{D}$ be the linear span of the e_n, which is a dense core of self-adjointness for N. Then there exists no symmetric operator Θ, whose domain includes $\mathcal{D}$, which maps $\mathcal{D}$ into the domain of N, and which satisfies the commutation relation

$$N\Theta f - \Theta N f = -if \tag{1.6}$$

for all $f \in \mathcal{D}$.

Proof: The commutation relation implies that

$$-i = \langle e_0, N\Theta e_0 - \Theta N e_0 \rangle = \langle N e_0, \Theta e_0 \rangle - \langle \Theta e_0, N e_0 \rangle = 0,$$

which is absurd. ■

We shall have more to say about this in future Chapters, but it is already clear that insofar as quantum phase phenomena exist, they are going to have to be described by non-canonical operators.

Historically, Dirac first introduced θ_r, but realized this inconsistency shortly thereafter. Although he published no acknowledgement, he did make the point in his Cambridge lectures on quantum mechanics [48], saying that if one worked with E without assuming it unitary, no error would result, which is true.

It is a pity that this was not more widely known, as other authors have also made this error, including Heitler [105], who subsequently also realized the problem. It was not until the papers of Louisell [154] and of Susskind & Glogower [218] that this became known to the physics community at large.

Roughly speaking, the response to this problem falls into several categories. One response is that we must make do with what might be called "functions of the phase". Susskind and Glogower took E, for instance, and broke it up into its real and imaginary parts, C and S, which are likened to the cosine and sine of what would be a phase operator if only there were one. In any event C and S are perfectly well defined. As E is a shift operator, weighted shift operators were then considered from the same point of view [149].

A second category of response is to introduce an operator with a phase-like connection more or less by fiat. This can be done by selecting a Hilbert space of functions of an angle, and the operator of multiplication by the angle on it. By choosing the Hilbert space to be "naturally equivalent" to the usual space of states, this operator can be transported to act there. By construction, the spectrum of this operator is the interval $[-\pi, \pi]$, say. As far as we know, the first example of such an operator, one that we shall denote by X throughout this book, was first introduced by Garrison & Wong [68]. Mathematically speaking, it is what is known as a Toeplitz operator. Usually considered as part of complex function theory, the theory of Toeplitz operators is a well developed mathematical topic. The operator X has been independently rediscovered a number of times, as evidenced by the multitude of references made to it, a selection of which are [66], [87], [171], [182].

An interesting curiosity is that Galindo (*ibid*) discovered a dense domain in the usual Hilbert space, $L^2(\mathbb{R})$, on which N and X were canonically conjugate. Which should be impossible, as E is not unitary — what is going on? Since this domain does not contain any of the eigenvectors of N, the usual Hermite-Gauss states h_n (the nth harmonic oscillator eigenvector), we are able to bypass the No-Go Theorem. However, the domain is unstable under the action of almost any operator of interest in quantum mechanics. For example, it turns out that $f = h_0 + h_1$ is in this domain, but $Af = h_0$ is not, where A is the lowering operator, so that while

$$[X, N]_- f = if,$$

we see that

$$[X, N]_{-} Af \neq iAf ;$$

indeed this last statement is meaningless, since Af does not belong to the domain being considered. So it seems to us that this domain is not useful for the usual formulation of quantum mechanics, and in this respect the No-Go Theorem has not been invalidated. But Galindo's work is interesting in its own right, and makes us think hard about how we phrase the No-Go Theorem.

Quantum mechanics tells us that any self-adjoint operator is observable. It can be questioned as to whether or not this is precisely true, something we shall discuss when setting up the quantum formalism, but something like it certainly is. This means that C, S, X, and the other "phase operators" that have been variously proposed, will be observable. Hence, subject to the technical problems of measuring observables with a continuous component in their spectrum, an experimental arrangement can in principle be devised which will measure each such operator and prepare its eigenstates. There is nothing wrong with these operators from the point of view of general quantum theory, therefore. The relevant question is whether or not they describe a phase property for the phenomena measured in some particular setup. To answer this, the theoretical description of the arrangement has to be written down and analyzed, as we do for a laser model in Chapter 11.

A third category of response does not introduce a single operator, although in the literature this approach is described as presenting "a hermitian phase operator"[14]. Rather this theory is based on a sequence of finite rank operators X_s associated with the $(s + 1)$-dimensional subspace of $L^2(\mathbb{R})$ which is the linear span of the first $s + 1$ number operator eigenvectors (the Hermite-Gauss functions). In this method, physical quantities such as expectations or variances are to be calculated for the operator X_s, and then the limit of the results of these calculations are taken as $s \to \infty$, and these limits are then interpreted as being the corresponding expectations and variances for the putative hermitian phase operator. It turns out that the sequence (X_s) converges weakly to the Toeplitz operator X, but not strongly, in that (for $k \geqslant 2$ the sequence of powers (X_s^k) converges weakly to an operator which is not the corresponding power X^k of X. Thus the limit as $s \to \infty$ of the variance of X_s in some state is not equal to the variance of X in that state; nor is it equal to the variance of any other self-adjoint operator in that state. Consequently, it is rather difficult to give a useful physical interpretation to the outcomes of this type of calculation.

However, this theory has many adherents, who point out that certain physically measurable quantities calculated with X_s for finite s have the proper form when the limit $s \to \infty$ is taken—after the calculations. (There is nothing mysterious about having to take the limit after the calculations; it would be simpler to say that the limits are being taken with respect to the weak operator topology.) This statement is hard to justify or refute with certainty, for on general principles it will be true for certain experimental arrangements, but not for others. The difficulty is that this theory is accompanied by assertions about what the state of the electromagnetic field is, often based on semi-classical or even classical considerations. As we shall argue in Chapter 11, the connection between the behaviour of operators on $L^2(\mathbb{R})$ and the properties of coherent laser light are problematical. A complete such connection requires a solution of the problem of the quantized electromagnetic field in interaction with cavity atoms, and no such solution exists. Failing that, aspects of this connection require a model which approximates the full electrodynamical problem adequately. If this requirement includes treating the field as a system with infinitely many degrees of freedom for which we can obtain a nonperturbative solution, again no such solution exists. If it is acceptable to truncate the field to a finite number of degrees of freedom and to describe the cavity atoms as finite level systems, a solution is possible, and this is the basis of the model treated in Chapter 11. However, when the operators (X_s) are considered within the framework of this model, they do not describe the phase of the coherent radiation, with or without the limit $s \to \infty$.

In the words of Sherlock Holmes, *"when you have eliminated the impossible whatever remains, however improbable, must be the truth"*. We do not claim to have eliminated all impossibilities — in the field of mathematical and physical research, that is not possible — but the comments made above concerning the various responses to this problem (and dealt with at greater length later on in this book) indicate to us that there are sufficient concerns about the validity of each that another approach should at least be attempted. When all else fails, one must return to first principles. Of course the problem is to identify the principle! We feel that *Weyl quantization* is such a principle, and in a series of papers, we used it to quantize the angle function in phase space. The operator that comes out of this approach which we denote by $\mathbf{\Delta}[\varphi]$ and call the phase operator (granting ourselves some "authors' license") will be shown to be bounded and self-adjoint, and to be the unique operator on $L^2(\mathbb{R})$ which describes the phase of the coherent radiation in the laser model of Chapter 11. Moreover, it is the only one of the phase operator proposals whose commutator with the number operator exactly mirrors the Poisson bracket of their corresponding phase space symbols. (The symbol of an operator will be taken to mean the phase space function whose quantization reproduces the original operator.) This last comment deserves some explanation, since it would seem to indicate that we are claiming that the number operator N and our phase operator $\mathbf{\Delta}[\varphi]$ are canonically conjugate, and since it is widely believed that the Poisson bracket of half the square $\frac{1}{2}r^2$ of the radius and the angle function φ in the plane are canonically conjugate as classical observables. However, this is not the case — the fact that the function φ is discontinuous on the plane is sufficient to ensure that the Poisson bracket of $\frac{1}{2}r^2$ and φ is equal to the sum of 1 and an additional distributional factor.

It is curious that quantum mechanics courses usually do not include the work of Weyl on quantization, which, in its basic form, is certainly quite simple. Similarly, you would be hard pressed to find any account of Weyl quantization in an ordinary book on quantum mechanics but see[1] [19], [63], [146]. So we have taken this opportunity to provide an account of Weyl quantization in the phase plane both using plane coordinates (p, q), and polar coordinates (r, β). The latter formalism provides us with the tools to define $\mathbf{\Delta}[\varphi]$ and to quantize general functions of the angle or the radius in phase space.

A unexpected by-product of considering the laser model turns out to be a new approximation method for dequantization, which is discussed in Chapter 12. By the *dequantization* of a given operator B we mean that phase space function T whose quantization is B. As mentioned above, T is known as the *symbol* of B to experts in pseudodifferential operator theory. It has other names as well, but we are going to stick with these two.

Independently of our work, Royer proposed consideration of $\mathbf{\Delta}[\varphi]$, and also quantizations of the angle function other than that of Weyl. His original proposal was reported at the Wigner conference held in Oxford in 1993.[2] While Royer's work has followed paths different from ours, it spurred us to look at other quantization schemes, and we have developed that topic from

[1]On second thought, these are not ordinary books on quantum mechanics.

[2]It seems less and less likely that the proceedings of that conference will be published.

our point of view, as discussed in Chapter 14.

This does not exhaust the topic of quantum phase. For one thing, there are proposals to consider other sorts of operators — relative phase operators and operators associated with another degree of freedom such as charge, for example.

Before leaving this introduction and getting down to business, we feel that we should say a word about the vexed question of mathematics and rigour. Certain operations that are needed in quantum phase theory need to be performed with a degree of care. For example, asymptotic expansions to which a limiting process (such as integration or infinite summation) will be applied *term-by-term to the expansion*, require the expansion to be *uniform* with respect to the expansion parameter (or some similar condition) — if this is not done, the results are not necessarily true. Another example where care is needed comes from the fact that not all operators on Hilbert space have traces. There is a particular relation, which involves taking traces, which is often used to determine the symbol of an operator. To apply this "familiar formula", as we shall describe it, to operators which are not trace class is clearly questionable, and so any use of this "familiar formula" requires care. There are a number of other topics where rigour is important, or at least where we feel it to be so, and we shall discuss these as they occur. We have also endeavoured to explain in each case where things can go wrong if this care is not expended, and more interestingly, why the formal calculational method can sometimes give the correct answer when it might not be expected to. (The familiar formula is a case in point.) While physical motivation is often a good insurance against incorrect results, we do not believe that nature has arranged things so that a physicist does not need a mathematical insurance policy against the day when things do go wrong.

But rigour is only one part of the mathematical art. Another part is *structure*, which involves context. It is the difference between determining all the properties of a given function, such as the Gaussian, and determining the properties common to all functions of a prescribed class (such as infinitely differentiable functions). In using the results in these two approaches, in the former detailed results will be obtained, but holding only for the Gaussian. In the latter approach, fewer detailed results can be obtained, but they are immediately applicable to any function of the class. But more: only in this way can it be asserted with confidence that something is always true, or never true, or true under such and such conditions.

An example of this is a result from Weyl quantization : any tempered distribution may be quantized, and the result is necessarily a continuous linear map from Schwartz space to its dual.

We often emphasize structure, for we believe that it offers a way to invest the results of calculations with significant meaning. In particular, structure is a necessary ingredient in the examination of the relation between the mathematical symbols and physical qualities and quantities. Faced with the question of whether or not a particular operator is measurable, one checks first whether or not it is self-adjoint, and similarly we can find out if an operator preserves probabilities by seeing whether or not it is unitary. In this way we can take advantage of various structural theorems (in the above two cases, concerning all self-adjoint operators or all unitary operators). We go so far as to say that structure is a necessary component of scientific progress.

Acknowledgements

We would like to thank John Bolton, David Clover, Rob Griffiths, Jon Hall, Chris Rowley, Chris Wigglesworth and Mark Woodford about various matters concerning LaTeX 2_ε and various matters pertaining to computers. Geoffrey Sewell was generous with his time and advice about statistical mechanics in general and his work on the laser model in particular. But most of all we are grateful to Diana, Lynne and Susie, who have put up with us for a period even longer than it took to write this book — no mean feat.

The quotation that begins this Chapter comes from the Book of Job. It exhorts us to be modest in the face of the wonders of creation. We are therefore chastened to understand that we are expounding no more than a fraction of what is known about any of the themes in this book: quantum theory, quantization and quantum phase. As most authors feel, we shall be satisfied if the reader has gained something from reading this book, as we have gained a measure of understanding from writing it.

CHAPTER 2

SOME REMARKS ON CLASSICAL MECHANICS

The classical tradition has been to consider the world to be an association of observable objects (particles, fluids, fields, etc.) moving about according to definite laws of force, so that one could form a mental picture in space and time of the whole scheme.

– P. A. M. Dirac, *Quantum Mechanics*

2.1 Introduction

The relation between classical and quantum mechanics is rather subtle, and by no means fully understood. In the most familiar mathematical sense, the relation seems rather obvious: in some cases quantum matrix elements, properly scaled with factors of $\hbar$, converge to their classical counterparts in the formal limit as $\hbar \to 0$.

But this does not really tell us how to construct macroscopic objects. A macroscopic object consists of a large number of microscopic subsystems acting in concert, and the classical behaviour we observe is some sort of average of their quantum interactions. The above limit says nothing about this. The sort of limit that is needed has been rigorously constructed in only a very limited sense. By a rigorous construction we mean setting up a fully quantum system involving N atoms (or whatever constitutes the system) with appropriate interactions, and then demonstrating that the limit $N \to \infty$ exists and exhibits classical behaviour. An example of such a derivation was given some years ago by Hepp [114], who emphasized the limitations of his construction, cf [18].

But why be concerned about this? After all, large N limits are notoriously difficult to do rigorously, and it is intuitively clear that some such construction must be possible if enough hard work were expended to do it. The answer in principle is that the recording devices for quantum events must behave classically in order that (a) we can make sense of the results, and (b) a measurement can be construed as having taken place. This requirement is part of the standard interpretation of quantum mechanics according to Bohr. Unless it is accepted that such devices can be constructed within quantum theory, we are faced with a logical inconsistency.

Perhaps this problem is just a holdover from the early days of quantum mechanics, and advances in technology now allow us to use small quantum systems as measurement devices, so doing away with the need for classical registrations? While experimentalists have certainly made significant progress in developing quantum probes, to consider them as providing a complete measurement would be the result of a faulty appreciation of how we must divide the experimental arrangement into a measuring device and the system being measured. That such a cut must always be made is an observation we owe to Heisenberg ([104], page 58) and Bohr (in *Discussions with Einstein*, [237]), and has been emphasized in recent years by Haag in connection with the problem of harmonizing the principles of relativity and quantum theory [94].

A number of instructive examples of the use of small quantum probes may be found in the book of Braginsky & Khalili [27]. For example, they consider the use of an electron as a probe to measure the charge on an LC circuit viewed as a quantum system. After detailing what happens to the electron as a result of this usage, we are told that "the experimenter sends the electron through an electronic lens and onto a photographic plate situated at the lens's focal plane". Hence, in the final analysis, the measuring apparatus here is the electron, lens and photographic plate together; it is the classical nature of this recording arrangement that enables us to draw conclusions.

There is considerably more to the classical limit than these remarks indicate, but they are enough in our view to justify the time honoured tradition of beginning a study of quantum mechanics with at least a few brief remarks on classical mechanics. In keeping to this tradition, our intention is no more than to emphasize certain similarities and differences between the two theories. In particular, we want to bring out the facts that they are both observable-state systems, and that there are mixed states in both theories. We also take this opportunity briefly to discuss the symplectic geometry underlying classical mechanics, so providing a contrast with the unitary geometry of quantum mechanics later on.

2.2 Axiomatics

Theoretical models provide us with a framework to organize experimental data. The purpose is to organize the widest range of phenomena with the minimum of assumptions, which themselves should as nearly as possible be operationally justified.

The ultimate goal of such a process is to base the system on a few axioms, held to be sacrosanct and "self-evident" verities, as in the *Elements* of Euclid. But even here, in the purest of mathematical schemes, we do not have a perfect representation of physical reality, rather a logico-deductive scheme in which interesting tautologies, known as theorems, follow from the axioms and postulates. Since Euclid, many other geometries have been proposed and studied, each having properties quite different from the others. In the sense of mathematical logic, they are all equally valid: the criterion is consistency. Once one would have added completeness as a criterion, but we now know (Gödel's Theorem) that even set theory itself is incomplete, and so every mathematical theory must be incomplete as well.

But Euclidean geometry continues to hold a great fascination for theoretical physicists as a model of clarity and organization. For Euclid had taken hundreds of apparently disconnected geometrical results and derived them all from a list of ten postulates, certainly an intellectual achievement of the first magnitude. For centuries since, there has been an aura of perfection about this that many have sought to emulate.

Even the great Newton felt this way, and organized his work along the lines of the *Elements*, to the point of obscuring the straightforward algebra with a cloak of geometry. The Greeks could be forgiven this, as they had too cumbrous a number system effectively to do algebra, and it is said by some classical scholars that they were disinclined to remedy this shortcoming on the socio-political grounds that while geometry was intellectually pure and fit for the aristocracy, algebra arose from calculations, which were associated with craft and commerce — the province of plebians. But all Newton succeeded in doing was to make his work relatively inaccessible to students.

So great was Newton's intellectual authority, however, that the axiomatic method has retained its siren call ever since. But there is a critical and essential difference between this method in pure mathematics and physics. For the mathematician, legitimate manipulations from the axioms constitute the necessary and sufficient condition for a statement to stand forever as a theorem; this is mathematical truth (subject to the foundations of set theory remaining fixed). But scientists must then compare the resulting theorems with experimental reality. If the results do not tally, the so-called axioms must be changed until the disparity is overcome. This is the process we may call falsification, after Popper.

Every physicist is aware that there are technical and philosophical difficulties associated with quantum mechanics, but the same is true of classical mechanics, for it fails at small distances and high velocities and energies. This creates an essential difficulty in applying axiomatic systems to physics: all current physical theories are models, and so have a limit to their validity, even if that limit is not precisely clear[1]. In addition, for neither theory is there a choice of observables and states which yields precisely those needed for physics and no others. By choice here we mean a "class" declaration of the form: *The observables are those functions in some specified class, and the states are those functionals on this class which are continuous in a specified topology.* Such a specification in advance is a necessity for a coherent theory, which cannot consist merely of a collection of isolated problems and special cases.

What we are proposing is that in spite of this, it is worth declaring a choice and using it as a template to refer to. So we shall proceed with the axiomatic method, and then comment on the necessary exceptions as we go.

If your inclination is not to worry about such matters much anyway, we envy you your attitude. But if you feel disquiet at the need for working with meta-quantities in the theory, we can offer you no solace beyond the fact that, models notwithstanding, nature cannot have inconsistencies or paradoxes, as Feynman has emphasized; and as far as is known, the models of classical and quantum mechanics are mathematically self-consistent.

2.3 Classical States And Observables

Not only quantum mechanics, but classical mechanics, too, is an observable-state system. The advantage of considering classical mechanics in this way is that doing so helps to clarify the intrinsic structure of the theory,

---

[1]We do not know what status to assign to quantum field theory. The rigorous version is starved of realistic models, and the calculational version is starved of rigour — the Planck length could well usher in a new story anyway.

supplementing the determination of the dynamical orbits.

For classical systems of idealized point particles it is axiomatic that all knowledge of the system at any instant is determined by the values of the generalized coordinates and momenta. In the case of a holonomic system with a finite number of degrees of freedom, these quantities are the coordinate functions of a differentiable manifold Π known as phase space, of even dimension $2d$; the number d is known as the number of degrees of freedom of the system. Only those systems for which phase space is $\mathbb{R}^2$ will be discussed in this book, in which case a (pure) state of the system may be identified with a point of Π. (Exceptionally, the laser model employs more general systems.)

If the forces are regular enough, they may be treated through Hamilton's equations, (2.5.4) below, the solutions of which exist and define a (Hamiltonian) flow in phase space, wherein each point lies on a unique dynamical orbit. Hence, knowledge of the forces acting in the system will serve to specify the future states of the system completely. Notably, Hamilton's equations are invariant under time reversal.

Under certain technical conditions, the Hamiltonian can be used to define a Lagrangian, and *vice versa*, through Legendre duality,

$$L(x,y) = \sup_{p\in\mathbb{R}} [px - H(p,y)] , \qquad H(p,q) = \sup_{x\in\mathbb{R}} [px - L(x,q)] . \quad (2.3.1)$$

In these circumstances, Hamilton's equations are equivalent to the Euler-Lagrange equations. Moreover, both can be obtained from variational principles, cf [67], [82], [31].

The second component of the classical observable-state system is the set of observables. These are taken to be functions on phase space, limited by whatever conditions of regularity are required by the physics of the problem.

Such basic systems do not begin to describe all the interesting problems covered by classical mechanics. These include systems with constraints, in which case the coordinate space is a more general manifold than $\mathbb{R}$, and phase space is its cotangent bundle. Then there are time and velocity dependent forces, forces where the particles run off to infinity and return in a finite time, attractors and repellors, and the like. See the references at the end of the Chapter for texts dealing with such systems.

Thus the definition of the (pure) states, observables and dynamics of a classical system have been given. Quantum mechanics is also described through states, observables and dynamics, but the two theories differ in that (amongst other things) in classical mechanics, Newton's Principle of Determinism and the Principle of Complete Knowledge hold:

> **Axiom 2.1** ***(Newton's Principle of Determinism)*** *Given the forces acting on the system, the initial (pure) state of the system determines all future states of the system uniquely.*

> **Axiom 2.2** ***(The Principle of Complete Knowledge)*** *All classical observables have a definite value in every pure state, and in principle these values may all be known simultaneously, with complete accuracy and without altering the state.*

Axiom 2.2 is not true in quantum mechanics. There is an axiom in quantum mechanics which at first sight looks very much like Axiom 2.1, however. But Axiom 2.1 must be understood in the light of the Principle of Complete Knowledge. In particular, this principle makes a distinct separation between the system and any measuring apparatus. It is a fundamental aspect of quantum mechanics that this is not possible.

There is another principle which must be mentioned here, although its purpose is to restrict consideration to non-relativistic systems, rather than being a principle of quantum theory.

> **Axiom 2.3** ***(Galileo's Principle of Relativity)*** *There is a collection of reference frames for space (the inertial frames) such that*
>
> - *the laws of nature are the same at all times in all inertial frames,*
> - *all inertial frames are in uniform rectilinear motion with respect to one another,*
> - *the invariance group for space and time is the inhomogenous Euclidean group (the Galilean transformations),*
> - *in principle, there is no upper limit to the speed of any object, so every point of space is in causal contact with every other.*

Before any more can be said about the structure of classical mechanics a commitment to a definite model must be made, and this is our next consideration.

2.4 The Formalism

As stated previously, constraints are never considered in this book and the phase space Π for the system is the Euclidean plane $\mathbb{R}^2$. By convention, the Euclidean coordinate system in Π has the momentum as the abscissa and the position as the ordinate. Then

Axiom 2.4 ***(Pure States)*** *A pure state is a point* $(p, q) \in \Pi$.

An observable F is a function on Π and its value in the state (p, q) is $F(p, q)$. It must therefore be assumed that the functions considered as observables are, at the least, defined everywhere. The inclusion of the dynamics will require that the Poisson bracket between observables results in an observable. In particular, in order that the dynamical flow be real analytic in the time variable, which is the usual formalism of dynamics, it is necessary to be able to take the Poisson bracket of an observable with the Hamiltonian function as often as we please. This forces an observable to be a function on phase space which is infinitely differentiable.

Axiom 2.5 ***(Observables)*** *The set of observables is the algebra* $\mathcal{C}^\infty(\Pi)$.

Remark It should be emphasized that there are other ways of axiomatizing classical mechanics. It is possible, for example, to develop a theory of classical mechanics for which the collection of observables is the space $\mathcal{C}(\Pi)$ of all continuous functions on Π. In such a formulation, there are time translations, but it is no longer possible to develop differential equations for the time evolution of observables. This seems to us to be too high a price to pay for being able to consider a greater number of observables.

Contrarily, not allowing discontinuous observables has the disappointing consequence that the angle function, φ, on phase space is not an observable, given that we are primarily interested (in this book) in studying the quantum analogue of this discontinuous function. Note that if discontinuous functions are allowed as observables, and if the collection of observables is to be stable under the Poisson bracket, distributions must also be allowed. For example, the derivative of the step function is a delta function. More pertinent to phase theory would be the Poisson bracket of φ with the classical analogue of the quantum number operator, $\nu = \frac{1}{2}(p^2 + q^2 - 1)$. If done carefully [53], this bracket is a distribution and not equal to 1 as a cursory calculation might suggest.

Without going into details, it is possible to define the Poisson bracket of a function and a distribution, the result being in general another distribution. This makes defining the Poisson bracket of two distributions problematic [135]. Even this problem can be overcome to a certain extent, in that a certain limited space of distributions can be found on which there is an extension of the Poisson bracket, with respect to which this space of distributions becomes a Lie algebra.

However, if this is the choice for observables, there cannot be even one pure state in the above sense, since that would require every distributional observable to be well defined at a given phase space point, which is impossible.

Therefore, if we wish to retain the notion of a pure state as a point in phase space and if we wish to retain the Poisson bracket as acting unrestrictedly between observables, the above Axiom for the nature of the observables is almost forced. If the axiom about pure states is dropped, the whole picture of phase space as the arena for classical mechanics goes. And if the Poisson bracket no longer gives a Lie algebra structure to the collection of all observables, the connection between classical and quantum mechanics is greatly weakened, if not severed. In our opinion, these arguments serve to justify the structure chosen here. These points will be considered again in Chapter 8.

A similar problem is encountered in the axiomatization of quantum mechanics, and results in the construction of two different models, to be referred to as the bounded and smooth models. The bounded model of quantum mechanics allows only bounded observables. It is completely functional, but does not permit study of time evolution in terms of the Hamiltonian directly, since the Hamiltonian operator is invariably unbounded. In contrast, the smooth model allows unbounded operators as observables, including the Hamiltonian. This permits us to develop equations of motion of a familiar type involving commutators. However, even with this wider latitude, certain other would-be observables do not fit into the formalism.

In Chapter 8, the connection between classical and quantum mechanics will be revealed as imperfect, in that it is not possible to construct a bijective correspondence between classical and quantum observables — for both technical and philosophical reasons. One source of imperfection is that it is desirable to consider quantum mechanical quantities corresponding to classical objects which are not observables in the sense of Axiom 2.5. This does not pose a mathematical problem, though it does cause us to think carefully about what physical interpretation should be placed on such quantum mechanical quantities. In the case of the quantum phase, in particular, such a pause for thought is both highly desirable and unavoidable. ∎

The algebraic operations on $\mathcal{C}^\infty(\Pi)$ of addition, scaling, and multiplication are the obvious pointwise ones. The algebra has an identity, the function[2] $\imath$ which takes the value 1 for all p and q; and complex conjugation defines an algebra involution, or *-operation. Evidently this algebra is commutative, in contrast to the analogous quantum algebra, which is not. There is a natural topology for this space with respect to which all of the algebraic operations are continuous, and with respect to which it is complete, a technical detail that will not be used.

In an observable-state formulation, as well as the algebra of observables, consideration must be given to its dual space of continuous linear functionals. But before doing so, there is an important question of notation to be settled.

Notation Since most of the vector spaces in the book are complex spaces, complex linear and antilinear operations must be distinguished.

---

[2]We shall use the symbol $\imath$ to represent a function which takes the value 1 on its domain. In different contexts throughout the book, different domains will be implied.

- For example, the inner product on a complex Hilbert space is linear in the right variable but antilinear in the left variable. Our notation for the inner product of the vectors ϕ and ψ in a Hilbert space $\mathcal{H}$ is $\langle \phi, \psi \rangle$, so the linearity relations are

$$\langle \overline{z}\,\phi, \psi \rangle \;=\; \langle \phi, z\psi \rangle \;=\; z\,\langle \phi, \psi \rangle\,, \qquad \phi,\, \psi \in \mathcal{H}, \quad (2.4.1.a)$$

where $z \in \mathbb{C}$ is a complex number. In the usual way, the norm of a vector $\phi \in \mathcal{H}$ is then

$$\|\,\phi\,\| \;=\; \langle \phi\,,\, \phi \rangle^{\frac{1}{2}}. \quad (2.4.1.b)$$

- It is also necessary to consider two-variable forms which are complex linear in both variables, especially in connection with topological vector spaces, such as spaces of test functions and distributions. To be specific, when considering a locally convex space $\mathcal{X}$ and its topological dual $\mathcal{X}'$, the notation for the duality pairing of an element $f \in \mathcal{X}$ with an element $T \in \mathcal{X}'$ is $[\![\, T\,,\, f\,]\!]$. The linearity conditions here are

$$[\![\, T\,,\, z\,f\,]\!] \;=\; [\![\, zT\,,\, f\,]\!] \;=\; z[\![\, T\,,\, f\,]\!]\,, \quad (2.4.2)$$

for all $f \in \mathcal{X}$, $T \in \mathcal{X}'$ and $z \in \mathbb{C}$.

■

Returning to the question of the dual of the algebra of classical observables, the following is standard distribution theory. For a detailed explanation of the terminology, see the books of Trèves [224].

Proposition 2.1 *The dual of $C^{\infty}(\Pi)$ is the space $\mathcal{E}'(\Pi)$ of distributions of compact support.*

It can be seen that the dual space $\mathcal{E}'(\Pi)$ contains the pure states by identifying the point $(p, q) \in \Pi$ with the Dirac delta function concentrated at that point, $\delta_{(p,q)} \in \mathcal{E}'(\Pi)$. For

$$[\![\, \delta_{(p,q)}\,,\, F\,]\!] \;=\; F(p, q), \quad (2.4.3.a)$$

as expected. In symbolic notation,

$$[\![\delta_{(p,q)} \, , F]\!] \; = \; \iint_{\Pi} \delta(p-p')\,\delta(q-q')\,F(p',q')\,dp'\,dq'. \tag{2.4.3.b}$$

Experience from the theory of algebras shows that the following two conditions characterize an important subset of functionals.

Definition 2.2 *A functional $T \in \mathcal{E}'(\Pi)$ is said to be* positive *if*

$$[\![T \, , F]\!] \; \geqslant \; 0 \tag{2.4.4}$$

for all $F \in \mathcal{C}^\infty(\Pi)$ whose values are positive at all points of phase space.

A functional $T \in \mathcal{E}'(\Pi)$ is said to be normalized *if*

$$[\![T \, , \mathrm{i}]\!] \; = \; 1, \tag{2.4.5}$$

where $\mathrm{i}(p,q) = 1$ *for all p, $q \in \mathbb{R}$, is the identity function on phase space. An example of such a normalized functional is $\delta_{(p,q)}$.*

We shall extend the term *state* to cover these functionals, on mathematical grounds. Thus we come to the following Axiom:

> **Axiom 2.6** ***(States)*** *A state is a distribution in $\mathcal{E}'(\Pi)$ which is positive and is normalized.*

It is an easy calculation to show that the pure states are states in this sense, but there are states which are not pure, as will be seen below. Then what distinguishes the pure states from the others? Although it could simply be said that the pure states are the delta functions, there is a geometric characterization of pure states, one which carries over to other observable-state systems when suitably interpreted. For the definitions of the terms used, see Peressini [177] or Schaefer [201].

Proposition 2.3 *The set of positive functionals in $\mathcal{E}'(\Pi)$ is a convex cone, and the set of states (as defined above) is a base for that cone. The extreme points of the base are precisely the delta distributions $\delta_{(p,q)}$, so that the pure states are the extreme points of the set of states. A mixed state is a state which is not pure.*

Having acquired the notion of mixed states, are there any? And if so, what do they represent physically?

Let ρ be a nonnegative function on Π which is integrable almost everywhere over every compact subset of Π, normalized so that

$$\int_{\Pi} \rho(p,q)\, dp\, dq \;=\; 1. \tag{2.4.6.a}$$

If ρ is of compact support and is not a delta function (and there are many such functions), then the formula

$$[\![T_\rho\, , F]\!] \;=\; \int_{\Pi} F(p,q)\, \rho(p,q)\, dp\, dq, \qquad F \in \mathcal{C}^\infty(\Pi), \tag{2.4.6.b}$$

defines a state T_ρ which is not pure.

As far as the physical meaning of mixed states goes, they are associated with imperfect information about the system, either because of complexity or possibly because some special circumstances make complete measurements impossible. The principal area of application of these ideas is statistical mechanics, see Section 2.5.2, but note carefully that this ignorance must be distinguished from the lack of definiteness in the values of a quantum observable in a general state.

Remark The model of classical mechanics above was constructed by first choosing the observables, and defining the states through mathematical duality. Some authors do not like to work with the resulting class of states, so they turn the problem around and choose the states first, usually taking them to be Borel probability measures. This has the merit of being a more natural choice from the point of view of the incomplete information interpretation for mixed states. The observables are then chosen by a form of duality (pre-duality is the mathematical term), in that they are taken to be those functions which are integrable against the states — Borel functions in this case. This leads to the problem stressed previously: the set of observables will not be a Lie algebra with respect to the Poisson bracket. While it is possible to define a Poisson bracket between two Borel functions in this formalism, it has to be done weakly (namely by integrating against a Borel measure) and consequently a significant part of the differential-geometrical structure of classical mechanics is lost. ■

2.5 Dynamics

2.5.1 *Hamiltonian Dynamics And Liouville's Theorem*

Classical dynamics is determined by a distinguished observable, the energy function or Hamiltonian H, which has a dual rôle: its value $H(p,q)$ gives the energy for a system in the pure state (p,q), and it also governs the time-evolution of a classical system, in that a system in the pure state (p_0,q_0) at time 0 will subsequently be in the pure state (p_t,q_t) at later time t, where the functions p_t, q_t are determined through Hamilton's equations ,

$$\frac{dp_t}{dt} = -\frac{\partial H}{\partial q}(p_t,q_t), \qquad \frac{dq_t}{dt} = +\frac{\partial H}{\partial p}(p_t,q_t). \tag{2.5.1}$$

It is sometimes convenient to use a vector notation for phase space points, so we write $\xi = (p,q)^{\mathrm{T}}$ for a general point, and $\xi_t = (p_t,q_t)^{\mathrm{T}}$ for its time evolutes. Along with this, Hamilton's equations can be written in vector form by introducing the two dimensional gradient,

$$\nabla H = \left(\frac{\partial H}{\partial p}, \frac{\partial H}{\partial q}\right)^{\mathrm{T}}. \tag{2.5.2}$$

In view of the opposite signs in Hamilton's equations, consider the matrix

$$J = \begin{pmatrix} 0 & -1 \\ 1 & 0 \end{pmatrix}, \tag{2.5.3}$$

in terms of which Hamilton's equations take the form

$$\frac{d\xi_t}{dt} = J\nabla H(\xi_t). \tag{2.5.4}$$

In Section 2.6, it will be shown that this formalism is more than a convenience: it leads the way to the natural geometry of classical mechanics.

A standard result of differential equation theory implies that Hamilton's equations have a unique local solution: from each initial point there is a unique trajectory. These curves are labelled by the time, which increases steadily, but may end at a finite value if the forces of interaction cause the motion to run off to spatial infinity. If not, the solution is global, but the trajectory may be quite complicated. The various possibilities are recounted in the theory of differentiable dynamics. For brevity and simplicity it will be assumed here that the trajectories are all well defined and global.

Given an observable F, if the system was in the state ξ_0 at time 0, the value of the observable in that state at time 0 is $F(\xi_0)$. If the time evolution of that pure state is described by the trajectory $t \to \xi_t$, then the value that the observable F takes at time t is $F(\xi_t)$. The $F(\xi_t)$ obey differential equations resulting from the equations of motion (2.5.4), as will now be shown.

In choosing to regard the observables as functions of p and q alone, the tacit assumption has been made that they depend on time only through the time dependence of the phase space coordinate functions p and q. This assumption can easily be modified if required. Observables are infinitely differentiable, so it is legitimate to calculate that the total time derivative of any observable is

$$\frac{d}{dt}F(\xi_t) = \frac{\partial F}{\partial p}(\xi_t)\frac{dp_t}{dt} + \frac{\partial F}{\partial q}(\xi_t)\frac{dq_t}{dt} = \left[\frac{\partial H}{\partial p}\frac{\partial F}{\partial q} - \frac{\partial H}{\partial q}\frac{\partial F}{\partial p}\right](\xi_t). \tag{2.5.5}$$

Introducing the Poisson bracket between any two observables by setting

$$\{F, G\} = \frac{\partial F}{\partial p}\frac{\partial G}{\partial q} - \frac{\partial F}{\partial q}\frac{\partial G}{\partial p}, \tag{2.5.6}$$

the time derivative can be written as

$$\frac{d}{dt}F(\xi_t) = \{H, F\}(\xi_t). \tag{2.5.7}$$

(Our sign convention for the Poisson bracket is not universal; we agree with Gallavotti [67] and disagree with Goldstein [82], for instance.)

In particular, the Hamiltonian of a system with no external forces is constant in time,

$$\frac{d}{dt}H(\xi_t) = 0, \tag{2.5.8}$$

as expected for the generator of an autonomous dynamics. Hence the Hamiltonian is time independent, and so its value at one time gives the system energy at all times (conservation of energy).

It should be noted that, with respect to the Poisson bracket, the space $\mathcal{C}^\infty(\Pi)$ is a Lie algebra. Moreover, this Lie algebra is related to another one, as follows. For any observable F define the continuous linear map $\mathfrak{L}_F$ on the algebra of observables by setting

$$\mathfrak{L}_F(G) = \{F, G\}, \qquad G \in \mathcal{C}^\infty(\Pi), \tag{2.5.9.a}$$

so that $\mathcal{L}_F$ can be written as the partial differential operator

$$\mathcal{L}_F = \left(-\frac{\partial F}{\partial q}\right)\frac{\partial}{\partial p} + \left(\frac{\partial F}{\partial p}\right)\frac{\partial}{\partial q}. \tag{2.5.9.b}$$

Because of the Leibnitz identity for the Poisson bracket,

$$\mathcal{L}_F(GK) = \mathcal{L}_F(G)K + G\mathcal{L}_F(K), \qquad G,\, K \in \mathcal{C}^\infty(\Pi), \tag{2.5.10}$$

(pointwise products in phase space are meant here), the observable $\mathcal{L}_F(G)$ is said to be the Lie derivative of G with respect to F.

The Lie derivative is an example of a vector field acting on the algebra $\mathcal{C}^\infty(\Pi)$ of observables. To be specific, a vector field on $\mathcal{C}^\infty(\Pi)$ is a linear map $\mathcal{L}$ from $\mathcal{C}^\infty(\Pi)$ to itself which satisfies the Leibnitz identity

$$\mathcal{L}(FG) = \mathcal{L}(F)G + F\mathcal{L}(G), \qquad F,\, G \in \mathcal{C}^\infty(\Pi). \tag{2.5.11}$$

The collection of all vector fields on $\mathcal{C}^\infty(\Pi)$ forms an infinite dimensional Lie algebra under the Lie bracket $[\,,\,]$ defined by

$$[\mathcal{L},\mathcal{M}](F) = \mathcal{L}(\mathcal{M}\,F) - \mathcal{M}(\mathcal{L}\,F) \tag{2.5.12}$$

for any vector fields $\mathcal{L}$, $\mathcal{M}$ and any $F \in \mathcal{C}^\infty(\Pi)$. To show that $[\mathcal{L},\mathcal{M}]$ is a vector field whenever $\mathcal{L}$ and $\mathcal{M}$ are is a relatively straightforward application of the properties of the vector fields $\mathcal{L}$, $\mathcal{M}$. Once this has been done, to show that we have defined a Lie bracket is elementary.

In terms of this structure, the map $F \mapsto \mathcal{L}_F$ from the space $\mathcal{C}^\infty(\Pi)$ of observables to the space of vector fields on $\mathcal{C}^\infty(\Pi)$ which sends to observable F to the vector field $\mathcal{L}_F$, is a Lie algebra homomorphism, since

$$[\mathcal{L}_F, \mathcal{L}_G] = \mathcal{L}_{\{F,G\}}, \qquad F,\, G \in \mathcal{C}^\infty(\Pi). \tag{2.5.13}$$

The proof of this identity follows from the fact that the Poisson bracket satisfies the Jacobi identity.

Returning to the subject of the time development, equation (2.5.7) can be rewritten in terms of the Lie derivative with respect to the Hamiltonian:

$$\frac{d}{dt}F(\xi_t) = \left[\mathcal{L}_H(F)\right](\xi_t). \tag{2.5.14}$$

The vector field $\mathcal{L}_H$ is often called the Liouville operator.

The higher Lie derivatives are

$$\frac{d^n}{dt^n}F(\xi_t) = \left[\mathcal{L}_H^n F\right](\xi_t)$$

for any $n \in \mathbb{N}$. It follows from this that, subject to conditions assuring convergence, the time evolution of any observable may be expressed in terms of a one parameter semigroup $\mathcal{T}$ of endomorphisms (linear operators) of $\mathcal{C}^{\infty}(\Pi)$ via the formula

$$F(\xi_t) = \left[\mathcal{T}_t F\right](\xi_0) = \left[\exp(t\mathcal{L}_H)F\right](\xi_0), \qquad t \geqslant 0. \tag{2.5.15}$$

Now that it has been shown how the Hamiltonian effects the time evolution of the observables, duality will determine the evolution of the elements of $\mathcal{E}'(\Pi)$, and hence of the states.

In this way a one parameter semigroup $\mathcal{T}^{\mathrm{tr}}$ of endomorphisms of $\mathcal{E}'(\Pi)$ is obtained via the formula

$$[\![\, \mathcal{T}_t^{\mathrm{tr}}(T)\, ,\, F\,]\!] = [\![\, T\, ,\, \mathcal{T}_t(F)\,]\!], \qquad F \in \mathcal{C}^{\infty}(\Pi),\, T \in \mathcal{E}'(\Pi), \tag{2.5.16}$$

for all $t \geqslant 0$, where $\mathcal{T}_t^{\mathrm{tr}}$ is the transpose of $\mathcal{T}_t$. Then if $\omega \in \mathcal{E}'(\Pi)$ is the state of the system at time $t = 0$, the state of the system at time t is $\omega_t = \mathcal{T}_t^{\mathrm{tr}}\omega$.

This construction extends the previous definition for the time evolution of pure states $\xi_0 \to \xi_t$. To see this, identify any pure state $\xi \in \Pi$ with the delta distribution $\delta_\xi \in \mathcal{E}'(\Pi)$ as usual. With respect to this identification, the trajectory ξ_t of a pure state, as governed by Hamilton's equations (2.5.1), is determined by the formula $\xi_t = \mathcal{T}_t^{\mathrm{tr}}\xi_0$ for all $t \geqslant 0$.

To summarize,

Axiom 2.7 ***(Dynamics)*** *There exists a distinguished observable, H, the Hamiltonian, which determines the Liouville operator $\mathcal{L}_H$, the time translation semigroup $\mathcal{T}$ of the observables, and that of the states, $\mathcal{T}^{\mathrm{tr}}$.*

Remark In many cases the system is closed and the internal forces are such that the dynamics are reversible. In such cases the time takes values in all of $\mathbb{R}$, and the semigroups become groups.

Another possible modification of these results is necessary if the observables are allowed to be explicitly time dependent. This would describe more complicated systems than have been considered here. Mathematically there is no difficulty in extending the theory, see Gallavotti (*ibid*). ■

It is a consequence of the assumptions that we have made that the well known Theorem of Liouville holds:

Theorem 2.4 ***(Liouville)*** *If V is a Lebesgue measurable set in phase space at time $t = 0$, in the course of time the set of points that constitute it move about under the influence of the dynamical forces, and at time t these points comprise the set*

$$V(t) = \{ (p_t, q_t) : (p, q) \in V \}. \tag{2.5.17}$$

While $V(t)$ is not the same as V in general, it has the same Lebesgue measure (phase space volume):

$$\int_{V(t)} dp\, dq = \int_V dp\, dq \tag{2.5.18}$$

for all $t \in \mathbb{R}$.

If F is any observable, it follows that

$$\int_{V(t)} F(p, q)\, dp\, dq = \int_V F(p_t, q_t)\, dp\, dq. \tag{2.5.19}$$

The proof is not difficult, and is constructed by calculating the effect of a time translation on the Jacobian, using Hamilton's equations and the equality of mixed second partial derivatives of the Hamiltonian [138].

2.5.2 *Mixed States And Statistical Mechanics*

It is true that there need be no mixed states in classical mechanics, provided that we are prepared to solve Hamilton's equations in all circumstances, and have obtained perfect and complete information from our measurements. While this is theoretically (or at least axiomatically) possible, such assumptions are unrealistic. Typically we must make do with incomplete knowledge of the system. Nevertheless, in such cases there may be enough information for us to be able to predict the system's probable future behaviour, subject to the laws of physics. In order for this to be successful, the complexity of the system must be of such a nature that we can employ the theory and methods of probability theory. In effect, the system must consist of a large number of particles, interacting regularly enough so that the notion of mean values of the dynamical quantities makes sense. Gibbs approached this problem by constructing what he called ensembles [75].

The results of this approach present us with mixed states characterizing systems in thermal equilibrium under different circumstances. We shall not belabour the point, but will briefly describe three such states that do describe interesting physical situations.

The microcanonical Gibbs state, given by

$$d\mu_{mic}(E, V, N; p, q) = \frac{\chi_V(q)}{Z_{mic}(E, V, N)\, N!}\, \delta\left(H_N(p, q) - E\right)\, d^N p\, d^N q, \tag{2.5.20.a}$$

where $Z_{mic}(E, V, N)$ is the microcanonical partition function,

$$Z_{mic}(E, V, N) = \frac{1}{N!} \int_\Pi \chi_V(q)\, \delta\left(H_N(p, q) - E\right)\, d^N p\, d^N q, \tag{2.5.20.b}$$

associated with confinement to the constant energy hypersurface $H_N(\xi) = E$ and spatial volume V. In the usual way, this state is appropriate only for systems with many degrees of freedom, here N, and ultimately $N \to \infty$.

This state describes autonomous systems which are energetically isolated, so that their energy hypersurfaces are the effective phase spaces for a system in this state.

Another state of interest for statistical mechanics is the canonical state, defined by

$$d\mu_{can}(\beta, V, N; p, q) = \frac{\chi_V(q)}{Z_{can}(\beta, V, N)\, N!} \exp\left[-\beta H_N(p, q)\right]\, d^N p\, d^N q, \tag{2.5.21.a}$$

where $Z_{can}(\beta, V, N)$ is the canonical partition function,

$$Z_{can}(\beta, V, N) = \frac{1}{N!} \int_\Pi \chi_V(q) \exp\left[-\beta H_N(p, q)\right]\, d^N p\, d^N q. \tag{2.5.21.b}$$

The third state usually considered in this context is the grand canonical state, defined by

$$d\mu_{gc}(\beta, V, \mu; p, q) = \frac{1}{Z_{gc}(\beta, V, \mu)} \sum_{n=0}^{\infty} \frac{1}{n!} \exp\left[-\beta H_n(p, q) + \beta\mu n\right]\, d^n p\, d^n q, \tag{2.5.22.a}$$

with grand partition function

$$Z_{gc}(\beta, V, \mu) = \sum_{n=0}^{\infty} e^{\beta\mu n} Z_{can}(\beta, V, n). \tag{2.5.22.b}$$

The canonical state is appropriate for systems with a fixed number of identical particles and at a fixed temperature β^{-1}. The grand canonical state is appropriate to a system containing arbitrary numbers of identical particles at fixed temperature β^{-1} and chemical potential μ.

Following Gibbs, we define thermodynamic functions

$$\begin{aligned} S_{mic}(E,V,N) &= \log Z_{mic}(E,V,N), && (2.5.23.a)\\ -\beta A_{can}(\beta,V,N) &= \log Z_{can}(\beta,V,N), && (2.5.23.b)\\ \beta\,|\,V\,|\,p_{gc}(\beta,V,\mu) &= \log Z_{gc}(\beta,V,\mu). && (2.5.23.c) \end{aligned}$$

Starting from the microcanonical entropy, we may compute the Helmholtz free energy A_{mic}, but it will not be equal to A_{can}. Similarly, the pressure densities p_{mic}, p_{can} and p_{gc} will all be different. It was a supposition of Gibbs that they would have the same (convergent) thermodynamical limits:

$$s(u,v) = \lim_{V\to\infty} \frac{1}{|\,V\,|} S_{mic}(U,V,N) \tag{2.5.24}$$

subject to

$$u = \lim_{V\to\infty} \frac{U}{|\,V\,|}, \qquad v = \lim_{V\to\infty} \frac{|\,V\,|}{N} \tag{2.5.25}$$

for the entropy density;

$$a(\beta,v) = \lim_{V\to\infty} \frac{1}{|\,V\,|} A_{can}(\beta,V,N) \tag{2.5.26}$$

for the free energy per unit volume; and

$$p(\beta,\mu) = \lim_{V\to\infty} p_{gc}(\beta,V,\mu) \tag{2.5.27}$$

for the equilibrium pressure.

We do not wish to embark on a further discussion of thermal equilibrium, nor on conditions under which these limit suppositions are valid. With varying degrees of reliability [138], they can be found in many places. However, we wish to bring to your attention the overview of classical thermodynamics and statistical mechanics in the preface by Wightman to the book on Lattice Gases by Israel [125].

For a classical version of the condition for thermal equilibrium discovered in quantum theory independently by Kubo, Martin & Schwinger (the

KMS condition, [143], [167]), see Katz [136]. This condition was later shown [2] to have the canonical state as a solution.

Finally, if it is desired to do classical statistical mechanics in an entirely algebraic fashion, the imposition of the thermodynamic limit requires that we combine together sets of observables with 1, 2, 3, ... degrees of freedom, in a graded structure. This leads into the theory of systems with infinitely many degrees of freedom.

2.6 Symplectic Geometry

For the simple classical systems under consideration, there is a natural symmetry scheme which explains neatly the choice of signs in Hamilton's equations, (2.5.4).

The matrix J is no idle curiosity. For suppose $W : \Pi \to \Pi$ is a smooth change of coordinates in phase space,

$$\zeta = W(\xi). \tag{2.6.1}$$

Then the equations of motion in the new coordinates take Hamiltonian form,

$$\frac{d\zeta_t}{dt} = J\nabla K \tag{2.6.2}$$

where $K(\zeta) = H(\xi)$, provided W satisfies the equation

$$\mathcal{J}(W)^{\mathrm{T}}\, J\, \mathcal{J}(W) = J, \tag{2.6.3}$$

where

$$\mathcal{J}(W) = \frac{\partial\left(\zeta^1, \zeta^2\right)}{\partial\left(\xi^1, \xi^2\right)} = \begin{bmatrix} \frac{\partial\zeta^1}{\partial\xi^1} & \frac{\partial\zeta^2}{\partial\xi^1} \\ \frac{\partial\zeta^1}{\partial\xi^2} & \frac{\partial\zeta^2}{\partial\xi^2} \end{bmatrix} \tag{2.6.4}$$

is the Jacobian matrix of W, and the superscript ^{T} denotes transpose. Such transformations are said to be canonical.

It can be shown fairly easily that W is canonical if and only if its Jacobian satisfies the condition

$$\det[\mathcal{J}(W)] = \left\{\zeta^1(\xi), \zeta^2(\xi)\right\} = 1, \tag{2.6.5}$$

where the Poisson bracket is taken using the old coordinates as independent variables. In words, the condition that the generalized position and momentum have a Poisson bracket equal to unity is a necessary and sufficient condition for Hamilton's equations to have their usual form. This underlines the special nature of this particular Poisson bracket, which has its counterpart in the importance of the commutator of the corresponding operators in quantum mechanics.

Another geometric quantity which is of importance in Hamiltonian mechanics is the closed 2-form

$$\omega \;=\; d\xi^1 \wedge d\xi^2. \tag{2.6.6}$$

In particular, it is true that the transformation W is canonical if and only if

$$\omega \;=\; d\zeta^1 \wedge d\zeta^2. \tag{2.6.7}$$

The 2-form ω can be used to put the vector fields on Π into one to one correspondence with the 1-forms. It is this correspondence that is behind the relation between the Lie and Poisson brackets considered earlier.

The equation (2.6.3) satisfied by the Jacobian of W can be considered as a mathematical relation in general, without reference to mechanics. That is, we consider the set of all matrices leaving J invariant:

$$\mathrm{Sp}\,(2;\mathbb{R}) \;=\; \left\{ A \in M_2(\mathbb{R}) \,:\, A^{\mathrm{T}} J A = J \right\}. \tag{2.6.8.a}$$

This turns out to be a group, known as the symplectic group of dimension 2. A simple calculation shows that for two dimensions, and only for two dimensions,

$$\mathrm{Sp}\,(2;\mathbb{R}) \;=\; \left\{ A \in M_2(\mathbb{R}) \,:\, \det A = 1 \right\}, \tag{2.6.8.b}$$

and so $\mathrm{Sp}\,(2;\mathbb{R})$ is identical to the group $\mathrm{SL}\,(2;\mathbb{R})$ of real matrices with unit determinant. (But note that $\mathrm{Sp}\,(2n;\mathbb{R})$ is not isomorphic to $\mathrm{SL}\,(2n;\mathbb{R})$ for $n > 1$.)

In mechanics, a smooth change of coordinates is canonical if and only if its Jacobian belongs to $\mathrm{Sp}\,(2;\mathbb{R})$. The identification between $\mathrm{Sp}\,(2;\mathbb{R})$ and $\mathrm{SL}\,(2;\mathbb{R})$ can be seen in equation (2.6.5).

The symplectic group has a related notion, that of symplectic forms. Using the matrix J to define the bilinear form

$$\Omega(\xi,\zeta) \;=\; \zeta^{\mathrm{T}} J \xi \;=\; \xi_1\zeta_2 - \xi_2\zeta_1, \tag{2.6.9}$$

a matrix A will belong to the symplectic group if and only if it leaves the form Ω invariant,

$$\Omega(A\xi, A\zeta) = \Omega(\xi, \zeta) \tag{2.6.10}$$

for all ξ, $\zeta \in \Pi$. Hence Ω is said to be a symplectic form.

Because the dynamics will be the same in all coordinate systems which can be obtained by canonical transformations, we may say that the geometry of classical mechanics is symplectic.

In contrast, the geometry of quantum mechanics is unitary, as the dynamics is invariant under unitary transformations. Moreover, while the invariance of the symplectic form leads to the finite dimensional symmetry group $\mathrm{Sp}\,(2;\mathbb{R})$, the invariance of the inner product leads to the infinite dimensional group of all unitary operators on Hilbert space.

2.7 Notes

Classical mechanics is important not solely as the precursor of quantum mechanics. It is still an active area of research, and by using modern global methods, mathematicians and physicists have made considerable progress in recent decades. The material above introduces some terminology and a particular point of view, but is not really a discussion of the modern theory of mechanics. For that the reader has to become used to the formalism of differential geometry: it is a curious turn of events that while classical mechanics is considered no more than background in a physicist's education these days, most books on differential geometry have considerable space devoted to it, although in an opaque language due to the emphasis on coordinate free global methods. Geometers are interested in mechanics because for complex systems, generalized coordinate space is a manifold and phase space a cotangent bundle. But there is more to mechanics than that, as may be seen from the text of Gallavotti.

There are far too many good books on mechanics for us to be able to offer a comprehensive list. The following is a very small selection, not including texts cited elsewhere in the chapter: [1], [7], [11], [32], [62], [93], [107], [214], [238].

Chapter 3
The Bounded Model

Another fine mess you've gotten me into!

– Hardy to Laurel

3.1 Introduction

This and the next two Chapters contain a description of the *kinematical* aspects of quantum mechanics, in an observable-state framework suitable for a reader with a working familiarity of quantum mechanics at the level of, say, Bohm [23], or anything similar.

A wholly pragmatic point of view would not require a full declaration of a model, but would simply agree that pure states are vectors in Hilbert space, that mixed states are density matrices, that observables are hermitian operators together with the identification of the position, momentum and energy operators, that the canonical commutation relations (CCR hereafter) underpin the uncertainty relations, and that dynamics comes from a prescription for Hamiltonian operators. Indeed, most of a first course in quantum theory is usually spent finding the energy spectrum for different model Hamiltonians. To be fair, consideration of the physical content of the theory is given to some extent or another, and the student is brought to an awareness of the probability aspects of the theory. From the point of view of the physical content of the theory this is probably sufficient.

But looking deeper, there are a number of points, both physical and mathematical, which should be pursued. The physical points involve the interpretation, and nowadays there are a number of specialized textbook treatments of these matters, including those by people who disagree with, or are simply unhappy with, the usual interpretation. Mostly, the standard interpretation will be assumed in this book.

There are mathematical difficulties even for the simplest system of a free massive point particle moving in one dimension. For one thing, the principal operators (position, momentum and energy) are unbounded, and so cannot act on every Hilbert space function. For example, if f is a normalized square integrable function of $x \in \mathbb{R}$, which is continuous everywhere and differentiable almost nowhere (such functions do exist), then $Pf = -i\hbar f'$ is not a square integrable function, where P is the usual representation for the momentum as a differentiation operator.

Should this point be addressed, or can it just be dismissed as a mathematical nicety? Here is the first decision to be made. The expectation and variance (and all the other moments) of P in the state determined by f will be infinite. We do not believe that the laboratory will blow up if it could be arranged that P be perfectly measured in the state determined by f. Rather, as the results of individual experimental runs are accumulated and a picture of the distribution begins to take form, the values obtained for the average of any power of P will grow without limit. There is, therefore, no probability distribution in this case, assuming the frequency interpretation of probabilities. Since the interpretation of the theory of quantum mechanics is based on probability distributions, allowing P and f together has violated the precepts of the theory. There is nothing mathematically special about P and f in this regard; there are infinitely many such mismatches between functions and operators, so this class of violation occurs infinitely often. That is unacceptable if quantum mechanics is to be a proper theory and not simply an *ad hoc* collection of rules dealing with problems on an individual basis.

It would be nice to say that there is an entirely satisfactory solution to this mathematical problem, but that would be untrue. There are two natural ways to proceed. The first is to work only with bounded operators. In this case all normalized vectors in Hilbert space determine states of the system. The drawback here is that the principal quantities noted above, position, momentum and energy amongst others, are not officially observables of such a theory. It will be shown, however, that they can be approximated as closely as needed by bounded operators. This approach leads to what we call the *bounded model*, which is discussed in this Chapter.

The second natural choice is to allow all the principal quantities as observables, which then requires that certain normalized vectors not be accepted as states. This possibility works because the states that remain form an adequately large collection in which every excluded vector is as near as required to a legitimate state. The condition specifying which vectors determine states is called smoothness, and the resulting model will be known as the *smooth model*, and will be discussed in the next Chapter.

Quantum theory for the systems considered in this book is characterized by the CCR, almost invariably for one degree of freedom. The CCR appears in both models, though in different forms. For the smooth model it appears as the commutator of the position and momentum, free to act on the smooth states. For the bounded model it is the exponentials of the position and momentum that appear. After both models have been discussed separately it will be shown how they are connected mathematically, whereupon both are available as necessary. For example, when working with the bounded model it is all right to consider the momentum operator directly, since its domain is available as part of the smooth model, and the relation of that domain to the bounded model is furnished by the connection theorem. The way this works will become clear with usage.

A note about the units employed in this book. Quantum mechanics is often presented with Planck's constant $\hbar$ explicitly included in the formulæ. Important as Planck's constant may be as a measure of the deviation of the quantum from the classical domain, the standard value of

$$\hbar = 1.0545\ldots \times 10^{-34} \quad \mathrm{J\,s}$$

is only specific to the SI system of units. However, any consistent system of units of measurement is equally valid, and we generally choose to work with atomic units, in which $\hbar = 1$. So $\hbar$ will not appear in the equations — except on those occasions where there is a particular reason to note its effect.

It will be taken as known that the first requirement of quantum theory is a Hilbert space $\mathcal{H}$, which (if quantum field theory is explicitly excluded from consideration) will be separable. This will be an unstated assumption from now on.

3.2 Bounded Approximations

Reference has been made to the fact that it is possible to approximate any unbounded operator arbitrarily closely by a bounded operator. It is this result which makes this model physically acceptable, and in view of the importance of this fact, it is worth outlining the proof.

Proposition 3.1 ***(Bounded Approximations)*** *For any unbounded self-adjoint operator A on a Hilbert space $\mathcal{H}$, we can find a sequence $\left(A_n\right)_n$ of bounded self-adjoint operators on $\mathcal{H}$ such that*

$$\lim_{n\to\infty} \| A\phi - A_n\phi \| = 0, \qquad \phi \in \mathcal{D}(A). \tag{3.2.1}$$

In other words, the sequence of operators $\left(A_n\right)_n$ converges strongly to A.

Proof: Let E_A be the spectral projection of the self-adjoint operator A, so that

$$A = \int_{\mathbb{R}} \lambda \, dE_A(\lambda),$$

and define the operators A_n by the formula

$$A_n = \int_{-n}^{n} \lambda \, dE_A(\lambda), \qquad n \in \mathbb{N}.$$

Standard functional analysis gives the desired convergence. ∎

It is important to realize that the convergence in this theorem is strong rather than merely weak. That is, for any ψ in the domain of A, the sequence $\| A\psi - A_n\psi \|$ converges to zero. In order for a convergence result to be of any use in quantum mechanics, it must hold true in this sense. For example, if $B_n \to B$ only weakly, meaning that the sequence of matrix elements $\langle \phi, (B - B_n)\psi \rangle$ converges to zero for all vectors ϕ, $\psi \in \mathcal{H}$, it cannot be concluded (without extra information) that any function of B_n converges. In particular, B_n^2 might not converge weakly to B^2. Physically, this means that the probability distributions obtained with the B_n do not converge to those of B. One cannot even conclude that the uncertainties converge. Weak convergence is simply too weak for quantum theory.

When applying the result of Proposition 3.1, it must be remembered that the phenomena it can be expected to describe are nonrelativistic and limited in energy and distance. If the energy is too high, the distances too small, the momentum too high, and so on, relativistic quantum field theory becomes applicable, and such phenomena as particle production and polarization of the vacuum can no longer be neglected.

With this in mind, suppose that A is an unbounded operator which might be expected to represent an observable. A typical state of the system is going to be such that the probability of recording values of A of modulus greater than some fixed value N (which should be imagined to be very large indeed), is negligibly small. Since the operator A_n is essentially what is left if values of A greater than n in modulus are ignored, A_n will give results that cannot be distinguished in a given experiment from that of A if n is sufficiently large (how large depends on both N and the experimental arrangements). So this sort of approximative technique is not only mathematically, but physically justifiable.

3.3 Observables And The Weyl Group

Although position and momentum are excluded from direct consideration in this model, it is important to retain some analogue of their commutation properties. The way to do this is to replace P and Q by the one parameter unitary groups $U(a) = \exp(ia\overline{P})$ and $V(b) = \exp(ib\overline{Q})$ that their closures generate. The commutation relation between Q and P,

$$QP - PQ = iI,$$

(as mentioned above, we choose to work with atomic units, so that $\hbar = 1$) which is strictly formal until the domain is specified, implies a commutation relation between U and V. This can be worked out by using the Baker-Campbell-Hausdorff formula [186], [99]. This formula provides a formal expression for the product of the exponentials of two operators. In particular, if the operators A and B both commute with their commutator $[A, B] = AB - BA$, then

$$e^A e^B = e^{A+B+\frac{1}{2}[A,B]}.$$

It turns out that the most convenient way to analyze the problem is to consider the operators $W(a, b) = \exp i\left(\overline{aP + bQ}\right)$ (where $a, b \in \mathbb{R}$). Disregarding closures at this formal level, substituting $A = aP + bQ$ and $B = cP + dQ$ (so that the commutator $[A, B]$ is a multiple of the identity) enables the product of $W(a, b)$ with $W(c, d)$ to be written as $W(a + c, b + d)$ multiplied by a phase factor. Having obtained this expression, the motivation for it in terms of P and Q can be put into the background, and the operators $W(a, b)$ considered in their own right. This is the kinematical foundation of the bounded model. Since it seems as if H. Weyl was the first person to propose doing this, this part of the theory bears his name.

3.3.1 *The Weyl Group*

The mathematical version of the CCR in the bounded model is realized through the Weyl group.

The set of all bounded operators on a Hilbert space $\mathcal{H}$ will be denoted by $\mathbb{B}(\mathcal{H})$.

Definition 3.2 ***(The Weyl Group)*** *A* representation of the Weyl group *on a Hilbert space $\mathcal{H}$ is a map W from $\mathbb{R}^2$ to $\mathbb{B}(\mathcal{H})$ such that*

$$W(a,b)W(c,d) = e^{\frac{1}{2}i(ad-bc)}\, W(a+c,b+d), \qquad (a,b),\,(c,d) \in \mathbb{R}^2. \quad (3.3.1)$$

In vector notation this becomes

$$W(\xi)W(\zeta) = W(\xi+\zeta)\, e^{\frac{1}{2}i\Omega(\xi,\zeta)}, \qquad \xi,\,\zeta \in \mathbb{R}^2, \qquad (3.3.2.a)$$

where Ω is the symplectic form introduced in Chapter 2:

$$\Omega(\xi,\zeta) = \xi_1\zeta_2 - \xi_2\zeta_1, \qquad \xi,\,\zeta \in \mathbb{R}^2. \qquad (3.3.2.b)$$

Moreover, the operators $W(a,b)$ are unitary, with

$$W(a,b)^{-1} = W(a,b)^* = W(-a,-b), \qquad a,\,b \in \mathbb{R}, \qquad (3.3.3)$$

and $W(0,0) = I$ is the identity operator. The operators $W(a,b)$ are known as Weyl operators, and constitute a projective unitary representation of the Weyl form of the CCR.

The reason for calling this a projective representation will be explained in Section 8.7 of Chapter 8.

Any such representation can also be expressed as a projective unitary representation of $\mathbb{C}$ instead of $\mathbb{R}^2$ by setting

$$W(\sqrt{2}a,\sqrt{2}b) = W[b-ia], \qquad a,\,b \in \mathbb{R}. \qquad (3.3.4)$$

Then the complex form satisfies the group law for $\mathbb{C}$,

$$W[z]W[w] = e^{i\,\mathrm{Im}(\bar{z}w)}\, W[z+w], \qquad z,\,w \in \mathbb{C}. \qquad (3.3.5)$$

The representation W of $\mathbb{R}^2$ will be distinguished from that of $\mathbb{C}$ by using round brackets in the first instance, and square brackets in the second. The $W[z]$ are also known as Weyl operators.

Since the Weyl group encapsulates quantum kinematics in bounded form, it is worth considering it from as many different points of view as possible.

Example 3.3 (The Schrödinger Representation) Perhaps the most important representation of the Weyl group is the familiar one for which the carrier space $\mathcal{H}$ is $L^2(\mathbb{R})$, and the action of the Weyl operators on functions is given by the rule

$$\big[W(a,b)\phi\big](x) \;=\; e^{\frac{1}{2}iab}\, e^{ibx}\, \phi(x+a), \qquad \phi \in L^2(\mathbb{R})\,. \tag{3.3.6}$$

It is then a simple calculation to show that these are indeed Weyl operators for a representation of the Weyl group on $L^2(\mathbb{R})$. (The complex form does not have a simple formula in this representation.)

This very important representation is the one implicit in Schrödinger's wave mechanics, and will be discussed and used throughout this book under the name *Schrödinger representation.*

3.3.2 *The Group Algebra*

There is a deep connection between the Weyl group and what is known to mathematicians as the group algebra. This concept was originally developed by Frobenius as a tool in the study of finite groups. Given a finite group G, consider the vector space $\mathbb{C}G$ of all formal linear combinations of the form

$$x \;=\; \sum_{g\in G} \alpha(g)\, g,$$

(technically $\mathbb{C}G$ is the free $\mathbb{C}$-module generated by G). The group structure of G turns $\mathbb{C}G$ into an algebra, with product defined by the expression

$$xy \;=\; \sum_{g\in G} \big(\alpha * \beta\big)(g)\, g, \qquad x = \sum_{g\in G} \alpha(g) g, \;\; y = \sum_{g\in G} \beta(g) g \,\in \mathbb{C}G,$$

where

$$\big(\alpha * \beta\big)(g) \;=\; \sum_{h\in G} \alpha(h)\, \beta(h^{-1}g), \qquad g \in G.$$

It is then the case that the algebra $\mathbb{C}G$ is commutative if G is, and that the identity of the group is also the identity of the algebra.

When continuous groups were first studied, it was soon discovered that there was a volume element (more precisely, the Haar measure) on any locally compact group which was invariant under left multiplication by group elements[1].

Consequently the sums in the formulation of the group algebra $\mathbb{C}G$ could be replaced by integrals, obtaining a group algebra of functions on the group. It was further found that the natural class of functions to be considered was the class of functions that were Lebesgue integrable with respect to this Haar measure μ. Thus, for any locally compact group G, the resulting space $L^1(G)$ is a Banach algebra with respect to the convolution product formula

$$(\alpha * \beta)(g) = \int_G \alpha(h)\,\beta(h^{-1}g)\,d\mu(h), \qquad \alpha,\, \beta \in L^1(G),$$

and, equipped with the involution

$$\alpha^*(g) = \overline{\alpha(g^{-1})}, \qquad \alpha \in L^1(G),$$

$L^1(G)$ becomes a Banach *-algebra.

Extending this construction further leads to the enveloping C*-algebra $C^*(G)$ of the Banach *-algebra $L^1(G)$. The collection $\mathcal{R}$ of all continuous *-algebra representations of $L^1(G)$ is used to define the seminorm[2]

$$\|\,\alpha\,\|_0 = \sup_{\pi \in \mathcal{R}} \|\,\pi(\alpha)\,\|, \qquad \alpha \in L^1(G).$$

The left regular representation L of $L^1(G)$ is defined on the Hilbert space $L^2(G)$ where, for any $\alpha \in L^1(G)$, L_α is the bounded operator on $L^2(G)$ given by the formula

$$L_\alpha(\phi) = \alpha * \phi, \qquad \phi \in L^2(G).$$

[1] Although the Haar measure is invariant under left translations, it is not generally invariant under right translations. But it is for a certain class of groups, the unimodular groups. Compact groups and abelian locally compact groups (for example, the real line) are all unimodular. The following material can readily be generalized to handle all locally compact groups, but will be written in terms of unimodular groups for simplicity of notation.

[2] Since $\|\,\pi(\alpha)\,\| \leqslant \|\,\alpha\,\|$ for all $\pi \in \mathcal{R}$ and $\alpha \in L^1(G)$, this seminorm is well-defined, with $\|\,\alpha\,\|_0 \leqslant \|\,\alpha\,\|$ for all $\alpha \in L^1(G)$.

Now L is an injective element of $\mathcal{R}$, which implies that $\|\cdot\|_0$ is a norm. Indeed, $\|\cdot\|_0$ is a C*-norm, and so the completion $C^*(G)$ of $L^1(G)$ with respect to $\|\cdot\|_0$ is a C*-algebra, called the *group C*-algebra*.

The representation theory of this C*-algebra is particularly simple, since there is a close correspondence between representations of $C^*(G)$ and the unitary representations of the group G.

Theorem 3.4 *To every nondegenerate* **-representation* π *of* $C^*(G)$ *there corresponds a weakly continuous unitary representation* U *of the group* G*, with both* π *and* U *acting on the same Hilbert space. Conversely, every weakly continuous unitary representation* U *of the group* G *yields a nondegenerate* **-representation* π *of the algebra* $C^*(G)$*, acting on the same Hilbert space, via the formula*

$$\pi(\alpha) = \int_G \alpha(g)\, U_g\, d\mu(g), \qquad \alpha \in C^*(G). \tag{3.3.7}$$

There is a similar relationship between the nondegenerate *-representations of the Banach *-algebra $L^1(G)$ and the weakly continuous unitary representations of G.

It turns out that neither the Banach *-algebra $L^1(G)$ nor the C*-algebra $C^*(G)$ contains an identity element unless the group G is discrete. However, the lack of an identity does not present a problem, since it is always possible to adjoin one. To be specific, we can consider the set

$$C_e^*(G) = C^*(G) \oplus \mathbb{C},$$

which becomes a *-algebra with respect to the product and involution

$$\begin{aligned} (\alpha, \lambda) \cdot (\beta, \mu) &= (\alpha * \beta + \mu\alpha + \lambda\beta, \lambda\mu), \\ (\alpha, \lambda)^* &= (\alpha^*, \overline{\lambda}), \end{aligned}$$

and this *-algebra is unital with identity $e = (0, 1)$. It is a standard result of the theory of C*-algebras that it is possible to equip $C_e^*(G)$ with a C*-norm $\|\cdot\|_0$ which extends[3] the norm $\|\cdot\|_0$ on $C^*(G)$ and for which $\| e \|_0 = 1$. The C*-algebra $C_e^*(G)$ is called the *unitization* of $C^*(G)$. Any representation of $C^*(G)$ can be extended naturally to $C_e^*(G)$ simply by sending the identity element e to the identity operator I.

[3] $C^*(G)$ can be regarded naturally as a subalgebra of $C_e^*(G)$ by identifying $\alpha \in C^*(G)$ with $(\alpha, 0) \in C_e^*(G)$.

The methods outlined above are standard fare of functional analysis, and can be found described in many books. For details of these matters, the reader is referred to [146], [174], [200].

The construction of the group algebra is highly suggestive in the context of the Weyl group, in that it indicates how to extend a group representation to a representation of an algebra consisting of functions on that group. The main difference to handle comes from the fact that a representation of the Weyl group is a projective group representation, and so the corresponding group algebra needs a slightly different algebraic structure to that outlined above.

3.3.3 *The Weyl Group C^*-Algebra*

Following this lead, consider a representation of the Weyl group on the Hilbert space $\mathcal{H}$. Arguing by analogy leads to consideration of the map $W : L^1(\mathbb{R}^2) \to \mathbb{B}(\mathcal{H})$ given by the formula

$$W[\![\alpha]\!] = \iint_{\mathbb{R}^2} \alpha(a,b) W(a,b)\, da\, db, \qquad \alpha \in L^1(\mathbb{R}^2). \tag{3.3.8}$$

The following properties can be shown to hold:

- The mapping W is a continuous linear injection from $L^1(\mathbb{R}^2)$ into $\mathbb{B}(\mathcal{H})$, with the following norm estimate

$$\| W[\![\alpha]\!] \| \leqslant \| \alpha \|_1 . \tag{3.3.9}$$

- If we define a product $\circ$ on $L^1(\mathbb{R}^2)$ by the rule

$$(\alpha \circ \beta)(a,b) = \iint_{\mathbb{R}^2} \alpha(a-x, b-y)\beta(x,y) e^{\frac{1}{2}i(ay-bx)}\, dx\, dy, \tag{3.3.10.a}$$

for $\alpha, \beta \in L^1(\mathbb{R}^2)$, then $L^1(\mathbb{R}^2)$ becomes an algebra and W an algebra homomorphism:

$$W[\![\alpha \circ \beta]\!] = W[\![\alpha]\!]\, W[\![\beta]\!], \qquad \alpha, \beta \in L^1(\mathbb{R}^2). \tag{3.3.10.b}$$

- $L^1(\mathbb{R}^2)$ becomes a *-algebra with respect to the involution

$$\alpha^*(a,b) = \overline{\alpha(-a,-b)}, \qquad \alpha \in L^1(\mathbb{R}^2), \tag{3.3.11.a}$$

and W is a *-algebra homomorphism:

$$W[\![\alpha^*]\!] = W[\![\alpha]\!]^*, \qquad \alpha \in L^1(\mathbb{R}^2). \tag{3.3.11.b}$$

- The two dimensional Gaussian function

$$H_0(a,b) = \frac{1}{2\pi}\, e^{-\frac{1}{4}(a^2+b^2)} \tag{3.3.12.a}$$

has the special property that $W[\![H_0]\!]$ is an orthogonal projection operator and satisfies the equation

$$W[\![H_0]\!]\, W(a,b)\, W[\![H_0]\!] = e^{-\frac{1}{4}(a^2+b^2)}\, W[\![H_0]\!], \tag{3.3.12.b}$$

for any $a, b \in \mathbb{R}$. In consequence, if $\phi,\ \psi \in \mathcal{H}$ are in the range of the projection $W[\![H_0]\!]$, then

$$\langle \phi\,,\, W(a,b)\psi\rangle = e^{-\frac{1}{4}(a^2+b^2)}\langle \phi\,,\, \psi\rangle, \qquad a,\, b \in \mathbb{R}. \tag{3.3.12.c}$$

Summarizing these ideas in terms of the previous Subsection, we note that equations (3.3.10.a) and (3.3.11.a) give $L^1(\mathbb{R}^2)$ the structure of a twisted nonunital Banach *-algebra, and that any representation W of the Weyl group generates a representation of this *-algebra.

Definition 3.5 ***(The Weyl Group C^*-Algebra)*** *The unitization of the enveloping C^*-algebra of the twisted Banach *-algebra $L^1(\mathbb{R}^2)$ (with product $\circ$ and involution *) is called the* Weyl group C*-algebra.

The operator $W[\![\alpha]\!]$ will play a significant rôle in quantization, and we shall find in Chapter 8 that it is essentially the *quantization of the classical observable which is the Fourier transform of α.*

3.3.4 *The von Neumann Uniqueness Theorem*

It is useful at this point to introduce the algebra $\mathfrak{B}$, consisting of all operators in $\mathbb{B}(\mathcal{H})$ which can be expressed as a polynomial in operators of the form $W(a,b)$ for $a,\, b \in \mathbb{R}$. For obvious reasons $\mathfrak{B}$ is called the polynomial algebra.

In order to do analysis with the Weyl group we require a notion of continuity, and experience shows that the proper one to choose is strong continuity: a representation of the Weyl group is *strongly continuous* if the function

$$(a, b) \mapsto W(a, b)\phi$$

is continuous from $\mathbb{R}^2$ to $\mathcal{H}$ for any $\phi \in \mathcal{H}$.

The simplest representations of the Weyl group are the so-called irreducible ones — these are the representations of the Weyl group for which there exists no nontrivial proper closed linear subspace of the Hilbert space $\mathcal{H}$ which is invariant under all of the operators $W(a, b)$. Schur's Lemma [126] provides a necessary and sufficient condition for a representation to be irreducible. For operational simplicity, we shall treat the result of Schur's Lemma as a method for defining which representations are irreducible, so we shall note that a representation of the Weyl group is *irreducible* if the only bounded operators on $\mathcal{H}$ which commute with all the operators $W(a, b)$ are multiples of the identity. For the purposes of quantum mechanics, the key property concerning representations of the Weyl group can be summarized by the following justly famous result of von Neumann [230], [184].

Theorem 3.6 ***(von Neumann's Uniqueness Theorem)*** *All strongly continuous irreducible representations of the Weyl group are unitarily equivalent.*

The irreducible representations of the Weyl group are characterized by the fact that the range of the projection operator $W[\![H_0]\!]$ *is one-dimensional. If* Ω_0 *is a unit vector in the range of this projection, then*

$$\langle \Omega_0 \, , \, W(a, b)\Omega_0 \rangle = e^{-\frac{1}{4}(a^2+b^2)}, \qquad a, \, b \in \mathbb{R}. \tag{3.3.13}$$

The algebra of polynomials $\mathfrak{B}$ *is weakly dense in* $\mathbb{B}(\mathcal{H})$. *Moreover the vector* Ω_0 *is cyclic in that* $\mathfrak{B}\Omega_0$ *is a dense linear subspace of* $\mathcal{H}$.

In general any strongly continuous representation of the Weyl group, whether irreducible or reducible, can be written as an orthogonal direct sum of a family of irreducible representations.

This result is clearly much more than a uniqueness theorem.

As a matter of terminology, if a representation of the Weyl group has a given property (such as irreducibility), the representation of the CCR it determines is said to have that property.

3.3.5 *Observables*

To complete the description of the kinematics of the bounded model, a precise choice of the states and observables must be made. Until further notice, the assumption is that the system Hilbert space $\mathcal{H}$ carries a strongly continuous and irreducible representation of the Weyl group, thereby incorporating the Weyl version of the CCR.

Although this model is based on the premise that observables be represented by bounded operators, this does not determine just which bounded operators should be chosen. Indeed, it is hard to ascertain precisely what requirements would result in a unique choice. To retain a possible connection to classical mechanics, most would agree that i times the commutator of any two observables should be an observable. In other words, the observables should constitute a Lie algebra.

For most purposes this is not enough and most physicists make the further assumption that the observables are a subset of a larger algebra of operators, with the algebra structure of this super-algebra being used to define the Lie algebra structure on the observables — this is our assumption as well.

But this still does not determine which larger algebra to choose. In order to make the necessary connections with the CCR, it is necessary to require that all the operators $W(a, b)$ belong to this larger algebra, and consequently it follows that the polynomial algebra $\mathfrak{B}$ is a subalgebra of it. However $\mathfrak{B}$ is itself too small.

To enlarge this set, the obvious thing to do is complete $\mathfrak{B}$ in some natural topology. For example, the closure in the operator norm topology could be considered. The result is a C*-algebra, which is an attraction, but this is still too small, since it does not even contain all bounded functions of P and Q.

Remark At this point the operators P and Q have not been defined, so a rule must be given to do so, as well as how to construct functions of them. The first step is to consider the strongly continuous one-parameter unitary groups

$$a \mapsto U(a) = W(a, 0), \qquad b \mapsto V(b) = W(0, b), \tag{3.3.14}$$

respectively. A theorem of Stone [186] says that such groups are differentiable, and their self-adjoint generators are P and Q, respectively.

This procedure is essentially one of retracing the motivational steps leading to the definition of W, and is well-defined once W is given. Bounded functions of these operators can now be constructed via the spectral theorem. This procedure does not determine the domains of P and Q, nor has any evidence been given to support the interpretation of P and Q as momentum and position operators, so this discussion must draw on the reader's prior knowledge (including that of spectral theory). ∎

The completion of $\mathfrak{B}$ in the strong operator topology on $\mathbb{B}(\mathcal{H})$ might be considered. This closure is certainly an algebra large enough to contain all bounded functions of P and Q, and some authors, notably Thirring [221], recommend its use. However, we choose to work with the closure of $\mathfrak{B}$ in the weak operator topology on $\mathbb{B}(\mathcal{H})$. For the weak closure of $\mathfrak{B}$ is its double commutant by the theorem of that name due to von Neumann, see [57]. The irreducibility of our representation of the Weyl group implies that the first commutant of $\mathfrak{B}$ is $\mathbb{C}I$, and hence the double commutant is the whole of $\mathbb{B}(\mathcal{H})$. In other words, we are choosing to work with the whole algebra $\mathbb{B}(\mathcal{H})$ of all bounded operators on $\mathcal{H}$. Summarizing,

Axiom 3.1 ***(Observables - Bounded Model)*** *Any bounded self-adjoint operator on a separable Hilbert space $\mathcal{H}$ which carries a strongly continuous irreducible representation of the Weyl group is an observable, and so the set of observables is the real Lie algebra $\mathbb{B}(\mathcal{H})_h$ of self-adjoint bounded operators on $\mathcal{H}$. This is a Lie subalgebra of the W^*-algebra of all bounded operators on $\mathcal{H}$, written $\mathbb{B}(\mathcal{H})$, where the Lie product on $\mathbb{B}(\mathcal{H})$ is given by the formula*

$$[A, B] = i\,(AB - BA), \qquad A,\, B \in \mathbb{B}(\mathcal{H}). \tag{3.3.15}$$

The following observations are pertinent:

- It is not necessary to specify $\mathcal{H}$, as all representations of this class are unitarily equivalent by von Neumann's uniqueness theorem.
- The term *algebra of observables* should be considered as a flag of convenience, which is sometimes used (somewhat inaccurately) to denote the Lie algebra $\mathbb{B}(\mathcal{H})_h$ of observables, while at other times is

used to describe the larger algebra $\mathbb{B}(\mathcal{H})$ of all bounded operators (which are not all self-adjoint and hence are not all observables) — the context in which this terminology is being employed should make its usage clear at any time.

- In Chapter 5, various concrete representations with these properties will be considered. Amongst these is the Schrödinger representation, which has been previously singled out, see Example 3.3. In this representation, the position observable Q is diagonal. Other representations include the so-called momentum representation in which the momentum observable P is diagonal, and the representation where the harmonic oscillator Hamiltonian is diagonal.
- The requirement that the representation of the Weyl group be strongly continuous is crucial. Without this condition, infinitely many pathological representations would be permitted.
- It is worth emphasizing that this model assumes that the system being considered has only a finite number of degrees of freedom — in fact, for most of this book, it will be assumed that the system has only 1 degree of freedom. The formalism here is not sufficient to deal with systems with infinitely many degrees of freedom.
- It is possible to restrict the collection of observables somewhat by requiring it to possess some weakened form of a product itself, rather than simply assuming that the observables form a subcollection of a larger algebra. This idea leads (amongst others) to the concept of a Jordan algebra, which approach is discussed at length in the book by Emch [57].

3.4 States In The Bounded Model

Having decided on the observables, the states must be chosen. Before doing so, a brief word about the physical meaning of a quantum state is in order.

Quantum and classical states have one thing in common: they carry all the information about the instantaneous situation of the system, though what this actually means is very different in the two cases, since for a quantum state the proviso *without recourse to a measurement* must be added.

According to the usual interpretation, the act of measurement must be treated separately, *ab extra* as it were, and is non-deterministic in nature.

Conventionally, the act of measurement is not connected to the dynamical evolution of the state. A measurement will result in a spectral value only. Immediately and uncontrollably, the act of measurement is supposed to collapse the state to the corresponding eigenstate. For spectral values in the continuum this needs a careful fine tuning, as a continuum eigenfunction is an eigendistribution which cannot be normalized, and so is not a state. These matters will be discussed in detail elsewhere, but let this serve as a reminder of how remote a quantum state is from direct experience (which can be forgotten in the welter of mathematics).

Before getting to that mathematics, it is interesting to learn why someone as committed to the standard interpretation as David Bohm came to reject it in favour of a hidden variable theory — the belief that there is a refined theory employing variables (in some sense) not yet known, which subsumes quantum theory and yet is deterministic. In the book *Quantum Implications*, Bohm states that it was not the fact that the results of measurements could be predicted only statistically that created his doubts about the completeness of the standard theory, but rather that the theory has no place in it for an "adequate notion of an independent actuality" [24]. To paraphrase Bohr's answer to Einstein on this very point, while it is true that there is no place for a classical actuality in quantum theory, usage of such an expression presupposes the sort of reality there is in the universe. The test of quantum mechanics is its internal mathematical consistency, its ability correctly to describe the phenomena in its domain without any failures, and its ability to reproduce the results of the coarser domain of classical mechanics. These are extremely complex issues which have been argued over continually ever since the Bohr-Einstein debates [237], and have not yet been resolved to everyone's satisfaction. In this book, the standard, or Copenhagen, interpretation is accepted as a working hypothesis. So,

> Other than in exceptional circumstances, a measurement will inevitably alter a state. The results of measurements are statistical in character, and physical quantities can only be assigned values by measuring them. This latter procedure is known as preparation.

3.4.1 *States As Functionals*

Recall that, in classical mechanics, states were defined as positive linear functionals on the space of observables. As the observables in this model are a subset of (and, indeed, span) the algebra $\mathbb{B}(\mathcal{H})$, the same definition of states can be used here. However, it turns out that there are functionals which are not wanted on physical grounds, and must be excluded by imposing an additional continuity condition. The aim is to end up with density matrices only, and that comes out of Gleason's Theorem below.

Definition 3.7 *Let ω be a linear functional on $\mathbb{B}(\mathcal{H})$. It is said to be* positive *if $\omega(A) \geqslant 0$ for all positive bounded operators A. It is said to be* normalized *if $\omega(I) = 1$. It is said to be* normal *if whenever $\big(A_n\big)_n$ is a sequence in $\mathbb{B}(\mathcal{H})$ which converges strongly to $A \in \mathbb{B}(\mathcal{H})$, then the sequence $\big(\omega(A_n)\big)_n$ converges to $\omega(A)$.*

With these concepts in hand the precise definition of states in the bounded model can now be given.

Axiom 3.2.a ***(States - Bounded Model)*** *A state of the system is a normalized positive linear functional ω on $\mathbb{B}(\mathcal{H})$ which is normal.*

3.4.2 *States As Density Matrices*

This Axiom, at first sight, gives us little idea about the exact nature of states. Moreover, it is not clear why the less than obvious continuity condition of normality was chosen. Gleason's Theorem resolves these difficulties. This Theorem refers to trace class operators, and requires the following material in its formulation.

Remark A positive trace class operator ρ has a complete orthonormal set of eigenvectors

$$\rho\phi_n = r_n\phi_n, \tag{3.4.1}$$

so that its spectral representation is

$$\rho = \sum_{n=0}^{\infty} r_n Q_n, \tag{3.4.2}$$

where Q_n is the projection operator onto the subspace spanned by ϕ_n. The eigenvalues are all positive and are related to the trace by

$$\text{Tr}\,(\rho) = \sum_{n=0}^{\infty} r_n. \tag{3.4.3}$$

If σ is a self-adjoint trace class operator, it has a complete set of eigenvectors ψ_n, (by the Hilbert-Schmidt Theorem [186]), so

$$\sigma = \sum_{n=0}^{\infty} s_n Q_n, \qquad Q_n \phi = \langle \psi_n, \phi \rangle \, \psi_n \,. \tag{3.4.4}$$

Hence every self-adjoint trace class operator is the difference of two positive trace class operators:

$$\sigma = \sigma_+ - \sigma_-, \qquad \sigma_\pm = \sum_{n \geqslant 0, \pm s_n > 0} s_n Q_n \,; \tag{3.4.5}$$

the eigenvalue 0 (if present) is omitted. By Lidskii's Theorem [211],

$$\text{Tr}\,(\sigma) = \sum_{n=0}^{\infty} s_n \tag{3.4.6}$$

is an absolutely convergent series.

An arbitrary trace class operator has a unique expression as a sum of self-adjoint trace class operators, $\tau = \tau_1 + i\tau_2$, with

$$\tau_1 = \frac{1}{2}\,(\tau + \tau^*)\,, \qquad \tau_2 = \frac{1}{2i}\,(\tau - \tau^*)\,. \tag{3.4.7}$$

An important observation is that

$$\| \tau \|_1 := \text{Tr}\big([\tau^* \tau]^{1/2}\big) \tag{3.4.8}$$

is a norm on the set of all trace class operators, under which it is a Banach space, denoted by $\mathcal{T}_1(\mathcal{H})$, whose dual is

$$\mathcal{T}_1(\mathcal{H})^* = \mathbb{B}(\mathcal{H})\,. \tag{3.4.9}$$

For an obvious reason, $\mathcal{T}_1(\mathcal{H})$ is known as the pre-dual of $\mathbb{B}(\mathcal{H})$. ■

In Chapter 8, further details concerning trace class operators will be considered. Now for Gleason's Theorem [79].

Theorem 3.8 ***(Gleason)*** *By a density matrix is meant a positive trace class operator on $\mathcal{H}$, whose trace is equal to unity. Every density matrix ρ determines a state ω_ρ through the formula*

$$\omega_\rho(A) \;=\; \mathrm{Tr}\;(\,\rho A\,)\,, \qquad\qquad A \in \mathbb{B}(\mathcal{H}). \tag{3.4.10}$$

Conversely, every state determines a density matrix through this formula.

Remark Without the condition of normality this theorem is not true. A version of this theorem is also true for nonseparable Hilbert spaces, provided the normality condition is revised to demand convergence of increasing positive nets of operators on $\mathbb{B}(\mathcal{H})$.

If we combine Gleason's Theorem with the eigenfunction decomposition for positive trace class operators, the result is that if ρ is a density matrix with eigenvalues r_n and corresponding eigenfunctions ϕ_n, and if Q_n is the projection along ϕ_n, then

$$\omega_\rho(A) \;=\; \mathrm{Tr}\;(\,\rho A\,) \;=\; \sum_{n=0}^{\infty} r_n \,\mathrm{Tr}\;(\,Q_n A\,) \;=\; \sum_{n=0}^{\infty} r_n \;\langle \phi_n, A\phi_n\rangle\,. \tag{3.4.11}$$

■

3.4.3 *Pure And Mixed States*

The geometric structure of the set of states depends on the notion of convex subsets of a vector space, and a passing knowledge of this material will be assumed.

Consider two states ω_1 and ω_2. If t_1 and t_2 are real numbers, it is clear from the positivity and normality requirements that $t_1\omega_1 \;+\; t_2\omega_2$ will also be a state provided that $0 \leqslant t_1, t_2 \leqslant 1$ and $t_1 + t_2 \;=\; 1$. In other words, the convex linear sum of states is also a state, and hence the set of states is convex.

In the analysis of any convex set, it is important to be able to identify the extreme points of that set. An element in a convex set is called extreme if it cannot be written as a nontrivial convex combination of other elements of that set. To be specific, if K is a convex set, then $a \in K$ is extreme if whenever we can write $a = \lambda b + (1 - \lambda)c$ where $b, c \in K$ and $0 < \lambda < 1$, then $b = c = a$. As an example, we note that the set of extreme points of a convex polyhedron in $\mathbb{R}^3$ is the set of its vertices.

As might be expected, therefore, the collection of extreme points of the convex set of states is of particular interest to us. However, since these states were not originally studied from the point of view of the theory of convex sets, they are most often referred to by a different name, and the extreme states are called pure states in the literature.

Before we proceed to identify the pure states, we shall introduce an (apparently different) collection of states, the *vector states*. A vector state is one whose associated density matrix is a projection operator with a one-dimensional range. Thus a vector state is one of the form

$$\omega_\phi(A) \;=\; \mathrm{Tr}\big(P_\phi A\big) \;=\; \langle\phi\,,\, A\phi\rangle, \qquad A \in \mathbb{B}(\mathcal{H}), \tag{3.4.12}$$

for some unit vector $\phi \in \mathcal{H}$, where $P_\phi \in \mathbb{B}(\mathcal{H})$ is the orthogonal projection along ϕ . In this context it is sometimes useful to employ the Dirac notation

$$P_\phi \;=\; |\,\phi\,\rangle\,\langle\,\phi\,| \tag{3.4.13}$$

to describe this density matrix/projection operator.

Suppose now that ω is a state, and let ρ be its associated density matrix. We can determine the complete orthonormal set of eigenvectors for ρ as described in the Remark at the beginning of the previous subsection, noting that the projection Q_n is the orthogonal projection P_{ϕ_n} for each integer n, and that the positive eigenvalues r_n sum to 1. If more than one of the eigenvalues r_n is positive, without loss of generality we can assume that $0 < r_1 < 1$, in which case the functionals

$$\omega_1 \;=\; \omega_{\phi_1}, \tag{3.4.14.a}$$

$$\omega_2 \;=\; \frac{1}{1 - r_1}\big(\omega \;-\; r_1\omega_{\phi_1}\big), \tag{3.4.14.b}$$

are distinct (since $\omega_1(P_{\phi_1}) = 1$, while $\omega_2(P_{\phi_1}) = 0$) states such that $\omega = r_1\omega_1 + (1 - r_1)\omega_2$. Consequently we deduce that ω is not pure. From this we infer that any pure state is a vector state. On the other hand, if $\phi \in \mathcal{H}$ is a unit vector, and if ω_1 and ω_2 are states (determined by density matrices

ρ_1 and ρ_2), and $0 < \lambda < 1$ is such that $\omega_\phi = \lambda\omega_1 + (1-\lambda)\omega_2$, then

$$\lambda\omega_1(P_\psi) + (1-\lambda)\omega_2(P_\psi) = \omega_\phi(P_\psi) = |\langle\phi, \psi\rangle|^2 = 0$$

whenever $\psi \in \mathcal{H}$ is orthogonal to ϕ, and so $\omega_1(P_\psi) = \omega_2(P_\psi) = 0$ whenever ψ is orthogonal to ϕ. Recalling the eigenvector decompositions of ρ_1 and ρ_2, we deduce that any eigenvector of ρ_1 (or ρ_2) corresponding to a nonzero eigenvalue is orthogonal to every vector which is orthogonal to ϕ, and hence is a multiple of ϕ itself. Consequently it follows that $\rho_1 = \rho_2 = P_\phi$, and hence that $\omega_1 = \omega_2 = \omega_\phi$. Therefore ω_ϕ is pure. We conclude that the terms *extreme state*, *pure state* and *vector state* are synonymous.

By a unit ray in $\mathcal{H}$ is meant a set of vectors in $\mathcal{H}$ of the form

$$\underline{\Phi} = \{ z\Phi : |z| = 1 \}, \tag{3.4.15}$$

where $\Phi \in \mathcal{H}$ is a unit vector. It is immediate that if Φ and Ψ belong to the same unit ray, they have the same projection operator, and so define the same vector state. Hence there is a bijective correspondence between unit rays and vector states.

States which are not pure are known as *mixed states*. They can be decomposed into a convex linear combination of pure states in a nontrivial manner.

It should be noted that, while there is a correspondence between the unit rays in $\mathcal{H}$ and the pure states on $\mathbb{B}(\mathcal{H})$, there is no such correspondence between the operation of addition of vectors in $\mathcal{H}$ and the operation of addition of states. To see this, if ϕ, ψ are orthogonal unit vectors in $\mathcal{H}$, then $\Phi = (\phi + \psi)/\sqrt{2}$ is a unit vector in $\mathcal{H}$. However the pure state ω_Φ cannot be written as a convex combination of the pure states ω_ϕ and ω_ψ alone — the expression for ω_Φ also contains cross terms involving ϕ and ψ simultaneously. Cross terms of this form frequently appear when calculating probabilities, as in the two slit experiment, and they correspond to interference effects.

This discussion can be summarized as an axiom:

Axiom 3.2.b ***(States - Bounded Model)*** *The set of states in the bounded model is a convex set and the pure states are its extreme points. Hence a mixed state cannot be decomposed into a convex linear sum of other states in a nontrivial fashion. A mixed state is a state which is not a pure state.*
States are given by density matrices, which are normalized positive trace class operators, through the formula $\omega(A) = \mathrm{Tr}\,(\,\rho A\,)$*, for all* $A \in \mathbb{B}(\mathcal{H})$.
The pure states are identifiable with projection operators, with one-dimensional ranges, or with unit rays in $\mathcal{H}$*. Hence all vector states are pure, and conversely.*
The density matrix for a mixed state has an eigendecomposition consisting of at least two distinct terms. The decomposition of a mixed state into pure states is not unique.

It is standard usage to identify a state with a density matrix. For a pure state, that means an identification with a projection operator of one dimensional range, or the associated unit ray. But more, there is a standard abuse of terminology identifying a unit ray with a representative unit vector, and so pure states with unit vectors. There is no harm in doing so, and will be the usage in this book.

3.5 Additional Reading

It has been noted that working from an explicitly defined model is not usual. Hence most textbooks do not address the issues discussed above from a structural point of view. The emphasis on an observable-state formulation, while implicit in Dirac's book [49], is essential in quantum field theory and statistical mechanics. The arguments for the bounded model in that context are forcefully made by one of its originators, Haag, in his book *Local Quantum Physics*, [94].

There are texts that cover more specialized topics that either will not be covered in this book or touched only slightly. One such topic is the path integral, for which [61], [194] and [206] are standard references. Feynman also has a book based on his Messenger lecture at Cornell, in which he discusses the nature of physics and the laws which describe it, [60]. Much earlier Heisenberg gave his ideas of the meaning of quantum mechanics in [104]. Max Jammer has traced the history of quantum mechanics in two cognate works, [128] and [129], which repay reading. For a proper treatment of Hilbert space in the midst of quantum mechanics, the book of Prugovecki is recommended [183]. Other references will be given in other Chapters.

Convexity is a topic that is not as widely taught as it might be, making references for it somewhat thin on the ground. Perhaps the best place to start is the article on convexity in the Encyclopedic Dictionary of Mathematics of the Mathematical Society of Japan, [126], from which further references may be gleaned. A recent introductory textbook, which can be recommended, is that of Webster [234].

CHAPTER 4
THE SMOOTH MODEL

The application of the theory is well known... But the fundamentals of it are usually hidden in a cloud of uncertainty and confusion
– F. Rohrlich, *Classical Charged Particles*

4.1 Introduction

The bounded model is based on the premise that all observables should be defined on the whole Hilbert space $\mathcal{H}$, which has the consequence that all observables must be bounded. Consequently, while maximizing the collection of permissible (pure) states, the bounded model significantly restricts the collection of permissible observables.

Conversely, the smooth model is based on the premise that as large a collection of operators as possible should be accounted observable. This collection includes unbounded operators, each with its own domain, and so it is necessary to find a common domain of vectors on which this posited collection of operators may act. This requires a decision as to what the expressions "as large as possible" should mean. Since the larger the class of observables the smaller the class of allowed states, this is a joint observable-state decision. Although these competing factors may at first seem to render the problem insoluble, a rather satisfactory solution does exist.

The first requirement is that all the operators have a common domain; if an operator is presented to us on some larger domain, it is its restriction to this common domain that concerns us. But this is not the only condition that must be satisfied. As with the bounded model, rather than being content with the collection of observables forming a (real) Lie algebra, the collection of observables will be required to form a subset of a larger algebra of (unbounded) operators, all of whose elements share the same common domain. The algebraic structure of this larger collection of operators demands this common domain be stable under the action of all of these operators. The Lie bracket of two observables will then be i times their commutator — a definition which makes sense on the common domain at least. The vectors in this common domain will then comprise the pure states.

On reflection, then, the possibility of an effective smooth model hinges on the possibility of finding some dense linear subspace $\mathcal{S}$ of $\mathcal{H}$ (density is a modest requirement on a domain), which satisfies at least these requirements:

- $\mathcal{S}$ must be small enough to be stable under the chosen collection of observables. As the largest set of operators for which this is so is the collection End($\mathcal{S}$) of all endomorphisms of $\mathcal{S}$ (linear maps from $\mathcal{S}$ to itself), the observables constitute a subalgebra of End($\mathcal{S}$). In this way, all the usual algebraic manipulations with observables needed for physics will be valid.
- $\mathcal{S}$ must be large enough so that the usual observables of quantum mechanics are essentially self-adjoint when restricted to it. This is a technical condition, but one necessary to exclude various undesirable (and non-physical) pathologies.
- On physical grounds, $\mathcal{S}$ is also required to be invariant under the action of the Hilbert space adjoints of all the operators considered, so if B is an element of the larger algebra (and thus maps $\mathcal{S}$ to itself), it is required that B^* maps $\mathcal{S}$ to itself.

It was promised above that at least one such domain exists, which is by no means obvious. But if there should be several, it will require some further principle to select one. Another worry is that unless these conditions are just right, the new formalism will have a significant number of mathematical pathologies. However, as for the bounded model, it is possible to impose topological requirements on the space $\mathcal{S}$ in order to obtain a structure theorem analogous to von Neumann's uniqueness theorem, so that these concerns can be set to rest.

4.2 The CCR On The Smooth Domain

The first task in building up the smooth model is to consider how the canonical commutation relation (CCR) will be reflected in its structure.

4.2.1 *The CCR In Heisenberg Form*

In this model the CCR refer directly to the commutators of P and Q or of the lowering and raising operators A and A^+. It is technically easier to start with the latter.

Definition 4.1 *To say that a separable Hilbert space $\mathcal{H}$ carries a representation of the canonical commutation relation in Heisenberg form means that there can be found on $\mathcal{H}$ a closed and densely defined operator A such that the domains of the operators A^*A and AA^* are equal and dense in $\mathcal{H}$ (this common dense domain will be denoted by $\mathcal{D}$, but is not the domain $\mathcal{S}$ discussed above), and such that*

$$AA^*\phi \;-\; A^*A\phi \;=\; \phi, \qquad\qquad \phi \in \mathcal{D}. \tag{4.2.1}$$

It is a standard result in operator theory that the *number operator*

$$N \;=\; A^*A, \tag{4.2.2}$$

defined on its natural domain $\mathcal{D}(N) \;=\; \mathcal{D}$, is then self-adjoint.

4.2.2 *The Common Domain*

Thus, even before we have introduced any common domain $\mathcal{S}$, we have the domain $\mathcal{D}$ to keep track of. Indeed, the above definition implicitly introduces two other domains, these being the domains of the unbounded operators A and A^*, both of which are, in principle, different subsets of $\mathcal{H}$ which properly contain $\mathcal{D}$. This proliferation of domains is characteristic of any analysis of unbounded operators — as more and more operators are introduced, more and more domains become defined. This would become burdensome, but fortunately it is unnecessary, provided that the smooth domain is introduced at this stage.

Consider how any element of the (as yet unknown) algebra of operators within which the observables are to lie relates to such a common domain. If B is such an operator, then $\mathcal{S}$ must be contained in the domain $\mathcal{D}(B)$ of B. Certainly, then, $\mathcal{S}$ must be contained in the intersection of all the domains $\mathcal{D}(B^n)$ of the operators B^n for any integer $n \geqslant 0$.

We must take account of all operators B, and since the greater the number of operators B, the smaller the intersection, it is possible that each new operator will diminish the size of $\mathcal{S}$. This would be undesirable and would, moreover, result in an unworkable theory. Fortunately, it is possible to avoid this difficulty, and to define a smooth domain in a simple manner which, moreover, explicitly takes account of the nature of the representation of the CCR (in Heisenberg form) carried by the Hilbert space $\mathcal{H}$. The key to the problem is the number operator N. Admittedly, the following definition comes *ab supra*, but the importance of the number operator in quantum mechanics should provide some grounds for its consideration until the utility of the definition provides its own justification.

Definition 4.2 *The common domain (called the smooth domain in what follows) of the smooth model is*

$$\mathcal{S} = \mathcal{D}^\infty(N) = \bigcap_{n \geqslant 0} \mathcal{D}(\overline{N^n}), \tag{4.2.3}$$

Moreover, in view of the identity in equation (4.2.1), *it is also the case that*

$$\mathcal{S} = \bigcap_{n \geqslant 0} \mathcal{D}(\overline{(AA^*)^n}).$$

The notation $\mathcal{D}^\infty(N)$ may be unfamiliar to the reader, but it is standard in the theory of unbounded operator algebras, although the notation $\mathcal{C}^\infty(N)$ is also used.

Notation At this stage it is appropriate to introduce a notational convention that will be adopted in what follows. Up to this point, elements of the Hilbert space $\mathcal{H}$ have been denoted by a lower case Greek symbol such as ϕ. As a subspace of $\mathcal{H}$, it would be perfectly proper to refer to elements of $\mathcal{S}$ using Greek symbols. However, the space $\mathcal{S}$ is not simply a subspace of $\mathcal{H}$, but is also a locally convex topological space in its own right, and it is questions of continuity of endomorphisms of $\mathcal{S}$ with respect to its own topology that are often of interest. Consequently, it is convenient to adopt a notational device to distinguish formulæ which are discussing endomorphisms of $\mathcal{S}$ from those formulæ which are about operators on $\mathcal{H}$. This will be done by writing elements of $\mathcal{S}$ using lower case Roman symbols like f, as in the above equation. This simple convention should cause no confusion, and acts as a useful *aide-mémoire* for the analyst.

This convention will be extended, when needed, by writing elements of the dual space $\mathcal{S}'$ of $\mathcal{S}$ using upper case Roman symbols, such as T. Hence $[\![T, f]\!]$ is the bilinear duality pairing of $f \in \mathcal{S}$ with $T \in \mathcal{S}'$. ■

4.2.3 *Kinematic Observables On $\mathcal{S}$*

So far, we have only presented the CCR in term of the raising and lowering operators A and A^*. We now define the operators P and Q, and show how these operators are related to the smooth domain $\mathcal{S}$.

First observe that because of the two characterizations of $\mathcal{S}$, both A and A^* leave $\mathcal{S}$ invariant. This enables the endomorphisms P and Q of $\mathcal{S}$ to be defined by the formulæ

$$Qf = \frac{1}{\sqrt{2}}(A + A^*)f, \qquad Pf = \frac{1}{i\sqrt{2}}(A - A^*)f, \qquad f \in \mathcal{S}. \quad (4.2.4)$$

The relation between these operators, the CCR and the space $\mathcal{S}$ can be summarized as follows. (The reader is referred to Putnam [184] for details of the proof.)

Theorem 4.3 ***(Properties Of The Smooth Domain)*** *The space $\mathcal{S}$ of equation* (4.2.3) *is a dense linear subspace of $\mathcal{H}$ with the following properties:*

1. *$\mathcal{S}$ is a core for N,*
2. *$\mathcal{S}$ is invariant under both A and A^*, and hence also under P, Q and N,*
3. *P, Q and the restriction of N to $\mathcal{S}$ are all essentially self-adjoint on $\mathcal{S}$, and the closures of the restrictions of A and A^* to $\mathcal{S}$ are A and A^* respectively,*
4. *$\mathcal{D}(A) = \mathcal{D}(A^*) = \mathcal{D}(\overline{P}) \cap \mathcal{D}(\overline{Q})$, on which space $A = \frac{1}{\sqrt{2}}(\overline{Q} + i\overline{P})$, $A^* = \frac{1}{\sqrt{2}}(\overline{Q} - i\overline{P})$,*
5. *the self-adjoint operator N has completely discrete spectrum $\mathbb{N} \cup \{0\}$, and $\mathcal{S}$ contains all the eigenvectors of N.*

Moreover $\mathcal{S}$ is the largest subspace of $\mathcal{H}$ which has these properties.

Hence, the CCR are satisfied on $\mathcal{S}$ either in the form of equation (4.2.1),

$$AA^*f - A^*Af = f, \quad (4.2.1)$$

or else in the form

$$QPf - PQf = if\,, \tag{4.2.5}$$

for all $f \in \mathcal{S}$.

Since $\mathcal{S}$ is central to the construction of the smooth model, any new characterization of it is useful. Particularly interesting are the two following representations, which do not involve N directly.

Proposition 4.4 *The two identities*

$$\mathcal{S} = \bigcap_{j,k=0}^{\infty} \mathcal{D}(\overline{Q}^j\, \overline{P}^k) = \bigcap_{j,k=0}^{\infty} \mathcal{D}\left(\overline{A^j(A^*)^k}\right). \tag{4.2.6}$$

hold, defining $\mathcal{S}$ *in terms of* P *and* Q *on the one hand, and* A *and* A^* *on the other.*

In the next Chapter, the representation in terms of P and Q will lead to a function space characterization of $\mathcal{S}$ in terms of smoothness and fall-off properties for functions which is easy to picture, and which gives the model its name.

4.2.4 *Topological Vector Spaces*

The space $\mathcal{S}$ is not a Hilbert space, nor even a Banach space. It is a special sort of topological vector space. In this Subsection, certain aspects of $\mathcal{S}$ will be considered from that point of view. Some general references on topological vector spaces and related topics are cited at the end of the Chapter.

Proposition 4.5 ***(Smooth Domain Topology)*** *The space* $\mathcal{S}$ *can be given a topological structure in the following fashion: the expressions*

$$\mathfrak{d}_n(f) = \|\, N^n f\, \|\,, \qquad n \in \mathbb{N} \cup \{0\},\ f \in \mathcal{S}, \tag{4.2.7}$$

define (Hilbertian) seminorms on $\mathcal{S}$ *which equip it with a Fréchet topology. If the kernel of* N *is finite-dimensional (which will be so in the cases of interest), then this topology is nuclear.*

The relationship between the Hilbert space $\mathcal{H}$ and the Fréchet space $\mathcal{S}$ forms an example of what is termed a rigged triple, or rigged Hilbert space [73]. For $\mathcal{S}$ has been defined so that it is a dense linear subspace of the Hilbert space $\mathcal{H}$, and since the Hilbert space norm $\|\cdot\|$ on $\mathcal{H}$ is equal to $\mathfrak{d}_0$ when restricted to $\mathcal{S}$, the embedding map from $\mathcal{S}$ into $\mathcal{H}$ is continuous. Additionally, every Hilbert space is self-dual — this is part of the Riesz Representation Theorem. Consequently, every element of $\mathcal{H}$ can be regarded as (identified with, more precisely) an element of the dual $\mathcal{S}'$ of $\mathcal{S}$. To be specific, an element $\phi \in \mathcal{H}$ defines a functional $\phi^\dagger$ in $\mathcal{S}'$ via the formula

$$[\![\phi^\dagger , f]\!] = \langle \phi, f \rangle , \qquad f \in \mathcal{S}. \tag{4.2.8}$$

It should be noted that the map $^\dagger : \mathcal{H} \to \mathcal{S}'$ is antilinear (rather than linear) injection, but it is still possible for us to regard its image as a dense subspace of $\mathcal{S}'$ in the strong dual topology. Thus there is a pair of continuous dense inclusions

$$\mathcal{S} \subseteq \mathcal{H} \subseteq \mathcal{S}' . \tag{4.2.9}$$

Gel'fand & Vilenkin (*ibid*) specify a number of technical requirements on the nature of the seminorms which define the topology of $\mathcal{S}$ in addition to these dense inclusions in order that the three spaces comprise a rigged triple. It turns out that as the topology on $\mathcal{S}$ is nuclear and Fréchet, these other conditions are satisfied automatically — consequently we have not felt it necessary to list them explicitly.

In all of the examples in this book, the æsthetic problems caused by the antilinearity of the embedding $^\dagger : \mathcal{H} \to \mathcal{S}'$ can be overcome by the fact that $\mathcal{H}$ possesses what is called a complex structure, namely an antilinear bijection[1] $J : \mathcal{H} \to \mathcal{H}$ which has the following properties:

$$J^2\phi = \phi, \tag{4.2.10.a}$$
$$\langle J\phi , J\psi \rangle = \langle \psi , \phi \rangle , \tag{4.2.10.b}$$

for all ϕ, $\psi \in \mathcal{H}$. In such a case, the mapping $\phi \mapsto (J\phi)^\dagger$ is a continuous linear map from $\mathcal{H}$ to $\mathcal{S}'$, and it will then be natural to regard $(J\phi)^\dagger$ as the image of $\phi \in \mathcal{H}$ in $\mathcal{S}'$.

[1]This map J should not be confused with the symplectic matrix in Chapter 2. That matrix also defines a complex structure, but one on phase space.

Notation (Bras And Kets) In the Dirac formalism, the pure states of the system are said to be denoted by so-called *kets*, and so we may tentatively identify $|\psi\rangle$ with $\psi \in \mathcal{H}$ (working for the while with the bounded model). Then there are *bras*, denoted $\langle\psi|$, which are thought of as dual vectors. The system works by having the bras act on the kets as antilinear functionals, so that $\langle\psi|\phi\rangle$ is the inner product $\langle\psi, \phi\rangle$.

Dirac tells us that the bras are defined only to the extent that the numbers $\langle\psi|\phi\rangle$ are finite; and that there is a one-to-one correspondence between bras and kets [49]. This is a reasonable statement in the context of the bounded model, where the pure states are characterized by the elements of $\mathcal{H}$, and hence both the bras and the kets can be indexed by the elements of $\mathcal{H}$, since any Hilbert space is self-dual.

However, it is not true for the smooth model, where our interpretation is that the pure states, the kets, are the elements of $\mathcal{S}$, whereas the bras are elements of the dual space $\mathcal{S}'$ — which is not a subspace of $\mathcal{H}$. In this way, a bra-ket pairing, formally written as $\langle T \mid f\rangle$, always makes sense as a duality pairing,

$$[\![T^*, f]\!] = \overline{[\![T, \overline{f}]\!]}, \qquad T \in \mathcal{S}',\ f \in \mathcal{S},$$

where a complex conjugation must be put in by hand to compensate for the sesquilinearity of the inner product.

This interpretation is consistent with a rigorous version of Dirac's treatment of operators with continuous spectra. In his book, Dirac tells us that if B is an observable with a continuous spectrum (which we shall take to be absolutely continuous and equal to $\mathbb{R}$ for definiteness) it has eigenvectors $|t\rangle$ such that

$$B\,|t\rangle = \varepsilon(t)\,|t\rangle, \qquad \langle s\,|\,t\rangle = \delta(s-t), \qquad s,\, t \in \mathbb{R},$$

where ε is some function from $\mathbb{R}$ to the spectrum of B. Clearly these bras and kets are not Hilbert space vectors. They are no longer labelled by elements of $\mathcal{H}$, and the evaluation $\langle s\,|\,t\rangle$ is no longer a finite complex number, given $s,\, t \in \mathbb{R}$. What are we to make of this?

As will be discussed in our treatment of generalized eigenvector expansions, applying the spectral calculus to an observable B (whose domain contains $\mathcal{S}$) with an absolutely continuous and nondegenerate spectrum means that there is a Hilbert space $L^2[\mathbb{R}, d\mu]$ of functions, a map ε from $\mathbb{R}$ to the spectrum of B and a unitary map U from $\mathcal{H}$ into $L^2[\mathbb{R}, d\mu]$ for some measure μ on $\mathbb{R}$ with respect to which B can be made diagonal[2],

$$[UBf](t) = \varepsilon(t)\,[Uf](t), \qquad f \in \mathcal{S},\ t \in \mathbb{R}.$$

Using the interpretation of kets and bras given above, the transformed kets are the elements Uf of $U[\mathcal{S}]$, and the bras are the elements of $U[\mathcal{S}']$. Amongst the latter are the delta distributions δ_t, so that Dirac's $\langle t \,|\, f\rangle$ is to be interpreted as

$$[\![\, \delta_t\,, Uf \,]\!] = [Uf](t), \qquad t \in \mathbb{R},\ f \in \mathcal{S}.$$

Then

$$[\![\, \delta_t\,, UBf \,]\!] = \varepsilon(t)[Uf](t), \qquad t \in \mathbb{R},\ f \in \mathcal{S}\,,$$

is the eigenvalue equation, meant weakly. There is no meaning to the ket $|\, s\rangle$; the delta function is a distribution and always a bra. The "normalization" condition $\langle t \,|\, s\rangle = \delta(t-s)$ is an unnecessary solecism which can be dispensed with.

Our conclusion is that the ideas of Dirac can be interpreted in a consistent fashion through rigorous distribution theory and spectral theory, but in order to interpret this very appealing notation and formalism requires a substantial amount of technical work; the very simplicity of the bra-ket formalism obscures the necessary precision which is needed to justify it. For that reason alone, we shall use the Dirac notation only very occasionally in what follows. ■

Before leaving this section, we introduce a concrete example of a topological vector space which will be of signal importance in what follows. As is well-known, the choice of an orthonormal basis for a (separable) Hilbert space enables us to identify that Hilbert space with the Hilbert sequence space ℓ^2 of square-summable sequences. While, in most cases of interest, a similar identification of the space $\mathcal{S}$ is possible, a less familiar sequence space than ℓ^2 is required. Consequently it is worth mentioning the sequence space involved, while omitting proofs. Although the material below is not essential for an understanding of what follows, it is nonetheless useful.

[2]The general case can be reduced to this by means of a Hilbert space-valued direct integral.

Definition 4.6 *A complex sequence* $(a_n)_{n\geqslant 0}$ *is called* rapidly decreasing *if the infinite set of series*

$$\sum_{n=0}^{\infty} n^{2k} \,|\, a_n \,|^2, \qquad k \in \mathbb{N}, \tag{4.2.11}$$

all converge. The set $\mathfrak{s}$ *of rapidly decreasing sequences is a vector space under component-wise operations. Equipped with the family of seminorms* $\{\mathfrak{q}_k \,:\, k \in \mathbb{N}\}$, *where*

$$\mathfrak{q}_k(a)^2 \;=\; \sum_{j=0}^{\infty} j^{2k} |\, a_j \,|^2, \qquad k \in \mathbb{N}, \tag{4.2.12}$$

it is a (reflexive and countably Hilbert) nuclear Fréchet space. Its dual space $\mathfrak{s}'$ *consists of those sequences* $(t_n)_{n\geqslant 0}$ *such that*

$$[\![\, t \,,\, a \,]\!] \;=\; \sum_{n=0}^{\infty} t_n\, a_n \tag{4.2.13}$$

converges absolutely for all $a \in \mathfrak{s}$, *with* $[\![\, t \,,\, a \,]\!]$ *serving as the bilinear duality pairing. A necessary and sufficient condition for this to be the case is that there exists some constant* $C > 0$ *and an integer* $r \in \mathbb{N}$ *such that*

$$|\, t_n \,| \;\leqslant\; C\,(n+1)^r, \qquad n \geqslant 0. \tag{4.2.14}$$

Such sequences are said to be slowly growing.

For example, the sequence $\big(1/(n{+}1)^5\big)$ is not rapidly decreasing, as equation (4.2.11) is violated for all $k \geqslant 6$. The sequence $\big(e^{-n}\big)$ is rapidly decreasing as the exponential decrease swamps any monomial growth. The sequence $\big(n^7\big)$ is slowly growing, whereas $\big(e^n\big)$ is not. Of course every rapidly decreasing sequence $\big(a_n\big)$ is p-summable, meaning that $\big(\,|\, a_n \,|^p\,\big)$ is an absolutely convergent series, and every p-summable series is slowly growing. For discussions of such sequences see [142], [52]. In Propositions 4.17 and 4.18 below, it will be shown when and how $\mathfrak{s}$ is equivalent, as a topological vector space, to $\mathcal{S}$.

4.3 Algebraic Structure Of The CCR

At an equivalent stage in the development of the bounded model, it was found that the Weyl form of the CCR could be couched in algebraic terms, as representations of the Weyl group. Questions of an algebraic nature were then considered, such as irreducibility, and used to determine the physically relevant representations. Subject to some technical complications, the same can be done for the smooth model.

4.3.1 *Unbounded Operator Algebras And Representations*

A *- algebra, $\mathcal{A}$, in this book, will be a complex vector space equipped with an associative (but not generally commutative) product that is distributive over addition. There will also be an involution, which is a self-inverse antilinear map $x \to x^*$ on $\mathcal{A}$ reversing the product, $(xy)^* = y^*x^*$. There may or may not be an identity; if there is, it is unique.

At this abstract level, there is no restriction on the nature of the elements of an algebra, but of course they may be algebras of unbounded operators on a Hilbert space. And for us the most important such algebra is of this type.

Definition 4.7 *Let $\mathcal{D}$ be a dense subspace of a (separable) Hilbert space $\mathcal{H}$. By $\mathcal{L}^+(\mathcal{D})$ is meant the collection of all linear endomorphisms a of $\mathcal{D}$ whose adjoint a^* is densely defined and which satisfy the following two conditions:*

$$\mathcal{D} \subseteq \mathcal{D}(a^*), \qquad a^*(\mathcal{D}) \subseteq \mathcal{D}.$$

*If $a \in \mathcal{L}^+(\mathcal{D})$, then the restriction of its adjoint to the domain $\mathcal{D}$ defines an element $a^+ = a\big|_{\mathcal{D}}$ of $\mathcal{L}^+(\mathcal{D})$, and $\mathcal{L}^+(\mathcal{D})$ becomes a unital *-algebra with composition of operators as product, the operator $^+$ as involution, and the identity map on $\mathcal{D}$ as unit.*

The algebra $\mathcal{L}^+(\mathcal{D})$ is a fundamental example of what is termed an unbounded operator algebra, and a comprehensive treatment of the theory of such algebras can be found in the work of Schmüdgen [203]. It is important to note that the *-algebra $\mathcal{L}^+(\mathcal{D})$ is truly an algebra of unbounded operators. To see why this is so, we note that an application of the Hellinger-Toeplitz Theorem [186] implies that $\mathcal{L}^+(\mathcal{H})$ is simply the algebra $\mathbb{B}(\mathcal{H})$ of bounded operators on $\mathcal{H}$, and it can be shown that if $\mathcal{L}^+(\mathcal{D})$ contains a closed operator then $\mathcal{D} = \mathcal{H}$ (and hence $\mathcal{L}^+(\mathcal{D}) = \mathbb{B}(\mathcal{H})$). Thus the *-algebra $\mathcal{L}^+(\mathcal{D})$ contains no closed operators whenever $\mathcal{D}$ is a proper subspace of $\mathcal{H}$.

We also note the important fact that if $\mathcal{A}$ is a division *-subalgebra of $\mathcal{L}^+(\mathcal{D})$, namely a *-subalgebra which possesses an identity and within which every nonzero element is invertible, then every element of $\mathcal{A}$ is a multiple of the identity.

The algebra $\mathcal{L}^+(\mathcal{D})$ is the backdrop for representation theory.

Definition 4.8 *If $\mathcal{A}$ is an abstract *-algebra with an identity, then by a* *-representation *of $\mathcal{A}$ on a dense subspace $\mathcal{D}$ of a Hilbert space $\mathcal{H}$ is meant a *-homomorphism π of $\mathcal{A}$ into $\mathcal{L}^+(\mathcal{D})$ which preserves the identity. The space $\mathcal{D}$ is said to carry the representation.*

Thus we have that $\pi(ab) = \pi(a)\pi(b)$ and $\pi(a^*) = \pi(a)^+$ for all $a, b \in \mathcal{A}$. This definition is, of course, extremely general, and it proves necessary to distinguish special cases of *-representations of *-algebras which enjoy particular properties. There are many possible definitions, but the only one in which we shall be interested is the following:

Definition 4.9 *A *-representation π of the *-algebra $\mathcal{A}$ acting on the dense subspace $\mathcal{D}$ of the (separable) Hilbert space $\mathcal{H}$ is called* self-adjoint *if the following condition holds:*

$$\mathcal{D} = \bigcap_{a\in\mathcal{A}} \mathcal{D}\big(\pi(a)^*\big). \tag{4.3.1}$$

It should be noted that this condition is one which requires information concerning the domains of the adjoints of all operators in the algebra, and does not of itself provide information about any one such domain in particular. It therefore makes no claims about the self-adjointness or otherwise of any of the operators $\pi(a)$. Indeed, it cannot, for since no element of $\mathcal{L}^+(\mathcal{D})$ can be closed, no element can be self-adjoint.

It is worth noting that any *-representation of $\mathcal{A}$ on the space $\mathcal{D}$ provides $\mathcal{D}$ with a locally convex topology, called the *graph topology*. This is the weakest locally convex topology on $\mathcal{D}$ for which the map $\pi(a)$ is a continuous map from $\mathcal{D}$ to $\mathcal{H}$ for all $a \in \mathcal{A}$, and as such is defined by the family of seminorms $\{\mathfrak{p}_a : a \in \mathcal{A}\}$, where

$$\mathfrak{p}_a(f) = \|af\|, \qquad a \in \mathcal{A},\, f \in \mathcal{D}. \tag{4.3.2}$$

Thus, although our definitions of algebras and representations to date have been purely algebraic, there is a mechanism whereby topological considerations can be introduced automatically.

4.3.2 *The Abstract CCR Algebra*

The Heisenberg form of the CCR can be turned into a representation of an abstract algebra in the above sense as follows.

Definition 4.10 *By the* abstract CCR algebra *(for one degree of freedom) is meant the noncommutative* **-algebra* $\mathbb{A}[p,q]$ *of all polynomials in two indeterminates* p *and* q*, which satisfy the relation*

$$qp - pq = i1\,, \tag{4.3.3}$$

where 1 *is the identity and the involution, denoted* **, is defined by* $p^* = p$*,* $q^* = q$ *and* $1^* = 1$*.*

As all representations of $\mathbb{A}[p,q]$ satisfy the Heisenberg form of the CCR, the representations of relevance to physics must be amongst them.

The first point to note is that no elementary *-representations of $\mathbb{A}[p,q]$ exist. For example, there are no finite-dimensional ones, for if there were we could find $k \times k$ matrices P and Q such that $[Q,P] = iI$. But this would imply that

$$0 = \mathrm{Tr}\big([Q,P]\big) = i\mathrm{Tr}(I) = i\,k,$$

which is absurd. More generally, there is no *-representation of $\mathbb{A}[p,q]$ by bounded operators, as the following theorem shows [184].

Theorem 4.11 ***(Winter-Wieland)*** *For any Hilbert space* $\mathcal{H}$*, there is no pair* P *and* Q *of bounded operators such that* $QP - PQ = i\,I$*.*

Proof: By an inductive argument, we see that if such operators existed, we would have that

$$i(n+1)P^n = QP^{n+1} - P^{n+1}Q, \qquad n \in \mathbb{N}.$$

Taking norms, this equality leads to the inequality

$$(n+1)\,\|\,P^n\,\| \leqslant 2\,\|\,Q\,\|\,\|\,P\,\|\,\|\,P^n\,\|\,, \qquad n \in \mathbb{N},$$

which would imply that $P^n = 0$ for large enough n. However the above identity shows us that $P^{n-1} = 0$ whenever $P^n = 0$, and so we would deduce that $P = 0$, which is impossible. ■

However, we can find interesting *-representations of the abstract CCR algebra $\mathbb{A}[p,q]$, and the one with which we shall be most concerned is the one carried by the smooth domain $\mathcal{S}$.

Proposition 4.12 *The *-representation $\pi : \mathbb{A}[p,q] \to \mathcal{L}^+(\mathcal{S})$ with*

$$\pi(1) = I, \qquad \pi(p) = P, \qquad \pi(q) = Q, \tag{4.3.4}$$

is a self-adjoint representation, and the standard nuclear Fréchet topology on $\mathcal{S}$ is the graph topology for this representation.

Notation Up until now the operators A and N have been considered as closed operators with their respective domains. In view of the importance that we choose to assign to the algebra $\mathcal{L}^+(\mathcal{S})$, and for reasons of simplicity (so as to be consistent with the definitions already given for the operators P and Q), we shall henceforth use the symbols A and N to denote the restrictions of these operators to the smooth domain $\mathcal{S}$, so that A and N may now be considered elements of $\mathcal{L}^+(\mathcal{S})$. Recalling the results of Theorem 4.3, we see that the operators previously denoted A and N will now be denoted $\overline{A}$ and $\overline{N}$ respectively, and moreover that the meaning of the statement $A^* = \overline{A^+}$ is unchanged by this change of convention.

Adopting this convention, it is possible to write

$$N = A^+A, \tag{4.3.5.a}$$

and also

$$A = \frac{1}{\sqrt{2}}(Q + iP), \quad A^+ = \frac{1}{\sqrt{2}}(Q - iP), \quad N = \frac{1}{2}(P^2 + Q^2 - 1), \tag{4.3.5.b}$$

consonant with equation (4.2.4). ■

4.3.3 *Gauge Invariant Representations*

Evidently, any vector in $\mathcal{D}(\overline{A})$ which is annihilated by $\overline{A}$ is also in the domain of $\overline{N}$ and is annihilated by $\overline{N}$ as well, from which it is clear that it also belongs to the smooth domain $\mathcal{S}$. The elements of the kernel of the operator $\overline{A}$ have a special rôle to play in our theory, and are called *Fock vectors.*

Since the closure $\overline{N}$ of the number operator N is self-adjoint, it can be used to generate a strongly continuous one-parameter unitary group on $\mathcal{H}$. Action by this group induces what are known as gauge transformations, more precisely global gauge transformations, sometimes known as gauge transformations of the first kind.

Definition 4.13 *By the* gauge group *is meant the strongly continuous one-parameter unitary group generated by the number operator $\overline{N}$,*

$$\Gamma(t) = e^{it\overline{N}}, \qquad t \in \mathbb{R}. \tag{4.3.6}$$

A vector $\psi \in \mathcal{H}$ is said to be a gauge invariant vector *if it is invariant under the action of the gauge group, so that*

$$\Gamma(t)\psi = \psi, \qquad t \in \mathbb{R}. \tag{4.3.7}$$

Given a representation of the canonical commutation relation on the Hilbert space $\mathcal{H}$, the fact that the spectrum of $\overline{N}$ consists of non-negative integers alone (see Theorem 4.3 above) implies that the gauge group Γ can be factored to provide a strongly continuous one parameter unitary group of the unit circle $\mathbb{T}$, rather than of $\mathbb{R}$. By a standard abuse of notation, the gauge group is frequently expressed in the factored form

$$\Gamma\left(e^{i\vartheta}\right) = e^{i\vartheta\overline{N}}, \qquad -\pi < \vartheta \leqslant \pi. \tag{4.3.8}$$

The gauge concept can be applied to the canonical commutation relation through this further definition.

Definition 4.14 *To say that a *-representation of the canonical commutation relation is* gauge invariant *is to say that the vector space of gauge invariant vectors is one-dimensional.*

There are a number of properties that can readily be established concerning Hilbert spaces which carry a gauge invariant representation of the canonical commutation relation.

Theorem 4.15 ***(Gauge Invariant Representations Revisited)*** *If the separable Hilbert space $\mathcal{H}$ carries a gauge invariant representation of the canonical commutation relation, then:*

1. *All of the eigenvalues of N are nondegenerate.*
2. *Letting Ω_k denote the (unique up to a phase factor) eigenvector of N with eigenvalue k, the set $(\Omega_k)_{k \geqslant 0}$ is an orthonormal basis for $\mathcal{H}$. These vectors will be referred as the Hermite-Gauss vectors*[3]*.*
3. *The vectors $\{\Omega_n : n \geqslant 0\}$ can be expressed in the form*

$$\Omega_n = \frac{1}{\sqrt{n!}} (A^*)^n \Omega_0, \qquad n \in \mathbb{N}. \tag{4.3.9}$$

The Hermite-Gauss vectors are known to physicists as the (harmonic) oscillator eigenfunctions and often denoted by the Dirac ket symbols $|n\rangle$. Their explicit form (as functions) depends on which representation is used, of course.

The following important uniqueness result holds, showing the power of gauge invariance.

Theorem 4.16 *Any two gauge invariant representations of the canonical commutation relation are unitarily equivalent.*

Proof: Suppose that the Hilbert spaces $\mathcal{H}$ and $\mathcal{K}$ carry gauge invariant representations of the canonical commutation relation, with respective operators $A_{\mathcal{H}}$, $A^*_{\mathcal{H}}$ and $A_{\mathcal{K}}$, $A^*_{\mathcal{K}}$, and respective families $(\Omega_n^{\mathcal{H}})$ and $(\Omega_n^{\mathcal{K}})$ of Hermite-Gauss vectors. Then it is easy to see that the map $\mathcal{U} : \mathcal{H} \to \mathcal{K}$, given by

$$\mathcal{U}\left(\Omega_n^{\mathcal{H}}\right) = \Omega_n^{\mathcal{K}}, \qquad n \geqslant 0,$$

is a unitary equivalence between these representations, since

$$A_{\mathcal{K}} = \mathcal{U} \cdot A_{\mathcal{H}} \cdot \mathcal{U}^{-1}, \qquad A^*_{\mathcal{K}} = \mathcal{U} \cdot A^*_{\mathcal{H}} \cdot \mathcal{U}^{-1}.$$

■

[3]This is because, as will be seen in the following Chapter, in the Schrödinger representation, they are precisely the classical Hermite polynomials multiplied by the Gaussian.

It is now possible, given the above results, to identify the space $\mathcal{S}$ more precisely.

Proposition 4.17 *For a gauge invariant representation of the canonical commutation relation on the Hilbert space $\mathcal{H}$, it can be proved that the space $\mathcal{S}$ consists precisely of those vectors f in $\mathcal{H}$ for which the sequence $\big(\langle \Omega_n\,,\, f\rangle\big)_{n\geqslant 0}$ belongs to $\mathfrak{s}$, the space of rapidly decreasing sequences introduced in Definition 4.6.*

One consequence of this Proposition (since $\mathfrak{s} \subseteq \ell^2$) is that

$$f \;=\; \sum_{n=0}^{\infty} \langle \Omega_n\,,\, f\rangle\, \Omega_n \tag{4.3.10}$$

for any $f \in \mathcal{S}$, where the convergence of this series is assured in $\mathcal{H}$. However, $\mathcal{S}$ has its own topology, and the Hermite-Gauss vectors all belong to $\mathcal{S}$, and so it can be asked whether the above series makes sense in the topology of $\mathcal{S}$. Moreover, as has been observed, the collection of Hermite-Gauss vectors forms an orthonormal basis for $\mathcal{H}$, and we would like to know to what extent, and in what manner, the collection of Hermite-Gauss vectors is a basis for $\mathcal{S}$. There is a theory concerning bases in topological vector spaces [166], and we have the best possible result in this case, as is enunciated in the following result.

Proposition 4.18 *With respect to the standard nuclear Fréchet topology on the space $\mathcal{S}$ defined by the seminorms introduced in equation (4.2.7), the Hermite-Gauss vectors $(\Omega_n)_{n\geqslant 0}$ form a Schauder basis, in the sense that for any vector $f \in \mathcal{S}$ a unique sequence $(a_n)_{n\geqslant 0}$ of complex coefficients can be found such that the infinite series*

$$\sum_{n=0}^{\infty} a_n\, \Omega_n \tag{4.3.11.a}$$

converges to f with respect to the topology on $\mathcal{S}$, and moreover that the linear functional

$$f \;\mapsto\; a_n\,, \qquad\qquad f \in \mathcal{S}, \tag{4.3.11.b}$$

is a continuous linear functional on $\mathcal{S}$ for any $n \geqslant 0$. In view of the fact that the Hermite-Gauss vectors form an orthonormal basis for $\mathcal{H}$, these *functionals are determined by the formulæ:*

$$a_n = \langle \Omega_n , f \rangle, \qquad n \geqslant 0, \ f \in \mathcal{S}. \qquad (4.3.11.c)$$

The map sending the vector f in $\mathcal{S}$ to the sequence $(a_n)_{n \geqslant 0}$ is a linear and topological isomorphism from $\mathcal{S}$ to $\mathfrak{s}$.

This gives a complete description of the basic properties of gauge invariant representations of the CCR. Further evidence of the importance of gauge invariant representations of the CCR can be found in the fact that all representations of the CCR can be expressed in terms of gauge invariant ones.

Proposition 4.19 *A representation of the canonical commutation relation can be written as an orthogonal direct sum of at most a countable number of gauge invariant representations [184], [52].*

4.3.4 *Irreducibility*

When considering the bounded model, there was a natural definition of an irreducible representation. One of the technical difficulties inherent in the analysis of unbounded operator algebras is that there is no single concept of irreducibility which will suffice in all cases [203]. However, in the case of *-representations of the CCR, only one of these definitions is necessary, and the concept of irreducibility introduced by this definition is closely related to that of gauge invariance. We make the following definition:

Definition 4.20 *A representation of the canonical commutation relation is* irreducible *if there is no closed subspace of $\mathcal{S}$ which is invariant under A and A^+.*

To understand, and interpret, this definition, we need some further terminology.

Definition 4.21 *A bounded operator $C \in \mathbb{B}(\mathcal{H})$ is said to* commute weakly *with a *-representation of the canonical commutation relation if*

$$\langle Bg , Cf \rangle = \langle C^* g , B^* f \rangle, \qquad f, g \in \mathcal{S}, \qquad (4.3.12)$$

for all operators $B \in \mathrm{End}(\mathcal{S})$ which are polynomials in A and A^+. The collection of all such operators C constitutes the weak commutant *of the representation.*

Just as there are many versions of irreducibility, there are other definitions of commutant that can be used in the theory of unbounded operator algebras (such as strong commutativity), but only this notion is used in this book.

Given these definitions, the following result relates them in a satisfactory manner.

Proposition 4.22 *For a *-representation of the canonical commutation relation, the following properties are equivalent:*

1. *the *-representation is gauge invariant,*
2. *the *-representation is irreducible,*
3. *the weak commutant of the *-representation is the set consisting of scalar multiples of the identity operator.*

It will be a basic assumption that, on physical grounds, these equivalent properties must be satisfied by any representation of the canonical commutation relation employed in the smooth model. This assumption is best justified by considering what reducible representations mean physically. Suppose, for example, that a representation is wanted which describes a particle carrying a generalized charge which can take the values ± 1. There will be two charge states, and to accommodate this the system Hilbert space must include an extra degree of freedom. One way to do this is to consider two-component Hilbert space of vectors of the form (ϕ_-, ϕ_+), where ϕ_-, $\phi_+ \in \mathcal{H}$. In other words, the usual system Hilbert space $\mathcal{H}$ has been replaced by the direct sum Hilbert space $\mathcal{H} \oplus \mathcal{H}$.

Not only the Hilbert space vectors, but the raising and lowering operators must also carry a charge label. It may be checked (and this is done later on) that the correct lowering operator for this system is the operator $A \oplus A$. This formalism leads to a (two-fold) reducible representation of the canonical commutation relation with two orthogonal unit Fock vectors.

This is a typical example, as reducibility is always associated with extra degrees of freedom, with these further degrees of freedom being associated with one or more additional physical quantities not previously described by the theory, such as the concept of generalized charge mentioned above. To make sense of this situation requires the notion of a *super-selection rule.* Following Streater and Wightman, a super-selection rule is a principle that rules out certain states as not being physically admissible, in such as way as to ensure that the representation carried by the subspace of physically admissible states is irreducible. One consequence is that not all self-adjoint operators are physically observable, since the only physically observable operators are those which map physically observable states to physically observable states. It might be a principle of nature, for example, that the only states of nature have generalized charge either -1 or $+1$. Then an admissible pure state is either of the form $(\phi_-, 0)$ or $(0, \phi_+)$, but cannot have both components nonzero. It is easy to construct examples of operators which map states of the form $(\phi_-, 0)$ to states of the form (ψ_-, ψ_+), and these would not be observable.

None of the systems considered in this book have such super-selection rules, so the interested reader is referred to Streater and Wightman [215] for further information. For a proof of the charge superselection rule within the framework of relativistic quantum field theory in the rigorous sense, see [217].

4.4 Axioms For The Smooth Model

We now have sufficient information to give a precise specification of the observables and states for the smooth model. In what follows, $\mathcal{H}$ is a Hilbert space which carries a gauge invariant representation of the CCR in Heisenberg form. We thus have the associated nuclear Fréchet space $\mathcal{S}$ as smooth domain, and the self-adjoint *-representation of the abstract CCR algebra $\mathbb{A}[p, q]$ carried by the space $\mathcal{S}$.

4.4.1 *Smooth Observables*

The construction of the smooth model was undertaken with the aim of choosing the largest collection of observables, given the technical requirements we have imposed. It is then clear that we need to choose the largest possible subcollection of $\mathcal{L}^+(\mathcal{S})$ to be the set of observables. In the bounded model, the set of observables was chosen to be the collection $\mathbb{B}(\mathcal{H})_h$ of self-adjoint elements of the algebra $\mathbb{B}(\mathcal{H})$ of bounded operators on $\mathcal{H}$, and we are seeking an analogous definition for the smooth model. However, as has been noted already, there are no self-adjoint elements of the algebra $\mathcal{L}^+(\mathcal{S})$, and so we must adopt a different definition. We recall that an element $B \in \mathcal{L}^+(\mathcal{S})$ is called *symmetric* if $B = B^+$.

> **Axiom 4.1** ***(Observables - Smooth Model)*** *Any symmetric element of the algebra $\mathcal{L}^+(\mathcal{S})$ is a (smooth) observable, and the set $\mathcal{L}^+(\mathcal{S})_h$ of smooth observables is a real Lie subalgebra of the (Lie) algebra $\mathcal{L}^+(\mathcal{S})$, where the Lie bracket of two elements in $\mathcal{L}^+(\mathcal{S})$ is defined to be i times the commutator of these elements.*

Note that the condition of symmetry ($B = B^+$) for an element B in $\mathcal{L}^+(\mathcal{S})$ is very similar to the condition for self-adjointness ($B = B^*$) for an element B in $\mathbb{B}(\mathcal{H})$, but is subtly different. It is not necessarily true, for example, even that the closure $\overline{B}$ of a symmetric element B in $\mathcal{L}^+(\mathcal{S})$ is self-adjoint (as an unbounded operator). We have adopted this definition, however, not only because it is the simplest possible but also because it provides us with the largest possible collection of observables.

Therefore P and Q are observables but A and A^+ are not (they are not symmetric), although they do belong to the algebra $\mathcal{L}^+(\mathcal{S})$ (see Section 4.6.1). It will prove possible, once the full terminology of the smooth model is established, to extend the notion of observables to incorporate operators other than those covered by the above Axiom (essentially by regarding these more general operators as idealizations, or approximations, of more complicated smooth observables), but we defer discussion of this matter to a later stage[4].

4.4.2 *Smooth States*

The definition of a general state, hence of the mixed states, follows the general considerations discussed in connection with states of the bounded model, in that the states of the smooth model will be positive normalized continuous linear functionals on $\mathcal{L}^+(\mathcal{S})$.

However, to interpret this statement, it is necessary to attribute meanings to the terms used. Since there was a unique meaning to the notion of a positive operator in $\mathbb{B}(\mathcal{H})$, there was no confusion as to what was meant

---

[4]The phase operator $\mathbf{\Delta}[\varphi]$ to be defined later is a bounded but not a smooth observable.

by a positive linear functional on $\mathbb{B}(\mathcal{H})$. However, there are several possible definitions of positivity of elements in $\mathcal{L}^+(\mathcal{S})$ [203], and so it is first necessary to specify the notion of positivity that we prefer. Our choice is the following — an operator $B \in \mathcal{L}^+(\mathcal{S})$ is said to be *positive* if

$$\langle f\,,\, Bf\rangle \geqslant 0\,, \qquad f \in \mathcal{S}, \tag{4.4.1}$$

and a positive linear functional on $\mathcal{L}^+(\mathcal{S})$ is then a linear functional on $\mathcal{L}^+(\mathcal{S})$ which takes positive real values on all positive elements of $\mathcal{L}^+(\mathcal{S})$. With this convention, we make the following statement:

> **Axiom 4.2.a *(States - Smooth Model)*** *A state of the system is a linear functional ω on the algebra of observables $\mathcal{L}^+(\mathcal{S})$ which is positive and normalized such that $\omega(I) = 1$.*

It would seem that this Axiom has ignored the question of continuity for states, which was explicitly included in our previous Axiom describing the states of the bounded model. There are reasons for this. To begin with, we have not yet assigned any particular topology to $\mathcal{L}^+(\mathcal{S})$, and without a topology any discussion of continuity is meaningless. However, there is a natural topology that can be defined on $\mathcal{L}^+(\mathcal{S})$ with respect to which $\mathcal{L}^+(\mathcal{S})$ becomes a locally convex topological *-algebra with separately continuous multiplication and continuous involution — technically speaking, this is the topology of uniform convergence on bounded subsets of $\mathcal{S}$ — but it is not necessary for us to be overly specific about the exact nature of this topology. This is because all positive linear functionals on $\mathcal{L}^+(\mathcal{S})$ are automatically continuous with respect to this topology.

As with the bounded model, it is convenient to have concrete representations of states in the smooth model. To that end, we have the following analogue of Gleason's Theorem 3.8:

Proposition 4.23 *Every state ω on $\mathcal{L}^+(\mathcal{S})$ is necessarily continuous, and has a representation*

$$\omega(B) = [\![\omega\,,\,B]\!] = \mathrm{Tr}\,(\rho B)\,, \qquad B \in \mathcal{L}^+(\mathcal{S}), \tag{4.4.2.a}$$

where ρ is a density matrix in the sense of Theorem 3.8 with the additional *property that the operators*

$$(N+1)^n \rho (N+1)^n \tag{4.4.2.b}$$

are well-defined and trace class for all positive integers n. Hence a state in the smooth model is known as a smooth state.

It is a consequence of these conditions that a density matrix ρ representing a smooth state ω has the property of mapping $\mathcal{H}$ into $\mathcal{S}$. Hence, for any positive integer n, the operator $(N+1)^n \rho (N+1)^n$ is an endomorphism of $\mathcal{S}$ as well as being a trace-class operator on $\mathcal{H}$. In particular, this means that the eigenvector associated with any nonzero eigenvalue of ρ belongs to $\mathcal{S}$.

We noted above that there was a convenient characterization of the smooth domain $\mathcal{S}$ in terms of the sequence space $\mathfrak{s}$. There is, unfortunately, no such simple description of the space $\mathcal{L}^+(\mathcal{S})$, but there such a characterization of the strong dual of $\mathcal{L}^+(\mathcal{S})$ (with respect to the natural topology on $\mathcal{L}^+(\mathcal{S})$ mentioned above).

Proposition 4.24 *The space $\mathfrak{s}^{(2)}$ of rapidly decreasing double sequences is the vector space of all double sequences $a = \big(a_{m,n}\big)_{m,n\geqslant 0}$ such that the series*

$$\mathfrak{q}^{(2)}_{j,k}(a)^2 = \sum_{m,n=0}^{\infty} m^{2j} n^{2k} |\, a_{m,n}\,|^2 \,, \qquad j,k \geqslant 0\,, \tag{4.4.3.a}$$

all converge. With respect to the topology defined by the countable family of seminorms $\{\mathfrak{q}^{(2)}_{j,k} : j,k \geqslant 0\}$, $\mathfrak{s}^{(2)}$ is a nuclear Fréchet space. The topological dual of $\mathcal{L}^+(\mathcal{S})$ can be identified with $\mathfrak{s}^{(2)}$, with an element $a \in \mathfrak{s}^{(2)}$ determining a continuous linear functional on $\mathcal{L}^+(\mathcal{S})$ via the formula:

$$[\![\, a \,,\, B \,]\!] = \sum_{m,n=0}^{\infty} a_{m,n} \langle \Omega_m \,,\, B\Omega_n \rangle \,, \qquad a \in \mathfrak{s}^{(2)},\ B \in \mathcal{L}^+(\mathcal{S}), \tag{4.4.3.b}$$

and this identification between $\mathfrak{s}^{(2)}$ and the strong dual of $\mathcal{L}^+(\mathcal{S})$ is a topological isomorphism.

The connection between these different characterizations of the dual of $\mathcal{L}^+(\mathcal{S})$ is as follows. If ω is an element of the dual of $\mathcal{L}^+(\mathcal{S})$ which is described by the double sequence $a \in \mathfrak{s}^{(2)}$, then there exists a trace-class operator ρ on $\mathcal{H}$ which can be defined in terms of its matrix coefficients with respect to the Hermite-Gauss functions as follows:

$$\langle \Omega_m \, , \, \rho \Omega_n \rangle \; = \; a_{n,m} \, , \qquad\qquad m, n \geqslant 0, \tag{4.4.4}$$

which satisfies the mollifying property (4.4.2.b) that $(N+1)^n \rho (N+1)^n$ be well-defined and trace-class for all positive integers n, and we then have that $\omega(B) = \mathrm{Tr}(\rho B)$ for all $B \in \mathcal{L}^+(\mathcal{S})$. Of course, if ω is a state on $\mathcal{L}^+(\mathcal{S})$, then ρ is the density matrix of Proposition 4.23.

Axiom 4.2.b ***(Pure & Mixed States - Smooth Model)*** *The set of states in the smooth model is a convex set and the pure states are its extreme points, in that a pure state cannot be decomposed into a convex linear sum of other states in a nontrivial fashion. A mixed state is a state which is not a pure state. States are given by density matrices which satisfy the additional mollifying condition in equation (4.4.2.b).*

The pure states are those whose density matrices are projections with one-dimensional ranges. Hence the terms pure state, extreme state *and* vector state *are synonymous. Pure states of the system may thus also be characterized by unit rays associated with $\mathcal{S}$.*

The density matrix for a mixed state has an eigendecomposition consisting of at least two distinct terms. The decomposition of a mixed state into pure states is not unique.

One of the reasons why there is no simple characterization of $\mathcal{L}^+(\mathcal{S})$, either as a tensor product or as a dual space, is that $\mathcal{L}^+(\mathcal{S})$ is not complete in its topology. It can be shown, however, that the completion of $\mathcal{L}^+(\mathcal{S})$ is the space $\mathcal{L}(\mathcal{S}, \mathcal{S}')$ of continuous linear maps from $\mathcal{S}$ to its strong dual $\mathcal{S}'$, and this space can be identified with the sequence space $\mathfrak{s}^{(2)'}$ of slowly growing double sequences. However, since we have not yet indicated the exact nature of the Schauder basis in $\mathcal{S}'$, we shall not be more specific about this identification at present. It is worth taking note of the space $\mathcal{L}(\mathcal{S}, \mathcal{S}')$, as it will be of significance in the discussion of Weyl quantization. Exactly as for the bounded model, we observe that the states form a convex set, and we define the pure states to be the extreme points of the set of states. Again we can characterize extreme states in terms of density matrices which are projections with one-dimensional range. The only difference now is that the a unit vector which defines such a projection must be in $\mathcal{S}$.

4.5 The Round-Off Approximation

A shortcoming of the smooth model is that certain operators which are natural candidates to be observables will not map the domain $\mathcal{S}$ into itself. An approximation method is available for dealing with this, analogous to the cut-off approximation method of the bounded model.

To illustrate the problem for this model in more detail, and how it can be overcome, consider the position operator Q on $L^2(\mathbb{R})$ as it is normally defined, that is, in the Schrödinger representation. By a standard result in spectral theory, the spectral measure E_Q for Q is given by the formula

$$E_Q(\Delta)\phi = \chi_\Delta \phi, \qquad \phi \in L^2(\mathbb{R}), \tag{4.5.1}$$

where the operator $E_Q(\Delta)$ consists of multiplication by the characteristic function χ_Δ of the Borel set Δ. As $E_Q(\Delta)$ is a projection operator it is bounded, and so $E_Q(\Delta)\phi$ belongs to $L^2(\mathbb{R})$ for all $\phi \in L^2(\mathbb{R})$. Unfortunately, $E_Q(\Delta)f$ may not belong to $\mathcal{S}$ even for $f \in \mathcal{S}$, as the sharp cut-off at the boundaries of Δ introduces discontinuities, and it will turn out that infinite differentiability is a necessary (but not sufficient) condition for membership of $\mathcal{S}$ (in this representation, at least).

This would seem to be a significant problem since, for numerous reasons, we would wish to be able to analyze spectral measures. However, why should we be required to class spectral measures as observables? To do so would be to say that, if J_1 and J_2 were distinct bounded intervals, the operators $E_Q(J_1)$ and $E_Q(J_2)$ were experimentally distinguishable observables. However this is not a reasonable assumption. To see this, consider the operators $E_Q(J)$ and $E_Q(K)$, where $J = [a, b]$ and $K = [a, b+\varepsilon]$, where $\varepsilon > 0$. Although J and K are distinct intervals, no matter how small ε is, were ε to be an order of magnitude less than the Planck length then no experimental device would be able to register a measurement outcome which belonged to the interval K, but not to J. Consequently there would be no operational difference between the *soi-disant* observables $E_Q(J)$ and $E_Q(K)$. For the same reasons, there would be no operational difference if, instead of operators such as $E_Q(J)$, we considered instead operators of the form

$$\phi \mapsto f_J \phi, \qquad \phi \in L^2(\mathbb{R}),$$

where f_J is a smooth function which is a close approximation to the characteristic function χ_J, but where the sharp edges of the discontinuities have been smoothed away. The resulting operators are then indeed smooth observables, and all is well. Mathematically, this results in replacing Q, for the purposes of its spectral analysis, with a smoothed version whose spectral decomposition involves a positive operator valued measure, rather than a projection valued measure. Similar considerations hold for any symmetric operator B which maps the domain $\mathcal{S}$ to itself.

The same difficulty arises in connection with certain Hamiltonians. One of many examples would be that of the finite square well potential problem. The discontinuity in the potential results in the Hamiltonian not leaving $\mathcal{S}$ invariant. However, for the same sort of reasons as those outlined above, it is not reasonable to require that a potential well have discontinuities — it would not be possible physically to create such forces. When all is said and done, the square well potential is just a mathematical idealization of the situation in which the value of the potential changes smoothly from a nonzero to a zero value in a very small, but nonzero, length. Replacing the "ideal" square well potential by such a smoothly-changing one, the Hamiltonian for the system preserves $\mathcal{S}$, and becomes a smooth observable. Similar arguments can be made for other physical systems.

To indicate that this type of replacement can always be done successfully, we make the following observations. Let A be a (not necessarily bounded) operator on $\mathcal{H}$ with domain containing $\mathcal{S}$, whose restriction to $\mathcal{S}$ is essentially self-adjoint. If (A_N) is any sequence of self-adjoint bounded operators on $\mathcal{H}$ such that $A_N f \to Af$ for all $f \in \mathcal{S}$, then a theorem of Rellich [55] shows that $g(A_N)f \to g(A)f$ for all $f \in \mathcal{S}$ and any continuous function g on $\mathbb{R}$. It is always possible to choose the operators (A_N) such that they (or, at least, their restrictions to $\mathcal{S}$) belong to $\mathcal{L}^+(\mathcal{S})$. One way to do this is by the method of truncation, so that A_N is the operator whose matrix coefficients with respect to the Hermite-Gauss functions are as follows:

$$\langle \Omega_m\,,\, A_N \Omega_n \rangle = \begin{cases} \langle \Omega_m\,,\, A\Omega_n \rangle\,, & 0 \leqslant m, n \leqslant N\,, \\ 0\,, & \text{otherwise.} \end{cases}$$

Consequently, by choosing N sufficiently large, we can replace the original operator A with a smooth observable in such a way that all calculated expectations, variances, and so on are as close as we choose (namely, to within experimental error) of the "true" expectations and variances that would result from the original operator A.

One advantage of the truncation approach is that not only the truncated operators A_N, but also their spectral functions $g(A_N)$, are smooth operators. However, this procedure may not always be regarded as being physical, and it may be necessary to find an alternate approximation scheme which has a more obvious physical (rather than purely mathematical) justification. An example of such an approach can be found in Dubin & Hennings [52] where, *inter alia*, it is shown that smoothing the Coulomb potential for small radii has the required result. This smoothing can be justified physically, since the model being discussed is non-relativistic, and hence the energies of any particles being considered are comparatively small, with the result that the Coulomb potential for small radii has no operational effect on the behaviour of particles, and so may reasonably be adjusted without substantially affecting the outcome of experiments.

Thus, at a number of stages, we are required to replace the ideal observables of elementary quantum theory with smoothed variants in the above fashion — which we call the round-off approximation for obvious reasons.

Just as the cut-off approximation lent the bounded model physical respectability, so this round-off method lends physical respectability to the smooth model.

4.6 Connecting The Models

4.6.1 *Common Terminology*

There are many common aspects to the bounded and smooth models; it will be useful for subsequent discussion to establish some common terminology and notation. To begin with, in either model we are provided with the (separable) Hilbert space $\mathcal{H}$, which carries a representation of the CCR, either in Weyl or in Heisenberg form (as we shall see below, a Hilbert space which carries a representation of the CCR in one form carries a representation in the other form — this is what unites the two models). In both cases, the collection of observables forms a real Lie subalgebra of some larger algebra. We shall denote this larger algebra by $\mathfrak{A}$, and call it the *algebra of observables* — a convenient term, if inaccurate, since not all of its elements are observables. Thus we have

$$\mathfrak{A} = \mathbb{B}(\mathcal{H}), \tag{4.6.1.a}$$

or

$$\mathfrak{A} = \mathcal{L}^{+}(\mathcal{S}), \tag{4.6.1.b}$$

according as we are working the bounded or smooth model, respectively. The observables in either model are then the symmetric (in the bounded model, therefore self-adjoint) elements of $\mathfrak{A}$, and we shall denote this collection by $\mathfrak{A}_h$.

In both models, the *states* of the system comprise the positive normalized elements of some larger collection $\mathfrak{A}_*$ of linear functionals[5] on $\mathfrak{A}$ for which there is a nonsingular bilinear duality pairing between $\mathfrak{A}$ and $\mathfrak{A}_*$. For the bounded model this collection is

$$\mathfrak{A}_* = \mathfrak{T}_1(\mathcal{H}), \tag{4.6.2.a}$$

the trace-class operators on $\mathcal{H}$, while for the smooth model we have

$$\mathfrak{A}_* = \mathcal{L}^{+}(\mathcal{S})', \tag{4.6.2.b}$$

the collection of all continuous linear functionals on $\mathcal{L}^{+}(\mathcal{S})$. Note that the relationship between the spaces $\mathfrak{A}$ and $\mathfrak{A}_*$ is different for the two models — $\mathfrak{A}$ is the dual of $\mathfrak{A}_*$ (but not conversely) in the bounded model; the reverse is true in the smooth model — this difference is forced upon us by the continuity properties we desire of states. In either model we may use the formula

$$[\![\rho, A]\!] = \mathrm{Tr}\big(AB_\rho\big), \qquad \rho \in \mathfrak{A}_*,\ A \in \mathfrak{A}, \tag{4.6.3}$$

to describe the nonsingular bilinear duality pairing between $\mathfrak{A}_*$ and $\mathfrak{A}$, where B_ρ is the density operator associated with $\rho \in \mathfrak{A}_*$ — in the bounded model B_ρ is equal to ρ itself, whereas in the smooth model B_ρ is the density matrix obtained from ρ via Theorem 4.23, the smooth analogue of Gleason's Theorem.

The states in either model are then the positive normalized (in the sense of the relevant Axioms) elements of $\mathfrak{A}_*$, and will be denoted by $\mathfrak{S}$. The terms *pure state* and *vector state* have the same definition in both models.

---

[5]This is a non-standard notation

Remark Streater and Wightman [215] quote Res Jost as saying that "In the thirties, under the demoralizing influence of perturbation theory, the mathematics required of a theoretical physicist was reduced to a rudimentary knowledge of the Latin and Greek alphabets." Things have certainly improved in the last sixty or so years: a vague acquaintance with Cyrillic script and the old German Fraktur alphabet is now also required[6]. ■

4.6.2 *The Connection Theorem*

The fact that the CCR in Weyl and Heisenberg form both have strong uniqueness theorems leads to a suspicion that there is a connection between them. This suspicion is well-founded, and this next theorem shows. The length and technicalities of the proof are somewhat surprising, and has been included because we have not found this theorem (in precisely this form) elsewhere. Although we will define the space again in subsequent Chapters, this proof requires some knowledge of the Schwartz space $\mathcal{S}(\mathbb{R}^2)$ of smooth functions on $\mathbb{R}^2$ all of whose partial derivatives are of rapid decrease — the reader is referred to Section 5.3 for a detailed definition.

Theorem 4.25 *A strongly continuous representation of the CCR in Weyl form can be used to generate a representation of the canonical commutation relation in the Heisenberg form. If the Weyl representation is irreducible, then so is the Heisenberg representation.*

Conversely, any representation of the CCR in Heisenberg form generates a representation in Weyl form, and the latter is irreducible if the former is. Moreover, these two processes are mutually inverse.

Proof: We shall only sketch some of the details. Given a strongly continuous and irreducible representation of the CCR in Weyl form, the representations (3.3.14)

$$a \mapsto U(a) = W(a,0), \qquad a \mapsto V(a) = W(0,a),$$

are strongly continuous unitary groups, and as such have self-

[6]Some knowledge of hieroglyphs enlivens the Bibliography, but is optional!

adjoint generators $\overline{P}$, $\overline{Q}$, respectively, whose domains are

$$\begin{aligned}\mathcal{D}(\overline{P}) &= \{\phi \in \mathcal{H} : \lim_{a\to 0} a^{-1}[U(a)\phi - \phi] \text{ exists}\},\\ \mathcal{D}(\overline{Q}) &= \{\phi \in \mathcal{H} : \lim_{a\to 0} a^{-1}[V(a)\phi - \phi] \text{ exists}\},\end{aligned}$$

with the definitions

$$\begin{aligned}\overline{P}\phi &= -i\lim_{a\to 0} a^{-1}[U(a)\phi - \phi], & \phi \in \mathcal{D}(\overline{P}),\\ \overline{Q}\phi &= -i\lim_{a\to 0} a^{-1}[V(a)\phi - \phi], & \phi \in \mathcal{D}(\overline{Q}).\end{aligned}$$

Moreover, we can show that $W[\![F]\!]\phi \in \mathcal{D}(\overline{P}) \cap \mathcal{D}(\overline{Q})$ for any F in $\mathcal{S}(\mathbb{R}^2)$ and $\phi \in \mathcal{H}$, with

$$\begin{aligned}\overline{P}\,W[\![F]\!]\phi &= W[\![\mathcal{L}_P F]\!]\phi,\\ \overline{Q}\,W[\![F]\!]\phi &= W[\![\mathcal{L}_Q F]\!]\phi,\end{aligned}$$

where $\mathcal{L}_P F$, $\mathcal{L}_Q F \in \mathcal{S}(\mathbb{R}^2)$ are given by the formulæ:

$$\begin{aligned}(\mathcal{L}_P F)(x,y) &= i(\partial_1 F)(x,y) + \frac{1}{2}yF(x,y),\\ (\mathcal{L}_Q F)(x,y) &= i(\partial_2 F)(x,y) - \frac{1}{2}xF(x,y).\end{aligned}$$

If we define $\mathcal{X}$ to be the subspace of $\mathcal{H}$ spanned by all vectors of the form $W[\![F]\!]\phi$, where $F \in \mathcal{S}(\mathbb{R}^2)$ and $\phi \in \mathcal{H}$, then $\mathcal{X}$ is a dense linear subspace of $\mathcal{H}$ contained in $\mathcal{D}(\overline{P}) \cap \mathcal{D}(\overline{Q})$, which is invariant under both $\overline{P}$ and $\overline{Q}$. Thus we can define endomorphisms A and A^+ of $\mathcal{X}$ by setting

$$\begin{aligned}Af &= \frac{1}{\sqrt{2}}\left(\overline{Q}f + i\overline{P}f\right),\\ A^+f &= \frac{1}{\sqrt{2}}\left(\overline{Q}f - i\overline{P}f\right).\end{aligned}$$

for any $f \in \mathcal{X}$, and direct calculation shows us that

$$AA^+f - A^+Af = f, \qquad f \in \mathcal{X}.$$

We can also calculate that

$$\langle Ag, f\rangle = \langle g, A^+f\rangle, \qquad f, g \in \mathcal{X},$$

so that both operators A and A^+ are closable, with $A^+ \subseteq A^*$ and $A \subseteq (A^+)^*$.

Standard functional analysis shows us that $B = (2I + A^*\overline{A})^{-1}$ and $C = (I + \overline{A}A^*)^{-1}$ both exist, belong to $\mathbb{B}(\mathcal{H})$, and have ranges $\mathcal{D}\left(A^*\overline{A}\right)$ and $\mathcal{D}\left(\overline{A}A^*\right)$ respectively, and we have that

$$(I + AA^+)f = (2I + A^+A)f, \qquad f \in \mathcal{X},$$

so we deduce that

$$B(I + AA^+)f = B(2I + A^+A)f = f = C(I + AA^+)f$$

for any $f \in \mathcal{X}$. If we consider the elliptic differential operator $\mathfrak{L}$ on $\mathcal{S}(\mathbb{R}^2)$ given by the formula

$$\mathfrak{L} = -\left(\partial_1^2 + \partial_2^2\right) + \frac{1}{4}\left(x^2 + y^2\right) - i\left(x\partial_2 - y\partial_1\right),$$

then we can show that

$$(I + AA^+)W[\![F]\!]\phi = \tfrac{1}{2}W[\![3F + 2\mathfrak{L}F]\!]\phi, \qquad F \in \mathcal{S}(\mathbb{R}^2), \ \phi \in \mathcal{H}.$$

Now it can be shown [220] that the operator $3I + 2\mathfrak{L}$ is a linear bijection from $\mathcal{S}(\mathbb{R}^2)$ to itself, and hence we deduce that $Bf = Cf$ for all $f \in \mathcal{X}$, and so $B = C$. Consequently $\mathcal{D}(A^*\overline{A}) = \mathcal{D}(\overline{A}A^*)$. Moreover, we can show that

$$\overline{A}A^*\phi - A^*\overline{A}\phi = \phi$$

for all ϕ in this common domain. Therefore the operator $\overline{A}$ provides the desired representation of the CCR in Heisenberg form. Conversely, if we have a representation of the canonical commutation relation in Heisenberg form provided by the closable operator A, then we can define the space $\mathcal{S}$ and the endomorphisms P and Q of $\mathcal{S}$ as above. Since the operators $\overline{P}$ and $\overline{Q}$ are self-adjoint, they can be used to generate strongly continuous one parameter unitary groups U and V respectively. We can show that these groups satisfy the so-called Weyl relation

$$U(a)V(b) = e^{iab}\,V(b)U(a), \qquad a,\, b \in \mathbb{R}.$$

Details of this argument are given by Putnam [184]. If we now define

$$W(a,b) = e^{-\frac{1}{2}iab}U(a)V(b), \qquad a,\, b \in \mathbb{R},$$

then it is easy to see that the map $W : \mathbb{R}^2 \to \mathbb{B}(\mathcal{H})$ is a strongly continuous representation of the Weyl group, and so we have a strongly continuous representation of the CCR in Weyl form. Moreover, it can be shown that these two constructions are mutually inverse, so we obtain a one-to-one correspondence between strongly continuous representations of the CCR in Weyl form and representations of the CCR in Heisenberg form.

Suppose now that W is a strongly continuous representation of the CCR in Weyl form, and let $\overline{A}$ be the closed operator, constructed as above, which implements the associated representation of the CCR in Heisenberg form. If $\psi \in \mathcal{H}$ belongs to the image of the projection $W[\![H_0]\!]$, then $\psi = W[\![H_0]\!]\psi$, and hence

$$A\psi = A\,W[\![H_0]\!]\psi = \frac{1}{\sqrt{2}}W\big[\!\big[(\mathcal{L}_Q + i\mathcal{L}_P)H_0\big]\!\big]\psi.$$

However, direct calculation shows us that $(\mathcal{L}_Q + i\mathcal{L}_P)H_0 = 0$, and hence $A\psi = 0$. Thus we deduce that $\psi \in \mathcal{D}(\overline{N})$ and that $\overline{N}\psi = 0$, so that $\Gamma(t)\psi = \psi$ for all $t \in \mathbb{R}$, and hence ψ is a gauge invariant vector.

Conversely, let us pick a gauge invariant vector $\psi \in \mathcal{H}$. Thus, if we define the self-adjoint operator $\overline{N} = A^*\overline{A}$, then we have that

$$\Gamma(t)\psi = e^{i\overline{N}t}\psi = \psi, \qquad t \in \mathbb{R},$$

so we deduce that $\psi \in \mathcal{D}(\overline{N})$ and that $\overline{N}\psi = 0$. Since the self-adjoint operator $\overline{N}$ is positive, it is clear that it generates a strongly continuous one parameter semigroup $\{e^{-\overline{N}t} : t \geqslant 0\}$, and it is also clear that $e^{-\overline{N}t}\psi = \psi$ for all $t \geqslant 0$. Since we can calculate that

$$\overline{N}\,W[\![F]\!]\phi = \frac{1}{2}W[\![(\mathcal{L} - I)F]\!]\phi, \qquad F \in \mathcal{S}(\mathbb{R}^2)\,,\, \phi \in \mathcal{H},$$

we deduce that

$$e^{-\overline{N}t}W[\![F]\!]\phi = e^{\frac{1}{2}t}W\big[\!\big[e^{-\frac{1}{2}\mathcal{L}t}F\big]\!\big]\phi, \qquad F \in \mathcal{S}(\mathbb{R}^2)\,,\, \phi \in \mathcal{H},\, t \geqslant 0.$$

Now it can be shown that [220]

$$W\big[\!\big[e^{-\frac{1}{2}\mathcal{L}t}F\big]\!\big] = W[\![K_t]\!]\,W[\![F]\!], \qquad F \in \mathcal{S}(\mathbb{R}^2)\,,\, t > 0,$$

where

$$K_t(x,y) = \frac{1}{4\pi \sinh\left(\frac{1}{2}t\right)} \exp\left[-\frac{1}{4}\coth\left(\frac{1}{2}t\right)(x^2+y^2)\right], \ t>0.$$

Hence we deduce that $e^{-\overline{N}t} = e^{\frac{1}{2}t}W[\![K_t]\!]$ for $t>0$, and so that

$$W[\![K_t]\!]\psi = e^{-\frac{1}{2}t}\psi, \qquad t>0.$$

Introducing more convenient notation, if we define the functions $H_\lambda \in \mathcal{S}(\mathbb{R}^2)$ for any $\lambda \geqslant 0$ by the formula

$$H_\lambda(x,y) = \frac{1}{2\pi}e^{-\frac{1}{4}(1+\lambda)(x^2+y^2)},$$

then we note that this definition for H_0 is the same as the old one, and moreover that

$$W[\![H_\lambda]\!]\psi = \frac{2}{2+\lambda}\psi, \qquad \lambda>0.$$

It is elementary to show that the map sending $\lambda \in [0,\infty)$ to $W[\![H_\lambda]\!] \in \mathbb{B}(\mathcal{H})$ is strongly continuous, and so we deduce (letting $\lambda \to 0$) that $W[\![H_0]\!]\psi = \psi$, so that ψ belongs to the image of the projection $W[\![H_0]\!]$. Thus the image of the projection $W[\![H_0]\!]$ is exactly equal to the space of gauge invariant vectors, from which we deduce that a representation of the CCR in Weyl form is irreducible if and only its associated representation of the canonical commutation relation in Heisenberg form is irreducible. ■

The connection theorem allows the Weyl operators to be written in terms of P and Q (or A and A^+) as follows.

Corollary 4.26 *The Weyl operators can be written as*

$$W(a,b) = e^{i\overline{(aP+bQ)}}, \tag{4.6.4}$$

or, in complex form,

$$W[z] = e^{i\overline{(zA+\overline{z}A^+)}}, \qquad z \in \mathbb{C}. \tag{4.6.5}$$

Proof: By the Trotter product formula [186],

$$e^{i\overline{(aP+bQ)}}\phi = \lim_{n\to\infty}\left[W(\tfrac{a}{n},0)W(0,\tfrac{b}{n})\right]^n\phi, \qquad \phi\in\mathcal{H}.$$

Now

$$\left[W(\tfrac{a}{n},0)W(0,\tfrac{b}{n})\right]^n = \left[e^{iab/2n^2}W(\tfrac{a}{n},\tfrac{b}{n})\right]^n = e^{iab/n}W(a,b)$$

so the result is immediate. ∎

4.7 Unitary Equivalence

At various stages in what has gone before, different representations of the canonical commutation relation (in either form) have been referred to as unitarily equivalent or not, as the case may be. The meaning of this is intuitively clear: they will be unitarily equivalent if there is a unitary operator taking one to the other.

This is bound up with the notion of the unitary equivalence of Hilbert spaces: any two Hilbert spaces of the same dimension are unitarily equivalent. But this is not the same as physical equivalence, a point worth discussing.

Definition 4.27 *A representation W_1 of the Weyl group on the Hilbert space $\mathcal{H}_1$ is said to be* unitarily equivalent *to a second representation W_2 of the Weyl group on the Hilbert space $\mathcal{H}_2$ if there exists a unitary operator $U : \mathcal{H}_1 \to \mathcal{H}_2$ such that*

$$W_2(a,b) \;=\; U\,W_1(a,b)\,U^{-1}, \qquad a,\, b \in \mathbb{R}. \tag{4.7.1}$$

It is clear that if W_1 and W_2 are unitarily equivalent representations of the Weyl group, then W_1 is strongly continuous (respectively irreducible) if and only if W_2 is. The unitary operator U can be used to transform all the calculations concerning the bounded model for the representation W_1 to the equivalent calculations for the representation W_2 — for example,

$$W_2[\![\alpha]\!] \;=\; U\,W_1[\![\alpha]\!]\,U^{-1}, \qquad \alpha \in L^1(\mathbb{R}^2)\,. \tag{4.7.2}$$

It is worth noting that this process goes both ways, in that any unitary operator U from the Hilbert space $\mathcal{H}$ to another Hilbert space $\mathcal{K}$ can be used to transform a representation W of the Weyl group on $\mathcal{H}$ to a representation of the Weyl group on $\mathcal{K}$ which is unitarily equivalent to W. This is done by noting that the definition

$$V(a,b) \;=\; U\,W(a,b)\,U^{-1}, \qquad a,\, b \in \mathbb{R}, \tag{4.7.3}$$

yields a representation V of the Weyl group on $\mathcal{K}$ which is unitarily equivalent to W. Thus, all unitary transformations of the Hilbert space $\mathcal{H}$ determine mathematically equivalent representations of the observables and states of the quantum system. Adopting the terminology of Halmos [97], we describe this by saying that any unitary transformation of the Hilbert space $\mathcal{H}$ determines a different *manifestation* of the (bounded) model.

A similar definition, with similar consequences, can be given for the smooth model.

Definition 4.28 *If the densely defined closable operators A_1 and A_2 provide representations of the canonical commutation relation (in Heisenberg form) on the Hilbert spaces $\mathcal{H}_1$ and $\mathcal{H}_2$ respectively, then these two representations are said to be* unitarily equivalent *if there exists a unitary operator $U : \mathcal{H}_1 \to \mathcal{H}_2$ such that $\mathcal{D}(A_2) = U\mathcal{D}(A_1)$ and such that*

$$A_2\phi = U A_1 U^{-1} \phi, \qquad \phi \in \mathcal{D}(A_1). \tag{4.7.4}$$

Thus the unitary map U must interpolate the domains as well as the values of the unbounded operators A_1 and A_2. As a notational shorthand, the identity

$$A_2 = U A_1 U^{-1} \tag{4.7.5}$$

shall be understand to imply both the correspondence between domains, as well as the equality of values required by the definition. This shorthand is a standard form of notation in unbounded operator theory.

It is clear that the map U also interpolates the other domains of importance in the smooth model, so that, for example, $N_2 = UN_1U^{-1}$. This implies the following relationship between the gauge groups,

$$\Gamma_2(t) = U\Gamma_1(t)U^{-1}, \qquad t \in \mathbb{R}, \tag{4.7.6}$$

from which is it easy to see that the property of gauge invariance is preserved under unitary equivalence. Moreover, the following identity

$$\mathcal{S}_2 = U\mathcal{S}_1, \tag{4.7.7}$$

holds between the common dense domains, and the representations π_1 and π_2 of the abstract Weyl algebra $\mathbb{A}[p,q]$ are unitarily equivalent in the sense that

$$\pi_2(a)Uf = U\pi_1(a)f, \qquad a \in \mathbb{A}[p,q],\ f \in \mathcal{S}_1. \tag{4.7.8}$$

Again, this procedure works both ways. Given any representation of the canonical commutation relation (in Heisenberg form) implemented by the closable densely defined operator A on the Hilbert space $\mathcal{H}$, and given any unitary map U from this Hilbert space $\mathcal{H}$ to another space $\mathcal{K}$, define an operator $B : \mathcal{D}(B) \to \mathcal{K}$ by setting $\mathcal{D}(B) = U\mathcal{D}(\overline{A})$ and requiring that

$$B\phi = U\overline{A}U^{-1}\phi, \qquad \phi \in \mathcal{D}(B). \tag{4.7.9}$$

Then B is closed and densely defined on $\mathcal{K}$, and moreover determines (or rather its restriction to USU^{-1} does) a representation of the CCR on $\mathcal{K}$ which is unitarily equivalent to that on $\mathcal{H}$. Thus, any unitary transformation of the Hilbert space $\mathcal{H}$ determines a different manifestation of the (smooth) model.

Having seen what does work, consider what does not, by returning to the problem of a quantum mechanical particle carrying a generalized charge. In the terminology of representation theory, if the appropriate representation of the canonical commutation relation (in Heisenberg form) for an uncharged particle is given by the closable densely defined operator A on the (separable) Hilbert space $\mathcal{H}$, then the appropriate representation of the CCR (in Heisenberg form) for the charged particle is given by the closable densely defined operator $A \oplus A$ on the direct sum Hilbert space $\mathcal{H} \oplus \mathcal{H}$. However, both $\mathcal{H}$ and $\mathcal{H} \oplus \mathcal{H}$ are separable Hilbert spaces, and hence are unitarily equivalent, yet define quite different physics!

That they define different physics is obvious in terms of the charge. That they define inequivalent representations is seen by transporting the charged representation to the original Hilbert space, it does not matter how, and then comparing the two representations there.

For example, if $\{\phi_n : n \in \mathbb{N}\}$ is an orthonormal basis for $\mathcal{H}$, then the set $\{(\phi_n, 0), (0, \phi_n) : n \in \mathbb{N}\}$ is an orthonormal basis for $\mathcal{H} \oplus \mathcal{H}$, and so $U : \mathcal{H} \oplus \mathcal{H} \to \mathcal{H}$ defines a unitary map such that

$$\begin{aligned} U(\phi_n, 0) &= \phi_{2n-1}, \\ U(0, \phi_n) &= \phi_{2n}, \end{aligned} \qquad n \in \mathbb{N}. \tag{4.7.10}$$

Defining the closable densely defined operator B on $\mathcal{H}$ by the formula

$$B = U(A \oplus A)U^{-1}, \tag{4.7.11}$$

it determines a representation of the CCR (in Heisenberg form) on $\mathcal{H}$ which is unitarily equivalent to that provided by $A \oplus A$ acting on $\mathcal{H} \oplus \mathcal{H}$.

Thus we have found two representations of the CCR (in Heisenberg form) on $\mathcal{H}$ which are not equal, or even unitarily equivalent. For example, if the representation provided by A were gauge invariant, then the representation provided by B would not be, since the space of gauge invariant vectors would in that case be two-dimensional.

The conclusion is that even though all separable Hilbert spaces are unitarily equivalent, representations of the CCR (on separable Hilbert spaces) are not all unitarily equivalent. However, the fact that all separable Hilbert spaces are unitarily equivalent implies that every representation of the CCR has a manifestation on any one given (separable) Hilbert space. In other words, any one Hilbert space can be used to describe all quantum physics that does not need an inseparable space.

4.8 Meaning And Form

4.8.1 *On Mathematical Quantization*

Any process of obtaining the associated quantum mechanical observable from its classical analogue is called quantization. Taking into account all of its aspects, it is an extensive theory, encompassing a number of different strands. Historically, it was arrived at empirically (in the original papers of Heisenberg and Schrödinger [103, 204]), and a working hypothesis was adopted which stated that there was a standard pair of operators P and Q which were the quantum mechanical analogues of the momentum and position coordinates p and q in Π. The quantization of some more complicated (but still relatively simple) function $f(p,q)$ on Π was simply assumed to be $f(P,Q)$. But in order for this to work, the function f has to be sufficiently simple that the function $f(P,Q)$ can be defined unambiguously. When this is the case, *mirabile dictu*, this approach worked — a tribute to the deep physical intuition of the creators of quantum mechanics.

Any more general process of quantization will have to deliver these particular results, as well as providing a method for quantizing more complicated functions f, for which the meaning of $f(P,Q)$ is not self-evidently defined. Thus any system of quantization is the process of assigning an unambiguous mathematical meaning to the expression $f(P,Q)$ for any suitable function f so that, in some sense, the quantum variable $f(P,Q)$ can then be interpreted as the quantum mechanical analogue of the classical quantity $f(p,q)$.

Amongst the various approaches to quantization, certain geometers would take quantum mechanics to be a deformation of classical mechanics. Consequently classical equations should have deformed analogues in quantum mechanics. To this end, recall that the time-evolution of a classical system is determined by its Hamiltonian function. As well, there is a self-adjoint observable H which determines the time-evolution of the quantum mechanical system. The originators of quantum mechanics made the assumption that H should be the quantum mechanical analogue of the classical Hamiltonian. Again this usually works, although there is no rigorous proof that it must be the case, although Schrödinger was strongly guided by the Hamilton-Jacobi equations in his derivation, and that equation is useful in describing the régime lying between classical mechanics and quantum mechanics proper.

If the classical equation of motion is compared with the quantum mechanical equation of motion, it is observed that the Poisson bracket (with the classical Hamiltonian) has been replaced by i times the commutator (with the Hamiltonian operator). The question the geometers now address is to what extent a quantization scheme preserves this connection between the Poisson bracket and the commutator. It is evident that this connection will not be total (after all, classical and quantum mechanics are not the same), and so some form of partial result is the best that can be expected. In practice, what is done is to look for a preferred subcollection of classical observables for which the connection between the Poisson bracket and (i times) the commutator can be made perfect, and then find what quantization schemes (if any) enable such a correspondence for these preferred observables. There is then a delicate balance between the collection of observables, the subcollection of preferred observables, and the exact nature of the Hilbert space $\mathcal{H}$ on which the quantization can be achieved (if at all). Since it is known in most cases which Hilbert space is required, a different route will be followed here.

That route, which is more traditional historically, is to find a direct method for assigning a quantization to any suitable classical observable. The process is designed in such a way that as large a class of what might be regarded as fundamental classical observables are quantized in the customary manner. After defining such a quantization scheme, its properties can be determined mathematically.

Even here, there is more than one way of doing things, but there is a preferred scheme which seems both to be simplest, yet provides the best results in as wide a context as possible. This is the proposal of Weyl, briefly expounded in his book on quantum mechanics and group theory, [236], which has the distinction of being the oldest formal quantization procedure. Weyl's scheme will be adopted throughout this book except in Chapter 11, Ordered Quantization, where some variants of Weyl quantization are considered.

Even before considering any details, it is clear that Weyl quantization (or any variant) is not going to be a *-algebra isomorphism from the algebra of classical observables to the algebra of quantum observables, again because classical and quantum mechanics are not the same. And because there is no such map, there will be some perfectly reasonable classical mechanical observables which do not possess a quantum analogue, and perfectly reasonable quantum mechanical observables whose classical analogue is not smooth, or even a function.

Some references are given at the end of the Chapter to discussions of the history of the various models of quantum theory, of the philosophical consequences of the theory, and of approaches that do not postulate a Hilbert space.

4.8.2 *The Correspondence Principle*

Mathematical formulation aside, the main problem of quantization lies in the interpretation of what it means. Suppose some classical observable is quantized, resulting in an operator representing a quantum mechanical observable. An allowed measurement of that quantum observable is then made. To what extent, if any, is the associated classical observable being measured? This is an extremely difficult question, and one to which there is no simple answer. We content ourselves with making the following comments.

Since, for example, observables such as position, momentum, orbital angular momentum and energy are the generators of (translation, rotation or time-evolution) symmetry groups in both classical and quantum mechanics, it is entirely reasonable that we should interpret the quantum mechanical position, momentum, orbital angular momentum and energy observables as representing the position, momentum, orbital angular momentum and energy of the system in question.

However even this reasonable viewpoint presents problems, since there many examples of quantum mechanical systems for which the spectrum of the Hamiltonian is discrete, indicating that the system can only be found with one of a discrete collection of energies. How is this to be reconciled with the fact that the "corresponding" classical Hamiltonian can in general take values lying in a continuous range? One explanation often given is that classical mechanics is the limit of quantum mechanics in the formal limit as $\hbar$ tends to 0, the physical rationale being that the effect of letting $\hbar$ tend to 0 is that particles approach their ionization energies in all states, and consequently behave ever more classically. However, any limiting procedure which will result in an observable with discrete spectrum changing into one with continuous spectrum would need to be handled carefully.

One approach to this limiting procedure, which has the requisite mathematical rigour, is the so-called *classical limit* [221]. In essence, this limit is summarized by the limiting formulæ

$$\lim_{\hbar \to 0} W\left(p/\sqrt{\hbar}, q/\sqrt{\hbar}\right) e^{i\sqrt{\hbar}aP}\, W\left(p/\sqrt{\hbar}, q/\sqrt{\hbar}\right)^* = e^{-iap}, \qquad (4.8.1.a)$$

$$\lim_{\hbar \to 0} W\left(p/\sqrt{\hbar}, q/\sqrt{\hbar}\right) e^{i\sqrt{\hbar}bQ}\, W\left(p/\sqrt{\hbar}, q/\sqrt{\hbar}\right)^* = e^{ibq}. \qquad (4.8.1.b)$$

It should be noted that the above limits are ones of strong convergence, and show how the (exponentials) of classical momentum and position can be regained from the (exponentials) of quantum mechanical momentum and position in the limit as $\hbar$ tends to 0. In this way, the classical limit is an example of a dequantization procedure which allows us to regain the originating classical observable from its quantum mechanical analogue.

It is sometimes said that the correspondence between classical and quantum mechanics lies in the Theorem of Ehrenfest, which is claimed to state that the expectations of quantum mechanical observables satisfy the corresponding classical equations of motion. However, this is not truly the case. In Theorem 3.3.15 of Thirring [221], Volume 3, it is shown that this interpretation can only be made for systems with special forms of potential.

Finally, it must be noted that there are quantum mechanical observables which have no direct classical analogues. For example, Dirac defined the spin of a system as the difference between the total angular momentum of that system (a constant of the motion) and the orbital angular momentum of that system (which is not a constant of the motion, but has a classical analogue), which definition results in a quantity which does not have a classical limit. Examples such as this tend to reinforce Bohr's view that human understanding only comes in classical terms.

4.9 Additional Reading

Some books which touch on a more geometrical approach to the theory than we take are [243] and [229]. Considerations of the balance between bounded and unbounded operators will be found in the short book of Isham [124].

Rigorous treatment of the so-called semi-classical region is technically difficult, requiring a broad range of mathematical techniques. A recent exposition is that of Landsman [146].

In the other direction, non-commutative geometry may be characterized, literally, as hyper-quantum mechanics, particularly if it is taken to include quantum groups. Some references, from which others may be gleaned, are [35], [33], [65], [133], [140], [159], [165]. References to topological vector spaces are [26], [70], [72], [71], [73], [69], [91], [130], [142], [190], [201], [207], [224] and [242]. Books on locally convex algebras not cited in the text are [106] and [163].

Chapter 5

Representations Of The CCR

If all this damned quantum jumping were really here to stay then I should be sorry I ever got involved with quantum theory.

– E. Schrödinger

But the rest of us are extremely grateful that you did.

– N. Bohr

5.1 Introduction

Having obtained the general structure of the sets of observables and states in a quantum mechanical system, and a few properties of these collections of objects, it is time to consider some familiar and some not so familiar realizations of these structures. The emphasis will be on how these representations reflect the general axiomatics of the previous two Chapters.

All the representations of the CCR considered in this Chapter are irreducible, and consequently isomorphic. In a mathematical sense, therefore, there is no difference between them, and anything that can be proved for one representation must be true for any of the others. However, there are often good physical reasons for wanting to consider a representation of the CCR in a certain form. In particular, it is often the case that some important operator is diagonal in a given representation, and so the quality it represents is easy to describe.

5.2 The Schrödinger Representation

The best known and most important representation of the CCR is the Schrödinger representation. With its natural interpretation, it describes a particle of nonzero mass moving on a line, and with the position operator diagonalized. It is irreducible in both the bounded and smooth models, and gauge invariant in the latter.

The system Hilbert space for this representation is $L^2(\mathbb{R})$. For the bounded model, the algebra of observables is $\mathbb{B}\left[L^2(\mathbb{R})\right]$, the set of all bounded operators on $L^2(\mathbb{R})$, and the states are given by the density matrices on $L^2(\mathbb{R})$. The action of the Weyl group in this representation is

$$\left[W(a,b)\phi\right](x) = e^{\frac{1}{2}iab}e^{ibx}\,\phi(x+a), \qquad \phi \in L^2(\mathbb{R})\,, \tag{5.2.1}$$

which was previously given in equation (3.3.6).

We can now identify the smooth domain $\mathcal{S}$ for the smooth model associated with this representation of the CCR. Following the details of Theorem 4.25, the unitary groups U and V derived from this representation W of the Weyl group are given by the formulæ:

$$\left[U(a)\phi\right](x) \;=\; \phi(x+a)\,, \tag{5.2.2.a}$$
$$\left[V(a)\phi\right](x) \;=\; e^{iax}\phi(x)\,, \tag{5.2.2.b}$$

for $\phi \in L^2(\mathbb{R})$ and $a \in \mathbb{R}$, and their self-adjoint generators are the operators $\overline{P}$ and $\overline{Q}$ respectively, which are defined on their respective domains by the formulæ:

$$\left[\overline{P}\phi\right](x) \;=\; -i\phi'(x)\,, \qquad \phi \in \mathcal{D}(\overline{P})\,, \tag{5.2.3.a}$$
$$\left[\overline{Q}\phi\right](x) \;=\; x\phi(x)\,, \qquad \phi \in \mathcal{D}(\overline{Q})\,, \tag{5.2.3.b}$$

and so $\overline{P}$ and $\overline{Q}$ are seen to be the standard operators for position and momentum originally proposed by Schrödinger. As in Proposition 4.4, we can now identify the smooth domain $\mathcal{S}$ explicitly as

$$\mathcal{S}(\mathbb{R}) \;=\; \left\{f \in \mathcal{C}^\infty(\mathbb{R}) \;:\; \int_{\mathbb{R}} \left|x^j f^{(k)}(x)\right|^2 dx \;<\; \infty\,,\; \forall j,k \geqslant 0\right\}. \tag{5.2.4.a}$$

This space has the alternate characterization

$$\mathcal{S}(\mathbb{R}) \;=\; \left\{f \in \mathcal{C}^\infty(\mathbb{R}) \;:\; \lim_{|\,x\,|\to\infty} x^j f^{(k)}(x) \;=\; 0\,, \quad \forall j,k \geqslant 0\right\}, \tag{5.2.4.b}$$

so that a function belongs to $\mathcal{S}(\mathbb{R})$ if and only if it is infinitely differentiable and it, and all of its derivatives, converge to zero at infinity faster than any polynomial. We also remember that the seminorms defined in equation (4.2.7) equip $\mathcal{S}(\mathbb{R})$ with a nuclear Fréchet locally convex topology.

Functions in $\mathcal{S}(\mathbb{R})$ will be referred to as *test functions* from time to time, but this usage is incorrect, strictly speaking, for an analyst expects a test function to be smooth and of compact support. But the term test function is sufficiently evocative that we choose to use it; as the term is never used differently in this book, no confusion should occur. The space $\mathcal{S}(\mathbb{R})$ was first described by Laurent Schwartz as one of the basic test function spaces in his theory of distributions. For this reason, functions in $\mathcal{S}(\mathbb{R})$ are conveniently known as *Schwartz functions.*

Equations (5.2.3.a) and (5.2.3.b) imply that the lowering and raising operators take the explicit form

$$[Af](x) = \frac{1}{\sqrt{2}}\left(\frac{df(x)}{dx} + xf(x)\right), \qquad (5.2.5.\text{a})$$

and

$$[A^+f](x) = \frac{1}{\sqrt{2}}\left(xf(x) - \frac{df(x)}{dx}\right), \qquad (5.2.5.\text{b})$$

respectively, for $f \in \mathcal{S}(\mathbb{R})$. The closure of A has domain $\mathcal{D}\left(\overline{P}\right) \cap \mathcal{D}\left(\overline{Q}\right)$, and is the closed operator[1] which provides our representation of the canonical commutation relation in Heisenberg form. The closure of A^+ has the same domain, and is the adjoint of A.

Given these differential operators, it is now possible to consider the number operator N, given by

$$N = \frac{1}{2}\left(-\frac{d^2}{dx^2} + x^2 - 1\right) \qquad (5.2.6)$$

on its domain $\mathcal{S}(\mathbb{R})$, and to show that the Hermite-Gauss vectors are the classical Hermite polynomials multiplied by the Gaussian, with appropriate normalizations; hence the terminology[2].

Proposition 5.1 *The Schrödinger representation of the CCR is gauge invariant, and the gauge invariant Fock vector h_0 is the Gaussian function*

$$h_0(x) = \pi^{-\frac{1}{4}}e^{-\frac{1}{2}x^2}. \qquad (5.2.7.\text{a})$$

[1] This is a point at which the distinction between A and A^+, defined on $\mathcal{S}(\mathbb{R})$, and their closures is important.

[2] In the previous Chapter, we employed the symbol Ω_k to denote the kth Hermite-Gauss vector. In each particular representation of the CCR, it will be convenient to introduce a distinct symbol to represent the particular form of the Hermite-Gauss vectors in that representation, leaving Ω_k to represent a generic Hermite-Gauss vector in any discussion which is representation-independent.

The remaining Hermite-Gauss vectors may be written in the form

$$h_k(x) = \frac{1}{\sqrt{2^k\, k!}} H_k(x) h_0(x)\,, \tag{5.2.7.b}$$

where H_k is the Hermite polynomial[3] *of degree k.*

The generating function for this orthonormal basis is

$$G_t(x) = \sum_{k=0}^{\infty} \frac{t^k}{\sqrt{2^k\, k!}} h_k(x) = \pi^{-\frac{1}{4}} \exp\left(-\frac{1}{4}t^2 + xt - \frac{1}{2}x^2\right)\,. \tag{5.2.8}$$

It is often extremely useful in calculating matrix elements of operators with respect to this basis to use the generating function, a technique we shall employ frequently.

Proposition 5.2 *The Hermite-Gauss functions are the normalized eigenvectors of the closure $\overline{N}$ of the number operator,*

$$\overline{N}\, h_k = k\, h_k\,, \qquad k \geqslant 0\,. \tag{5.2.9}$$

In order to rewrite this as the spectral decomposition of $\overline{N}$, introduce the projection operator P_k along h_k,

$$P_k \phi = \langle h_k, \phi \rangle\, h_k, \qquad k \geqslant 0,\ \phi \in \mathcal{H}. \tag{5.2.10.a}$$

The symbol P_k will be reserved for this operator.

From the orthonormality of the h_k it follows that

$$P_j P_k = \delta_{jk} P_k\,, \qquad j,\, k \geqslant 0, \tag{5.2.10.b}$$

and from the completeness of the Hermite-Gauss functions it follows that they give a decomposition of the identity,

$$\sum_{k=0}^{\infty} P_k = I\,. \tag{5.2.11}$$

Then

$$\overline{N} = \sum_{k=0}^{\infty} k P_k \tag{5.2.12}$$

---

[3]In other words the Hermite-Gauss *vectors*, which are an abstract construct to be found in any smooth representation of the CCR, are here represented by the concrete Hermite-Gauss *functions*. This (not such a) coincidence explains our choice of terminology.

is the spectral decomposition for $\overline{N}$ in terms of the P_k.

Remark As everyone knows (including $\hbar$ explicitly here),

$$H = \left(\overline{N} + \frac{1}{2}\right)\hbar\omega \tag{5.2.13}$$

is the Hamiltonian operator for the quantum simple harmonic oscillator of natural frequency ω. The spectral values of $\overline{N}$ thus count the excitations of the oscillator. When the oscillator represents a mode of the quantized electromagnetic field, these excitations can be identified with photons of frequency ω. But this interpretation must be applied carefully. Photons are relativistic particles of zero mass; they have no rest frame, are not strictly localizable, and are not conserved. This is why there are apparently spatially nonlocal effects for electromagnetic fields in cavities. If we consider that the field is established throughout the cavity, the occurrence of phase relations at different points is natural. Still, we shall use the term *number operator* as a convenience. ■

As in the previous Chapter, the smooth model automatically yields a rigged triple structure,

$$\mathcal{S}(\mathbb{R}) \subseteq L^2(\mathbb{R}) \subseteq \mathcal{S}'(\mathbb{R}), \tag{5.2.14}$$

where $\mathcal{S}'(\mathbb{R})$ is the topological dual of $\mathcal{S}(\mathbb{R})$, called the space of *tempered distributions.*[4] As remarked before, the embedding of $L^2(\mathbb{R})$ into $\mathcal{S}'(\mathbb{R})$ follows from the self-duality of $L^2(\mathbb{R})$, and hence is naturally antilinear rather than linear. However, since $L^2(\mathbb{R})$ possesses a complex structure J given by the formula

$$\big[J\phi\big](x) = \overline{\phi(x)}\,, \qquad \phi \in L^2(\mathbb{R}), \tag{5.2.15}$$

it is sensible, as was mentioned previously, to identify each element ϕ in $L^2(\mathbb{R})$ with the element $(J\phi)^\dagger$ in $\mathcal{S}'(\mathbb{R})$, so that the embedding of $L^2(\mathbb{R})$ into $\mathcal{S}'(\mathbb{R})$ regards $\phi \in L^2(\mathbb{R})$ as an element of $\mathcal{S}'(\mathbb{R})$ via the formula

$$[\![\phi\,,f]\!] = \int_{\mathbb{R}} \phi(x)f(x)\,dx\,, \qquad f \in \mathcal{S}(\mathbb{R}). \tag{5.2.16}$$

[4]The term *tempered* indicates that these distributions are less singular than others.

As was remarked in the previous Chapter, this rigged triple structure has turned up unannounced, so to speak, arising from the general structure of the representation of the CCR. This particular rigged triple has the further merit that each of the three spaces involved in it are invariant under the action of the Fourier transform, a property which is of considerable use in quantum mechanics. While appearing on the doorstep uninvited might be poor manners in polite society, serendipitous benefits such as these are most welcome in mathematics and physics.

From the fact that the functions in $\mathcal{S}(\mathbb{R})$ are infinitely differentiable and of rapid decrease at infinity, a dual characterization of tempered distributions can be determined.

Proposition 5.3 *A tempered distribution T may be characterized by a finite sequence $(w_j)_{0\leqslant j\leqslant n}$ of continuous functions for which there exists a sequence $(C_j)_{0\leqslant j\leqslant n}$ of positive constants such that*

$$|\, w_j(x)\,| \leqslant C_j \left(1 + |\, x\,|^j\right), \qquad x \in \mathbb{R},\ 0 \leqslant j \leqslant n, \tag{5.2.17.a}$$

through the symbolic formula

$$T = \sum_{j=0}^{n} (-1)^j \frac{d^j}{dx^j} w_j(x). \tag{5.2.17.b}$$

This expression is meant to hold in the distributional sense that for all $f \in \mathcal{S}(\mathbb{R})$,

$$[\![T\,,\, f]\!] = \sum_{j=0}^{n} \int_{\mathbb{R}} w_j(x)\, \frac{d^j f(x)}{dx^j}\, dx\,. \tag{5.2.17.c}$$

In other words, the formula (5.2.17.b) is to be interpreted weakly.

In the discussion of the bounded model, it was pointed out that the Gaussian function

$$H_0(a, b) \;=\; \frac{1}{2\pi} e^{-\frac{1}{4}(a^2+b^2)} \tag{3.3.12.a}$$

has the property that $W[\![H_0]\!]$ is a projection operator. In the Schrödinger representation, working out the Gaussian integral

$$W[\![H_0]\!](G_t) \;=\; h_0 \tag{5.2.18.a}$$

leads to the conclusion that

$$W[\![H_0]\!](\psi) = \langle h_0, \psi \rangle \, h_0 \,, \qquad \psi \in L^2(\mathbb{R}), \tag{5.2.18.b}$$

and that

$$W[\![H_0]\!] = P_0 \,; \tag{5.2.18.c}$$

that is, $W[\![H_0]\!]$ is the projection onto the subspace spanned by the Hermite-Gauss function h_0.

We have already noted in the previous Chapter that the space $\mathcal{S}(\mathbb{R})$ of test functions is topologically isomorphic to the space $\mathfrak{s}$ of sequences of rapid decrease and, consequently, the dual space $\mathcal{S}'(\mathbb{R})$ of tempered distributions will be isomorphic to the space $\mathfrak{s}'$ of sequences of slow increase. It is important to see how this latter identification is possible.

Remark The term Schauder basis has been mentioned previously without comment, but it is now necessary to define it precisely, and to introduce some related concepts.

A *topological basis* for a Hausdorff locally convex vector space E is a sequence (e_n) of vectors in E such that for any $x \in E$ there is a unique sequence of (complex) coefficients (x_n), the series $\sum_n x_n e_n$ converging in E to the element x. Thus any topological basis for E defines a sequence (u_n) of *coordinate functionals*, these being the linear functionals on E defined by the formula

$$u_n(x) = x_n \,, \qquad x \in E \,, \; n \in \mathbb{N}.$$

The topological basis (e_n) is called a *Schauder basis* if all of the coordinate functionals are continuous on E. In view of the identity

$$[\![\, u_m \,, e_n \,]\!] = \delta_{mn} \,, \qquad m, n \in \mathbb{N},$$

is it convenient to describe the basis vectors and coordinate functionals in a Schauder basis as being *orthogonal*, although no geometric meaning is implied by the use of that term.

A Schauder basis (e_n) is said to be *equicontinuous* if the collection of coordinate functionals (u_n) forms an equicontinuous collection in E', and it is said to be *absolute* if, for every continuous seminorm p

on E, there exists another continuous seminorm q on E such that

$$\sum_{n\in\mathbb{N}} |u_n(x)|\,p(e_n) \leqslant q(x)\,, \qquad x \in E.$$

For a nuclear Fréchet space E, a topological basis is necessarily both equicontinuous and absolute. Moreover, the coefficient functionals (u_n) are an absolute and equicontinuous basis for the dual space E', known as the *dual basis*. See Jarchow [130]. ■

As was seen in Proposition 4.18, the Hermite-Gauss functions $\big(h_n\big)$ form a Schauder basis for the nuclear Fréchet space $\mathcal{S}(\mathbb{R})$. For the moment, let us denote the dual basis for $\mathcal{S}'(\mathbb{R})$ by $\big(\mathfrak{h}_n\big)$. Then any element $f \in \mathcal{S}(\mathbb{R})$ has a unique expansion of the form

$$f = \sum_{n=0}^{\infty} a_n h_n\,, \tag{5.2.19.a}$$

where the sequence $a = \big(a_n\big)$ belongs to $\mathfrak{s}$, and any element $T \in \mathcal{S}'(\mathbb{R})$ has a unique expansion of the form

$$T = \sum_{n=0}^{\infty} t_n \mathfrak{h}_n\,, \tag{5.2.19.b}$$

where the sequence $t = \big(t_n\big)$ belongs to $\mathfrak{s}'$, and the pairing between T and f can be expressed in terms of the coordinate functionals as follows:

$$[\![T\,, f]\!] = \sum_{n=0}^{\infty} t_n a_n. \tag{5.2.19.c}$$

However, Proposition 4.18 tells us more about the dual basis $\big(\mathfrak{h}_n\big)$, since we know that

$$\mathfrak{h}_n(f) = a_n = \langle h_n, f\rangle = \int_{\mathbb{R}} h_n(x) f(x)\,dx \tag{5.2.20.a}$$

for any $f \in \mathcal{S}(\mathbb{R})$ and $n \geqslant 0$ (since the functions h_n are real-valued), and so we see that

$$\mathfrak{h}_n = (Jh_n)^\dagger\,, \qquad n \geqslant 0. \tag{5.2.20.b}$$

Consequently we make the usual identification between $(Jh_n)^\dagger$ and h_n, and so simply write $\mathfrak{h}_n = h_n$, and regard $\big(h_n\big)$ as the dual absolute Schauder basis for $\mathcal{S}'(\mathbb{R})$, as well as being the Schauder basis for $\mathcal{S}(\mathbb{R})$. This abuse of notation is standard.

It should be noted that the real-valued nature of the Hermite-Gauss functions in the Schrödinger representation is what provides us with such a simple description of the dual basis for $\mathcal{S}'(\mathbb{R})$ — in other representations of the CCR this will not be the case, and the complex structure will need to be incorporated into the comparable description explicitly.

5.2.1 *Approximate Position Operators*

In Section 4.5, The Round-Off Approximation, it was observed that, even for smooth operators, the associated spectral measures were not themselves smooth observables. It was commented that it was possible to adjust the spectral description of such observables in a manner which avoids this problem, and we now proceed to demonstrate how this can be done in the case of the position operator Q. To begin with, we remind the reader of the spectral theory of this operator. These results are (of course) standard, but are frequently omitted from mathematics books — Akhiezer & Glazman [3] being an admirable exception in this respect.

If we reserve the symbol Q to denote the operator with $\mathcal{S}(\mathbb{R})$ as its domain, the closure $\overline{Q}$ of Q has domain

$$\mathcal{D}(\overline{Q}) \;=\; \Big\{\phi \in L^2(\mathbb{R}) \,:\, \int_{\mathbb{R}} |\, x\phi(x)\,|^2\,dx \,<\, \infty\Big\}, \tag{5.2.21.a}$$

and on this domain $\overline{Q}$ is defined by the formula

$$(\overline{Q}\phi)(x) \;=\; x\phi(x)\,, \qquad\qquad \phi \in \mathcal{S}(\overline{Q})\,. \tag{5.2.21.b}$$

It is clear that $\overline{Q}$ is an extension of Q. The operator Q is essentially self-adjoint on $\mathcal{S}(\mathbb{R})$, and its unique self-adjoint extension is $\overline{Q}$. The spectrum of $\overline{Q}$ is absolutely continuous, and is equal to $\mathbb{R}$. As noted previously, the spectral projections are defined by the formula

$$E_Q(\Delta)\phi \;=\; \chi_{\Delta}\,\phi, \qquad\qquad \phi \in L^2(\mathbb{R}), \tag{4.5.1}$$

for any Borel subset Δ of $\mathbb{R}$.

The spectral calculus is implemented for $\overline{Q}$ as follows. For any Borel function F on $\mathbb{R}$ we define the dense domain

$$\mathcal{D}_F \;=\; \Big\{\phi \in L^2(\mathbb{R}) \,:\, \int_{\mathbb{R}} |\, F(x)\phi(x)\,|^2\,dx \,<\, \infty\Big\}. \tag{5.2.22.a}$$

Then $F(\overline{Q})$ is the closed operator defined on the domain $\mathcal{D}_F$ by the formula

$$\left[F(\overline{Q})\phi\right](x) = F(x)\phi(x)\,, \qquad \phi \in \mathcal{D}_F. \tag{5.2.22.b}$$

The round-off approximation takes advantage of the fact that convolution is a smoothing operation. In particular, it is not hard to show that $g * \chi_\Delta$ belongs to $\mathcal{S}(\mathbb{R})$ for any function $g \in \mathcal{S}(\mathbb{R})$ and any Borel subset Δ of $\mathbb{R}$. (Here and below, $f * g$ denotes the convolution product of the functions f, g.) If we choose $g \in \mathcal{S}(\mathbb{R})$ to be a function such that

$$\int_{\mathbb{R}} |\, g(x)\,|^2\, dx = 1\,, \tag{5.2.23.a}$$

$$\int_{\mathbb{R}} x|\, g(x)\,|^2\, dx = 0\,, \tag{5.2.23.b}$$

then, for any Borel subset Δ of $\mathbb{R}$, we define the bounded operator $E_{Q,g}(\Delta)$ by the formula

$$E_{Q,g}(\Delta) = \left(|\, g\,|^2 * \chi_\Delta\right)(\overline{Q})\,. \tag{5.2.24.a}$$

Note that $E_{Q,g}$ defines a positive operator valued measure on $\mathbb{R}$ such that

$$\overline{Q} = \int_{\mathbb{R}} x dE_{Q,g}(x)\,, \tag{5.2.24.b}$$

and which has the property that $E_{Q,g}(\Delta)$ maps $\mathcal{S}(\mathbb{R})$ continuously into itself for each Borel subset Δ of $\mathbb{R}$.

A number of authors, including Davies [41] and Busch, Grabowski & Lahti [30], find it advantageous to take the notion of an observable to mean a positive operator valued measure such as $E_{Q,g}$. Consequently, a positive operator valued measure is often called an *approximate position operator*. The above authors' work is done with the bounded model in mind, but there is no particular problem in presenting an analogous theory for the smooth model, and this is done in Dubin & Hennings [52]. In that context, the positive operator valued measure $E_{Q,g}$ is called a *question* about the observable Q — the distinction between an observable and a question about it being that the smoothing factor g introduces and (to some extent) quantifies the inevitable uncertainties that will arise in any measurement of Q.

This smoothing procedure is not just a mathematical nicety, it has deep implications for the theory of measurement in quantum mechanics. In the standard theory of the bounded model, if a system were in the pure state $|\,\phi\,\rangle\,\langle\,\phi\,|$ described by the unit vector $\phi \in L^2(\mathbb{R})$ (using the Dirac notation), and if a measurement for Q were made which registered a value lying in the Borel subset Δ of $\mathbb{R}$, then it would be said that the system state collapsed after that measurement to the pure state

$$\|\, E_Q(\Delta)\phi \,\|^{-2}\,|\, E_Q(\Delta)\phi\,\rangle\,\langle\, E_Q(\Delta)\phi\,|. \tag{5.2.25}$$

However, as has already been indicated, to imply that we can make such measurements of Q implies that there is a perfect experiment with which to undertake such measurements, and this is not possible. A question about Q incorporates into its definition some acceptance of the inevitable inaccuracies of the information it provides about Q, and so it is reasonable to suppose that there would be a number of different experimental devices which could be contrived which would provide information about Q within these tolerances. The mathematical construct which represents an experimental set-up providing information concerning a given question about Q is called an *instrument observable* by Dubin & Hennings [52] and, as indicated above, there are many possible instruments for any one given question. However, once a particular instrument has been chosen, there is again a definite rule for the collapse of the state of the system once a measurement has been conducted. For the simplest instrument which can be constructed for the question about Q given above, this rule states that, were the system were initially in the pure state $|\,f\,\rangle\,\langle\,f\,|$ given by the unit vector $f \in \mathcal{S}(\mathbb{R})$, and were a measurement for Q were made by this instrument which gave an outcome lying in the Borel subset Δ of $\mathbb{R}$, then the output state of the system would be the mixed state

$$K^{-1}\int_{\Delta} |\, g_s(\overline{Q})f\,\rangle\,\langle\, g_s(\overline{Q})f\,|\, ds\,, \tag{5.2.26.a}$$

where K is the normalization factor

$$K \;=\; \int_{\Delta} \|\, g_s(\overline{Q})f\,\|^2\, ds \;=\; \int_{\mathbb{R}} (|\,g\,|^2 * \chi_{\Delta})(x)|\, f(x)\,|^2\, dx\,, \tag{5.2.26.b}$$

and $g_s(\overline{Q})$ is the operator obtained from $\overline{Q}$ by applying the spectral calculus to the Borel function

$$g_s(x) \;=\; g(x-s). \tag{5.2.26.c}$$

It is important to note that the uncertainties in the measurement of Q have resulted in the introduction of mixed states into the mathematics. This is to be expected, since mixed states are precisely those for which surety about spectral values is not present.

It should be also noted that, formally, the ideal (standard) theory of measurement results from replacing the Schwartz function g in the above theory by the Dirac delta distribution δ. Close approximations to the ideal measurement situation can be made by choosing the function g to approximate that distribution. In this sense, we can choose a function g so that our general measurement apparatus can mirror the ideal situation to within any desired degree of accuracy.

It is also worth noting one further implication of the differences between the ideal and the general formulæ of wave-packet collapse. In the ideal case, in order for formula (5.2.25) for the collapsed state to make sense, it must be that it is only be possible to register a measurement for Q lying in the Borel set Δ if it is the case that $\| E_Q(\Delta)\phi \| > 0$, in other words if the probability of Q taking a value in the Borel set Δ is nonzero. The comparable observation for the smoothed case is that, in order to be able to register a value for Q in the Borel set Δ, it is required that

$$\int_{\mathbb{R}} (|g|^2 * \chi_\Delta)(x) |f(x)|^2 \, dx > 0.$$

Since the function $|g|^2 * \chi_\Delta$ is an approximation for χ_Δ, but a smooth one, this observation again indicates how the function g measures the level of uncertainty that arises in the making of, and interpretation of, measurements of Q.

It is not the purpose of this book to expound a detailed survey of the quantum theory of measurement — we have simply raised the issues to point out in the first place that these are issues which need to be addressed, and moreover that there are mathematically satisfactory ways in which they can be dealt with.

5.3 The Fourier Transform

Before studying any more representations of the CCR, we must specify the conventions that we shall be using concerning the Fourier transform. The form of the Fourier transform with which we shall work is the one given by the following definition.

Definition 5.4 *There exists a unique unitary map $\mathcal{F} : L^2(\mathbb{R}^d) \to L^2(\mathbb{R}^d)$ such that whenever ϕ is in the dense subspace $L^1(\mathbb{R}^d) \cap L^2(\mathbb{R}^d)$,*

$$[\mathcal{F}\phi](k) = (2\pi)^{-\frac{1}{2}d} \int_{\mathbb{R}^d} \phi(x)\, e^{-ix\cdot k}\, dx\,. \tag{5.3.1.a}$$

Its inverse is the map $\mathcal{F}^{-1} : L^2(\mathbb{R}^d) \to L^2(\mathbb{R}^d)$, where

$$[\mathcal{F}^{-1}\phi](x) = (2\pi)^{-\frac{1}{2}d} \int_{\mathbb{R}^d} \phi(k)\, e^{ix\cdot k}\, dk \tag{5.3.1.b}$$

whenever $\phi \in L^1(\mathbb{R}^d) \cap L^2(\mathbb{R}^d)$.

These are vector formulæ, so that $x \cdot k$ denotes the standard inner product of x, $k \in \mathbb{R}^d$, and dx, dk are both the d-dimensional Lebesgue measure.

As for every unitary operator, $\mathcal{F}$ has a spectral representation. Restricting attention to the case $d = 1$, this was found by Wiener, although in a different terminology [240]. By classical methods he showed that

$$\mathcal{F}h_n = i^{-n}h_n\,, \qquad n \geqslant 0\,. \tag{5.3.2.a}$$

Thus the eigenvalues of $\mathcal{F}$ are 1, -1, i and $-i$, and the spectral decomposition of $\mathcal{F}$ can be written as

$$\mathcal{F} = \sum_{n=0}^{\infty} i^{-n} P_n\,, \tag{5.3.2.b}$$

where P_n is the projection operator along h_n already introduced. We note the curiosity that the Fourier transform is just one of the unitary operators generated by the gauge group of the Schrödinger representation,

$$\mathcal{F} = \Gamma\big(-\tfrac{1}{2}\pi\big). \tag{5.3.2.c}$$

Equation (5.3.2.a) implies that the Fourier transform and its inverse leave $\mathcal{S}(\mathbb{R})$ invariant.

This result can be extended. For any integer $d \in \mathbb{N}$, we define the space $\mathcal{S}(\mathbb{R}^d)$ to be the space of all smooth functions f on $\mathbb{R}^d$ such that f and all of its partial derivatives (of all orders) tend to zero at infinity faster than any polynomial in the coordinate variables. In other words, the smooth function f belongs to $\mathcal{S}(\mathbb{R}^d)$ if and only if the functions

$$x \mapsto x_1^{a_1} \cdots x_d^{a_d} \frac{\partial f^{b_1+\cdots+b_d}}{\partial x_1^{b_1} \cdots \partial x_d^{b_d}}(x_1, \ldots, x_d)$$

are bounded on $\mathbb{R}^d$ for any $a_1, \ldots, a_d$, $b_1 \ldots, b_d \geqslant 0$. It is clear that, when $d = 1$, this definition coincides with that already given for $\mathcal{S}(\mathbb{R})$. We can equip $\mathcal{S}(\mathbb{R}^d)$ with a nuclear Fréchet topology (just as was done for $\mathcal{S}(\mathbb{R})$), and the interested reader is referred to the work of Reed & Simon [186] for further details. Elements of $\mathcal{S}(\mathbb{R}^d)$ are often referred to as *Schwartz functions*, or *test functions*, as was the case when $d = 1$. The topological dual $\mathcal{S}'(\mathbb{R}^d)$ of $\mathcal{S}(\mathbb{R}^d)$ is called the space of *tempered distributions*, as before, and the three spaces $\mathcal{S}(\mathbb{R}^d)$, $L^2(\mathbb{R}^d)$ and $\mathcal{S}'(\mathbb{R}^d)$ form a rigged triple, with $L^2(\mathbb{R}^d)$ equipped with a complex structure J (of complex conjugation) which linearizes the natural antilinear embedding † of $L^2(\mathbb{R}^d)$ in $\mathcal{S}'(\mathbb{R}^d)$.

Proposition 5.5 *Given a function $f \in \mathcal{S}(\mathbb{R}^d)$, its Fourier transform $\mathcal{F}f$ and its Fourier inverse transform $\mathcal{F}^{-1}f$ both belong to $\mathcal{S}(\mathbb{R}^d)$. Moreover, the restrictions of $\mathcal{F}$ and $\mathcal{F}^{-1}$ to $\mathcal{S}(\mathbb{R}^d)$ are mutually inverse continuous endomorphisms (and hence topological isomorphisms) of the space of test functions $\mathcal{S}(\mathbb{R}^d)$.*

Since the Fourier transform is a continuous endomorphism of $\mathcal{S}(\mathbb{R}^d)$, it defines a continuous endomorphism $\mathcal{F}$ of the space $\mathcal{S}'(\mathbb{R}^d)$ of tempered distributions via the formulæ

$$\begin{aligned} [\![\mathcal{F}T, f]\!] &= [\![T, \mathcal{F}f]\!], \\ [\![\mathcal{F}^{-1}T, f]\!] &= [\![T, \mathcal{F}^{-1}f]\!], \end{aligned} \qquad T \in \mathcal{S}'(\mathbb{R}^d),\ f \in \mathcal{S}(\mathbb{R}^d). \quad (5.3.3)$$

so that the Fourier transform on $\mathcal{S}'(\mathbb{R}^d)$ is simply the transpose of the Fourier transform on $\mathcal{S}(\mathbb{R}^d)$. There is no need to distinguish between the Fourier transforms on $\mathcal{S}(\mathbb{R}^d)$, $L^2(\mathbb{R}^d)$ and on $\mathcal{S}'(\mathbb{R}^d)$. This is because it can be shown that

$$\begin{aligned} [\![(J\phi)^\dagger, \mathcal{F}f]\!] &= \int_{\mathbb{R}^d} \phi(x)(\mathcal{F}f)(x)\,dx = \int_{\mathbb{R}^d} (\mathcal{F}\phi)(x)f(x)\,dx \\ &= [\![\big(J(\mathcal{F}\phi)\big)^\dagger, f]\!] \end{aligned} \quad (5.3.4)$$

for any $\phi \in L^2(\mathbb{R}^d)$ and $f \in \mathcal{S}(\mathbb{R}^d)$. Consequently the Fourier transform of the tempered distribution $(J\phi)^\dagger \in \mathcal{S}'(\mathbb{R}^d)$ is the same as the tempered distribution $\big(J(\mathcal{F}\phi)\big)^\dagger$. Since our standard convention is to identify an element $\psi \in L^2(\mathbb{R}^d)$ with the element $(J\psi)^\dagger \in \mathcal{S}'(\mathbb{R}^d)$, the above identity shows that, with respect to this convention, the Fourier transform on $\mathcal{S}'(\mathbb{R}^d)$ is simply an extension of the Fourier transform on $L^2(\mathbb{R}^d)$.

5.4 The Momentum Representation

The Schrödinger representation is important because it diagonalizes the position operator, while the momentum representation is particularly important because it diagonalizes the momentum operator. It is obtained from the Schrödinger representation by the action of the Fourier transform operator, as will now be shown.

Consider the strongly continuous irreducible representation $\mathcal{W}$ of the Weyl group on $L^2(\mathbb{R})$ defined by the formula

$$\mathcal{W}(a,b) = \mathcal{F}W(a,b)\mathcal{F}^{-1}, \qquad a,\, b \in \mathbb{R}\,, \tag{5.4.1}$$

where W is the representation of the Weyl group defined in equation (3.3.6) in reference to the Schrödinger representation. Direct calculation then shows that

$$\mathcal{W}(a,b) = W(-b,a), \qquad a,\, b \in \mathbb{R}. \tag{5.4.2}$$

In order to interpret this new representation of the CCR, it is enough to observe that its one-parameter unitary subgroups $\mathcal{U}(a) = \mathcal{W}(a,0)$ and $\mathcal{V}(a) = \mathcal{W}(0,a)$ have the form

$$\mathcal{U}(a) = V(a)\,, \qquad \mathcal{V}(a) = U(-a)\,, \qquad a \in \mathbb{R}\,, \tag{5.4.3.a}$$

where U and V are the unitary subgroups of the representation W. These unitary groups therefore have infinitesimal generators $\overline{\mathcal{P}}$ and $\overline{\mathcal{Q}}$, respectively, where $\mathcal{P}$, $\mathcal{Q} \in \mathcal{L}^+(\mathcal{S}(\mathbb{R}))$ are the smooth observables

$$\mathcal{P} = Q\,, \qquad \mathcal{Q} = -P\,. \tag{5.4.3.b}$$

In other words, $\mathcal{P}$ and $\mathcal{Q}$ are given by the formula

$$\begin{aligned} (\mathcal{P}g)(k) &= kg(k)\,, \\ (\mathcal{Q}g)(k) &= i\frac{dg}{dk}(k)\,, \end{aligned} \qquad g \in \mathcal{S}(\mathbb{R})\,, \tag{5.4.4}$$

and hence we see that the representation $\mathcal{W}$ of the Weyl group defines an irreducible representation of the CCR acting on $L^2(\mathbb{R})$ with respect to which the momentum operator $\mathcal{P}$ is diagonal — it is for this reason that this representation is called the momentum representation. Moreover it is clear that the common smooth domain of this representation is, once again, Schwartz space $\mathcal{S}(\mathbb{R})$.

If we consider the lowering and raising operators $\mathcal{A}$ and $\mathcal{A}^+$ for this representation of the smooth model, then we calculate that

$$\mathcal{A} = \frac{1}{\sqrt{2}}(\mathcal{Q} + i\mathcal{P}) = iA\,, \qquad \mathcal{A}^+ = \frac{1}{\sqrt{2}}(\mathcal{Q} - i\mathcal{P}) = -iA^+\,, \tag{5.4.5}$$

and hence the number operator $\mathcal{N}$ for the momentum representation is the same as the number operator for the Schrödinger representation,

$$\mathcal{N} = N\,. \tag{5.4.6}$$

Thus, if we denote the Hermite-Gauss functions for the momentum representation by $(\mathbf{h}_n)_{n\geqslant 0}$, then it is clear that the unique Fock vector is $\mathbf{h}_0 = h_0$, and moreover that

$$\mathbf{h}_n = \frac{1}{\sqrt{n!}}(\mathcal{A}^+)^n\mathbf{h}_0 = i^{-n}h_n\,, \qquad n \geqslant 0\,. \tag{5.4.7}$$

Hence the Fourier transform $\mathcal{F} : L^2(\mathbb{R}) \to L^2(\mathbb{R})$ is the unitary map which provides the equivalence between the Schrödinger representation and the momentum representation of the CCR. In summary, we see that, mathematically, the momentum representation is no more than a trivial relabelling of the Schrödinger representation — however, the physical interpretation of these two representations is very different.

5.5 The Heisenberg Representation

In 1925, in the first paper on what we now call quantum mechanics, Heisenberg presented the following representation. Heisenberg's construction of this representation was derived from his analysis of spectroscopic data, and was somewhat involved (*cf* [103]). An extensive and careful account of the construction from a historical point of view is given by Tomonaga [223]. If one is not interested in the original motivation for this representation, which involves the consideration of action-angle variables, an appeal to double indices, and the like, the construction is not difficult to fit into the formalism we have been considering, where it appears as just another representation.

This representation is most readily accessible for the CCR in Heisenberg form. In the language of the smooth model, the system Hilbert space is ℓ^2, the space of square summable sequences. The closure $\overline{A}$ of the lowering operator is given by the formula

$$\overline{A}\left(c_0, c_1, c_2, \ldots, c_n, \ldots\right) = \left(c_1, \sqrt{2}c_2, \sqrt{3}c_3, \ldots, \sqrt{n+1}c_{n+1}, \ldots\right), \tag{5.5.1.a}$$

and its adjoint A^* is the raising operator

$$A^*\left(c_0, c_1, c_2, \ldots, c_n, \ldots\right) = \left(0, c_0, \sqrt{2}c_1, \ldots, \sqrt{n}c_{n-1} \ldots\right), \tag{5.5.1.b}$$

where both of these closed operators share the same domain

$$\mathcal{D}(\overline{A}) = \mathcal{D}(A^*) = \left\{ \left(c_n\right)_{n\geqslant 0} \in \ell^2 : \left(\sqrt{n+1}\, c_n\right)_{n\geqslant 0} \in \ell^2 \right\}. \tag{5.5.1.c}$$

It follows that the common dense domain $\mathcal{S}$ is the space $\mathfrak{s}$ of rapidly decreasing sequences (see Definition 4.6), and the lowering and raising operators are the restrictions of $\overline{A}$ and A^* to $\mathfrak{s}$, and are denoted A and A^+ as usual. The action of the number operator $N = A^+A$ on $\mathcal{S}$ can be calculated to be

$$N\left(c_0, c_1, c_2, \ldots\right) = \left(0, c_1, 2c_2, \ldots, nc_n, \ldots\right) \tag{5.5.2.a}$$

for $c \in \mathfrak{s}$. The unique self-adjoint extension $\overline{N}$ of N has domain

$$\mathcal{D}(\overline{N}) = \left\{ \left(c_n\right)_{n\geqslant 0} \in \ell^2 : \sum_{n=0}^{\infty} n^2 |\, c_n\,|^2 < \infty \right\}, \tag{5.5.2.b}$$

and formula (5.5.2.a) can be used to define the action of $\overline{N}$ on $\mathcal{D}(\overline{N})$. For obvious reasons, in the older literature, this representation of the CCR was known as *matrix mechanics.*

This representation of the CCR is gauge invariant, with unit Fock vector

$$e_0 = (1, 0, 0, \ldots, 0, \ldots), \tag{5.5.3}$$

unique up to a phase factor. The Hermite-Gauss elements for this representation are the sequences $(e_n)_{n\geqslant 0}$, where these sequences are defined by the formula

$$\left(e_n\right)_m = \delta_{mn}, \qquad m, n \geqslant 0. \tag{5.5.4}$$

By taking appropriate linear combinations of the raising and lowering operators defined in equations (5.5.1.a) and (5.5.1.b), we can obtain the matching expressions for the position and momentum operators P and Q for this representation. Written in terms of matrix coefficients, these were what Heisenberg found.

This representation is unitarily equivalent to the Schrödinger representation, as it must be, and the relevant unitary map $\mathfrak{U} : L^2(\mathbb{R}) \to \ell^2$ is given by linear extension from its action on the Hermite-Gauss elements of $L^2(\mathbb{R})$,

$$\mathfrak{U}h_n = e_n, \qquad n \geqslant 0. \tag{5.5.5}$$

We note that we have restricted our discussion of the representations of the CCR on ℓ^2 to the case of the smooth model. This is because it is not easy to write down an explicit formula for the representation of the Weyl group. If W is the standard representation of the Weyl group in the Schrödinger representation (see equation (3.3.6)), then we have a representation $\widehat{W}$ of the Weyl group on the Hilbert space ℓ^2 given by the formula

$$\widehat{W}(a,b)e_n = \sum_{m=0}^{\infty} \langle h_m\,, W(a,b)h_n\rangle\, e_m, \qquad n \geqslant 0,\ a,\, b \in \mathbb{R}. \tag{5.5.6}$$

and the unitary map $\mathfrak{U}$ defined above intertwines W and $\widehat{W}$. It is left to the interested reader to find explicit expressions for the matrix coefficients $\langle e_m\,, \widehat{W}(a,b)e_n\rangle = \langle h_m\,, W(a,b)h_n\rangle$ for $m,\, n \geqslant 0$.

5.6 The Bargmann-Segal Representation

Another gauge invariant representation of interest is due to Bargmann [12, 13] and Segal [209], and its main feature is that the raising operator is diagonal in this representation.[5] In this representation, the system Hilbert space, $\mathcal{B}$, consists of all functions which are analytic everywhere[6] and square

---

[5] This representation is sometimes also referred to as the Fock-Bargmann-Segal representation, but not by us. It is convenient to use the abbreviation BS in the index.

[6] Mathematicians prefer the term holomorphic to analytic nowadays, a usage that is not general amongst physicists. Functions which are everywhere holomorphic are called *entire*.

integrable in the sense that the integral

$$\| F \|^2 = \frac{1}{\pi} \int_{\mathbb{C}} | F(z) |^2 e^{-|z|^2} dA(z) \tag{5.6.1}$$

is well defined and finite, where $dA(z) = dx\,dy$ is the standard Lebesgue measure on $\mathbb{C}$. (The notation $dA(z)$ will be used throughout the book.) Then $\mathcal{B}$ is a Hilbert space with respect to the inner product

$$\langle F, G \rangle = \frac{1}{\pi} \int_{\mathbb{C}} \overline{F(z)}\, G(z)\, e^{-|z|^2} dA(z), \qquad F, G \in \mathcal{B}, \tag{5.6.2}$$

and the functions

$$E_k(z) = \frac{1}{\sqrt{k!}} z^k, \qquad k \geqslant 0, \tag{5.6.3}$$

constitute an orthonormal basis for $\mathcal{B}$.

The unitary map $\mathfrak{U}_{BS} : L^2(\mathbb{R}) \to \mathcal{B}$ given by the formula

$$\mathfrak{U}_{BS} h_k = E_k, \qquad k \geqslant 0, \tag{5.6.4.a}$$

can be written in terms of an integral kernel, in that

$$\left[\mathfrak{U}_{BS}\phi\right](z) = \int_{\mathbb{R}} \mathcal{U}(z, x)\, \phi(x)\, dx, \qquad \phi \in L^2(\mathbb{R}), \tag{5.6.4.b}$$

where $\mathcal{U}(z, x)$ is given by the explicit formula

$$\mathcal{U}(z, x) = \pi^{-\frac{1}{4}} \exp\left(-\frac{1}{2}(z^2 + x^2) + \sqrt{2}zx\right). \tag{5.6.4.c}$$

This unitary map can be used to carry the Schrödinger representation of the CCR from $L^2(\mathbb{R})$ to $\mathcal{B}$, obtaining an irreducible representation W_{BS} of the Weyl group on $\mathcal{B}$ afforded by the formula

$$W_{BS}(a, b) = \mathfrak{U}_{BS}\, W(a, b)\, \mathfrak{U}_{BS}^{-1}, \qquad a, b \in \mathbb{R}. \tag{5.6.5.a}$$

It is in fact simplest to present this representation of the Weyl group in complex form, so we consider the operators $W_{BS}[w] = W_{BS}(a, b)$ where $w = \frac{1}{\sqrt{2}}(b - ia)$ (see equation (3.3.4)), in which case we can write

$$\left(W_{\mathrm{BS}}[w]F\right)(z) = e^{-\frac{1}{2}|w|^2} e^{i\overline{w}z} F(z + iw), \qquad w \in \mathbb{C},\ F \in \mathcal{B}. \tag{5.6.5.b}$$

Direct calculation then shows us that

$$\begin{aligned}\big(W_{BS}[\![H_0]\!]F\big)(z) &= \frac{1}{\pi}\int_{\mathbb{C}} e^{-\frac{1}{2}|w|^2}\big(W_{BS}[w]F\big)(z)\,dA(w)\\ &= \frac{1}{\pi}\int_{\mathbb{C}} e^{-|w|^2}F(w)\,dA(w)\end{aligned} \tag{5.6.6}$$

for any $F \in \mathcal{B}$, which demonstrates the fact that $W_{BS}[\![H_0]\!]$ is the one-dimensional projection $|\,E_o\,\rangle\,\langle\,E_0\,|$, and hence that this representation of the Weyl group is irreducible.

To see the manner in which the smooth model is presented in this representation, we note that the one-parameter unitary subgroups $U_{BS}(a) = W_{BS}(a,0)$ and $V_{BS}(a) = W_{BS}(0,a)$ of W_{BS} have self-adjoint generators $\overline{P_{BS}}$ and $\overline{Q_{BS}}$, respectively, given by the formulæ

$$\big(\overline{P_{BS}}F\big)(z) = -\frac{i}{\sqrt{2}}\big(F'(z) - zF(z)\big)\,, \tag{5.6.7.a}$$

$$\big(\overline{Q_{BS}}F\big)(z) = \frac{1}{\sqrt{2}}\big(F'(z) + zF(z)\big)\,, \tag{5.6.7.b}$$

where the domains of these two operators are the largest possible subspaces of $\mathcal{B}$ for which the above definitions make sense. Consequently, then, the lowering and raising operators are the closed operators with the common domain

$$\begin{aligned}\mathcal{D} &= \mathcal{D}\big(\overline{A_{BS}}\big) = \big\{F \in \mathcal{B} : zF(z) \in \mathcal{B}\big\}\\ &= \mathcal{D}\big(A^*_{BS}\big) = \big\{F \in \mathcal{B} : F'(z) \in \mathcal{B}\big\}\,,\end{aligned} \tag{5.6.8}$$

and are there given by the formulæ

$$\big(\overline{A_{BS}}F\big)(z) = F'(z)\,, \tag{5.6.9.a}$$

$$\big(A^*_{BS}F\big)(z) = zF(z)\,, \tag{5.6.9.b}$$

for any $F \in \mathcal{D}$. It is in this sense that the raising operator $\overline{A_{BS}}$ is diagonal in this representation.

The gauge invariant vectors of this representation, which are the elements of the kernel of the number operator $N_{BS} = A^*_{BS}\overline{A_{BS}}$, are hence the elements of the kernel of the lowering operator. Thus the space of gauge invariant vectors in $\mathcal{B}$ is the space of constant functions, and consequently is one-dimensional, spanned by the unit vector E_0. Thus we deduce that the unique Fock vector for this gauge invariant representation is E_0, and it is clear that the Hermite-Gauss functions for this representation are the vectors $\{E_k : k \geqslant 0\}$.

Although there is no simple functional characterization of the elements of the smooth domain $\mathcal{S}_{BS}$ for this representation, it is easy to show that the elements of $\mathcal{S}_{BS}$ consist of those functions $F \in \mathcal{B}$ for which the coordinate sequence $(\langle E_k , F\rangle)_{k\geqslant 0}$ belongs to the sequence space $\mathfrak{s}$ of sequences of rapid decrease. As is now usual for us, we shall reserve the symbols P_{BS}, Q_{BS}, A_{BS}, A^+_{BS} and N_{BS} to denote the restrictions of the above momentum, position, lowering, raising and number operators to the smooth domain, which restrictions are all continuous endomorphisms of the nuclear Fréchet space $\mathcal{S}_{BS}$.

This representation is useful in that it points the way to the construction of certain field theory representations, since the infinite dimensional analogue of the Gaussian measure

$$d\gamma(z) \;=\; e^{-|z|^2}\, dA(z) \tag{5.6.10}$$

exists and is well behaved.

5.7 Hardy Space And Function Theory

A final, and for phase theory very important, representation of the CCR is that on Hardy space. This representation is sufficiently important to merit some discussion of the surrounding function theory, establishing notation and conventions in the process.

5.7.1 *Function Theory*

The material presented below is intended merely to remind the reader of the key details concerning Fourier analysis on the unit circle. It is not the aim here to present a detailed and rigorous exposition of these matters — the interested reader can find the details and proofs in many good books on Fourier analysis.

The complex unit circle $\mathbb{T}$ can be identified with the real interval $(-\pi, \pi]$ via the complex exponential function, so the point $-\pi < \vartheta \leqslant \pi$ corresponds in a unique way with the point $e^{i\vartheta}$ in $\mathbb{T}$. In this notation, the Lebesgue measure on $\mathbb{T}$ is written $d\vartheta$.

5.7.1.1 *Integration Over The Unit Circle*

For any $1 \leqslant p < \infty$, the space $L^p(\mathbb{T})$ denotes the linear space of all Lebesgue measurable functions f on $\mathbb{T}$ such that

$$\| f \|_p = \left(\frac{1}{2\pi} \int_{-\pi}^{\pi} |f(e^{i\vartheta})|^p \, d\vartheta \right)^{\frac{1}{p}} < \infty , \tag{5.7.1}$$

and $L^\infty(\mathbb{T})$ is the space of all essentially bounded functions[7] on $\mathbb{T}$. We let $\| f \|_\infty$ denote the essential supremum of $| f |$ for any $f \in L^\infty(\mathbb{T})$. Then the space $\left(L^p(\mathbb{T}), \| \cdot \|_p\right)$ is a complete normed space for any $1 \leqslant p \leqslant \infty$. Moreover, since the set $(-\pi, \pi]$ has finite measure, we have the inclusions

$$L^\infty(\mathbb{T}) \subseteq L^{p_2}(\mathbb{T}) \subseteq L^{p_1}(\mathbb{T}) \subseteq L^1(\mathbb{T}) \tag{5.7.2}$$

for any $1 \leqslant p_1 \leqslant p_2 \leqslant \infty$.

For any $f \in L^1(\mathbb{T})$ we can define its *classical Fourier coefficients* by the formula

$$\hat{f}_n = \frac{1}{2\pi} \int_{-\pi}^{\pi} f(e^{i\vartheta}) \, e^{-in\vartheta} \, d\vartheta , \qquad n \in \mathbb{Z}. \tag{5.7.3}$$

Among this infinity of integration spaces, the space $L^2(\mathbb{T})$ is of particular interest, since it is a Hilbert space with respect to the inner product

$$\langle f , g \rangle = \frac{1}{2\pi} \int_{-\pi}^{\pi} \overline{f(e^{i\vartheta})} g(e^{i\vartheta}) \, d\vartheta , \qquad f, g \in L^2(\mathbb{T}) , \tag{5.7.4}$$

and an orthonormal basis for this Hilbert space is the sequence $\left(\chi_n\right)_{n \in \mathbb{Z}}$, where

$$\chi_n(e^{i\vartheta}) = e^{in\vartheta} , \qquad n \in \mathbb{Z} . \tag{5.7.5}$$

We note that the classical Fourier coefficients of a function $f \in L^2(\mathbb{T})$ are precisely the expansion coefficients of f with respect to this orthonormal

[7] An essentially bounded function on $\mathbb{T}$ is a measurable function which is bounded on the complement of some set of measure zero. Given an essentially bounded function f on $\mathbb{T}$, its *essential supremum* is the infimum of the collection of suprema of f taken over the complements of sets of measure zero. In other words, K is the essential supremum of f if and only if the set $\{e^{i\vartheta} \in \mathbb{T} : f(e^{i\vartheta}) > K\}$ has measure zero, while the set $\{e^{i\vartheta} \in \mathbb{T} : f(e^{i\vartheta}) > K - \varepsilon\}$ has strictly positive measure for any $\varepsilon > 0$. The *essential infimum* of an essentially bounded function is defined similarly.

basis, in that

$$\hat{f}_n = \langle \chi_n , f \rangle , \qquad n \in \mathbb{Z} . \tag{5.7.6}$$

5.7.1.2 *Harmonic Extensions And Hardy Spaces*

A key feature of the analysis of integrable functions on $\mathbb{T}$ is the fact that they possess a *harmonic extension* to the open unit disc

$$\mathbb{D} = \{ z \in \mathbb{C} : |z| < 1 \} . \tag{5.7.7}$$

To be specific, any function $f \in L^1(\mathbb{T})$ defines a function $\tilde{f}$ on $\mathbb{D}$ via the formula

$$\tilde{f}(re^{i\vartheta}) = \sum_{n \in \mathbb{Z}} \hat{f}_n \, r^{|n|} e^{in\vartheta} , \tag{5.7.8.a}$$

for $0 \leqslant r < 1$ and $-\pi < \vartheta \leqslant \pi$. It is well-known that $\tilde{f}$ can be obtained from f via an integral expression involving the *Poisson kernel*,

$$\tilde{f}(re^{i\vartheta}) = \frac{1}{2\pi} \int_{-\pi}^{\pi} \frac{1 - r^2}{1 - 2r\cos(\vartheta - \beta) + r^2} \, f(e^{i\beta}) \, d\beta , \tag{5.7.8.b}$$

where $0 \leqslant r < 1$, $-\pi < \vartheta < \pi$, and it can be shown that each such function $\tilde{f}$ is harmonic, with $\nabla^2 \tilde{f} = 0$, in $\mathbb{D}$.

Of particular interest to us will be the subspaces $H^p(\mathbb{T})$ of the integration spaces given above, which are defined as follows,

$$H^p(\mathbb{T}) = \{ f \in L^p(\mathbb{T}) : \hat{f}_n = 0 \quad \forall n < 0 \} , \tag{5.7.9}$$

for any $1 \leqslant p \leqslant \infty$ (in this book, we shall be primarily interested in the spaces $H^p(\mathbb{T})$ for $p = 1, 2, \infty$. These spaces are known as Hardy spaces (on the unit circle[8]), after the mathematician and cricket lover G. H. Hardy [102]. The harmonic extension procedure, when applied to Hardy functions, gives excellent results. To be specific, if $f \in H^p(\mathbb{T})$ then its harmonic extension $\tilde{f}$ is not just harmonic, it is analytic in $\mathbb{D}$, with power series expansion

$$\tilde{f}(z) = \sum_{n=0}^{\infty} \hat{f}_n z^n , \qquad z \in \mathbb{D} , \tag{5.7.10}$$

[8]There are Hardy spaces which can be defined on domains other than $\mathbb{T}$, but these are not of importance to the theory presented in this book.

and $\tilde{f}$ satisfies the additional property that

$$\sup_{0 \leqslant r < 1} \frac{1}{2\pi} \int_{-\pi}^{\pi} \left|\tilde{f}(re^{i\vartheta})\right|^p d\vartheta \;=\; \frac{1}{2\pi} \int_{-\pi}^{\pi} \left|f(e^{i\vartheta})\right|^p d\vartheta. \tag{5.7.11}$$

Moreover, this process can be inverted — to be able to do so was indeed the original motivation for introducing the Hardy spaces. If F is a function which is analytic in $\mathbb{D}$ and which is such that

$$\sup_{0 \leqslant r < 1} \frac{1}{2\pi} \int_{-\pi}^{\pi} \left|F(re^{i\vartheta})\right|^p d\vartheta \;<\; \infty\,, \tag{5.7.12}$$

then the radial limit

$$\check{F}(e^{i\vartheta}) \;=\; \lim_{r\to 1} F(re^{i\vartheta})\,, \qquad\qquad \vartheta \in \mathbb{T}\,, \tag{5.7.13}$$

exists almost everywhere in $\mathbb{T}$, and moreover defines a function $\check{F} \in H^p(\mathbb{T})$. Additionally, these two processes are mutually inverse, in that $\big(\tilde{f}\big)\check{} = f$ for any function $f \in H^p(\mathbb{T})$, while $\big(\check{F}\big)\tilde{} = F$ for any function F which is analytic in $\mathbb{D}$ and which satisfies the boundedness condition (5.7.12).

This ability to characterize elements of the spaces $H^p(\mathbb{T})$ in terms of analytic functions is extremely useful in analysis, enabling us to use the results of complex function theory when studying these spaces.

5.7.1.3 *The Hardy Hilbert Space*

For quantum mechanics, the most important of the Hardy spaces is $H^2(\mathbb{T})$. Since $H^2(\mathbb{T})$ is a closed linear subspace of the Hilbert space $L^2(\mathbb{T})$, it is a Hilbert space in its own right. An orthonormal basis for the Hilbert space $H^2(\mathbb{T})$ is the subcollection $\big(\chi_n\big)_{n\geqslant 0}$ of the previously given orthonormal basis for $L^2(\mathbb{T})$.

It is important to bear in mind this key distinction between $L^2(\mathbb{T})$ and its subspace $H^2(\mathbb{T})$, namely that $\big(\chi_n\big)_{n\in\mathbb{Z}}$ is an orthonormal basis for the former space, while $\big(\chi_n\big)_{n\geqslant 0}$ is an orthonormal basis for the latter. We emphasize this point because the similarity between the two orthonormal bases means that calculations for $L^2(\mathbb{T})$ and $H^2(\mathbb{T})$ often look very similar, with the only difference lying in different ranges of summation for indices. However, this small difference has a significant impact on both the mathematical analysis and the physical interpretations that can be placed on such analysis.

Since $H^2(\mathbb{T})$ is a closed linear subspace of the Hilbert space $L^2(\mathbb{T})$, we can consider the (continuous) orthogonal projection $P_+ : L^2(\mathbb{T}) \to L^2(\mathbb{T})$ of $L^2(\mathbb{T})$ onto $H^2(\mathbb{T})$. It is clear that the action of P_+ is to extract from f that part of its Fourier expansion which involves the positively-indexed Fourier coefficients, so that

$$P_+ f \;=\; \sum_{n=0}^{\infty} \hat{f}_n \chi_n \,, \tag{5.7.14.a}$$

where $f \in L^2(\mathbb{T})$ has the expansion

$$f \;=\; \sum_{n \in \mathbb{Z}} \hat{f}_n \chi_n \,. \tag{5.7.14.b}$$

This orthogonal projection P_+ is known as the *Szëgo-Riesz projection.*

5.7.2 *Toeplitz Operators*

Since $H^2(\mathbb{T})$ is a Hilbert space, there are many operators one might consider on it, but one class is of particular importance for both function theory and phase theory — the multiplication operators.

For every function $\omega \in L^\infty(\mathbb{T})$, and function $f \in H^2(\mathbb{T})$, their pointwise product ωf is certainly in $L^2(\mathbb{T})$, but it may have nonzero Fourier coefficients with indices $n < 0$. We can project these out with P_+, and the result is a multiplication operator in $H^2(\mathbb{T})$ associated with the function ω. The notation

$$M(\omega) f \;=\; P_+(\omega f), \qquad\qquad f \in H^2(\mathbb{T}) \,, \tag{5.7.15}$$

will be used for this operator, and $M(\omega)$ will be known as the *Toeplitz operator* generated by ω. It is clear that $M(\omega)$ is a bounded operator on $H^2(\mathbb{T})$, with

$$\| M(\omega) \| \;\leqslant\; \| \omega \|_\infty \,, \tag{5.7.16}$$

for any $\omega \in L^\infty(\mathbb{T})$.

It is easy to overlook the important distinction between the operation of multiplication by ω in $L^2(\mathbb{T})$ and the Toeplitz operator $M(\omega)$ on $H^2(\mathbb{T})$. Although multiplication by ω is certainly a continuous linear operator on $L^2(\mathbb{T})$, it is not in general an operator which leaves the subspace $H^2(\mathbb{T})$ invariant. For example, if $\omega = \chi_{-1}$ and $f = \chi_0$, then $\omega f = \chi_{-1}$, which does not belong to $H^2(\mathbb{T})$, even though f does. The Szëgo-Riesz projection is vital in order to define operators from $H^2(\mathbb{T})$ to itself.

What do Toeplitz operators look like? The answer lies in the form of the matrix elements of $M(\omega)$ with respect to the natural basis $(\chi_n)_{n\geqslant 0}$. We see that

$$\langle \chi_k, M(\omega)\chi_j\rangle \;=\; \hat{\omega}_{k-j}, \qquad\qquad j,\, k \geqslant 0, \tag{5.7.17}$$

where the $\hat{\omega}_n$, as usual, denote the Fourier coefficients of w. Brown and Halmos [29] have shown that if A is any bounded operator on $H^2(\mathbb{T})$ whose matrix elements have this characteristic difference property, in that there exists a sequence $(a_n)_{n\in\mathbb{Z}}$ such that

$$\langle \chi_k, A\chi_j\rangle \;=\; a_{k-j}, \qquad\qquad j,\, k \geqslant 0, \tag{5.7.18}$$

then the sequence (a_n) belongs to $\ell^2(\mathbb{Z})$, and so the series

$$a \;=\; \sum_{n=-\infty}^{\infty} a_n\chi_n \tag{5.7.19}$$

converges to a function a in $L^2(\mathbb{T})$ such that $M(a) = A$.

The spectral theory of Toeplitz operators is well known. The particular results below will prove useful further on. Here ω is an arbitrary function in $L^\infty(\mathbb{T})$.

- $M(\omega)$ is self-adjoint if and only if ω is real almost everywhere, and positive if and only if ω is positive almost everywhere.
- If $M(\omega)$ is self-adjoint, then its spectrum $\sigma[M(\omega)]$ is the closed bounded interval

  $$\sigma[M(\omega)] \;=\; [\operatorname{ess\,inf}\,\omega, \operatorname{ess\,sup}\,\omega]\,. \tag{5.7.20}$$

  If, moreover, ω is not equal (almost everywhere) to a constant function, then $M(\omega)$ has no point spectrum, and its continuous spectrum is absolutely continuous.
- If ω is real valued and not equal to a constant function (almost everywhere), denote the spectrum of $M(\omega)$ by the interval $[c, d]$. If we assume the technical condition[9] that there are measurable

[9]This technical condition ensures that the operator $M(\omega)$ is of unit multiplicity, and hence is unitarily equivalent to a single multiplication operator, rather than a direct sum of a number of such operators — this is the continuum form of nondegeneracy.

functions $a, b : [c,d] \to \mathbb{R}$ such that $0 \leqslant b(t) - a(t) < 2\pi$, and

$$\{ e^{i\vartheta} : \omega(e^{i\vartheta}) \geqslant t \} = \{ e^{i\vartheta} : a(t) \leqslant \vartheta \leqslant b(t) \}, \qquad (5.7.21)$$

for all $t \in [c,d]$, then we can obtain a concrete spectral representation of $M(\omega)$ on the space $L^2([c,d];d\mu)$, where $d\mu$ is the measure

$$d\mu(t) = \frac{1}{\pi} \sin\left(\frac{a(t) - b(t)}{2}\right) dt, \qquad (5.7.22)$$

for there exists a unitary operator $V : H^2(\mathbb{T}) \to L^2([c,d];d\mu)$ such that

$$\left[VM(\omega)V^{-1}\phi\right](t) = t(V\phi)(t), \qquad \phi \in H^2(\mathbb{T}), \qquad (5.7.23)$$

so that $VM(\omega)V^{-1}$ is multiplication by t on $L^2([c,d];d\mu)$. An explicit formula for the unitary map V can be found — the reader is referred to the book of Rosenblum & Rovnyak [196] for details. A particular example of this unitary map will be considered in Chapter 10, when the Toeplitz phase operator is considered in detail.[10]

5.7.3 *The Representation Of The CCR On Hardy Space*

The Hilbert space $H^2(\mathbb{T})$ is isomorphic to $L^2(\mathbb{R})$ via the unitary map $\mathfrak{U}_T$ from $L^2(\mathbb{R})$ to $H^2(\mathbb{T})$ given by the formula

$$\mathfrak{U}_T h_n = \chi_n, \qquad n \geqslant 0, \qquad (5.7.24)$$

and so there is an irreducible representation W_T of the Weyl group on $H^2(\mathbb{T})$ defined by the formula

$$W_T(a,b) = \mathfrak{U}_T W(a,b) \mathfrak{U}_T^{-1}, \qquad a, b \in \mathbb{R}. \qquad (5.7.25)$$

This time, it is simplest to display the particular form of the smooth model. The (self-adjoint extensions of the) lowering and raising operators are the closed operators defined on the common domain

$$\mathcal{D}(\overline{A_T}) = \mathcal{D}(A_T^*) = \{ f \in H^2(\mathbb{T}) : (\sqrt{n}\hat{f}_n)_{n\geqslant 0} \in \ell^2 \} \qquad (5.7.26.a)$$

[10] The expression *Toeplitz phase operator* is preferred to "phase operator on Hardy space".

by the formulæ

$$\overline{A_T}\chi_n = \begin{cases} \sqrt{n}\chi_{n-1}, & n \geqslant 1, \\ 0, & n = 0, \end{cases} \qquad A_T^*\chi_n = \sqrt{n+1}\chi_{n+1}, \quad n \geqslant 0. \tag{5.7.26.b}$$

Thus $\overline{N_T} = A_T^*\overline{A_T}$, the (self-adjoint extension of the) number operator, has domain

$$\mathcal{D}(\overline{N_T}) = \{f \in H^2(\mathbb{T}) : (n\hat{f}_n)_{n\geqslant 0} \in \ell^2\}, \tag{5.7.27.a}$$

and is such that

$$\overline{N_T}\chi_n = n\chi_n, \qquad n \geqslant 0. \tag{5.7.27.b}$$

We deduce that the smooth domain $\mathcal{S}_T$ for this model is given by

$$\mathcal{S}_T = \{f \in H^2(\mathbb{T}) : (\hat{f}_n)_{n\geqslant 0} \in \mathfrak{s}\}. \tag{5.7.28}$$

It is clear that any function $f \in \mathcal{S}_T$ is infinitely differentiable. As usual, we shall refer to the restrictions to $\mathcal{S}_T$ of the lowering, raising and number operators by the symbols A_T, A_T^+ and N_T respectively — it is of course possible now to define the momentum and position operators P_T and Q_T as well.

From the above discussion, it is clear that the gauge group is defined on $\mathcal{S}_T$ by the formula

$$\big(\Gamma(t)f\big)(e^{i\vartheta}) = f\big(e^{i(\vartheta+t)}\big), \qquad f \in \mathcal{S}_T,\ t \in \mathbb{R}, \tag{5.7.29}$$

and thus the gauge invariant vectors for this representation are the constant functions, and hence form a one-dimensional subspace of $H^2(\mathbb{T})$ spanned by the unique Fock vector χ_0. Moreover, we deduce that the Hermite-Gauss functions for this representation are the vectors $(\chi_n)_{n\geqslant 0}$.

It would be useful if it were possible to express the various operators of this representation of the CCR in a closed form with respect to the functions on $H^2(\mathbb{T})$, but this does not seem to be practically possible. The number operator N_T is a rare exception to this observation, since we can write

$$\big(N_T f\big)(e^{i\vartheta}) = -i\frac{d}{d\vartheta}f(e^{i\vartheta}), \qquad f \in \mathcal{S}_T. \tag{5.7.30}$$

It should be noted that N_T is not a Toeplitz operator. In spite of the form of its functional definition, it is a positive operator and hence, for example the operator $I + N_T$ has a unique positive square root. Convenient functional representations for the lowering and raising operators cannot be found, it seems — the best possible being, apparently, the polar forms

$$\begin{aligned} A_T f &= (I + N_T)^{\frac{1}{2}} M(\chi_{-1}) f \,, \\ A_T^+ f &= M(\chi_1)(I + N_T)^{\frac{1}{2}} f \,, \end{aligned} \qquad f \in \mathcal{S}_T \,. \tag{5.7.31}$$

The difficulty in finding a closed form for the operators involved in this representation can best be illustrated in the following way. The unitary map $\mathfrak{V} = \mathfrak{U}_T \mathfrak{U}_{BS}^{-1} : \mathcal{B} \to H^2(\mathbb{T})$ intertwines the Bargmann-Segal and the Hardy space representations of the CCR. Since there are good closed form expressions for the various operators to be encountered in the Bargmann-Segal representation, we would be able to find closed form expressions for the operators in the Hardy space representation provided that we could find a closed form expression for this unitary map $\mathfrak{V}$. Now, we can express $\mathfrak{V}$ in terms of an integral kernel, writing

$$(\mathfrak{V}F)(e^{i\vartheta}) \;=\; \frac{1}{\pi} \int_{\mathbb{C}} G_{\frac{1}{2}}(e^{i\vartheta}\overline{z}) F(z) e^{-|z|^2} \, dA(z) \,, \qquad F \in \mathcal{B} \,, \tag{5.7.32.a}$$

where $G_{\frac{1}{2}}$ is the entire function

$$G_{\frac{1}{2}}(z) \;=\; \sum_{k=0}^{\infty} \frac{z^k}{\sqrt{k!}} \,, \qquad z \in \mathbb{C} \,. \tag{5.7.32.b}$$

Thus any closed form expression for the operators found in the Hardy space representation will be based upon properties of and expressions for this function $G_{\frac{1}{2}}$. Since, however, there is no simple expression for this function, we cannot obtain any results of practical use. However, the above discussion is interesting in its own right, not least because the function $G_{\frac{1}{2}}$ occurs elsewhere in phase theory. For this reason, detailed knowledge of the properties of the family of entire functions

$$G_\alpha(z) \;=\; \sum_{k=0}^{\infty} \frac{z^k}{(k!)^\alpha} \,, \qquad 0 < \alpha < 1 \,, \tag{5.7.33}$$

would be beneficial in a number of problems.

5.7.4 *The Wrong Phase Operator?*

For representations of the CCR to be physically equivalent, they must be unitarily equivalent. However, this implies not only that the underlying

Hilbert spaces are unitarily isomorphic, but also that the unitary isomorphism between these two spaces intertwines the representations of the Weyl group that these Hilbert spaces carry. In particular, this implies that properties such as gauge invariance are invariants of physical equivalence. We have observed previously (see page 75) how unitarily equivalent Hilbert spaces (or even the same Hilbert space) can carry physically inequivalent representations of the CCR.

This extra requirement of physical equivalence is easily overlooked, and a number of authors have been led to consider to consider a representation of the CCR on $L^2(\mathbb{T})$ with which can be associated an operator which is analogous to a number operator, and also another self-adjoint operator which is canonically conjugate to it. This has led these authors to suppose that they have found a quantum mechanical phase operator. However, they have not — their mathematics is correct, but solves a different problem.[11]

Define a (new) "number operator" $\mathcal{N}$ on the dense domain[12] $\mathcal{C}^\infty(\mathbb{T})$ of $L^2(\mathbb{T})$, which consists of the infinitely differentiable functions on $\mathbb{T}$, by

$$[\mathcal{N}f](e^{i\vartheta}) = -i\frac{d}{d\vartheta}f(e^{i\vartheta}), \qquad f \in L^2(\mathbb{T}) . \tag{5.7.34}$$

It is important that $\mathcal{N}$ should not be confused with N because of the high degree of similarity between their defining formulæ —the former acts on $L^2(\mathbb{T})$, the latter on $H^2(\mathbb{T})$.

An operator that has been proposed as a phase operator from time to time is defined on this domain by the formula

$$[\Phi f](e^{i\vartheta}) = \vartheta f(e^{i\vartheta}) . \tag{5.7.35}$$

In other words, Φ is simply the operator of multiplication by ϑ, and it is easy enough to see that

$$\mathcal{N}\Phi f - \Phi\mathcal{N}f = -i\,f \tag{5.7.36}$$

for all f *in this domain.*

It is tempting to say that, since the Hilbert space $L^2(\mathbb{T})$ is unitarily isomorphic to $L^2(\mathbb{R})$, and since $\mathcal{N}$ looks like the number operator N on $H^2(\mathbb{T})$, the images of the maps $\mathcal{N}$ and Φ will give a canonically conjugate

[11] It is convenient to refer to this in the index as the *circle representation.*

[12] The space $\mathcal{C}^\infty(\mathbb{T})$ consists of those elements of $L^2(\mathbb{T})$ whose (doubly infinite) sequence of Fourier coefficients is rapidly decreasing in both directions — we do not need the details.

phase-number pair when transported to $L^2(\mathbb{R})$. However, this argument fails because $\mathcal{N}$ is not the number operator coming from some gauge invariant representation of the CCR — the presence of the negative eigenvalues of $\mathcal{N}$ renders this impossible. In some sense, the negative eigenvalues of $\mathcal{N}$ correspond to excitations with an opposite generalized charge (whatever the physical meaning that might have). Since there was no such degree of freedom in the system we started with, $\mathcal{N}$ is not counting the same physical excitations as N. Thus, $\mathcal{N}$ and Φ are not a conjugate pair associated with an irreducible representation of the CCR.

One proposed solution to this problem is to incorporate the Szëgo-Riesz projection into the definitions, considering the operators $N = P_+\mathcal{N}$ and $X = P_+\Phi$, which can be regarded as operators on $H^2(\mathbb{T})$. This approach regains the true number operator N, and for this reason many authors posit $X = P_+\Phi$, the Toeplitz operator of multiplication by the angle function, as a quantum mechanical phase operator. However, as the No-Go Theorem would lead us to expect, N and X are no longer canonically conjugate on $\mathcal{S}_T$.

Attempts have been made to rectify this problem by finding an alternative domain to $\mathcal{S}_T$ on which N and $P_+\Phi$ are canonically conjugate. Just such a domain has been constructed by Galindo [66]. Unfortunately this domain, and any other on which canonicity holds, will not contain the linear span of the Hermite-Gauss vectors, and so will not be a common domain on which the standard quantum mechanical observables act and leave invariant.

5.8 The CCR: Dirac's Method

We end this Chapter with a section that could well have come at the beginning, namely a critique of Dirac's method for "deriving" the CCR, wherein he suggests the "connection" between the commutator and the Poisson bracket. The argument is of historical interest only, since the proper connection is between the commutator and the Moyal, and not the Poisson, bracket (see Chapter 13). Yet the attraction of the supposed connection remains a popular one, and has an important bearing on the nonexistence of a canonical phase operator.

It is worth following Dirac's book in this matter [49], since in fairness to him, he does not say what is often ascribed to him. Most of us are so familiar with Dirac's method, or more probably the folk-lore surrounding it, that the bracket-commutator connection seems almost obvious. But that is a result of the clarity of Dirac's exposition, and the apparent simplicity is misleading. At crucial points in the argument, decisions are made which are by no means obvious and should be considered with great care. Since this section is essentially a historical review, we shall temporarily drop atomic units and include Planck's constant $\hbar$ here, so that the equations have the more traditional form.

Dirac begins by noting that Poisson brackets are important in classical mechanics, so he exhorts us to

> *try to introduce a quantum Poisson bracket which shall be the analogue of the classical one.*

He demands that the quantum Poisson bracket, to be denoted $\{\ ,\ \}_Q$, have the algebraic properties of the classical Poisson bracket. Thus the quantum Poisson bracket should be a Lie bracket, but should also satisfy the additional identity

$$\{uv, w\}_Q = u\,\{v, w\}_Q + \{u, w\}_Q\, v \tag{5.8.1}$$

for any quantum observables u, v and w (Dirac called these *q-numbers*), which condition is necessary because the collection of quantum observables is to be a real Lie subalgebra of the larger algebra of observables (as is also the case classical observables). For the classical algebra, the ordering in this relation does not matter, since the algebra of observables is commutative, but in the quantum case it matters a great deal.

Dirac also assumes that every q-number is hermitian, and moreover that the quantum Poisson bracket is a hermitian operator, so that

$$\{u\,,\,v\}_Q^* = \{v\,,\,u\}_Q \tag{5.8.2}$$

for any two q-numbers u and v. These are not contentious assumptions, and can be taken to be unobjectionable. It follows from these assumptions that, for any four q-numbers u_1, u_2, v_1, and v_2 we have the identity

$$\{u_1, v_1\}_Q\,(u_2v_2 - v_2u_2) = (u_1v_1 - v_1u_1)\,\{u_2, v_2\}_Q\,. \tag{5.8.3}$$

Now comes the most crucial step. Remember, at this stage we do not know which q-number corresponds to a given classical observable — some additional input is necessary. Dirac's argument at this point is that

> *The strong analogy between the quantum Poisson bracket ··· and the classical Poisson bracket ··· leads us to make the assumption that quantum Poisson brackets, or at any rate the simpler ones of them, have the same values as the corresponding classical Poisson brackets*

In particular, since the classical position and momentum observables p and q have Poisson bracket $\{p, q\} = 1$, Dirac argues that we should expect their corresponding q-numbers P and Q to be such that $\{P, Q\}_Q = I$, the identity operator. The standard assignment of q-numbers for P and Q yields the commutator identity

$$[P, Q] = -i\hbar I,$$

where the real constant $\hbar$ is later identified, on the basis of analysis of the classical limit, to be equal to Planck's constant divided by 2π, in which case we deduce from equation (5.8.3) that the quantum Poisson bracket of any two q-numbers u and v is given by the formula

$$\{u, v\}_Q = \frac{i}{\hbar}(uv - vu). \tag{5.8.4}$$

The question then remains to what extent Dirac's last assumption is valid — to identify how "simple" the q-numbers u and v have to be so that $\{u, v\}_Q = \{u, v\}$. We know that this is true for the position and momentum observables, but the No-Go Theorem tells us that this is impossible for the number operator and any putative phase operator.

Concerning how frequently one is likely to come across such operator pairs, Dirac tells us that

> *A Poisson bracket in quantum mechanics is a purely algebraic notion and is thus a rather more fundamental concept than a classical Poisson bracket, which can only be defined with reference to a set of canonical coordinates and momenta*[13] *For this reason canonical coordinates and momenta are of less importance in quantum mechanics than in classical mechanics; in fact, we may have a system in quantum mechanics for which canonical coordinates and momenta do not exist and we can still give meaning to*

[13]It should be noted that it is now possible to define the classical Poisson bracket in a coordinate free manner, which fact moderates the validity of this argument.

> *Poisson brackets. Such a system would be one without a classical analogue and we should not be able to obtain its quantum conditions by the method described here.*

One of us took Dirac's course in Quantum Mechanics at Cambridge some years ago, based on his book, and remembers a bright spark asking him about the existence of canonical pairs other than position and momentum. After a long silence, the class was told that "we were not going to go into that here" or something very much like that, as Thucydides would have it.

We must emphasize that we are not casting aspersions at Dirac's work — we have the greatest respect for what he achieved. However, Dirac was not perfect, and our dispute is with those who take everything that Dirac wrote as being carved in stone, and subject to no critical analysis. Dirac himself was more robust. For example, he gave J. E. Roberts the thesis problem of casting his (Dirac's) bra and ket formalism into the rigged Hilbert space framework because, and we paraphrase him, *the time had come to do so*[14] [189, 188].

5.9 Additional Reading

Many textbooks consider the Schrödinger, Heisenberg and momentum representations of the CCR in some version or another. The Bargmann-Segal representation is less well covered, and often only in connection with normal-ordered quantization. This is the case in the excellent book of Berezin & Shubin [19]; but see Folland [63]. Other than the books on quantum mechanics previously cited, we recommend Kemble's text for its careful treatment of a number of topics otherwise overlooked, relying on methods of classical analysis [137]. For example, the reader's attention is brought to Kemble's explanation of why the bound state eigenfunctions of the hydrogen atom do not form a basis for $L^2(\mathbb{R}^3)$, despite the fact that the eigenfunctions of a Sturm-Liouville problem are complete.

There are a number of books on Hardy spaces and Toeplitz operators. Besides those noted in the text, some relevant references are [25], [56] and [118].

---

[14] J.E.Roberts, private communication.

Chapter 6

Probability in Quantum Mechanics

I am not saying this in order to criticize, but your argument is sheer nonsense.

– N. Bohr

Surely, after 62 years, we should have an exact formulation of some serious part of quantum mechanics?

– J .S. Bell

Probably never before has a theory been evolved which has given a key to the interpretation and calculation of such a heterogeneous group of phenomena of experience as has quantum theory. In spite of this, however, I believe that the theory is apt to beguile us into error in our search for a uniform basis for physics, because, in my belief, it is an incomplete representation of real things, although it is the only one that can be built out of the fundamental concepts of force and material points (quantum corrections to classical mechanics).

– A. Einstein, J. Franklin Inst. **221**, 1936

Grammatici certant et adhuc sub judice lis est (Scholars dispute, and the case is still before the courts).

– Quintus Horatius Flaccus, 65-8 BC

In this Chapter we consider the probabilistic aspects of quantum theory, which originate in Bohr's proposal that a wave function is to be interpreted as giving the statistical distribution of the allowed values of any quantity being measured. It should be noted that, throughout this book, we are adopting the conventional *frequency of occurrence* interpretation of probability, so that the probability of an event happening is understood as being equal to the theoretical limit of the proportion of times that this event occurs in a sequence of identical tests (in the limit as the number of such tests tends to infinity). That such a limit exists is a consequence of the Weak Law of Large Numbers.

During the birth pangs of quantum mechanics this probabilistic interpretation of quantum mechanics was a source of great contention (and still is, for some), since it implies that we cannot, in general, predict in advance what the outcome of the next measurement of an observable might be, even if we know all the constraints and forces acting upon a system — moreover, this indeterminacy is unavoidable. However, on the whole, we have grown used to these concepts, and now accept that the universe is ordered thus.

We note, before proceeding, that the formalisms in this Chapter are valid in both the bounded and the smooth models, provided that the states and observables being considered at any time are appropriate to the model in question.

6.1 Quantum Probability Distributions

Now that we have introduced both observables and states as quantum mechanical quantities, we need to know how to derive information from their mathematical structure concerning the measurements that might be made of them. We start with the interpretation of observables. This should present few conceptual difficulties, since the interpretation given below is entirely analogous with the interpretation normally made of classical observables.

> **Axiom 6.1** *(The Rôle Of The Spectrum) In an experiment to measure the values of an observable A, the only values that can occur are the numbers in its spectrum, whatever the state.*

Should A have a purely discrete spectrum, the only measured values will be its eigenvalues. For an observable A with a nonempty continuous spectrum, there is a continuum of possible outcomes.

While we were justifying both the smooth and the bounded models, we stressed that there had to be imperfections inherent in any piece of measuring apparatus, and that this implied that a perfect measurement of observables with continuous spectra was not possible. There are, essentially, two mathematical solutions to this problem. The first solution, to which we have already referred when discussing approximate position operators in the preceding Chapter, involves the use of approximate observables, questions, and instrument observables. We shall discuss this approach again further on in this Chapter. A second approach will be discussed in Chapter 16, when we discuss various aspects of quantum measurement. There we argue that any measuring device must only be able to register outcomes on some discrete scale associated with that device. Thus, irrespective of whether the observable (which the apparatus is seeking to measure) has continuous spectrum or not, any measurement process can be interpreted as registering the values of some observable with discrete spectrum. However, both of these approaches must be seen as refinements of the ideal probabilistic theory of measurement, and should follow that theory.

We therefore proceed by stating the standard probabilistic interpretation of states for quantum mechanics in a manner which is valid for all types of observable.

Axiom 6.2 ***(The Quantum Probability Distribution)*** *The values that result from a measurement do so randomly. Suppose a quantum system is in the state given by the density matrix* ρ*, and that the observable* A *has the spectral representation*

$$A = \int_{\sigma(A)} \lambda \, dE_A(\lambda). \tag{6.1.1.a}$$

Then, if V *is a Borel subset of the spectrum* $\sigma(A)$ *of* A*, the quantity*

$$\mathbf{Pr}\,[\rho, A\,;\,V] = \mathrm{Tr}\,(\,\rho E_A(V)\,)\,. \tag{6.1.1.b}$$

is the probability that upon measurement a value for A *is obtained which lies within the Borel set* V*. We therefore call this quantity a quantum probability distribution.*

It is sometimes convenient to consider all Borel subsets of $\mathbb{R}$ rather than only subsets of the spectrum. This change is harmless, since if V is a Borel subset of $\mathbb{R}$ such that $\overline{V} \cap \sigma(A) = \varnothing$, then $E_A(V) = 0$, and so $\mathbf{Pr}\,[\rho, A\,;\,V]$ is equal to 0. We have emphasized previously that spectral projections such as $E_A(V)$ are, in general, not smooth observables. However, this creates no problems for the application of Axiom 6.2 in the smooth model, since operators such as $\rho E_A(V)$ are trace class for all (smooth) states ρ, (smooth) observables A and Borel subsets V of $\mathbb{R}$.

Although Axiom 6.2 has been expressed in a formalism best suited for observables with continuous spectrum, there is no problem in applying it to observables with discrete spectrum. For suppose that the observable A has the nondegenerate eigenvalue α in its spectrum, with corresponding unit eigenvector ψ, and let V be an open interval of $\mathbb{R}$ which contains α, but no other point of the spectrum of A. Then the spectral measure $E_A(V)$ is the projection P_ψ onto the one-dimensional subspace of $\mathcal{H}$ spanned by ψ,

$$E_A(V)\phi \;=\; P_\psi \phi \;=\; \langle \psi \,,\, \phi \rangle \, \psi \,, \qquad \phi \in \mathcal{H} \,, \tag{6.1.2.a}$$

and so the probability of registering the value α as the outcome of a measurement of the observable A, when the system is in the state defined by the density matrix ρ, is

$$\mathbf{Pr}\,[\rho, A\,;\, V] \;=\; \mathrm{Tr}\;(\,\rho P_\psi\,) \;=\; \langle \psi \,,\, \rho\psi \rangle \,. \tag{6.1.2.b}$$

In particular, if $\rho \;=\; |\,\phi\,\rangle\,\langle\,\phi\,|$ is some pure state, then this probability is the transition probability

$$\mathbf{Pr}\,[\rho, A\,;\, V] \;=\; |\,\langle \phi \,,\, \psi \rangle\,|^2 \,. \tag{6.1.2.c}$$

Note the occurrence of the square of the modulus in this expression. In the paper introducing his path-sum integral, Feynman [59] emphasizes that this rule is what distinguishes quantum from classical probability.

There are a number points that should be emphasized in respect of this Axiom:

1. For any fixed state ρ and observable A, the map $V \to \mathbf{Pr}\,[\rho, A\,;\, V]$ is a probability measure when considered as a function of the (Borel) subsets of the real line. To emphasize this viewpoint, the notation $m_{\rho;A}$ is introduced for this measure,

   $$\mathbf{Pr}\,[\rho, A\,;\, V] \;=\; m_{\rho;A}(V), \qquad V \subseteq \mathrm{Bor}(\mathbb{R}). \tag{6.1.3}$$

   This is not just a mathematical observation, but is an essential part of the foundations of quantum theory. Certainly no such construction is possible for pure states in classical mechanics.

   It is important to note how observables with discrete spectrum are described in terms of these probability measures. For example, consider again our example of an observable A which possesses the nondegenerate eigenvalue α with associated unit eigenvector ψ. If ρ is equal to the pure state P_ψ, then

$$m_{P_\psi;A}(V) = \begin{cases} 1, & \alpha \in V, \\ 0, & \alpha \notin V, \end{cases} \tag{6.1.4}$$

for any Borel subset V of $\mathbb{R}$, and hence we see that $m_{P_\psi;A}$ is the atomic measure concentrated at α. Indeed Gleason's Theorem 3.8, or its smooth variant, can be used to imply that the real number α is an eigenvalue of the observable A if and only if there exists a state ρ for which $m_{\rho;A}(\{\alpha\}) > 0$.

2. Functions of the operator A can be dealt with very neatly within this formalism. Using the spectral calculus,

$$\mathrm{Tr}\,(\rho f(A)) = \int_{\sigma(A)} f(\lambda)\, dm_{\rho;A}(\lambda). \tag{6.1.5}$$

for any functions f for which $f(A)$ belongs to the observable algebra of the model in question.

3. In particular, the quantities

$$\mathrm{Tr}\,(\rho A^k) = \int_{\sigma(A)} \lambda^k\, dm_{\rho;A}(\lambda), \qquad k \in \mathbb{N}, \tag{6.1.6}$$

are the moments of A in the state ρ. The first two moments are of particular significance: the first moment is the expectation (or average) of the values obtained in measuring A in the state ρ:

$$\mathbf{Exp}_\rho[A] = \mathrm{Tr}\,(\rho A); \tag{6.1.7}$$

and from the first and second moments we construct the variance,

$$\mathbf{Var}_\rho[A] = \mathbf{Exp}_\rho[A^2] - (\mathbf{Exp}_\rho[A])^2. \tag{6.1.8}$$

The square root of the variance,

$$\mathbf{Unc}_\rho[A] = (\mathbf{Var}_\rho[A])^{\frac{1}{2}}, \tag{6.1.9}$$

is the uncertainty of A in that state.

One important consequence of this result is the following. If A is an observable and ρ a state such that $\mathbf{Unc}_\rho[A] = 0$, then since

$$\mathbf{Unc}_\rho[A] = \int_{\sigma(A)} (\lambda - \alpha)^2\, dm_{\rho;A}(\lambda) = 0,$$

where $\alpha = \mathbf{Exp}_\rho[A]$, it follows that $m_{\rho;A}$ is the atomic measure at α, and hence that α is an eigenvalue of A. Consequently we deduce that $\mathbf{Unc}_\rho[A] > 0$ for any state if A is an observable with no point spectrum.

4. In the bounded model, a measurement of a projection operator P in a pure state will result in the value 0 if the state vector is orthogonal to the closed subspace $P\mathcal{H}$ which is the range of P. It will result in the value 1 if the state vector is in the range.
   More generally, if a vector ψ has a nonzero component in both the kernel and the range of P, a measurement will give us either 1 (yes) or 0 (no) and a sequence of identical measurements of P will give us a distribution of 0s and 1s, with the probability of obtaining 1 equal to $\| P\psi \|^2$, and that for 0 equal to $\| (I - P)\psi \|^2$.
   Hence projection operators are sometimes known as questions, or propositions. Some years ago, Birkhoff and Von Neumann attempted to build up all of quantum theory from the geometry of projection operators [21]. Their analysis led them to a certain class of lattices which are now studied under the name of quantum logic. See the list at the end of the Chapter for additional references.

Feynman (*ibid*) has illustrated the rôle of interference as the basis of the structure of the theory through the so-called Young's Slit experiment, where an electron beam is incident on a two slit barrier, beyond which there is a registration screen. The pattern on the screen is quite different as one or the other slit is closed, or if both are open. The proviso is that the beam is allowed to reach the screen unhindered by anything but the barrier. The phenomenon involved can be described in general terms as follows.

Given two pure states, ϕ and ψ, and an observable A, each state will determine a probability distribution with respect to A. A third pure state can defined by forming their normalized sum ζ,

$$\zeta = \frac{\phi + \psi}{\| \phi + \psi \|}.$$

The probability distribution of A in the pure state ζ is given by the formula

$$\mathbf{Pr}[\zeta, A; V] = \frac{1}{\| \phi + \psi \|^2}\big(\mathbf{Pr}[\phi, A; V] + \mathbf{Pr}[\psi, A; V] + 2\mathrm{Re}\,\langle \phi, E_A(V)\psi \rangle\big),$$

for any Borel set V.

The probability distributions $\mathbf{Pr}[\zeta, A; V]$ consists of the sum of the probabilities for ϕ and ψ separately, plus a cross term (which is present even if ϕ and ψ are orthogonal). When referring to the two-slit arrangement, where we can take ϕ and ψ to represent the states describing the electron beams emanating from the slits separately, this cross term is what is responsible for the pattern not consisting simply of two slightly smeared spots. In an older terminology, this is a manifestation of the wave nature of matter.

In constructing the bounded model, the fact that a unit ray determines a state rather than a unit vector was discussed at length — the choice of representative from the unit ray is unimportant, and the absolute phase factor does not affect the physics. It was also pointed out that this was no longer the case for transition processes, and that can be seen here. It does not matter whether ζ or $e^{i\alpha}\zeta$ is chosen since they both give the same state and, in this application, the same interference pattern. But if ϕ is replaced by $e^{i\beta}\phi$ and ψ by $e^{i\gamma}\psi$, the result is quite different (assuming that β and γ are uncorrelated) — relative phase affects the physics. Note, however, that the phase of the state wave function is not the same thing as the phase of the quantized electromagnetic field, and it is this latter concept, or at least an aspect of it, that the term quantum phase operator refers to.

6.2 Uncertainty Relations

That observables are represented in quantum mechanics by operators has (amongst many others) two important consequences. The first is the one we have discussed in the previous section, namely that (in principle) any number lying in the spectrum of the observable may be registered as the outcome of some measurement process. The second consequence is usually expressed in terms of the so-called uncertainty relations. These relations impose limits on the degree of accuracy with which two observables can be measured simultaneously. In particular, even if both observables have a discrete spectrum, so that separate measurements can be made for them with complete accuracy, simultaneous measurements cannot be made for both of them, with complete accuracy, unless the two observables commute. More generally, for pairs of observables with continuous spectra, the uncertainty relations tell us that the levels of precision with which these two quantities can simultaneously be measured are inversely proportional (at best).

The standard example of an uncertainty relation is that for the position and momentum observables. To be able to deal with these observables without approximating them we shall work for a while in the smooth model of quantum mechanics. As is well-known, the standard position and momentum operators (5.2.3.b) and (5.2.3.a) of the Schrödinger representation on $L^2(\mathbb{R})$ satisfy the canonical commutation relation (4.2.5), and from this it is possible to deduce that

$$\mathfrak{Unc}_\rho[Q] \cdot \mathfrak{Unc}_\rho[P] \geqslant \frac{1}{2} \tag{6.2.1}$$

in any state ρ. If we restored Planck's constant explicitly, the right-hand side of this inequality would be $\frac{1}{2}\hbar$ instead of $\frac{1}{2}$. This inequality was originally discovered by Heisenberg, and his view of it is considered in detail in his book on the physical principles of quantum theory [104].

The uncertainty between P and Q can be generalized to yield a relation that must be satisfied by any pair of operators. The sharpest form was first derived by Robertson, apparently [191, 192, 193].

Theorem 6.1 *Let A and B be observables and ρ a density matrix. Then*

$$\begin{aligned}\mathfrak{Unc}_\rho[A]^2 \cdot \mathfrak{Unc}_\rho[B]^2 \quad \geqslant \quad & \left\{\frac{1}{2}\mathsf{Exp}_\rho[AB+BA] - \mathsf{Exp}_\rho[A]\,\mathsf{Exp}_\rho[B]\right\}^2 \\ & + \frac{1}{4}\left\{\mathsf{Exp}_\rho[i(AB-BA)]\right\}^2, \end{aligned} \tag{6.2.2}$$

which is known as the uncertainty relation[1] *for A and B.*

Proof:

In the first stage of the proof[2], the problem is reduced to the pure state case by means of the Gel'fand-Naimark-Segal (GNS) construction [57]. Denote the algebra of observables by $\mathfrak{A}$, as usual, with the density matrices interpreted correspondingly. The map

$$X, Y \mapsto \mathsf{Exp}_\rho[X^*Y], \qquad X, Y \in \mathfrak{A},$$

[1] This result can be seen as a non-commutative generalization of the probabilistic statement that the square of the covariance of two random variables is never greater than the product of the variances of these random variables. However, this non-commutative version has richer implications than has the simple probabilistic one.

[2] We are including a proof because the textbooks consider only pure states [222], [124].

is sesquilinear, hermitian and positive. Defining

$$\mathfrak{K} = \{ X \in \mathfrak{A} : \mathbf{Exp}_\rho [X^* X] = 0 \},$$

then $\mathfrak{K}$ is a linear subspace of $\mathfrak{A}$, and we can define an inner product on the quotient space $\mathcal{K} = \mathfrak{A}/\mathfrak{K}$ via the formula

$$\langle X + \mathfrak{K}, Y + \mathfrak{K} \rangle = \mathbf{Exp}_\rho [X^* Y], \qquad X, Y \in \mathfrak{A}.$$

Any $X \in \mathfrak{A}$ defines an endomorphism $\pi(X)$ of $\mathcal{K}$ by setting

$$\pi(X)(Y + \mathfrak{K}) = XY + \mathfrak{K}, \qquad Y \in \mathfrak{A},$$

and the resulting map $\pi : \mathfrak{A} \to \mathrm{End}(\mathcal{K})$ is a *-algebra representation. Finally, the vector

$$\Omega = I + \mathfrak{K}$$

in $\mathcal{K}$ is a cyclic vector for this *-representation.
Write $\hat{A} = \pi(A) - \mathbf{Exp}_\rho [A] I$ and $\hat{B} = \pi(B) - \mathbf{Exp}_\rho [B] I$ for notational simplicity. Working in the inner product space $\mathcal{K}$,

$$\mathbf{Var}_\rho [A] = \|\hat{A}\Omega\|^2, \qquad \mathbf{Var}_\rho [B] = \|\hat{B}\Omega\|^2,$$

and hence the Cauchy-Schwarz inequality yields the bound

$$\begin{aligned}
&\mathbf{Var}_\rho [A] \cdot \mathbf{Var}_\rho [B] \\
&\qquad \geqslant |\langle \hat{A}\Omega, \hat{B}\Omega \rangle|^2 \\
&\qquad = \langle \Omega, \hat{A}\hat{B}\Omega \rangle \langle \Omega, \hat{B}\hat{A}\Omega \rangle \\
&\qquad = \tfrac{1}{4} \langle \Omega, (\hat{A}\hat{B} + \hat{B}\hat{A})\Omega \rangle^2 + \tfrac{1}{4} \langle \Omega, i(\hat{A}\hat{B} - \hat{B}\hat{A})\Omega \rangle^2,
\end{aligned}$$

after some elementary calculation. Since

$$\begin{aligned}
&\langle \Omega, (\hat{A}\hat{B} + \hat{B}\hat{A})\Omega \rangle \\
&\qquad = \langle \Omega, (\pi(A)\pi(B) + \pi(B)\pi(A))\Omega \rangle - 2\,\mathbf{Exp}_\rho [A]\, \mathbf{Exp}_\rho [B] \\
&\qquad = \mathbf{Exp}_\rho [AB + BA] - 2\,\mathbf{Exp}_\rho [A]\, \mathbf{Exp}_\rho [B],
\end{aligned}$$

and

$$\begin{aligned}
\langle \Omega, i(\hat{A}\hat{B} - \hat{B}\hat{A})\Omega \rangle &= \langle \Omega, i(\pi(A)\pi(B) - \pi(B)\pi(A))\Omega \rangle \\
&= \mathbf{Exp}_\rho [i(AB - BA)],
\end{aligned}$$

the result follows. ■

The first term in the upper bound may be omitted, since it is non-negative, yielding the inequality

$$\mathbf{Unc}_\rho [A] \cdot \mathbf{Unc}_\rho [B] \geqslant \frac{1}{2} \left| \mathbf{Exp}_\rho [AB - BA] \right|, \tag{6.2.3}$$

which is perhaps a more familiar relation, reducing as it does to equation (6.2.1) when $A = Q$ and $B = P$.

The familiar fact that particles do not have well defined paths in space is often claimed as one of consequences of the Uncertainty Principle. This is only partially true. To know the trajectory of a particle in space would require exact knowledge of both the position and momentum of that particle at all times, which would require the uncertainties of both position and momentum to vanish. That position and momentum cannot have zero uncertainty is not a consequence of the Uncertainty Principle, however, but comes from the fact that the position and momentum operators have no point spectrum. The Uncertainty Principle can, however, be used to make an even bolder statement than the above — we cannot predict the trajectory of a particle to within any pre-assigned degree of accuracy, in that we cannot say that the uncertainty of position will remain small over all time. For if the uncertainty in position is small initially, then the Uncertainty Principle ensures that the corresponding uncertainty in momentum is large. Thus, if we know where the particle is (at one time) fairly well, we can know very little about in which direction and how fast it is moving at that time, and the consequence of this is that the uncertainty in position will increase for future times, and will stop being small. The reader is referred to the work of Heisenberg (*ibid*) for a discussion of these matters.

Estimating the order of magnitude of quantities is of considerable importance as a guide to the applicability and effect of physical laws, so it must be noted in passing that the nonexistence of classical paths for particles has significant content only for atomic particles. For billiard balls one can know both position and momentum with sufficient accuracy so that classical mechanics essentially holds. A rough estimate shows that quantum uncertainties are definitely not the source of the deviations from line experienced by amateur billiards players!

6.3 Wave Packet Collapse

Consider setting up an experiment to measure an observable A when the system is in the state ρ. The outcome of such a measurement is an element of the spectrum of A and, if it were possible to repeat this same experiment many times, information could be obtained about the probability measure $m_{\rho;A}$ which governs the distribution of the outcomes of such experiments. But what happens to the state of the system after any one of these measurements? Expressed in a manner which allows for degeneracy and continuous spectra, the official view is the following:

Axiom 6.3 ***(Collapse Of The Wave Packet)*** *If an experiment results in the recording of a certain eigenvalue, then immediately after registration, the state is uncontrollably transformed into the corresponding eigenstate.*
More generally, if the experiment detects the occurrence of values of the observable A in the Borel subset V of the real line (or whatever parameter space contains the spectrum), a positive response for the state ρ results in its immediate transformation (collapse) into the state

$$\rho_{out} = \frac{E_A(V)\rho E_A(V)}{\mathrm{Tr}\ (E_A(V)\rho E_A(V))}, \tag{6.3.1}$$

where $E_A(V)$ is the associated spectral projection.

This is the projection, or collapse, postulate of von Neumann [230] as modified by Lüders [156] to allow for degeneracy. We shall refer to equation (6.3.1) as Lüders' equation.

Some comment should be made about the use of the word "uncontrollably" in the above Axiom. There is first of all an uncontrollable element acting before the collapse insofar as it is not known in advance which particular spectral value will be registered by the measurement, and which one does appear is outside our control. The second uncontrollable element in the measurement, and the one that the Axiom is usually taken to refer to, is that, once a particular spectral value is registered, there is no way of stopping, slowing, or in any way affecting the collapse of the wave packet to the output state. One might say that it is an "instantaneous filter".

But the Axiom has a positive as well as a negative aspect, as Lüder's equation (6.3.1) prescribes a formula for obtaining a unique state from the initial state contingent upon the registration of a spectral value. Moreover, the normalization factor in the denominator gives the probability that a positive outcome will occur:

$$\mathbf{Pr}\,[\rho; A\,;\, V] \;=\; \mathrm{Tr}\,(\,\rho E_A(V)\,) \;=\; \mathrm{Tr}\,\big(\,\rho E_A(V)^2\,\big) \;=\; \mathrm{Tr}\,(\,E_A(V)\rho E_A(V)\,)\,. \tag{6.3.2}$$

In this regard, a cursory inspection of Lüders' equation might lead to the belief that it could be invalidated if one tried to calculate the output state ρ_{out} in the case when $\mathrm{Tr}\,(\,E_A(V)\rho E_A(V)\,) \;=\; 0$. But this cannot happen, as obtaining an outcome in the set V requires a strictly positive probability, and so the spectre of being required to divide by zero is illusory (as are all other spectres).

The following observations concerning this Axiom are worth noting:

1. Axiom 6.3 makes the usual predictions when applied to pure states. Indeed we see that if the initial state of the system is the pure state defined by the unit vector $\psi \in \mathcal{H}$, so that

$$\rho \;=\; |\,\psi\,\rangle\,\langle\,\psi\,|\,, \tag{6.3.3.a}$$

then the probability of recording a measurement for the observable A which lies in the Borel subset V of $\mathbb{R}$ is

$$\mathbf{Pr}\,[\rho, A\,;\, V] \;=\; \mathrm{Tr}\big[\rho E_A(V)\big] \;=\; \|\,E_A(V)\psi\,\|^2\,, \tag{6.3.3.b}$$

and moreover, if such a measurement is recorded, then the subsequent state of the system will be

$$\rho_{out} \;=\; \|\,E_A(V)\psi\,\|^{-2}\;|\,E_A(V)\psi\,\rangle\,\langle\,E_A(V)\psi\,|\,, \tag{6.3.3.c}$$

a pure state defined by the unit vector $\|\,E_A(V)\psi\,\|^{-1}\;E_A(V)\psi$.

2. For observables represented by positive operator valued measures, the Lüders formula must be modified. This is accomplished by abstracting the properties of the transformation $\rho \to \rho_{out}$ as a mapping, and declaring that all such mappings represent a wave packet collapse. There are technical difficulties to deal with (particularly for the smooth model) which are beyond the scope of this book, and we refer to the literature for details [41], [52]. A particular example is furnished by the approximate position observables considered in

Section 5.2.1. Using the example discussed there, were the system initially in the state ρ, then the apparatus designed to measure Q, described by the approximate position operator, would register an outcome lying in the Borel subset V of $\mathbb{R}$ with probability

$$K = [\![\mathcal{I}[\rho; V], I]\!], \tag{6.3.4.a}$$

and the subsequent state of the system (given such an experimental outcome) would then be the state

$$\rho_{out} = \frac{1}{K}\mathcal{I}[\rho; V], \tag{6.3.4.b}$$

where $\mathcal{I}[\rho; V]$ is the positive element of $\mathfrak{A}_*$ defined by the formula

$$\mathcal{I}[\rho; V] = \int_V g_s(\overline{Q})\rho g_s(\overline{Q})^* \, ds\,; \tag{6.3.4.c}$$

this last integral should be interpreted weakly. The mapping $\mathcal{I}$ is a particular example of what has been referred to as an instrument observable. This particular instrument observable has the property of being translationally covariant, in that

$$U(a)^* \mathcal{I}[\rho; V + a]\, U(a) = \mathcal{I}[U(a)^* \rho\, U(a); V], \qquad V \in \mathrm{Bor}(\mathbb{R}),$$

for all $a \in \mathbb{R}$, where $U(a) = W(a, 0)$ is the one parameter unitary group which implements spatial translations in $L^2(\mathbb{R})$. Davies (*ibid*) has shown how this concept can be generalized to cover the notion of invariance under a general class of groups (including the rotation group for systems moving in $\mathbb{R}^3$).

One important consequence of the theory of approximate observables and instrument observables is that it answers the problem of the non-repeatability of experiments. As an example of this, in the ideal situation described by the above Axioms, the probability of recording a measurement of the observable A in the Borel set $\mathbb{R}$ is 1, and the state of the system subsequent to such a (successful!) measurement will be $\rho_{out} = E_A(\mathbb{R})\rho = \rho$ for any input state ρ. In other words, an experiment which "does not care" what its outcome it does not affect the system. However, we should expect that any experimental intervention with a physical system will, to some extent, affect that system. In the case of an approximate observable, this is what happens. Again using our standard example from

Section 5.2.1, while the probability of recording a measurement in the Borel set $\mathbb{R}$ is still

$$K = [\![\mathcal{I}[\rho;\mathbb{R}], I]\!] = \int_{\mathbb{R}} [\![\rho, g_s(\overline{Q})^* g_s(\overline{Q})]\!] \, ds = [\![\rho, I]\!] = 1,$$

the state subsequent to such a measurement is

$$\rho_{out} = \mathcal{I}[\rho;\mathbb{R}] = \int_{\mathbb{R}} g_s(\overline{Q}) \rho g_s(\overline{Q})^* \, ds,$$

which is not the same as ρ. Thus even a trivial experiment affects the system. More generally, if an ideal experiment were made of a system in the state ρ and an outcome was registered lying in the Borel subset V of $\mathbb{R}$, so that the subsequent state of the system is the state ρ_{out} given by equation (6.3.1), then a repetition of this experiment would record an outcome lying in the Borel set V with probability one, since $E_A(V)\rho_{out} = \rho_{out}$. This is no longer the case with approximate observables — in general the probability of recording a second outcome lying in the Borel set V is strictly less than one.

3. In the Young's two slit experiment, we know that placing a detector in the region just beyond the slits but before the screen, in order to see through which hole the electrons pass, destroys the interference effects between the slits. It is consistent to interpret this by saying that a wave packet collapse occurs each time an electron is observed. In any event, the experiment with the detector is not the same experiment as if it were not there: observations affect the system in quantum theory.

6.3.1 *Reality*

It was noted in Chapter 3 that it is part of the standard interpretation of the theory that when a state is not an eigenstate of an observable it cannot be said to have a definite value for that observable. This is in sharp contrast to the situation in classical mechanics, and so the law of the excluded middle does not hold in quantum mechanics[3].

---

[3]The law of the excluded middle states that if p is a sentential variable such as *3 is greater than 4*, then either p is true or its negation is true. (In symbols, $\models p \vee \neg p$.)

If it is accepted that an observable does not have a value in a state which is not one of its eigenstates, and if a value of that observable is obtained as the result of an observation, the implication seems to be that the observation itself caused (created) the value of the observable. This places the act of measurement, and by implication the experimenter, as the creator of (macroscopic) reality. And this even though the experimenter appears nowhere in the equations of the theory.

Yet the entire scientific enterprise is devoted to the explication of an objective reality, so it is no wonder that this interpretation has been the subject of considerable attack. On page 49, a case for rejecting the standard interpretation in favour of a hidden variable theory as put by Bohm was mentioned. Anyone who does not accept Bohm's argument, or something similar, must disagree with the notion of an objective reality it requires.

We disagree with there being such an objective reality, and our position on the question is essentially the relatively modest one proposed by Wallace [233],

> *The most natural sense to give to the word interpretation is our manner of identifying the abstract mathematical symbols of theory with the concepts that we use to construct descriptions of our experience, in short, of making the link between the objective and the subjective.*
> *Though reality cannot be defined either in terms of our experiences themselves or of the mathematical structure of our theories, it is given substance by making workable (verifiable) identifications between the mathematical symbols and the totality of our experience of the physical world.*

Another part of what the quantum meaning of reality is, involves the delayed choice experiments, whose interpretation leads to an argument in which we do not wish to become embroiled. Suffice it to say that our position is that the quantized relativistic field description is primary, and a particle picture must be constructed from it. In Haag's book (*ibid*), the way to do this is discussed at length. Electromagnetic field experiments are the most difficult to analyze, due to the zero mass of the photon. The possibility of retrospective creation of reality in these cases can usually be traced to giving the particle picture primacy and making the assumption that photons are committed to some path as the experiment proceeds. Wallace (*ibid*) also discusses the Wheeler delayed choice experiment in a way that accords to our understanding of the matter.

In more general terms, we remind the reader that the experiment of Aspect [9, 8, 10] verifying Bell's inequalities [17] proves that quantum mechanics gives the accurate description of a certain delayed choice experiment, and that the results are inconsistent with any local reality theory (these terms requiring the precise definition Bell gives them).

6.3.2 *Consciousness*

This brings us to the curious case of the effect of consciousness on the measurement process. Curious we say, because we might have expected Schrödinger and Wigner, of all people, to have come to a different conclusion. The arrangement known as Schrödinger's cat is too well known to need sketching. (Originally this appeared in German in [205]. It was translated by J. D. Trimmer and printed in the Proceedings of the American Philosophical Society in 1980, and that translation is reprinted in [237].)

Wigner substituted a friend and made the triggered event non-lethal (see the reprint in [237] with the title *Remarks on the Mind-Body Question*), but let us not be coy. We may imagine the box expanded into an escape proof prison cell complete with the lethal cyanide release trigger of Schrödinger, used as a death penalty in cases of capital crime. Thus no individual is directly responsible for taking a life. In the box happens to be a physicist found guilty of peddling quantum paradoxes, a capital crime if there ever were one. The crux of the matter for the proponents of "consciousness in measurement" is that until the prison cell is opened, the peddler is in a state of being neither alive nor dead, but this state collapses to one of being either dead or alive immediately on our opening the cell door.

One snappy response to this, proposed by Omnès [172], is that we can perform an autopsy to determine when death occurred, and so when the state of the peddler actually collapsed, and hence there is no mystery. We agree, but wish to go further. When propounding a gedanken experiment, a strict rule is that it must conform to the laws of nature, and one of the laws of nature is that biological creatures such as physicists are either alive or dead, exclusively, and cannot be in a mixed state. (Our families and friends may disagree with this, but let us take that as a second order effect!) If we consider an observable we call life, we may suppose that it represents a super-selection rule, on empirical grounds (cf the discussion on page 76). This notion is compatible with theory since a physicist is, first of all, a macroscopic system, and the interaction between measurement devices and systems with infinitely many degrees of freedom is generally agreed to be the origin of irreversibility. This topic will come up again in the discussion of the laser. It might also be mentioned that there are indications that quantum mechanics can provide this without invoking classical mechanics as an outside agency. See Hepp [114] for an attempt in this direction, though not everyone finds that the attempt as successful as we do [18].

It might be objected that we have no right invoking a super-selection law without deriving it from the theory. But positing such a law on the basis of observation seems to us to be no different than accepting the conservation of strangeness without a full theoretic understanding of its origin (the symmetry considerations of the "eight-fold way" and its extensions do not constitute a dynamical explanation). So our position is that if it is really a physicist (or a cat) you are talking about, to suggest a mixed life-death state is plain nonsense. This does not suggest anything about the problem of describing life entirely within quantum theory, or any other reductionist goal. Nor does it answer the question of how complex a system must be before this super-selection rule holds for it; that is an empirical matter at present.

Proponents of the delayed choice interpretation run the risk of conceding some very curious interpretations of the behaviour of matter. Consider the following example. Suppose that a two slit experiment is run, with the results recorded on a video tape but not observed by any human being. These tapes are then placed in a film archive. Five years later, the administrators decree that it is necessary to get rid of a number of old tapes to make room for an additional administrator (as usual), but in order to avoid a public outcry, all the old tapes will be viewed by a management trainee before consignment to the pulper so as to choose a few deemed worth keeping. On viewing the tape of the two slit experiment, the run results are transferred to a human mind. Do we believe that only then does the wave function of the experimental system collapse? Suppose the viewer is not clever enough to understand the tapes; does the collapse still take place? Suppose in the meantime the two slit apparatus has been dismantled. Does this affect the collapse?

The honest position is that we cannot prove or disprove the delayed choice interpretation answers to these questions by direct observation. Nor can we deduce the answers by a law of nature such as a superselection rule of life (which enables autopsies to give definite conclusions, mostly). Under these circumstances, all that quantum theory tells us is that at the time of viewing by someone who understands the pictures, information about the two slit runs, made years ago, has subsequently become available.

We shall not discuss the possibility that the collapse postulate can be replaced by what is known as decoherence theory. This theory is in an early stage of development in any event, and we refer the reader to the book of Omnès [172], as well as the comprehensive review of that book by Faris [58].

6.4 Mixed States And The Universe

In the laser model discussed in Chapter 11, we shall have to consider a system consisting of several subsystems interacting with one another. The evolution of the full (closed) system is unitary, but by projecting on to any given subsystem, an irreversible dynamics is found. This technique of projection onto a subsystem is extremely important, and we have chosen to introduce it here under the disguise of answering the question: why are there mixed states in quantum mechanics?

6.4.1 *Compound Systems*

Our experience in classical mechanics is that there mixed states arise out of incomplete knowledge, and the same is true in quantum mechanics. We have already seen one way in which this happens, since mixed states were seen to arise naturally as the outcomes of measurements of approximate observables. Here is a simple and standard demonstration of another way in which mixed states can be created.

Notation The following terminology will make the ensuing discussion easier to describe. By a quantum mechanical system will be meant the ordered pair $\Sigma = (\mathcal{H}, \mathfrak{A})$, where $\mathcal{H}$ is a separable Hilbert space which carries a representation of the canonical commutation relation (either in Weyl or in Heisenberg form), $\mathfrak{A}$ is the corresponding algebra of observables. We denote the set of states by $\mathfrak{S}$. ■

6.4.1.1 *Tensor Products*

In the next few sections we shall indicated how two systems are compounded. When the systems consist of particles of the same type, questions of statistics arise, as in multi-electron systems. But if, as will be supposed here, the systems are not identical, the situation is simpler, and the composite system can be described in terms of tensor products.

Suppose that $\mathcal{H}$ and $\mathcal{K}$ are two given Hilbert spaces[4]. The algebraic tensor product $\mathcal{H} \otimes \mathcal{K}$ of $\mathcal{H}$ and $\mathcal{K}$ is an inner product space when equipped with the sesquilinear pairing

$$\langle \phi \otimes \alpha \,,\, \psi \otimes \beta \rangle \;=\; \langle \phi \,,\, \psi \rangle \, \langle \alpha \,,\, \beta \rangle \,, \qquad \phi,\, \psi \in \mathcal{H},\; \alpha,\, \beta \in \mathcal{K}. \qquad (6.4.1)$$

Denote by $\mathcal{H}\overline{\otimes}\mathcal{K}$ the Hilbert space completion of $\mathcal{H} \otimes \mathcal{K}$, known as the Hilbert tensor product of $\mathcal{H}$ and $\mathcal{K}$.

If $(\phi_n)_{n\geqslant 0}$ is an orthonormal basis for $\mathcal{H}$, while $(\alpha_n)_{n\geqslant 0}$ is an orthonormal basis for $\mathcal{K}$, then $(\phi_m \otimes \alpha_n)_{m,n\geqslant 0}$ is an orthonormal basis for $\mathcal{H}\overline{\otimes}\mathcal{K}$. Consequently the Hilbert space $\mathcal{H}\overline{\otimes}\mathcal{K}$ is separable.

There is a natural method for combining continuous linear operators on $\mathcal{H}$ and $\mathcal{K}$ to obtain a continuous linear operator on the tensor product Hilbert space $\mathcal{H}\overline{\otimes}\mathcal{K}$. Given $A \in \mathbb{B}(\mathcal{H})$ and $B \in \mathbb{B}(\mathcal{K})$, let $A \otimes B \in \mathbb{B}(\mathcal{H}\overline{\otimes}\mathcal{K})$ be the continuous linear map defined uniquely by the formula

$$\big[A \otimes B\big](\phi \otimes \alpha) \;=\; A\phi \otimes B\alpha, \qquad \phi \in \mathcal{H},\; \alpha \in \mathcal{K}. \qquad (6.4.2)$$

If instead A and B are densely defined unbounded linear operators on $\mathcal{H}$ and $\mathcal{K}$ respectively, they can be combined to form a linear operator $A \otimes B$ which is defined on the subspace $\mathcal{D}(A) \otimes \mathcal{D}(B)$ of $\mathcal{H}\overline{\otimes}\mathcal{K}$ (since $\mathcal{D}(A)$ is dense in $\mathcal{H}$, $\mathcal{D}(B)$ is dense in $\mathcal{K}$, and $\mathcal{H} \otimes \mathcal{K}$ is dense in $\mathcal{H}\overline{\otimes}\mathcal{K}$, it follows that $\mathcal{D}(A) \otimes \mathcal{D}(B)$ is dense in $\mathcal{H}\overline{\otimes}\mathcal{K}$) by the same formula restricted to the subdomain,

$$\big[A \otimes B\big](\phi \otimes \alpha) \;=\; A\phi \otimes B\alpha, \qquad \phi \in \mathcal{D}(A),\; \alpha \in \mathcal{D}(B), \qquad (6.4.3)$$

noting that $A \otimes B$ is closable if both A and B are.

6.4.1.2 *Compounding Bounded Models*

Consider how one might compound two given copies of the bounded model, Σ_1, Σ_2. The first step is to choose a Hilbert space, and that is clearly the

[4]As usual, all the Hilbert spaces being considered here are complex and separable.

tensor product $\mathcal{H} = \mathcal{H}_1 \overline{\otimes} \mathcal{H}_2$. From the strongly continuous representations of the Weyl group that are given, W_1 and W_2 respectively, the formula

$$W(a,b) = W_1\left(\frac{a}{\sqrt{2}}, \frac{b}{\sqrt{2}}\right) \otimes W_2\left(\frac{a}{\sqrt{2}}, \frac{b}{\sqrt{2}}\right), \qquad a, b \in \mathbb{R}. \qquad (6.4.4)$$

determines a strongly continuous representations of the Weyl group on $\mathcal{H}$. As usual, the algebra of observables for Σ is $\mathbb{B}(\mathcal{H})$. Technically, $\mathbb{B}(\mathcal{H})$ is equal to the completion of the algebraic tensor product of $\mathbb{B}(\mathcal{H}_1)$ with $\mathbb{B}(\mathcal{H}_2)$ in a certain topology (the W^*-tensor product topology will do, [200]). We write

$$\mathfrak{A} = \mathbb{B}(\mathcal{H}) = \mathbb{B}(\mathcal{H}_1) \overline{\otimes} \mathbb{B}(\mathcal{H}_2) . \qquad (6.4.5)$$

The pre-dual $\mathfrak{A}_*$ for the compound system, namely the trace class operators on $\mathcal{H}_1 \overline{\otimes} \mathcal{H}_2$, can be identified as the closure of the algebraic tensor product $(\mathfrak{A}_1)_* \otimes (\mathfrak{A}_2)_*$ of the pre-duals of the component subsystems[5].

6.4.1.3 *Compounding Smooth Models*

Suppose now that Σ_1 and Σ_2 are smooth models, with common domains $\mathcal{S}_1$ and $\mathcal{S}_2$. It is a standard construction in the theory of locally convex spaces to complete their algebraic tensor product in an appropriate topology (the three main topologies, projective, inductive and injective, all coincide in this case), indicated $\mathcal{S} = \mathcal{S}_1 \overline{\otimes} \mathcal{S}_2$. In the Schrödinger representation, $\mathcal{S}(\mathbb{R}) \overline{\otimes} \mathcal{S}(\mathbb{R})$ is none other than $\mathcal{S}(\mathbb{R}^2)$. In all cases, $\mathcal{S}$ is a nuclear Fréchet space, and will be taken to be the common domain for the compound system.

By hypothesis, Σ_k carries a representation of the CCR in Heisenberg form resulting from the lowering and raising operators, A_k and A_k^+ acting on $\mathcal{S}_k$, for $k = 1, 2$. By setting

$$A = \frac{1}{\sqrt{2}}\left[A_1 \otimes 1 + 1 \otimes A_2\right], \qquad (6.4.6)$$

the operator A and its adjoint A^* leave $\mathcal{S}$ invariant, and a simple calculation shows that A and A^+ satisfy the CCR on the common domain. Moreover, the common domain is equal to $\mathcal{S} = \mathcal{D}^\infty(\overline{N})$ as in equation (4.2.3), where $N = A^+ A$.

Proceeding as in Chapter 4, the algebra of observables associated to this construction will be taken to be $\mathfrak{A} = \mathcal{L}^+(\mathcal{S})$, which can be shown

[5] With respect to an appropriate topology.

to contain (properly) the algebraic tensor product of $\mathcal{L}^+(\mathcal{S}_1)$ and $\mathcal{L}^+(\mathcal{S}_2)$. None of these algebras is complete[6] since, as has been mentioned before, the completion of $\mathcal{L}^+(\mathcal{S})$ is the space $\mathcal{L}(\mathcal{S}, \mathcal{S}')$ of continuous linear maps from $\mathcal{S}$ to its dual space, which is not an algebra. Notwithstanding the incompleteness, the notation $\mathfrak{A} = \mathfrak{A}_1 \overline{\otimes} \mathfrak{A}_2$ will be used.

As observed above, it is known that the spaces $(\mathfrak{A}_1)_*$, $(\mathfrak{A}_2)_*$ and $\mathfrak{A}_*$ have natural nuclear Fréchet topologies, and it can be shown that $\mathfrak{A}_*$ is naturally topologically isomorphic to the completed tensor product $(\mathfrak{A}_1)_* \overline{\otimes} (\mathfrak{A}_2)_*$.

This completes the construction of the compound model $\Sigma = \Sigma_1 \overline{\otimes} \Sigma_2$ for the smooth model [7].

6.4.1.4 *Compound Systems — Summary*

We summarize the observations concerning compound systems as follows:

Axiom 6.4 ***(Compounding Systems With Non-Identical Particles)*** *Given two quantum mechanical systems of the same type (bounded or smooth), $\Sigma_j = [\mathcal{H}_j, \mathfrak{A}_j]$ for $j = 1, 2$, provided that the particles comprising the two systems are not identical,*

$$\Sigma = \left(\mathcal{H}_1 \overline{\otimes} \mathcal{H}_2, \mathfrak{A}_1 \overline{\otimes} \mathfrak{A}_2\right) \tag{6.4.7}$$

is the quantum mechanical compound system, and is of the same type as its constituent subsystems.

A few observations are in order concerning compound systems.

1. The Connection Theorem 4.25 states that any Hilbert space which carries a representation of the CCR in Weyl form also carries a representation of the CCR in Heisenberg form, and vice versa. Thus, if both of the Hilbert spaces $\mathcal{H}_1$ and $\mathcal{H}_2$ carry representations of the CCR, then the tensor product Hilbert space $\mathcal{H}_1 \overline{\otimes} \mathcal{H}_2$ carries representations of the CCR in both Weyl and Heisenberg form, through the constructions given above. These two representations of the

[6] More precisely, none of them are complete in the topologies being considered for them in this book.

[7] There would be no difficulty in compounding a bounded with a smooth model, but the need to do so does not arise in this book.

CCR on the tensor product Hilbert space are then equivalent to each other in the sense of the Connection Theorem.

2. The representation in the compound system is never irreducible, even when the component subsystem representations are. It is enough to show this for the smooth model. If f_1 and f_2 are vectors in the kernels of A_1 and A_2, respectively, then

$$f_1 \otimes f_2 \,, \qquad A_1^+ f_1 \otimes f_2 \; - \; f_1 \otimes A_2^+ f_2 \,,$$

are linearly independent elements of the kernel of A, so the representation is reducible (or, equivalently, not gauge invariant).

The states of the compound system $\Sigma = \Sigma_1 \overline{\otimes} \Sigma_2$ are, of course, the positive normalized elements of $\mathfrak{A}_*$. If $\omega_1 \in \mathfrak{S}_1$ and $\omega_2 \in \mathfrak{S}_2$, then we can define a state $\omega_1 \otimes \omega_2 \in \mathfrak{S}$ such that

$$\big(\omega_1 \otimes \omega_2\big)(B \otimes C) \;=\; \omega_1(B)\,\omega_2(C) \tag{6.4.8}$$

for all $B \in \mathfrak{A}_1$ and $C \in \mathfrak{A}_2$. It is important to note, however, that not all states in $\mathfrak{S}$ are of this type.

Observables in $\mathfrak{A}$ of the form $B \otimes I_2$ and $I_1 \otimes C$ are said to be *localized* in Σ_1 and Σ_2, respectively, and have the property that

$$\big(\omega_1 \otimes \omega_2\big)(B \otimes I_2) \;=\; \omega_1(B)\,, \qquad \big(\omega_1 \otimes \omega_2\big)(I_1 \otimes C) \;=\; \omega_2(C)\,, \tag{6.4.9}$$

for all $B \in \mathfrak{A}_1$ and $C \in \mathfrak{A}_2$. Note that the raising and lowering operators are not localized in this way.

Given a state $\omega \in \mathfrak{S}$ of the compound system, the formulæ

$$\omega_1(B) \;=\; \omega\big(B \otimes I_2\big)\,, \qquad \omega_2(C) \;=\; \omega\big(I_1 \otimes C\big)\,, \tag{6.4.10}$$

for $B \in \mathfrak{A}_1$ and $C \in \mathfrak{A}_2$, determine states ω_1 and ω_2 of the component subsystems Σ_1 and Σ_2 respectively. However, $\omega \neq \omega_1 \otimes \omega_2$, a fact which is physically significant, as will be discussed in the next Subsection.

6.4.2 *Mixed States*

In this subsection it will be supposed that the universe can be decomposed as the tensor product of the system of interest (called the system) and the remainder, hereafter called the reservoir. (There seems to be no natural way to avoid using the term *system* in these two different ways.) In an obvious notation, $\Sigma^{(u)} = \Sigma^{(s)}\overline{\otimes}\Sigma^{(r)}$, where u stands for "universe"[8].

From within the system, it is by definition not possible to measure what is going on in the reservoir (and vice versa), for having information about the rest of the universe would mean that at least some of what has been called the reservoir would be, in fact, part of the system. This means that given an observable $A \in \mathfrak{A}^{(s)}$, there is no automatic way to extend it to an observable in $\mathfrak{A}^{(u)}$. But there is a way that is most natural, one that we shall adopt, namely

$$A^{(u)} = A \otimes I^{(r)}. \tag{6.4.11}$$

The reason for the assertion that this is most natural will become clear below.

Now suppose that the universe is in a pure state represented by the unit vector $\Psi \in \mathcal{H}^{(u)}$, and consider the state $\omega_\Psi^{(s)} \in \mathfrak{S}^{(s)}$ it defines through its action on the extension of system observables given above:

$$\omega_\Psi^{(s)}(A) = \langle \Psi , A^{(u)}\Psi \rangle = \langle \Psi , (A \otimes I^{(r)})\Psi \rangle . \tag{6.4.12}$$

This state $\omega_\Psi^{(s)}$ is what an experimenter in the system can determine about the state of the universe. In general $\omega_\Psi^{(s)}$ will be a mixed state for the system, defined by a density matrix $\rho_\Psi^{(s)}$, so that

$$\omega_\Psi^{(s)}(A) = \mathrm{Tr}(\rho_\Psi^{(s)} A) , \qquad A \in \mathfrak{A}^{(s)} . \tag{6.4.13}$$

Similarly, a hypothetical experimenter within the reservoir would determine a (mixed) state $\omega_\Psi^{(r)} \in \mathfrak{S}^{(r)}$ for the reservoir determined by the formula

$$\omega_\Psi^{(r)}(B) = \langle \Psi , (I^{(s)} \otimes B)\Psi \rangle = \mathrm{Tr}(\rho_\Psi^{(r)} B) , \qquad B \in \mathfrak{A}^{(r)} . \tag{6.4.14}$$

Thus, from the universe pure state $|\,\Psi\,\rangle\langle\,\Psi\,|$ we can deduce a system state $\omega^{(s)}$ and a reservoir state $\omega^{(r)}$. However it is not possible, in general, to recreate the universe state $|\,\Psi\,\rangle\langle\,\Psi\,|$ from the component states $\omega^{(s)}$ and

[8] In general there would be particles of the same type, say electrons, in both the system and the reservoir, and that should be taken into account. To do so requires (i) decomposing the universe into particle types; and (ii) using symmetric tensor products for Bosons and antisymmetric ones for Fermions. There are no technical difficulties in doing so, but the result is a considerable increase in notational complexity, so we are simply going to ignore this complication.

$\omega^{(r)}$. In particular, it can be shown that (in general), the tensor product state $\omega^{(s)} \otimes \omega^{(r)}$ is a mixed state of the universe, and hence not equal to $|\,\Psi\,\rangle\,\langle\,\Psi\,|$.

Seemingly this is rather mysterious, since the impossibility of recovery is true even when there is no interaction between the system and reservoir. The solution to the mystery is that the splitting of a universe pure state into parts destroys the phase relations between the two parts. In fact, the universe state has the same fate as Humpty-Dumpty: all the Queen's horses and men cannot put it together again. This loss of information has interesting implications for physicists who believe that they have a "theory of everything" for, even if they do, how can they know it?

It is in this lack of ability to recreate universe (pure) states from system and reservoir states that the need for mixed (system) states lies — that a system state is mixed, in this context, results from the fact that we have (unavoidably) imperfect knowledge of the universe. We shall return to this topic when we discuss the dynamics of open systems in the next Chapter.

It might reasonably be asked what a section on mixed states is doing in a Chapter concerning probability in quantum mechanics. The answer is that $\omega_\Psi^{(s)}$ has an interpretation as a discrete random variable whose values are projection operators, and this will also provide justification for saying that $A \otimes I^{(r)}$ is the most natural way to extend A into the universe.

Let $\big(\phi_n^{(s)}\big)_{n\geqslant 0}$ be an orthonormal basis for the system Hilbert space $\mathcal{H}^{(s)}$. On general principles, a sequence $\big(\psi_n^{(r)}\big)_{n\geqslant 0}$ of vectors in the reservoir Hilbert space $\mathcal{H}^{(r)}$ can be found so that the identity

$$\Psi \;=\; \sum_{n\geqslant 0} \phi_n^{(s)} \otimes \psi_n^{(r)} \tag{6.4.15.a}$$

holds. Using the orthonormality of the $\mathcal{H}^{(s)}$ basis,

$$\sum_{n\geqslant 0} \big\|\psi_n^{(r)}\big\|^2 \;=\; \|\,\Psi\,\|^2 \;=\; 1\,. \tag{6.4.15.b}$$

Writing $p_n^{(r)} = \big\|\psi_n^{(r)}\big\|^2$, then we see that $0 \leqslant p_n^{(r)} \leqslant 1$, and $\sum p_n^{(r)} = 1$. As weights, these constants can be interpreted as probabilities.

Letting $P_n^{(s)} = |\,\phi_n^{(s)}\,\rangle\,\langle\,\phi_n^{(s)}\,|$ be the projection operator along $\phi_n^{(s)}$,

$$\omega_\Psi^{(s)}(A) \;=\; \sum_{n\geqslant 0} \big\|\psi_n^{(r)}\big\|^2 \,\langle \phi_n^{(s)}\,,\, A\phi_n^{(s)}\rangle \;=\; \sum_{n\geqslant 0} p_n^{(r)}\,\mathrm{Tr}\,\big(P_n^{(s)}A\big), \tag{6.4.16.a}$$

for any observable $A \in \mathfrak{A}^{(s)}$, and so

$$\rho_{\Psi}^{(s)} = \sum_{n \geqslant 0} p_n^{(r)} P_n^{(s)} . \tag{6.4.16.b}$$

In this light, the density matrix $\rho_{\Psi}^{(s)}$ could be interpreted as a discrete random variable, taking the projections $P_n^{(s)}$ as possible values, each with probability $p_n^{(r)}$. The expectation values calculated in equation (6.4.16.a) are then seen to have been calculated conditionally. Interestingly, the values of this random variable are determined by the system, while the associated probabilities determined by the reservoir.

This interpretation is possible for any choice of orthonormal basis for $\mathcal{H}^{(s)}$, and would result in different values and probabilities, although still with a pure state-valued random variable interpretation. On the one hand, because the random variable in question varies with basis, this representation does not have absolute significance. On the other hand, these observations do provide some justification for the rules we have adopted for extending system (and reservoir) observables to the universe, since these rules enable us to derive system (mixed) states from universe (pure) states which are capable of interpretation as random variables, and it should be recalled that mixed states in classical mechanics arose similarly from pure state-valued random variables.

6.5 Additional Reading

The original *vade mecum* on the theory of lattices and its applications is the monograph of Birkhoff [20]; a treatment for mathematics undergraduates from a more modern point of view is that of Davey & Priestley [39].

A few books on the foundations of quantum mechanics with emphasis on quantum logic in one form or another are those of Jauch [131], Piron [179], Varadarajan [229] and Mackey [161]. A more general treatment is found in Ludwig's approach [157], and a thorough treatment of the problem of objective reality may be found in Mittelstaedt [168]. Treatments where probability is placed in the forefront are Gudder [92] and Holevo [119].

Tensor analysis from the point of view of linear algebra can be found in many places these days, and we recommend Greub [89]. When topology is involved, the books previously noted on topological vector spaces contain everything that is needed.

CHAPTER 7

DYNAMICAL SYSTEMS

A systematic presentation of the general concepts of quantum mechanics requires rather extensive preliminary knowledge of general functional analysis . . . An attempt to skip this information by substituting it with a reference to similar information taken, for example, from finite dimensional linear algebra, would look like regular cheating to the reading mathematician.

– Berezin & Shubin

We may start off a particular state vector in Hilbert space at a particular time. Suppose that we then make it vary with time in accordance with the Schrödinger equation: what would happen to it? Roughly, what happens to it is that it gets knocked right out of Hilbert space in the shortest time interval possible.

– P. A. M. Dirac [50]

Everything so far discussed has given us no more than a snapshot of quantum theory at a given instant. The heart of a physical theory is in the way it includes the forces acting internally and externally, in other words, the dynamics, and that is what we consider in this Chapter. Interests of space have forced us to be selective in our choice of dynamical systems — if the particular topics discussed in this Chapter seem idiosyncratic, it is only because they have been chosen with later applications in mind. For example, scattering is not discussed, but the damped oscillator is.

7.1 Eigenfunction Expansions & Generalized Eigenvectors

There is no need to convince anyone of the utility of eigenvectors, but for observables with a continuous component to their spectrum there are no such entities to associate with the continuous spectral values. In place of eigenvectors there are tempered distributions which act very much like eigenvectors, only they are not elements of Hilbert space. Since the smooth model incorporates a rigged Hilbert space, a rigorous treatment of these eigendistributions (as we shall refer to them) is possible.

To avoid complications that would obscure the basic idea, it will be assumed that the continuous spectrum is separated from the eigenvalues. This means that every operator can be uniquely separated into a sum of two operators, one with a wholly discrete and one with a wholly continuous spectrum. Moreover these two operators commute (strongly) so for spectral purposes they can be considered separately. It is no real loss of generality, therefore, to assume in the rest of this section that the operators considered have a wholly continuous spectrum.

The position operator in the Schrödinger representation exemplifies the sort of operator that will be considered. By δ_y is meant the delta function concentrated at the point $y \in \mathbb{R}$, treated as a tempered distribution. Thus

$$[\![\delta_y , f]\!] = f(y), \qquad f \in \mathcal{S}(\mathbb{R}) . \tag{7.1.1.a}$$

In a more familiar notation, $\delta_y(x) = \delta(x - y)$, but it is precisely this sort of formal symbolism we are trying to avoid.

Replacing f by Qf yields

$$[\![\delta_y , Qf]\!] = (Qf)(y) = y\, f(y) = y\, [\![\delta_y , f]\!], \qquad f \in \mathcal{S}(\mathbb{R}), \tag{7.1.1.b}$$

for any $y \in \sigma(Q) = \mathbb{R}$, with everything well defined. To get to the eigenvalue equation, Q has to be moved across to act on δ_y. Recall that Q is here being regarded as a continuous endomorphism of $\mathcal{S}(\mathbb{R})$, and hence it defines a continuous endomorphism Q^{tr} of $\mathcal{S}'(\mathbb{R})$ via the formula

$$[\![Q^{\mathrm{tr}} T , f]\!] = [\![T , Qf]\!], \qquad T \in \mathcal{S}'(\mathbb{R}),\ f \in \mathcal{S}(\mathbb{R}) . \tag{7.1.1.c}$$

Hence it is clear that

$$Q^{\mathrm{tr}} \delta_y = y\, \delta_y \tag{7.1.1.d}$$

for any $y \in \mathbb{R}$, and this may fairly be called an eigenvalue equation for Q. However, the key fact to notice is that the distributions δ_y are not functions in $L^2(\mathbb{R})$, but are rather true distributions in $\mathcal{S}'(\mathbb{R})$.

It might be thought that this is a curious phenomenon associated with the fact that Q is not bounded, but that is not so. It is a consequence of the continuous nature of the spectrum of Q, as can be seen by considering the function $\tanh \overline{Q}$ obtained by applying the spectral functional calculus to the operator Q,

$$\big(\tanh \overline{Q}\, \phi\big)(x) \;=\; \tanh x\; \phi(x)\,, \qquad\qquad \phi \in L^2(\mathbb{R})\,. \tag{7.1.5.a}$$

We can see that the operator $\tanh \overline{Q}$ is bounded (with unit norm) and self-adjoint. Moreover, its spectrum is the interval $[-1, 1]$ and is wholly continuous. However it has eigendistributions δ_x for any $x \in \mathbb{R}$, with

$$\big(\tanh \overline{Q}\big)^{\mathrm{tr}} \delta_x \;=\; \tanh x\; \delta_x\,, \qquad\qquad x \in \mathbb{R}\,, \tag{7.1.5.b}$$

and hence we see that the need for a general theory of eigendistributions is intrinsic to the study of operators with continuous spectrum, and is not simply due to the introduction of the smooth model.

Consider another familiar example, the momentum operator P in the Schrödinger representation. Its eigendistributions are the family of complex exponentials $\{\, T_k \,:\, k \in \mathbb{R}\,\} \subseteq \mathcal{S}'(\mathbb{R})$ given by

$$T_k(x) \;=\; e^{-ikx}, \qquad\qquad k \in \mathbb{R}\,, \tag{7.1.6}$$

since

$$P^{\mathrm{tr}} T_k \;=\; k T_k, \qquad\qquad k \in \mathbb{R}\,. \tag{7.1.7}$$

Moreover

$$(P^n)^{\,\mathrm{tr}} T_k \;=\; k^n T_k, \qquad\qquad k \in \mathbb{R}\,, \tag{7.1.8}$$

for any positive integer n.

We wish to codify the principles indicated by the above examples. For the remainder of this analysis, we shall assume that we are working with some manifestation of the smooth model, so that we have a system Hilbert space which carries a gauge invariant representation of the CCR. In particular, we shall only consider smooth observables, namely symmetric elements of $\mathcal{L}^+(\mathcal{S})$. We now extract the following definition.

Definition 7.1 *Given a symmetric observable* $A \;=\; A^+ \;\in\; \mathcal{L}^+(\mathcal{S})$, *an element* $T \;\in\; \mathcal{S}'$ *is said to be a* generalized eigenfunction *(or* generalized eigendistribution*) for* A *associated with the (continuum) spectral value* $\lambda \in \sigma(A)$ *if*

$$[\![T\,,\, Af]\!] \;=\; \lambda [\![T\,,\, f]\!]\,, \qquad\qquad f \in \mathcal{S}\,, \tag{7.1.9.a}$$

which condition can be expressed more simply by saying

$$A^{\mathrm{tr}}\, T \;=\; \lambda T\,, \tag{7.1.9.b}$$

where A^{tr} *is the transpose of* A*, a continuous endomorphism of* $\mathcal{S}'$*.*

Having introduced the concept of a generalized eigendistribution, we also need the concept of a complete collection of such quantities. We introduce this concept most conveniently in the following manner:

Definition 7.2 *Given a symmetric operator* $A = A^{+} \in \mathcal{L}^{+}(\mathcal{S})$*, a* spectral function *for* A *is a surjective map* $\varepsilon : \mathbb{R} \to \sigma(A)$*. Given a spectral function* ε *for* A*, a* complete family of (generalized) eigendistributions *for* A *is a family* $\{T_\lambda : \lambda \in \mathbb{R}\} \subseteq \mathcal{S}'$ *such that*

$$A^{\mathrm{tr}} T_\lambda \;=\; \varepsilon(\lambda)\, T_\lambda\,, \qquad\qquad \lambda \in \mathbb{R}\,, \tag{7.1.10}$$

and such that, given $f \in \mathcal{S}$*,* $f = 0$ *whenever* $[\![\, T_\lambda\,,\, f\,]\!] = 0$ *for all* $\lambda \in \mathbb{R}$*.*

Thus the family $\{\delta_x : x \in \mathbb{R}\}$ forms a complete family of generalized eigendistributions for the position operator Q (with respect to the spectral function $\varepsilon(x) = x$), while the family of exponentials $\{T_k : k \in \mathbb{R}\}$ introduced in equation (7.1.6) forms a complete family of generalized eigendistributions for the operator P^n for any positive integer n (with respect to the spectral function $\varepsilon(k) = k^n$) — this last example shows the utility of our choosing to parametrize elements of the spectrum of A using a spectral function, since doing so enables the same family of eigendistributions to serve for more than one observable.

Although our discussion to date has been in terms of observables with wholly continuous spectrum, that has been simply for the purposes of clarity. It can be shown that any symmetric element of $\mathcal{L}^{+}(\mathcal{S})$ possesses a spectral function and a complete family of eigendistributions associated with that spectral function. However, describing how this is to be achieved can become rather complicated, notationally, through issues of degeneracy. We choose again to simplify our discussion by restricting our attention to the nondegenerate case. We shall therefore[1] only consider symmetric observables $A \in \mathcal{L}^{+}(\mathcal{S})$ which are *cyclic* in the sense that there exists a vector $\Omega \in \mathcal{S}$ such that the set $\{A^n\Omega : n \geqslant 0\}$ has dense linear span in $\mathcal{H}$.

[1] This is a simplification since every symmetric observable in $\mathcal{L}^{+}(\mathcal{S})$ can, in some sense, be written as a direct sum of cyclic symmetric observables [73].

Theorem 7.3 *Every cyclic symmetric operator $A = A^+ \in \mathcal{L}^+(\mathcal{S})$ possesses a complete family $\{T_\lambda : \lambda \in \mathbb{R}\}$ of generalized eigendistributions associated with some spectral function ε. If A is essentially self-adjoint on $\mathcal{S}$, then the generalized eigenvectors T_λ are uniquely defined.*

That an essentially self-adjoint element of $\mathcal{L}^+(\mathcal{S})$ possesses a unique complete family of generalized eigendistributions follows from the fact that such an operator possesses a unique self-adjoint extension — this is not true of a more general symmetric operator. The reader is referred to Dubin & Hennings [52] for a proof of this result.

When the momentum representation was discussed in Chapter 5, it was shown that the Fourier transform $\mathcal{F}$ implemented the unitary equivalence of the Schrödinger and momentum representations of the CCR. In particular it was noted that the position and momentum operators Q and P are related by the formula $Q = \mathcal{F}P\mathcal{F}^{-1}$. It can now also be seen that the Fourier transform (when extended to $\mathcal{S}'(\mathbb{R})$) also acts as a mapping between the families of generalized eigendistributions for P and Q, since $\mathcal{F}^{-1}T_k = \sqrt{2\pi}\delta_k$ for any $k \in \mathbb{R}$. In other words, the momentum operator P is unitarily equivalent to one whose generalized eigendistributions are delta distributions.

This is a general result, which can be shown using the functional form of the spectral theorem, a proof of which is given in Reed & Simon [186]. Given an essentially self-adjoint operator $A \in \mathcal{L}^+(\mathcal{S})$, there exists a unitary transformation $U : \mathcal{H} \to L^2(\sigma(A), dm)$, where m is a regular Borel measure on $\sigma(A)$, such that UAU^{-1} is the multiplication operator

$$\big(UAU^{-1}\phi\big)(\lambda) = \lambda\,\phi(\lambda)\,, \qquad \phi \in L^2(\sigma(A), dm),\ \lambda \in \sigma(A)\,. \tag{7.1.11}$$

Since $\mathcal{H}$ carries a representation of the CCR in Heisenberg form, so does $L^2(\sigma(A), dm)$, and this representation will be irreducible provided that A is cyclic. Consequently the measure space $L^2(\sigma(A), dm)$ possesses its own smooth domain and associated collections of test functions and tempered distributions. For any $\lambda \in \sigma(A)$, the distribution

$$T_\lambda = U^{\mathrm{tr}}\delta_\lambda \tag{7.1.12}$$

is a generalized eigendistribution for A, with

$$A^{\mathrm{tr}}T_\lambda = \lambda T_\lambda\,, \tag{7.1.13}$$

and the collection $\{T_\lambda : \lambda \in \sigma(A)\}$ is a complete family of generalized eigendistributions[2] for A. It should be noted that one consequence of these observations is the fact that

$$\langle f , g \rangle = \int_{\sigma(A)} \overline{[\![T_\lambda , f]\!]} \, [\![T_\lambda , g]\!] \, dm(\lambda) \,, \tag{7.1.14}$$

and, more generally,

$$\langle f , F(A) g \rangle = \int_{\sigma(A)} F(\lambda) \overline{[\![T_\lambda , f]\!]} \, [\![T_\lambda , g]\!] \, dm(\lambda) \,, \tag{7.1.15}$$

for all $f, g \in \mathcal{S}$ and all suitable functions F, which indicates that the family $\{T_\lambda : \lambda \in \sigma(A)\}$ constitutes a weak partition of the identity.

7.2 Dynamics Of Closed Systems

In the early days of quantum theory, the quantum dynamical systems considered were quite concrete: the simple harmonic oscillator, the Hydrogen atom, and so on. It was from these examples, with the correspondence principle as a rough guide, that the notion of a Hamiltonian operator as a sum of kinetic and potential operators emerged. It was quickly realized that Hamiltonian operators do not need to have this special form, enabling a more general presentation of the theory of quantum dynamics.

At this point it is necessary to distinguish between open and closed systems. Closed systems are ones which are energetically isolated; open systems are not. Closed systems are easier to describe: their Hamiltonians are free of any explicit time dependence and generate their dynamics in a particularly simple manner.

---

[2] Technically, a complete family of generalized eigendistributions should be parametrized by $\mathbb{R}$, and not by $\sigma(A)$, as above. More precisely, a spectral function ε should be introduced, and a family of eigendistributions $\{S_\lambda : \lambda \in \mathbb{R}\}$ considered, where S_λ is equal to $T_{\varepsilon(\lambda)}$. The above, incorrect, notation has been chosen in the interests of clarity.

Axiom 7.1 ***(Time Evolution, Closed Systems)*** *For a closed quantum mechanical system* $\Sigma = (\mathcal{H}, \mathfrak{A})$, *the dynamics are governed by a strongly continuous one-parameter unitary group* $\mathcal{U}$ *of* $\mathbb{R}$ *into* $\mathbb{B}(\mathcal{H})$. *The densely defined self-adjoint operator* H *which is the infinitesimal generator of the group* $\mathcal{U}$, *defined by the formula*

$$\mathcal{U}_t = e^{-itH}, \qquad t \in \mathbb{R}, \tag{7.2.1}$$

is called the Hamiltonian for the system. As an observable, the Hamiltonian H *represents the energy of the system.*

Axiom 7.1 is, as stated, valid for the bounded model[3]. In the smooth model it is also necessary to assert that the operators $\mathcal{U}_t$ and H are all continuous endomorphisms of the smooth domain $\mathcal{S}$, and moreover that $\mathcal{S}$ is a core of self-adjointness for H. It is clear that the additional requirements of the smooth model are fairly strong, and that some of the "standard" Hamiltonians of quantum mechanics (for example the Hydrogen atom Hamiltonian) do not satisfy them. In particular physical situations, therefore, it may be necessary to apply some form of the Round-Off Approximation to obtain a suitable smooth observable. That this can be done successfully is shown in Dubin & Hennings [52].

7.2.1 *The Schrödinger And Heisenberg Pictures*

Given a Hamiltonian observable H and its associated time-evolution unitary group $\mathcal{U}$, it is still necessary to explain how these quantities are used to implement the time-evolution of the system. Any such implementation is usually called a *picture* of quantum mechanics, and there are two particularly important ones[4].

[3]In many physical situations some symmetric operator will be presented as the Hamiltonian observable, and the unitary group $\mathcal{U}$ will not be known. Moreover, the operator thus presented will not automatically be self-adjoint. The prudent practitioner therefore first imposes the boundary conditions that the physics requires, thereby finding an appropriate self-adjoint extension of the given operator, which will then be used as the Hamiltonian observable.

[4]Pictures other than the Schrödinger and Heisenberg pictures are generally referred to as *interaction pictures*, and involve some sharing of the dynamical time-evolution of the system between both observables and states.

In the *Schrödinger picture* of quantum mechanics, it is assumed that observables (unless explicitly time-dependent) do not evolve with time, and that all the time-evolution implies by the dynamics of the system is carried by the states. Thus, if $\omega \in \mathfrak{S}$ is the state of the system at time 0, then the state of the system at time t is $\omega_t \in \mathfrak{S}$, where

$$\omega_t(A) = \omega\left(\mathfrak{U}_t^* A \mathfrak{U}_t\right) \tag{7.2.10}$$

for any $A \in \mathfrak{A}$ (provided that no measurement takes place, causing a collapse of the wave packet). If the state ω is represented by the density matrix ρ, then the time evolved state ω_t is represented by the density matrix

$$\rho_t = \mathfrak{U}_t \, \rho \, \mathfrak{U}_t^* . \tag{7.2.11.a}$$

In particular, if the state is pure, so that $\rho = P_\phi$ for some unit vector ϕ, it follows that $\rho_t = P_{\phi_t}$, so that the future state is also pure, determined by the unit vector ϕ_t, where

$$\phi_t = \mathfrak{U}_t \phi . \tag{7.2.11.b}$$

The opposite viewpoint is taken in the *Heisenberg picture*, in which all the dynamical time-evolution is to be carried by the observables of the system. An observable $A \in \mathfrak{A}$ at time 0 will then be interpreted as having evolved to the observable $A(t)$ at time t, where

$$A(t) = \mathfrak{U}_t^* \, A \, \mathfrak{U}_t \, , \tag{7.2.12}$$

and that a state of the system will not change with time (provided again that no measurement takes place). For any $t \in \mathbb{R}$ there is an automorphism τ_t of the algebra $\mathfrak{A}$ of observables such that $\tau_t(A) = A(t)$ for any observable A, and indeed $\{\tau_t : t \in \mathbb{R}\}$ is a one-parameter group of automorphisms of the algebra $\mathfrak{A}$.

It is important to notice that there is no physical difference between the two pictures, since

$$\omega_t(A) = \omega\big(\tau_t(A)\big) = \omega\big(\mathfrak{U}_t^* \, A \, \mathfrak{U}_t\big), \qquad \omega \in \mathfrak{S}, \; A \in \mathfrak{A}, \tag{7.2.13}$$

and hence all expectations between states and observables will be the same in both pictures.

7.2.2 *Equations Of Motion*

It is usual at this point to differentiate the above equations and obtain what are called the equations of motion. This may not, strictly speaking, be possible in the bounded model, since to do so will involve H explicitly in the equations, and H may not a bounded observable[5]. However there is no problem in the smooth model since, under the assumptions made, the time-evolved states (in the Schrödinger picture) are all smooth states, and the time-evolved observables (in the Heisenberg picture) are all smooth observables, and moreover time-differentiation is well-defined. Were we using the smooth model, under the assumptions made, not only would ρ_t and ϕ_t determine smooth states, but time differentiation would be well defined. In the Schrödinger picture, then, the equation of motion for a time-evolved state ω_t is

$$i\frac{d}{dt}\omega_t(A) = \omega_t(AH - HA) = \omega_t([A, H]), \qquad A \in \mathfrak{A}. \tag{7.2.14.a}$$

In terms of the representing density matrix ρ_t, this equation reads

$$i\frac{d}{dt}\rho_t = [H, \rho_t], \tag{7.2.14.b}$$

which is known as von Neumann's equation. For pure states $\rho_t = P_{\phi_t}$, this equation becomes

$$i\frac{d}{dt}\phi_t = H\phi_t, \tag{7.2.14.c}$$

which is the standard time dependent Schrödinger equation. On the other hand, in the Heisenberg picture, the equation of motion for a time-evolved observable $A(t)$ reads

$$i\frac{d}{dt}A(t) = [A(t), H]. \tag{7.2.15}$$

7.3 Dynamics Of Open Systems

One consequence of the energetic isolation of closed quantum mechanical systems is that their dynamics are reversible — changing the direction of

[5]What is intriguing, therefore, is the fact that formal calculations performed in the bounded model are often successful in giving the correct results. The question of how regular (while remaining unbounded) a Hamiltonian operator must be for these formal calculations to work is too complicated to go into here — suffice it to say that there is an extensive literature devoted to just this point.

time does not affect the validity of the dynamics. Mathematically, this effect is described by a time-reversal operator R in the following manner. Suppose, as usual, that the system Hilbert space $\mathcal{H}$ carries an irreducible representation of the CCR in Weyl form, with associated gauge invariant Fock vector Ω. Then, since

$$\overline{\langle W[u]\Omega\,,\,W[v]\Omega\rangle} \;=\; e^{i\mathrm{Im}(\overline{u}v)}\,e^{-\frac{1}{2}|\,u-v\,|^2} \;=\; \langle W[\overline{u}]\Omega\,,\,W[\overline{v}]\Omega\rangle$$

for all $u, v \in \mathbb{C}$, it is possible to use the cyclicity of the Fock vector Ω to prove the existence of a bounded antilinear self-inverse map[6] $R : \mathcal{H} \to \mathcal{H}$ such that

$$\langle R\phi\,,\,\psi\rangle \;=\; \overline{\langle \phi\,,\,R\psi\rangle}\,, \qquad\qquad \phi, \psi \in \mathcal{H}\,, \qquad\qquad (7.3.1.a)$$

for which

$$RW[z]\Omega \;=\; W[\overline{z}]\Omega\,, \qquad\qquad z \in \mathbb{C}. \qquad\qquad (7.3.1.b)$$

The procedure of time-reversal is now achieved by conjugating both observables and states with the operator R. Thus any density matrix ρ is to be replaced by the density matrix

$$\rho^{\mathrm{TR}} \;=\; R\,\rho\,R\,, \qquad\qquad (7.3.2.a)$$

while any observable A is to be replaced by the observable

$$A^{\mathrm{TR}} \;=\; R\,A\,R. \qquad\qquad (7.3.2.b)$$

Since $W^{\mathrm{TR}}(a,b) \;=\; W(a,-b)$ for any $a, b \in \mathbb{R}$, evidently conjugation with R changes the sign of momentum while preserving the sign of position, and hence can be seen as effecting a reversal of the sign of time. It is now clear that the time evolution of time-reversed states (or observables, according to the picture used) is implemented by the time-reversed unitary group

$$\mathfrak{U}_t^{\mathrm{TR}} \;=\; R\,\mathfrak{U}_t\,R\,, \qquad\qquad t \in \mathbb{R}\,, \qquad\qquad (7.3.2.c)$$

[6] It is also elementary to show that this map is also a continuous antilinear endomorphism of the smooth domain $\mathcal{S}$, and so the time-reversal operator can be deployed meaningfully within either the bounded or the smooth model.

and to say that a system's dynamics are reversible[7] is to require that

$$R\mathcal{U}_t R\mathcal{U}_t \;=\; I\,, \qquad\qquad t \in \mathbb{R}\,, \tag{7.3.3}$$

so that $\mathcal{U}_t^{\mathrm{TR}} = \mathcal{U}_{-t}$ for all $t \in \mathbb{R}$. It is worth noting that if $\{\mathcal{U}_t \,:\, t \geqslant 0\}$ is a one-parameter unitary semigroup which satisfies equation (7.3.3) for all $t \geqslant 0$, then the semigroup can be extended to a one-parameter unitary group $\{\mathcal{U}_t \,:\, t \in \mathbb{R}\}$ which represents reversible dynamics.

Macroscopic matter does not behave like this, and we perceive it to be irreversible. The question of how this can come about when the dynamics of closed systems are reversible is the principal theme of statistical mechanics. To consider this question in any detail is well beyond the scope of this book, but as the laser is both an important example of a system exhibiting quantum behaviour on a macroscopic scale (and is relevant to quantum phase theory) some brief remarks, at least, must be made about the interplay of the physics and mathematics that must be brought to bear on such problems.

Macroscopic behaviour in any sense is a summation of the complex behaviour of systems with many degrees of freedom. Our position is that an explanation of systems with irreversible dynamics lies in the analysis of systems with infinitely many degrees of freedom, based on some version or another of the thermodynamic limit.

One way to treat the dynamics of an open system is to consider it as part of a larger closed system, driven by its interaction with the rest of the system, as well as evolving through its own dynamics. This description covers numerous types of phenomena, and often results in the system being driven irreversibly to a final state. The final states are frequently states of thermal equilibrium, but can (as is the case with the laser model) be stable states far from thermal equilibrium, being sustained by energy from the larger system.

While the mechanisms — and the physics — of these two types of equilibrium states are very different, they have a number of mathematical features in common. Certainly the closed system has to be one of infinitely many degrees of freedom. If the open subsystem is also infinite, then in some sense it must be small compared to the closed system, but such sys-

[7]Requiring that the dynamics be reversible is an additional requirement on the system, and restricts the class of observables that can be Hamiltonian operators for quantum mechanical systems.

tems will not be considered in this book, where the open subsystems will always be finite. This constraint on the closed system is necessary but not sufficient, and the particular nature of the interactions plays an essential part in determining the large time dynamics of the system.

Adopting this approach involves determining the equation of motion for the open system, which will incorporate the driving effect of the rest of the closed system. For nontrivial interactions there is no chance of solving this equation, which is usually of integro-differential type, non-Markovian and possibly with a stochastic noise term. Depending on the nature of the phenomena it may be possible (on physical grounds) to neglect certain terms, and in favourable circumstances to extract a solution based on the leading terms. This usually comes down to imposing some sort of limit or specialization on the true equation of motion, turning it into a Markovian equation which can be solved.

But in many cases even this programme can not be carried through. Instead one is forced to replace the effect of the full system by an empirical model. A familiar example of this is to impose a heat bath on the system to drive it into equilibrium. Another possibility is to add a term to the Hamiltonian of the system under discussion, simulating the effect of the environment. Equations of this type may then either be solved exactly or (if this is not possible) have their solutions approximated by perturbation techniques.

7.3.1 *System-Reservoir Dynamics*

Fundamentally, an open system is part of a closed system, which we are going to call the universe.[8] As a closed system, the universe will necessarily evolve through a unitary group generated by a time independent Hamiltonian. The nature of the dynamics this induces on an open subsystem will now be investigated.

Suppose, then, that the universe is the tensor product of a system and a reservoir as in Section 6.4.2 of the previous Chapter, $\Sigma^{(u)} = \Sigma^{(s)}\overline{\otimes}\Sigma^{(r)}$, and that the universe Hamiltonian is

$$H^{(u)} = H^{(s)} \otimes I^{(r)} + I^{(s)} \otimes H^{(r)} + \lambda H_I, \tag{7.3.4.a}$$

[8]It is not proposed that calling the closed system the universe is anything other than a mnemonic device, and must not be taken too seriously.

where $H^{(s)}$ and $H^{(r)}$ are the system and reservoir Hamiltonians, respectively, and λH_I is the system-reservoir interaction, with coupling constant λ. The universe Hamiltonian generates a strongly continuous one parameter unitary group $\{\mathfrak{U}_t^{(u)} : t \in \mathbb{R}\}$, so that after a time t, in the Heisenberg picture a universe observable B evolves to the observable

$$\tau_t^{(u)}[B] = \mathfrak{U}_{-t}^{(u)} B \mathfrak{U}_t^{(u)}, \qquad B \in \mathfrak{A}^{(u)},\ t \in \mathbb{R}, \tag{7.3.4.b}$$

and correspondingly, in the Schrödinger picture a universe state $\omega^{(u)}$ evolves to the state $\omega_t^{(u)}$ given by

$$\omega_t^{(u)}(B) = \omega\left(\tau_t^{(u)}[B]\right), \qquad B \in \mathfrak{A}^{(u)},\ t \in \mathbb{R}. \tag{7.3.4.c}$$

Of course, the maps $\tau_t^{(u)}$ are all continuous endomorphisms of the algebra $\mathfrak{A}^{(u)}$ of universe observables.

In Section 6.4.2 of the previous Chapter it was shown how a pure state of the universe can be used to define a (mixed) state on the system. The same mathematics can be performed for any universe state, with the same results.

Definition 7.4 *If the universe is in the state $\omega^{(u)} \in \mathfrak{S}^{(u)}$, then it induces the system state $\omega^{(s)}$ through the formula*

$$\omega^{(s)}(B^{(s)}) \; = \; \omega^{(u)}(B^{(s)} \otimes I^{(r)}), \tag{7.3.5}$$

for all $B^{(s)} \in \mathfrak{A}^{(s)}$. The partial trace $\mathfrak{S}^{(u)} \to \mathfrak{S}^{(s)}$ which sends $\omega^{(u)}$ to $\omega^{(s)}$ will be referred to as the projection from the universe states onto the system states.

As in the previous Chapter, the import of this is that if the universe is in the state $\omega^{(u)}$, an ideal observer within the system would identify the system state as $\omega^{(s)}$. (A similar, hypothetical, ideal observer in the reservoir would be able to identify an reservoir state $\omega^{(r)}$, defined analogously.) This would then determine the state of the system at time t to be $\omega_t^{(s)}$, where $\omega_t^{(s)}$ is determined in terms of the time-evolved universe state $\omega_t^{(u)}$ by the formula

$$\omega_t^{(s)} = \left(\omega_t^{(u)}\right)^{(s)}. \tag{7.3.6}$$

Since this *is* the dynamical evolution as determined in $\Sigma^{(s)}$, the problem is to find laws that govern the transformation $\omega^{(s)} \to \omega_t^{(s)}$ without requiring knowledge of the universe we do not have (that is, not available within the system).

This is more difficult than might be imagined at first. Given an initial system state $\omega^{(s)}$, a universe state $\omega^{(u)}$ must be found whose system projection is $\omega^{(s)}$. That universe state can then be time-evolved, and the resulting universe state can then be projected back to obtain a system state. However, it would be naïve to hope that such a prescription can be implemented without difficulty, since the state projection operator from $\mathfrak{S}^{(u)}$ to $\mathfrak{S}^{(s)}$ is not injective. For example, given any system state $\omega^{(s)}$, the universe state $\omega^{(s)} \otimes \omega^{(r)}$ projects to $\omega^{(s)}$ for *any* reservoir state $\omega^{(r)}$, and the system projections of the universe time evolutes of the universe states $\omega^{(s)} \otimes \omega^{(r)}$ will, in general, be different[9] for each choice of reservoir state $\omega^{(r)}$.

A definite rule for the extension of system states to universe states must therefore be adopted. The standard choice for this rule is based on physical grounds. As has already been indicated, it is presumed that the system is small compared to the reservoir, and hence, while the reservoir may have a significant effect on the system, it will be assumed that the system will have comparatively small impact on the reservoir, and that the reservoir is consequently (almost) in equilibrium. We therefore choose a particular reservoir state $\nu^{(r)}$ which is stationary with respect to the reservoir time evolution group $\mathfrak{U}^{(r)}$ determined by the reservoir Hamiltonian $H^{(r)}$, so that

$$\nu^{(r)}(B) = \nu^{(r)}\big(\mathfrak{U}_{-t}^{(r)} B \mathfrak{U}_t^{(r)}\big), \qquad B \in \mathfrak{A}^{(r)}, \ t \in \mathbb{R}. \tag{7.3.7}$$

Any system state $\omega^{(s)}$ then defines the universe state $\omega^{(u)} = \omega^{(s)} \otimes \nu^{(r)}$, and the time-evolute[10] (at time t) $\widetilde{\tau}_{\lambda;t}[\omega^{(s)}]$ of the system state $\omega^{(s)}$ will be defined to be the projection of the time-evolved universe state $\omega_t^{(u)}$, so that

$$\widetilde{\tau}_{\lambda;t}[\omega^{(s)}](B^{(s)}) = \big(\omega^{(s)} \otimes \nu^{(r)}\big)\big(\mathfrak{U}_{-t}^{(u)}(B^{(s)} \otimes I^{(r)})\mathfrak{U}_t^{(u)}\big), \tag{7.3.8}$$

for $B^{(s)} \in \mathfrak{A}^{(s)}$ and $t \in \mathbb{R}$. More generally, equation (7.3.8) can be extended to define an endomorphism $\widetilde{\tau}_{\lambda;t}$ of the space $\mathfrak{A}_*^{(s)}$. Under favourable circumstances (meaning that the interactions are not too violent) the mapping $\widetilde{\tau}_{\lambda;t}$ will be a *continuous* linear endomorphism of $\mathfrak{A}_*^{(s)}$ for all $t \in \mathbb{R}$. This will

---

[9] It is assumed that there is a nontrivial interaction between the system and the reservoir — otherwise there is no point in this discussion in the first place.

[10] The parameter λ is explicitly included in the notation as a reminder of the coupling.

now be assumed. Dually, in the bounded model, the formula

$$\omega^{(s)}\big(\tau_{\lambda;t}(B^{(s)})\big) \;=\; \widetilde{\tau}_{\lambda;t}[\omega^{(s)}](B^{(s)}) \tag{7.3.9}$$

(where $\omega^{(s)} \in \mathfrak{A}^{(s)}_*$ and $B^{(s)} \in \mathfrak{A}^{(s)}$) then defines a group $\{\tau_{\lambda;t} \,:\, t \in \mathbb{R}\}$ of continuous *-automorphisms of the algebra of system observables. For the smooth model, this requires an additional assumption of regularity, since the algebra of observables $\mathfrak{A}^{(s)}$ is not the full dual of $\mathfrak{A}^{(s)}_*$ — rather, the reverse is true. It will, however, be assumed that the endomorphisms $\tau_{\lambda;t}$ can be defined.

Although this is the correct way to proceed, the result is less satisfactory than might have been hoped for since, aside from substantial technical problems, the family $\{\,\widetilde{\tau}_{\lambda;t} \,:\, t \in \mathbb{R}\,\}$ does not satisfy the group law[11]. Even when restricted to positive times only, it does not satisfy the semigroup law, since $\widetilde{\tau}_{\lambda;s}\widetilde{\tau}_{\lambda;t} \neq \widetilde{\tau}_{\lambda;s+t}$ for all times $s,\, t \geqslant 0$.

That is not to say that the various states $\widetilde{\tau}_{\lambda;t}[\omega^{(s)}]$ are completely uncorrelated. Going over from linear functionals to density matrices for clarity, under the (vague but strong) assumptions that have been made, if $\rho^{(s)}_{\lambda;t}$ is the density matrix associated with $\widetilde{\tau}_{\lambda;t}[\omega^{(s)}]$ then $\rho^{(s)}_{\lambda;t}$ satisfies the linear integro-differential equation[12]

$$\frac{d\rho^{(s)}_{\lambda;t}}{dt} = K\rho^{(s)}_{\lambda;t} + \lambda^2 \int_0^t M(\lambda; t-u)\rho^{(s)}_{\lambda;u}\, du \,+ \xi(t) \tag{7.3.10}$$

for all $t \geqslant 0$, where K and $M(\lambda;t)$ are endomorphisms of $\mathfrak{A}^{(s)}_*$ and ξ is a $\mathfrak{A}^{(s)}_*$-valued function of $\mathbb{R}$. The form of these various terms, and indeed their presence, depends strongly on the nature of the full dynamics and on the choice of initial reservoir state. In the examples in this book, K will be the linear operator on $\mathfrak{A}^{(s)}_*$ given by commutation with the free Hamiltonian of the system,

$$K \;=\; -i[H^{(s)}, \cdot].$$

[11] That this is true follows, in essence, from the fact that the reservoir is affected (if only slightly) by the system. Thus, if the universe is in the state $\omega^{(s)} \otimes \nu^{(r)}$ at time 0, it is not in the state $\widetilde{\tau}_{\lambda;t}[\omega^{(s)}] \otimes \nu^{(r)}$ at time t, which would (essentially) be necessary were the endomorphisms $\widetilde{\tau}_{\lambda;t}$ to satisfy the group law.

[12] It does not seem appropriate to outline a derivation of this equation here, as whenever it is used in what follows, it will be derived *in situ* for the particular system being considered.

The effect of M, the memory kernel, is that the driven system dynamics does not satisfy the semigroup law in time (is not Markovian), since it implies that the future time-evolution of the system state depends not simply upon the current system state, but also upon the past system states. When it is present, $\xi(t)$ is a stochastic noise term constructed from reservoir variables. The noise term makes this equation a quantum analogue of Langevin's equation for the velocity of a particle in Brownian motion.

If the stochastic noise term is zero, this equation is usually termed a generalized master equation, and is derived by what is known as Zwanzig's projection technique [244]. With various degrees of mathematical rigour and physical intuition, derivations are recounted in various books and papers, eg, [4].

7.3.2 *Thermal Equilibrium*

As mentioned previously, the commonest examples of system-reservoir dynamics are those which drive a system into thermal equilibrium, and those which model the laser, where the final system state is far from thermal equilibrium. We shall primarily be interested in the latter examples, but a few remarks about thermal equilibrium states will serve to provide a contrast to the laser model.

A *finite* system is one with only finitely many degrees of freedom. Consider a finite system Σ with Hamiltonian H such that $\sigma_\beta = e^{-\beta H}$ is trace class. Normalizing σ_β defines the canonical density matrix ρ_β for inverse temperature β, and this is a thermal equilibrium state. Amongst other reasons for this identification, ρ_β is the state that minimizes the free energy F, which is defined as the linear functional on the states (given in terms of density matrices) by the formula

$$F(\rho) = \mathrm{Tr}\left(\rho H + \beta^{-1}\rho \ln \rho\right) . \tag{7.3.11}$$

This is equivalent to ρ_β satisfying the β-KMS condition,

$$\mathrm{Tr}\left(\rho A_t B\right) = \mathrm{Tr}\left(\rho B A_{t+i\beta}\right) \tag{7.3.12}$$

for all observables A, B, and all times $t \in \mathbb{R}$, where the operators $A_{t+i\beta}$ have been obtained from the time-evolutes A_t of the observable A by analytic continuation.

For an infinite system, things are more delicate. A state is said to be globally thermodynamically stable if it minimizes the free energy functional, and locally thermodynamically stable if no local modification of it will decrease the relative free energy. The KMS condition (7.3.12) is then equivalent to local (but, in general, not global) thermodynamic stability. A detailed discussion of these results can be found in Sewell [210].

If an infinite reservoir $\Sigma^{(r)}$ is weakly and locally coupled (see [141] for the meaning of these terms) to a finite system $\Sigma^{(s)}$, and if we assume that the reservoir state $\nu^{(r)}$ is a β-KMS state with respect to the reservoir Hamiltonian $H^{(r)}$, then the reservoir drives the system $\Sigma^{(s)}$ irreversibly into its (canonical) β-KMS state.

How are these dynamics achieved, in terms of the generalized master equation? The important observation is that there are two time scales in operation here. The dissipative motion in the reservoir is slow compared to the free motion of the finite subsystem. The natural time scale for $\Sigma^{(s)}$, due to $H^{(s)}$, is t, whereas the time scale for dissipation in $\Sigma^{(r)}$ is $T = \lambda^2 t$. Since typical diffusion effects take place over periods of seconds or minutes, while atomic interactions take nanoseconds, the physics implies that the constant λ is extremely small. Thus to drive $\Sigma^{(s)}$ into thermal equilibrium requires a weak interaction with $\Sigma^{(r)}$. Over the long term, as the effect of the reservoir dominates, the time scale at which the system should be viewed is T rather than t.

The standard mathematical model for realizing these ideas is to let the constant λ tend to zero in such a way that $T = \lambda^2 t$ remains constant. The end result of this procedure is called the *weak coupling limit*, and was originally proposed by van Hove [122], and made rigorous by Davies [40] (under the physically reasonable condition of the decay of the truncated multi-time correlation functions for the reservoir). In models for which this approach works, the weak coupling limit of the non-Markovian generalized master equation is a Markovian master equation which determines a collection of continuous endomorphisms $\{\,\widetilde{\tau}(T) \,:\, T \geqslant 0\,\}$ of $\mathfrak{A}_*^{(s)}$ (in the diffusive time variable), which satisfy the semigroup law $\widetilde{\tau}(S)\widetilde{\tau}(T) = \widetilde{\tau}(S+T)$ for all S, $T \geqslant 0$. Moreover, the limit of $\widetilde{\tau}(T)\omega^{(s)}$ as $T \to \infty$ is, for any system state $\omega^{(s)}$, a system KMS state (thus representing thermal equilibrium). This is in accord with the general statement above. A simple example illustrating these points can be found in Sewell §3.7.1, *ibid.*

7.3.3 *States Far From Equilibrium*

Quite different physical mechanisms come into play when a system is driven irreversibly into a final state far from equilibrium. In thermodynamic terms, this requires putting energy into the system and decreasing (extracting) entropy — precisely the opposite of what happens in the approach to thermal equilibrium. No longer is the system driven on a time scale slow compared to its atomic scale, and so the weak coupling limit is not applicable.

One method for achieving the desired results uses what is known as the singular coupling limit. A model with an infinite reservoir (a free Fermion or Bose field) is used, the coupling constant λ is 1, and the interaction Hamiltonian is linear in the field and its adjoint. Leaving details and justification until later, the singular coupling limit replaces the field $A(k)$ (where k is the momentum) by $\varepsilon^{-2}A(\varepsilon^{-1}k)$ and takes the limit $\varepsilon \to 0+$. The two-time correlation functions for the field are then either 0 or a delta function in time in this limit; hence the name. The resulting semigroup $\{\tau_t : t \geqslant 0\}$ of continuous endomorphisms of $\mathfrak{A}^{(s)}$, or the equivalent semigroup $\{\widetilde{\tau}_t : t \geqslant 0\}$ of continuous endomorphisms of $\mathfrak{A}_*^{(s)}$ are known as the *reduced system dynamics.*

The singular coupling limit can also be explained in terms of different time scales, but in this case the relative sizes of the system and reservoir time scales must be reversed. Details of this argument can be found in Palmer [173].

Due to the complexity of the equations involved, it may not be possible explicitly to perform the singular coupling limit for a generalized master equation. However, general theory indicates that this limit is theoretically possible, and moreover that the resulting reduced system dynamics must satisfy particular properties. These properties are fairly restrictive, and consequently it may be possible to model the driving effect of the reservoir empirically, by considering particular examples of dynamics which satisfy the necessary conditions. For the purposes of this book, it is not necessary to give a detailed discussion of the nature of the properties that reduced system dynamics must satisfy — it suffices simply to describe the exact form of such dynamics.

Proposition 7.5 *The reduced system dynamics $\{\tau_t : t \geqslant 0\}$ (of a finite system) is a one-parameter semigroup of endomorphisms of $\mathfrak{A}^{(s)}$*

$$\tau_t(B) = e^{tZ}B, \qquad B \in \mathfrak{A}^{(s)}, \tag{7.3.13.a}$$

where the generator Z is an endomorphism of $\mathfrak{A}^{(s)}$ of the form

$$Z(B) = i[K,B] + \sum_{j=1}^{J} \left(O_j^* B O_j - \frac{1}{2} \{O_j^* O_j, B\}_+ \right) . \qquad (7.3.13.b)$$

Here the symbol $\{\, , \,\}_+$ denotes the anticommutator of two observables,

$$\{A\,,\,B\}_+ = AB + BA. \qquad (7.3.13.c)$$

Moreover the operator K is a self-adjoint Hamiltonian for the system, while the operators O_j $(1 \leqslant j \leqslant J)$ are all bounded.

Thus we can study reduced system dynamics by choosing candidate operators K and O_j for $1 \leqslant j \leqslant J$ and investigating the properties of the resulting dynamical semigroup, with a view to most closely modelling the desired physical phenomena.

The form of the generator Z was discovered by Gorini, Kossakowski & Sudarshan [85] in the case where the system Hilbert space was finite dimensional. Lindblad [150] showed that an operator Z of the above form defines a valid reduced system dynamics for an infinite dimensional (separable) system Hilbert space, provided that all the operators (including K) are bounded. His theory included the possibility that J is infinite, provided that the sum $\sum_j O_j^* O_j$ converges to a bounded operator. For these reasons, generators of the above type will be termed *GKSL generators.* A more general result, allowing for unbounded operators, is not known. However Alli & Sewell [5] have constructed a reduced system dynamics applicable to the laser model whose generator is of the above type, but for which the constituent operators are unbounded — we shall begin to study part of this model in the next Section.

By transposition, a family τ of automorphisms of $\mathfrak{A}^{(s)}$ of the type described in this Proposition yields a one-parameter family $\tilde{\tau}$ of endomorphisms of the collection $\mathfrak{A}_*^{(s)}$ of density matrices. Moreover, in the special case that the system is in a pure state ω defined by the unit vector ϕ, the state $\tilde{\tau}(t)\omega$ is also a pure state, determined by the unit vector $T_t\phi$, where T is a differentiable contraction semigroup of isometries of $\mathcal{H}^{(s)}$. Its generator, L, satisfies the inequality

$$\langle L\phi, \phi \rangle + \langle \phi, L\phi \rangle \leqslant 0. \qquad (7.3.14)$$

This property of reduced system dynamics was discovered by Lumer &

Phillips [158], who termed it *dissipativity.* We have already used this term (and will continue to use it) to describe dynamics whose effect is to drive a system irreversibly into some final state. However it should be noted that in a conservative system equality holds in (7.3.14), in which case the reduced system dynamics is unitary, and hence reversible. Consequently, in this case the reduced dynamics have a dissipative generator (in the sense of Lumer & Phillips) without exhibiting what we have chosen to refer to as dissipative behaviour.

Reduced system dynamics for open dissipative systems were introduced as being a way of describing the behaviour (internal to the system) of a system-reservoir universe. It is interesting to note, therefore, that the usual methods of mathematical analysis for such dynamics involve constructing a reservoir and having it interact with the system. In this context, the reservoir here need have no physical significance — it is a mathematical artefact which enables us to extend the system dynamics to unitary dynamics on a larger space — this process is known as *dilation.* The self-adjoint generator of that unitary dynamics can then be identified, and can then be projected back to the system, thereby specifying the generator of the system dynamics. Once this has been, the reservoir can be discarded[13].

In the final two Sections of this Chapter examples of the two approaches to open dissipative systems outlined above will be given. In the next Section an example will be studied for which an explicit calculation of the singular coupling limit can be performed, and the final Section a model will be discussed entirely in terms of its reduced system dynamics, described in terms of a GKSL generator.

---

[13]The original idea for this approach seems to have been due to Naimark [170], who observed that a symmetric operator can be dilated to a self-adjoint operator on some larger Hilbert space. If the spectral measure for that self-adjoint dilation is projected back onto the original Hilbert space, the result is what is now referred to as a positive operator valued measure for the original symmetric operator. It should be noted, however, that there are in general many different possible self-adjoint dilations of a given symmetric operator, and hence there are many possible different positive operator valued measures associated with that operator — a fact that has been noted previously. The best brief account of the relevant spectral theory of these observations can be found in the Appendix to Riesz & Sz.Nagy [187].

7.4 The Damped Oscillator

The damped oscillator is the name given to the model compound system $\Sigma^{(u)}$ (the universe) comprising a simple harmonic oscillator, $\Sigma^{(s)}$ (the system), a free Bose field system, $\Sigma^{(r)}$ (the reservoir), with a quadratic interaction between them which is such that, after applying the singular coupling limit, the reservoir will drive the system irreversibly into a final state far from equilibrium.

7.4.1 *The Bose Field*

Field theory differs essentially from systems with finitely many degrees of freedom in a number of respects. Perhaps the most important of these is the occurrence of infinitely many inequivalent representations of the CCR in either Weyl or Heisenberg form. The field to be considered is the Bose field in one spatial dimension. A particular characteristic of the field is that its action changes the number of particles in the reservoir, and so the reservoir Hilbert space needs to account for the presence of any number of particles. It is most convenient to describe the field in terms of the momentum representation. Thus, for any positive integer n, the n-particle Hilbert space is $L^2_+(\mathbb{R}^n)$, consisting of all square-integrable functions $F_n(k_1, \ldots, k_n)$ of n momentum coordinates which are invariant under all momentum coordinate interchanges — this symmetry is what characterizes the Bose field. For notational simplicity, we shall adopt the notational convention that $L^2_+(\mathbb{R}^0)$ represents the one dimensional Hilbert space $\mathbb{C}$. The reservoir Hilbert space $\mathcal{H}^{(r)}$ is then the *Fock space*

$$\mathbb{F} = \widehat{\bigoplus}_{n\geqslant 0} L^2_+(\mathbb{R}^n) , \tag{7.4.1}$$

consisting of those sequences $\mathcal{F} = \big(F_n\big)_{n\geqslant 0}$ such that $F_n \in L^2_+(\mathbb{R}^n)$ for all $n \geqslant 0$ and for which the series $\sum_{n\geqslant 0} \| F_n \|_n^2$ converges. Then $\mathbb{F}$ is a Hilbert space with respect to the inner product

$$\langle \mathcal{F}, \mathcal{G}\rangle = \sum_{n=0}^{\infty} \langle F_n, G_n\rangle_n , \qquad \mathcal{F} = \big(F_n\big)_{n\geqslant 0}, \mathcal{G} = \big(G_n\big)_{n\geqslant 0} \in \mathbb{F} . \tag{7.4.2}$$

There is a bounded and a smooth model for fields, but as only the fields themselves will be needed, the smooth model will be used. Hence a common dense domain $\mathcal{S}^{(r)}$ for the algebra of smooth observables must be chosen.

This domain is obtained by restricting each F_n to belong to the n-variable Schwartz space $\mathcal{S}_+(\mathbb{R}^n)$ and then combining them by means of the algebraic direct sum[14],

$$\mathcal{S}^{(r)} = \bigoplus_{n\geqslant 0} \mathcal{S}_+(\mathbb{R}^n) . \tag{7.4.3}$$

The matrix elements for the free field are tempered distributions, but these singularities can be dealt with by smearing the fields with test functions from $\mathcal{S}(\mathbb{R})$. For technical reasons, it is easiest introduce the smeared field operators first, before considering the unsmeared fields.

Any function $\phi \in L^2(\mathbb{R})$ defines a continuous linear map $B_n(\phi)$ from $L^2_+(\mathbb{R}^n)$ to $L^2_+(\mathbb{R}^{n-1})$ for any $n \in \mathbb{N}$ by the formula

$$[B_n(\phi)F](k_1,\ldots,k_{n-1}) = \int_{\mathbb{R}} F(k_1,\ldots,k_{n-1},k)\phi(k)\,dk, \qquad F \in L^2_+(\mathbb{R}^n), \tag{7.4.4}$$

and also defines a continuous linear map $B_n^+(\phi)$ from $L^2_+(\mathbb{R}^n)$ to $L^2_+(\mathbb{R}^{n+1})$ for any $n \in \mathbb{N}\cup\{0\}$ by the formula

$$B_n^+(\phi)F = \sigma_{n+1}(F\otimes\phi), \qquad F \in L^2_+(\mathbb{R}^n), \tag{7.4.5.a}$$

where, for any $N \in \mathbb{N}$, σ_N is the orthogonal projection of $L^2(\mathbb{R}^N)$ onto $L^2_+(\mathbb{R}^N)$, so that

$$[\sigma_N F](k_1,\ldots,k_N) = \frac{1}{N!}\sum_{\pi\in S_N} F(k_{\pi 1},\ldots,k_{\pi N}), \qquad F \in L^2(\mathbb{R}^N), \tag{7.4.5.b}$$

the sum being taken over all permutations of N symbols. In other words,

$$\left[B_n^+(\phi)F\right](k_1,..,k_{n+1}) = \frac{1}{n+1}\sum_{j=1}^{n+1} F(k_1,..,k_{j-1},k_{j+1},..,k_{n+1})\,\phi(k_j), \tag{7.4.5.c}$$

for $F \in L^2_+(\mathbb{R}^n)$. We note that, whenever $f \in \mathcal{S}(\mathbb{R})$, the map $B_n(f)$ maps $\mathcal{S}_+(\mathbb{R}^n)$ continuously into $\mathcal{S}_+(\mathbb{R}^{n-1})$, and $B_n^+(f)$ maps $\mathcal{S}_+(\mathbb{R}^n)$ continuously into $\mathcal{S}_+(\mathbb{R}^{n+1})$.

For any $f \in \mathcal{S}(\mathbb{R})$, the lowering and raising fields $A[f]$ and $A^+[f]$ are defined as endomorphisms of $\mathcal{S}^{(r)}$ for any $f \in \mathcal{S}(\mathbb{R})$ in terms of the operators

[14] This choice is technically easier to work with than the locally convex direct sum, and will suffice here, even though it is not complete in the direct sum topology.

$B_n(f)$ and $B_n^+(f)$ via the formulæ

$$[A[f]\mathcal{G}]_n = \sqrt{n+1}\, B_{n+1}(f) G_{n+1}, \qquad n \geqslant 0, \tag{7.4.6.a}$$

$$[A^+[f]\mathcal{G}]_n = \begin{cases} 0, & n = 0, \\ \sqrt{n}\, B_{n-1}^+(f) G_{n-1}, & n \geqslant 1, \end{cases} \tag{7.4.6.b}$$

for any $\mathcal{G} = (G_n)_{n\geqslant 0} \in \mathcal{S}^{(r)}$. It can be shown that these fields satisfy the canonical commutation relations in smeared form,

$$\big(A[f]A[g] - A[g]A[f]\big)\mathcal{G} = 0, \tag{7.4.7.a}$$
$$\big(A^+[f]A^+[g] - A^+[g]A^+[f]\big)\mathcal{G} = 0, \tag{7.4.7.b}$$
$$\big(A[f]A^+[g] - A^+[g]A[f]\big)\mathcal{G} = \langle \overline{g}, f\rangle\, \mathcal{G}, \tag{7.4.7.c}$$

for any f, $g \in \mathcal{S}(\mathbb{R})$ and $\mathcal{G} = (G_n)_{n\geqslant 0} \in \mathcal{S}^{(r)}$. Moreover,

$$\big\langle A[f]\mathcal{G}_1\,,\, \mathcal{G}_2\big\rangle = \big\langle \mathcal{G}_1\,,\, A^+[\overline{f}]\mathcal{G}_2\big\rangle \tag{7.4.8}$$

for all $f \in \mathcal{S}(\mathbb{R})$ and $\mathcal{G}_1$, $\mathcal{G}_2 \in \mathcal{S}^{(r)}$, so it follows that the endomorphisms $A[f]$ and $A^+[f]$ are closable, with

$$\overline{A[f]} = \big(A^+[\,\overline{f}\,]\big)^*, \tag{7.4.9.a}$$
$$\overline{A^+[f]} = \big(A[\,\overline{f}\,]\big)^*, \tag{7.4.9.b}$$

for any $f \in \mathcal{S}(\mathbb{R})$.

The unsmeared Bose (lowering) field itself is the collection of operators $A(k)$, defined for each $k \in \mathbb{R}$, such that

$$A[f] = \int_{\mathbb{R}} f(k) A(k)\, dk, \qquad f \in \mathcal{S}(\mathbb{R}). \tag{7.4.10}$$

These operators $A(k)$ are perfectly well-defined, being given by the formula

$$[A(k)\mathcal{G}]_n(k_1, \ldots, k_n) = \sqrt{n+1} G_{n+1}(k_1, \ldots, k_n, k), \tag{7.4.11}$$

for $\mathcal{G} = (G_n)_{n\geqslant 0} \in \mathcal{S}^{(r)}$. If the unsmeared fields $A(k)$ are used, it will also be necessary to consider their adjoints $A^+(k)$, so that

$$A^+[f] = \int_{\mathbb{R}} f(k) A^+(k)\, dk, \qquad f \in \mathcal{S}(\mathbb{R}). \tag{7.4.12}$$

Unlike the unsmeared lowering fields, however, the unsmeared raising fields $A^+(k)$ are not well-defined, being distributional in their k-behaviour, and so must be handled with care. Formulæ such as

$$\left[\, A(k)\,,\, A^+(p) \,\right] \;=\; \delta(k-p), \qquad\qquad k,\, p \in \mathbb{R}, \tag{7.4.13}$$

can be established, but need to be interpreted weakly (that is, smeared with test functions). But with a little care such identities represent useful calculational shortcuts leading to correct results.

In a system with one degree of freedom, the Fock vector played a critical rôle in determining the representation of the CCR. The same rôle is played here by the vector $\Omega = (\Omega_n)_{n\geqslant 0}$ in $\mathcal{S}^{(r)}$ whose components are

$$\Omega_n \;=\; \begin{cases} 1, & n = 0, \\ 0, & n \in \mathbb{N}\,. \end{cases} \tag{7.4.14}$$

This vector has the property of being (up to a phase) the only normalized vector in Fock space which is annihilated by all lowering operators:

$$A[f]\Omega \;=\; 0, \qquad\qquad f \in \mathcal{S}(\mathbb{R}). \tag{7.4.15}$$

By analogy with our earlier terminology, we choose to call Ω a Fock vector. The set of all vectors which are polynomials in the $A[f]$ and $A^+[g]$ (for all $f,\, g \in \mathcal{S}(\mathbb{R})$) acting on Ω is dense in Fock space $\mathbb{F}$. In standard terminology, Ω is cyclic for the polynomials in the fields.

The reservoir smooth model $\Sigma^{(r)}$ will not have to be spelled out in detail beyond the choice of the smooth domain $\mathcal{S}^{(r)}$. The algebra of reservoir observables will be $\mathfrak{A}^{(r)} = \mathcal{L}^+\big(\mathcal{S}^{(r)}\big)$, and it contains the *-algebra of polynomials in the smeared lowering and raising fields[15].

7.4.2 *Equations Of Motion*

The system $\Sigma^{(s)}$ will be described by the usual smooth model for one degree of freedom, associated with a gauge invariant representation of the CCR. To distinguish them from the field operators in the reservoir, the lowering and raising operators for the system will be denoted a and a^+ respectively[16].

[15]Indeed, only the polynomial algebra will be needed for the calculations of this model

[16]When discussing an operator b, it is sometimes necessary to state results which are common to both b and b^+. It is therefore useful to adopt a notational convention which can be used to refer to both of these operators at the same time. To this end the symbol $b^\sharp$ will be used.

The universe is, as usual, a tensor product $\Sigma^{(s)} \otimes \Sigma^{(r)}$, with Hilbert space

$$\mathcal{H} \;=\; \mathcal{H}^{(s)} \overline{\otimes} \mathbb{F}\,, \tag{7.4.16}$$

and the algebra of observables for the universe will be denoted by $\mathfrak{A}$.

Henceforth no superscript will be used for universe variables, and sometimes $a^\sharp \otimes I^{(r)}$ and $I^{(s)} \otimes A^\sharp$ will be written simply as $a^\sharp$ and $A^\sharp$.

The Hamiltonian for the system is the operator

$$H^{(s)} \;=\; \omega\, a^+ a \tag{7.4.17}$$

for some real constant ω, and the reservoir Hamiltonian is

$$H^{(r)} \;=\; \int_{\mathbb{R}} \varepsilon(k)\, A^+(k) A(k)\, dk, \tag{7.4.18.a}$$

where ε is some infinitely differentiable polynomially bounded function.

At first sight, it might seem that there are problems with defining this operator, since it involves the unmeasured Bose fields $A^+(k)$. But, since

$$\big[H^{(r)}\mathcal{G}\big]_n(k_1,\ldots,k_n) \;=\; \Big(\sum_{j=1}^{n} \varepsilon(k_j)\Big) G_n(k_1,\ldots,k_n), \qquad n \geqslant 0, \tag{7.4.18.b}$$

for any $\mathcal{G} \;=\; (G_n)_{n\geqslant 0} \in \mathcal{S}^{(r)}$, it is clear that $H^{(r)}$ maps $\mathcal{S}^{(r)}$ to itself. Moreover, the unitary group $\mathcal{U}^{(r)}$ defined by $H^{(r)}$ is well-behaved, with

$$\big[\mathcal{U}_t^{(r)}\mathcal{G}\big]_n(k_1,\ldots,k_n) \;=\; \Big(\prod_{j=1}^{n} e^{-it\varepsilon(k_j)}\Big) G_n(k_1,\ldots,k_n), \qquad n \geqslant 0, \tag{7.4.18.c}$$

for any $\mathcal{G} = (G_n)_{n\geqslant 0} \in \mathcal{S}^{(r)}$; and is also an endomorphism of $\mathcal{S}^{(r)}$.

By using the spectral calculus for the momentum operator P, the operator $\varepsilon(\overline{P})$ can be defined by

$$\big[\varepsilon(\overline{P})f\big](k) \;=\; \varepsilon(k)f(k), \qquad f \in \mathcal{S}(\mathbb{R}). \tag{7.4.18.d}$$

Then

$$\mathcal{U}_t^{(r)*}\, A[f]\, \mathcal{U}_t^{(r)} \;=\; A\big[e^{-it\varepsilon(P)}f\big], \qquad f \in \mathcal{S}(\mathbb{R}),\ t \in \mathbb{R}\,. \tag{7.4.18.e}$$

It should be noted that the Fock vector Ω is a stationary vector for this reservoir Hamiltonian, being an eigenvector with eigenvalue 0.

The description of the damped oscillator dynamics is completed by defining the interaction Hamiltonian to be the quadratic expression

$$H_I \;=\; a^+ \otimes A[g] + a \otimes A^+[g] \tag{7.4.19}$$

for some real-valued "coupling" function $g \in \mathcal{S}(\mathbb{R})$. Further on a special choice for g will be made as a result of the singular coupling limit.

As usual, the Hamiltonian for the universe is given by the formula

$$H \;=\; H^{(s)} \otimes I^{(r)} \;+\; I^{(s)} \otimes H^{(r)} \;+\; H_I, \tag{7.4.20}$$

and the one parameter unitary group that it generates is denoted $\mathfrak{U}$.

The problem is to determine the time evolution of the observables and states with respect to the Hamiltonian H, and this will be done in the Heisenberg picture. Thus the quantities of interest are then the operator-valued functions

$$\begin{aligned} a(t) \;&=\; \mathfrak{U}_t^* \,(a \otimes I^{(r)})\, \mathfrak{U}_t\,, && (7.4.21)\\ A(k,t) \;&=\; \mathfrak{U}_t^* \,(I^{(s)} \otimes A(k))\, \mathfrak{U}_t\,. && (7.4.22) \end{aligned}$$

Given the above definition of H, the coupled evolution equations for $a(t)$ and $A(k,t)$ are

$$\frac{d}{dt}a(t) \;=\; -i\omega a(t) \;-\; i\int_{\mathbb{R}} g(k)A(k,t)\,dk \tag{7.4.23.a}$$

$$\frac{\partial}{\partial t}A(k,t) \;=\; -i\varepsilon(k)A(k,t) \;-\; ig(k)a(t). \tag{7.4.23.b}$$

This last equation for the field $A(k,t)$ is equivalent to the integro-differential equation

$$A(k,t) \;=\; e^{-it\varepsilon(k)}(I^{(r)} \otimes A(k)) \;-\; ig(k)\int_0^t e^{-i(t-s)\varepsilon(k)}a(s)\,ds, \tag{7.4.24}$$

which, after substitution into the differential equation for $a(t)$, leads to the following integro-differential equation for $a(t)$,

$$\frac{d}{dt}a(t) \;=\; -i\omega a(t) \;-\; \int_0^t M(t-s)a(s)\,ds \;+\; I^{(s)} \otimes W(t), \tag{7.4.25}$$

where the *memory kernel* M is given by

$$M(t) = \int_{\mathbb{R}} |\,g(k)\,|^2\, e^{-i\varepsilon(k)t}\,dk, \tag{7.4.26}$$

and the *stochastic noise field* W is

$$W(t) = -i \int_{\mathbb{R}} g(k) e^{-i\varepsilon(k)t} A(k)\, dk = -iA\big[e^{-i\varepsilon(\overline{P})t} g\big] = -i\mathfrak{U}_t^{(r)*} A[g]\, \mathfrak{U}_t^{(r)}. \tag{7.4.27}$$

Equation (7.4.25) is the equation of Langevin type for the damped oscillator, as anticipated in the discussion of open system dynamics.

7.4.3 *The Dynamical Solution*

One can see in advance of a solution what sort of time behaviour equation (7.4.25) represents. Fundamental is the fact that if g is square integrable, then M and W will approach zero as $t \to \infty$, by virtue of the Riemann-Lebesgue Lemma — the memory and noise effects decay with time. However, since the above equations for $a(t)$ and $A(k,t)$ have exact solutions, detailed information about the universe time evolution can be obtained.

Looking for a solution for $a(t)$ of the form

$$a(t) = G(t)\,(a \otimes I^{(r)}) + \int_{\mathbb{R}} g(k) H(k,t)\,(I^{(s)} \otimes A(k))\, dk\,, \tag{7.4.28}$$

for suitable functions $G(t)$ and $H(k,t)$, standard Laplace transform techniques indicate that G is determined by the formula

$$\hat{G}(z) = \Big[z + i\omega + \hat{M}(z)\Big]^{-1}, \tag{7.4.29.a}$$

in which case $H(k,t)$ is given by the formula

$$H(k,t) = -i \int_0^t e^{-i\varepsilon(k)(t-s)}\, G(s)\, ds. \tag{7.4.29.b}$$

Here $\hat{G}(z)$ denotes the Laplace transform of G,

$$\hat{G}(z) = \int_0^\infty G(t)\, e^{-zt}\, dt, \tag{7.4.30}$$

where $\hat{M}$ is the Laplace transform of the memory kernel M. Similarly, the differential equations can be solved to determine $A(k,t)$,

$$\begin{aligned} A(k,t) &= g(k)\, H(k,t)\, \big[a \otimes I^{(r)}\big] + e^{-i\varepsilon(k)t} \big[I^{(s)} \otimes A(k)\big] \\ &\quad + g(k) \int_{\mathbb{R}} g(p)\, \frac{H(k,t) - H(p,t)}{\varepsilon(k) - \varepsilon(p)}\, \big[I^{(s)} \otimes A(p)\big]\, dp. \end{aligned} \tag{7.4.31}$$

From the discussion of dynamics for system-reservoir models, we were led to the conclusion that the reduced system dynamics will depend on the choice of a reservoir state which is stationary under free reservoir dynamics. In this case that state is taken to be the pure state determined by the Fock vector Ω. The initial state of the system can be arbitrary, but it will be easier to understand the model if it is taken to be pure, determined by the normalized function $f \in \mathcal{S}^{(s)}$, say. Hence the initial universe state is pure, with unit vector $f \otimes \Omega$, corresponding to the density matrix[17] $\rho = P_f \otimes P_\Omega$.

If the various parameters of the Hamiltonian are sufficiently regular, then the matrix elements (also referred to as Wightman functions, n-point functions or correlation functions)

$$\mathcal{W}_n(k_1, t_1; \ldots ; k_n, t_n) = \operatorname{Tr}\left(\rho\, A(k_1, t_1) \ldots A(k_n, t_n)\right) \qquad (7.4.32)$$

will be tempered distributions in the momentum variables, $\mathcal{W}_n \in \mathcal{S}'(\mathbb{R}^n)$. Using the standard Wightman reconstruction procedure, these distributions can be used to construct the Hilbert space, ground state, and formulæ for the action of the fields in the representation (inequivalent to that of the free field) whose matrix elements are these Wightman functions. It is not difficult to do this, but the details will not be needed in what follows.

The Wightman functions for both the system and the reservoir can be calculated explicitly for this model, and the results are

$$\begin{aligned}\big\langle (f \otimes \Omega), a^+(s_m) \ldots a^+(s_1) a(t_1) \ldots a(t_n)(f \otimes \Omega)\big\rangle \\ = \Big(\prod_{j=1}^{m} \overline{G(s_j)}\Big)\Big(\prod_{k=1}^{n} G(t_k)\Big) \left\langle f, \left[a^+\right]^m a^n f\right\rangle, \qquad (7.4.33.a)\end{aligned}$$

and

$$\begin{aligned}\langle (f \otimes \Omega), A(k_1, t_1) \ldots A(k_n, t_n)(f \otimes \Omega)\rangle \\ = \Big(\prod_{j=1}^{n} g(k_j)\Big)\Big(\prod_{j=1}^{n} H(k_j, t_j)\Big) \langle f, a^n f\rangle . \qquad (7.4.33.b)\end{aligned}$$

The simplicity of these solutions is due to the quadratic nature of the coupling — and the choice of initial state of the reservoir.

[17]If we wanted to drive $\Sigma^{(s)}$ into thermal equilibrium we would use the (inequivalent) representation of the field appropriate to the β-KMS state, and take the ground state of that representation as the initial reservoir. For this representation see Dubin [51].

This solution represents non-Markovian evolution. In accordance with the discussion concerning irreversibility, the singular coupling limit can now be imposed to extract the irreversible semigroup behaviour. This could be done in terms of a limit of relative time scales, but the effect of this procedure is the same as is obtained by setting

$$g(k) = g_0, \qquad \varepsilon(k) = k, \tag{7.4.34}$$

where g_0 is some real constant. The solutions in this special case will be distinguished by the subscripts sc. The function g is no longer in $\mathcal{S}(\mathbb{R})$, and a bit of care must be exercised in subjecting the solutions to this limiting procedure.

The first consequence of this limit is that the memory term is no longer an integral over past times, but is distributional in nature,

$$M_{sc}(t) = 2\pi g_0^2 \delta(t)\,, \tag{7.4.35}$$

which is why the semigroup law will be satisfied. Using this to calculate the other functions that appear in the solution, we obtain

$$G_{sc}(t) = e^{-\zeta t}, \tag{7.4.36.a}$$

where

$$\zeta = \pi g_0^2 + i\omega \tag{7.4.36.b}$$

is the complex natural frequency for this problem. It then follows that

$$H_{sc}(k,t) = \frac{i}{\zeta - ik}\left(e^{-\zeta t} - e^{-ikt}\right), \tag{7.4.37}$$

so that the time dependence of the system lowering operator is

$$a_{sc}(t) = e^{-\zeta t}(a \otimes I^{(r)}) + i g_0\, I^{(s)} \otimes \int_{\mathbb{R}} \frac{e^{-\zeta t} - e^{-ikt}}{\zeta - ik} A(k)\, dk. \tag{7.4.38}$$

The one-point functions for the system in this limit are

$$\big\langle (f \otimes \Omega), a_{sc}(t)(f \otimes \Omega)\big\rangle = e^{-i\omega t} e^{-\pi g_0^2 t} \langle f, a\, f\rangle\,, \tag{7.4.39}$$

$$\big\langle (f \otimes \Omega), a_{sc}^{+}(t)(f \otimes \Omega)\big\rangle = e^{i\omega t} e^{-\pi g_0^2 t} \langle f, a^{+} f\rangle\,, \tag{7.4.40}$$

both of which decay to zero exponentially.

As a function of time the number operator yields

$$\big\langle (f \otimes \Omega), a_{sc}^{+}(t) a_{sc}(t)(f \otimes \Omega)\big\rangle = e^{-2\pi g_0^2 t} \langle f, a^{+} a\, f\rangle\,, \tag{7.4.41}$$

showing that the system excitations decay exponentially to zero. This means that the reservoir is a sink in this limit, driving the oscillator into a final ground state.

Considering the field in the singular coupling limit, the time dependent one-point function is

$$\langle (f \otimes \Omega), A_{sc}(k,t)(f \otimes \Omega)\rangle = \frac{ig_0}{\zeta - ik}\left[e^{-\zeta t} - e^{-ikt}\right], \tag{7.4.42}$$

and so by the Riemann-Lebesgue Lemma,

$$\lim_{t\to\infty}\langle f \otimes \Omega, A_{sc}[g,t] f \otimes \Omega\rangle = \lim_{t\to\infty}\int_{\mathbb{R}} f(k)\,\langle f \otimes \Omega, A_{sc}(k,t) f \otimes \Omega\rangle\, dk = 0 \tag{7.4.43}$$

for any $f \in \mathcal{S}(\mathbb{R})$. Thus the field effects also decay with time.

7.4.4 *The Generator Of The Irreversible Dynamics*

The final step in the analysis is to project the universe dynamics (in the singular coupling limit) down to the system, hence obtaining the reduced system dynamics τ.

To do this, note that equation (7.4.33.a), in the special case that all the time parameters are the same, implies that

$$\tau_t\left[(a_{sc}^+)^m a_{sc}^n\right] = \overline{G(t)^m} G(t)^n\,(a_{sc}^+)^m a_{sc}^n = e^{-m\bar{\zeta}-n\zeta}\,(a_{sc}^+)^m a_{sc}^n \tag{7.4.44}$$

for all nonnegative integers m, n. In particular,

$$\tau_t\, a_{sc} = e^{-\pi g_0^2 t}\, e^{-i\omega t}\, a_{sc}, \tag{7.4.45.a}$$

$$\tau_t\, a_{sc}^+ = e^{-\pi g_0^2 t}\, e^{i\omega t}\, a_{sc}^+, \tag{7.4.45.b}$$

$$\tau_t\, a_{sc}^+ a_{sc} = e^{-2\pi g_0^2 t}\, a_{sc}^+ a_{sc}, \tag{7.4.45.c}$$

and, with somewhat more work,

$$\tau_t\, W_{sc}[z] = \exp\left[-\frac{1}{2}|\,z\,|^2\left(1 - e^{-2\pi g_0^2 t}\right)\right] W_{sc}[e^{-\zeta t} z], \tag{7.4.45.d}$$

for the action on the Weyl group in complex form. The above equations should be compared with the identity

$$\tau_t\, a_{sc} a_{sc}^+ = e^{-2\pi g_0^2 t}\, a_{sc} a_{sc}^+ + \left(1 - e^{-2\pi g_0^2 t}\right) I, \tag{7.4.45.e}$$

from which it is clear that the map τ_t is not an algebra homomorphism of $\mathfrak{A}^{(s)}$.

Examination of these results tells us that τ is of the GKSL-form, with

$$\tau_t B = e^{tZ} B, \qquad B \in \mathfrak{A}^{(s)}, \tag{7.4.46.a}$$

and generator

$$Z(B) = i\omega\left[a^+ a, B\right] + 2\pi g_0^2\, a^+ B a - \pi g_0^2 \left\{a^+ a, B\right\}_+, \qquad B \in \mathfrak{A}^{(s)}. \tag{7.4.46.b}$$

7.5 Two Level Systems

At its deepest level, spin is a manifestation of relativistic covariance, as can be seen from the representation theory of the Lorentz group. There is also a non-relativistic theory of spin due to Pauli, the principal application of which is the classification of the shell system of the elements and the fine structure of atomic and molecular spectra. However, our interest with Pauli's model of spin lies not with multi-electron atomic systems and applications of the Pauli theory to atomic spectroscopy, but is simply because it provides a model of two-level systems, since in the laser model the cavity atoms will be so caricatured. Thus the use of the word *spin* in what follows is simply a mathematical association, and no suggestion is being made that the energy levels being discussed have anything to do with electron or atomic spin.

7.5.1 *One Free Spin*

Let $\Sigma^{(s)}$ model a two-level system. The Hilbert space is two dimensional, $\mathcal{H}^{(s)} = \mathbb{C}^2$. Vectors are pairs of complex numbers, $(z_1\,,\,z_2)^{\mathrm{T}}$, and the inner product is

$$\left\langle \begin{pmatrix} z_1 \\ z_2 \end{pmatrix}, \begin{pmatrix} w_1 \\ w_2 \end{pmatrix} \right\rangle = \overline{z_1} w_1 + \overline{z_2} w_2. \tag{7.5.1}$$

The vectors $e_+ = (1\,,\,0)^{\mathrm{T}}$ and $e_- = (0\,,\,1)^{\mathrm{T}}$ form the standard orthonormal basis, and represent the upper and lower energy states as spin up and down, respectively. As $\mathcal{H}^{(s)}$ is finite dimensional this is necessarily a bounded model, and the algebra of system observables is the algebra of all 2×2 matrices with complex coefficients, $\mathfrak{A}^{(s)} = M_2(\mathbb{C})$.

The Pauli matrices

$$\sigma_1 = \begin{pmatrix} 0 & 1 \\ 1 & 0 \end{pmatrix}, \qquad \sigma_2 = \begin{pmatrix} 0 & -i \\ i & 0 \end{pmatrix}, \qquad \sigma_3 = \begin{pmatrix} 1 & 0 \\ 0 & -1 \end{pmatrix},$$

together with the identity matrix I, are the basic physical observables for the system. Using the summation convention, they satisfy the identities:

$$\sigma_\alpha \, \sigma_\beta = \delta_{\alpha\beta} \, I + i \, \varepsilon_{\alpha\beta\gamma} \, \sigma_\gamma.$$

It is clear that I, σ_1, σ_2, and σ_3 form a basis for the 4-dimensional real vector space of self-adjoint elements of $M_2(\mathbb{C})$, the observables for the system.

It will also be useful to introduce the matrices

$$\sigma_+ = \frac{1}{2}(\sigma_1 + i\sigma_2) = \begin{pmatrix} 0 & 1 \\ 0 & 0 \end{pmatrix}, \quad \sigma_- = \frac{1}{2}(\sigma_1 - i\sigma_2) = \begin{pmatrix} 0 & 0 \\ 1 & 0 \end{pmatrix},$$

noting that $\sigma_+^2 = \sigma_-^2 = 0$, that $\left[\sigma_+, \sigma_-\right] = \sigma_3$, and that the anti-commutator $\left\{\sigma_+, \sigma_-\right\}_+$ is equal to I. Moreover,

$$\begin{aligned} \sigma_+ e_- &= e_+, \\ \sigma_- e_+ &= e_-, \end{aligned}$$

so the operator σ_+ acts to flip a down spin to an up spin, with the operator σ_- doing the reverse. The observable σ_3 measures the alignment of the spin. In this atomic caricature, spin flip corresponds to a transition between levels.

Vector notation can be useful when working with Pauli matrices. If $\vec{\sigma}$ denotes the matrix-valued vector $(\sigma_1, \sigma_2, \sigma_3)$, any real vector $\vec{a} = (a_1, a_2, a_3)$ can be used to describe the matrix

$$\vec{a} \cdot \vec{\sigma} = a_1 \sigma_1 + a_2 \sigma_2 + a_3 \sigma_3.$$

Then, since every (self-adjoint) observable ξ has a unique representation in the form

$$\xi = x \, I + \vec{a} \cdot \vec{\sigma}$$

for $x \in \mathbb{R}$ and $\vec{a} \in \mathbb{R}^3$, it is clear that the simplest way to express the action of any (linear) operator on observables is to find a compact formula for the action of that operator on the expression $\vec{a} \cdot \vec{\sigma}$.

To make good the interpretation as a model of a simple atom, the free atomic Hamiltonian is

$$K = \frac{1}{2}\varepsilon\,\sigma_3, \tag{7.5.2}$$

which has eigenvalues $\pm\frac{1}{2}\varepsilon$, with corresponding eigenvectors $e_\pm$. Thus there are these two nondegenerate energy levels, and that is about all there is to it. The free Hamiltonian generates the one parameter dynamical unitary group

$$\mathcal{U}_t = e^{-itK} = \begin{pmatrix} e^{-\frac{1}{2}it\varepsilon} & 0 \\ 0 & e^{\frac{1}{2}it\varepsilon} \end{pmatrix}. \tag{7.5.3}$$

The free time evolution for general observables can then be expressed by the formula

$$\mathcal{U}_t^*\,(\vec{a}\cdot\vec{\sigma})\,\mathcal{U}_t = \vec{a}_t\cdot\vec{\sigma}, \qquad \vec{a}\in\mathbb{R}^3,\ t\in\mathbb{R}, \tag{7.5.4}$$

where

$$\vec{a}_t = \begin{pmatrix} \cos(\varepsilon t) & \sin(\varepsilon t) & 0 \\ -\sin(\varepsilon t) & \cos(\varepsilon t) & 0 \\ 0 & 0 & 1 \end{pmatrix} \vec{a}. \tag{7.5.5}$$

This, together with the fact that $\mathcal{U}_t^*\,I\,\mathcal{U}_t = I$, gives a complete description of the evolution of one free spin.

7.5.2 *One Pumped Spin*

The system is now envisaged as being subject to external forces corresponding to a source and a sink. This is an open system, and the external forces will be modelled through a GKSL-generator. Our treatment follows that to be found in the paper of Alli and Sewell [5].

Accordingly, in the general expression (7.3.13.b) for the generator of reduced system dynamics, K is taken to be the free spin Hamiltonian $\frac{1}{2}\varepsilon\sigma_3$ just considered, and $J = 3$, $O_1 = \sqrt{b_+}\,\sigma_+$, $O_2 = \sqrt{b_-}\,\sigma_-$, $O_3 = \sqrt{b_3}\,\sigma_3$. The constants ε, b_+, b_- and b_3 are all strictly positive.

It is convenient to introduce the constants

$$u = \frac{1}{2}(b_+ + b_-) + 2b_3, \qquad v = b_+ + b_-, \qquad \eta = \frac{b_+ - b_-}{b_+ + b_-}, \tag{7.5.6}$$

noting that these constants are constrained by the relations

$$v < 2u, \qquad |\eta| < 1, \tag{7.5.7}$$

for then the action of Z on the generators is given by

$$\begin{aligned} Z(\sigma_+) &= (-u + i\varepsilon)\,\sigma_+, & (7.5.8.a)\\ Z(\sigma_-) &= (-u - i\varepsilon)\,\sigma_-, & (7.5.8.b)\\ Z(\sigma_3) &= -v\,(\sigma_3 - \eta\, I)\,. & (7.5.8.c)\end{aligned}$$

Hence the one parameter group $\tau_t = e^{tZ}$ is such that

$$\begin{aligned} \tau_t\,\sigma_+ &= e^{-ut}\,e^{i\varepsilon t}\,\sigma_+, & (7.5.9.a)\\ \tau_t\,\sigma_- &= e^{-ut}\,e^{-i\varepsilon t}\,\sigma_-, & (7.5.9.b)\\ \tau_t\,\sigma_3 &= e^{-vt}\,\sigma_3 + \eta(1 - e^{-vt})I, & (7.5.9.c)\end{aligned}$$

from which it follows that

$$\tau_t\,\vec{a}\cdot\vec{\sigma} = e^{-ut}\,\vec{a}_t\cdot\vec{\sigma} + a_3\left[(e^{-vt} - e^{-ut})\sigma_3 + \eta(1 - e^{-vt})I\right], \tag{7.5.9.d}$$

where the map $\vec{a} \mapsto \vec{a}_t$ implements the time evolution of the free Hamiltonian, as described in equation (7.5.5).

From the form of the generator and the consequent properties of τ it follows that the various coupling constants have physical significance, with b_+ coupling the atom to a pump, and b_- and b_3 coupling it to sinks. It is clear, therefore, that the character of the time evolution depends on the balance between these constants.

To complete the description of this model, we consider the time dependence of certain quantities whose physical interpretation is of interest. First note that for this system an arbitrary pure state can be parametrized by an angle, since a general unit vector can be written as

$$\psi_\theta = \begin{pmatrix}\cos\theta\\ \sin\theta\end{pmatrix}, \tag{7.5.10}$$

(a phase factor could be included, but doing so would not change anything substantial).

The operators

$$N_\pm = \frac{1}{2}\,(I \pm \sigma_3) \tag{7.5.11.a}$$

are the observables for the number of excitations in the upper and lower state, respectively. After a time t, these evolve to

$$\tau(t)N_{\pm} = \frac{1}{2}(1 \pm \eta)\left(1 - e^{-vt}\right) I + e^{-vt} N_{\pm} \,. \tag{7.5.11.b}$$

The expectation of $\tau(t)N_{\pm}$ in the state ψ_θ is then

$$\mathbf{Exp}_\theta\left[\tau(t)N_{\pm}\right] = e^{-vt}\left\{\begin{matrix}\cos^2\theta \\ \sin^2\theta\end{matrix}\right\} + \frac{1}{2}(1 \pm \eta)\left(1 - e^{-vt}\right) , \tag{7.5.12.a}$$

and so, asymptotically,

$$\lim_{t\to\infty} \mathbf{Exp}_\theta\left[\tau(t)N_{\pm}\right] = \frac{1}{2}(1 \pm \eta) \,. \tag{7.5.12.b}$$

The observable σ_3 can be interpreted as measuring the relative occupation levels of the upper and lower states of the system. In this context σ_3 could be referred to as the pumping observable, and its expectation in a given state would then be the pumping parameter for that state. For the state ψ_θ,

$$\mathbf{Exp}_\theta\left[\sigma_3\right] = \cos 2\theta \,, \tag{7.5.13.a}$$

with uncertainty

$$\mathbf{Unc}_\theta\left[\sigma_3\right] = \sin^2 2\theta \,. \tag{7.5.13.b}$$

After a time t, the pumping parameter evolves to the value

$$\mathbf{Exp}_\theta\left[\tau(t)\sigma_3\right] = e^{-vt}\cos 2\theta + \eta\left(1 - e^{-vt}\right) , \tag{7.5.14.a}$$

with uncertainty

$$\mathbf{Unc}_\theta\left[\tau(t)\sigma_3\right] = e^{-2vt}\sin^2 2\theta \,, \tag{7.5.14.b}$$

and the limiting pumping parameter is η. This is the limiting value of the excess of the population in the upper level over that in the lower level, so η is a pumping control parameter (fixed in the dynamical generator). The populations decay into their limiting values exponentially, with average lifetime v^{-1}, so v is the decay rate parameter.

Roughly speaking, the operators $\sigma_{\pm}$ represent the polarizability of the atom. For the above pure state we find

$$\mathbf{Exp}_\theta\left[\sigma_{\pm}\right] = \frac{1}{2}\sin 2\theta \,. \tag{7.5.15}$$

These expectations change with time to give

$$\mathsf{Exp}_\theta\left[\sigma_\pm(t)\right] \;=\; \frac{1}{2}\,e^{-ut}\,e^{\pm i\varepsilon t}\,\sin 2\theta\,. \tag{7.5.16}$$

Thus the polarizability of this atom decays exponentially under this external interaction, and the rate of decay is governed by the average lifetime for the decay, u^{-1}. Thus we now have an interpretation of all the parameters in the model.

It is worth noting one particular system state which is stationary with respect to the reduced system dynamics. Consider the mixed state ϖ whose density matrix is

$$\rho \;=\; \begin{pmatrix} \frac{1}{2}(1+\eta) & 0 \\ 0 & \frac{1}{2}(1-\eta) \end{pmatrix}. \tag{7.5.17}$$

The expectations

$$\varpi\left(\tau(t)\sigma_\pm\right) \;=\; 0, \tag{7.5.18.a}$$

$$\varpi\left(\tau(t)\sigma_3\right) \;=\; \eta. \tag{7.5.18.b}$$

show that ϖ corresponds to a state with no polarizability and an average pumped population equal to the Hamiltonian pumping parameter. Moreover these values are independent of time.

7.6 Further Reading

For further work on dynamical semigroups, see Vanheuverswijn [227, 228]; deMoen, Vanheuverwijn & Verbeure [43]; Kümmerer [144], Davies [41] and Alicki & Lendi [4], and further references there.

For references on field theory see Baumgärtel & Wollenberg [16], Bogolubov, Logunov, Oksak & Todorov [22], Glimm & Jaffe [80], Haag [94], Itzykson & Zuber [127], Jost [134], Schweber [208], Streater & Wightman [215], Strocchi [216], and the AMS/IAS course on Fields and Strings, by Deligne *et al* [45]. Works which cover the applications of field theory to statistical mechanics include Dubin [51], Emch [57], Sewell [210], Ruelle [199], amongst very many others.

CHAPTER 8
WEYL QUANTIZATION

Can two walk together, except they be agreed?

– Amos 3:3

8.1 Introduction

A quantization scheme is a prescription to assign a quantum mechanical observable to a given classical observable. Given that physical concepts are most readily expressed in classical terms, such a prescription is clearly desirable for a better understanding and interpretation of quantum mechanical phenomena. In this Chapter we discuss the theory of Weyl quantization, which is one of many possible such prescriptions — it is certainly the oldest, and simplest.

Is such a procedure possible? Is it unique? Is it even necessary? Before many advances in the experimental art that we now take for granted, Kemble [137] felt that it was, even though it might be a waste of time from the point of view of practicability:

> *It must be frankly admitted at the outset that an examination of the efforts which have been made to set up a general theory answering the above and related questions suggests at times the possibility that the game is not worth the candle. ... One's doubts are emphasized by an examination of the dynamical variables actually measured for atomic systems. As a matter of fact, position, or configuration, linear momentum, energy, and a single arbitrary component of magnetic moment are the only independent dynamical variables whose measurement can be carried out in principle with arbitrary precision for an individual atomic system. ... Nevertheless, it is to be remembered that the develop-*

ment of scientific theory is always conditioned by artistic considerations of simplicity and symmetry, and by the urge for unity and completeness.

Although open minded, this statement was unduly pessimistic. Ever improving experimental techniques mean that an increasing number of physical quantities can be measured with accuracy. In addition, it is our view that completeness is more than an æsthetic urge, not least because the debates of Bohr and Einstein revolved around this very issue of quantization.

We believe that a detailed analysis of this problem of quantization is justified. Moreover it is a necessary component of our theory of phase, and so its study is imperative for us. We shall see below that a comprehensive theory of quantization is indeed possible.

8.2 Quantization Heuristics

We are therefore interested in finding a prescription for associating some operator with a given function T on phase space Π, which operator $\boldsymbol{\Delta}[T]$ will then be the quantization of the classical observable T. Throughout this book the symbol $\boldsymbol{\Delta}[T]$ will be reserved for the operator to be assigned to the phase space function T according to the rule of Weyl quantization (even when T is a tempered distribution rather than a function). For the time being the discussion will be strictly heuristic.

8.2.1 *Position And Momentum*

The physical meaning of the Schrödinger representation is based on the identification of the operators $\overline{Q}$ and $\overline{P}$ as representing the quantum version of position and momentum respectively. Borel functions of either position alone or momentum alone are obtained from $\overline{Q}$ and $\overline{P}$ by means of the spectral theorem,

$$f(\overline{Q}) = \int_{\mathbb{R}} f(q)\, dE_Q(q) \qquad \text{and} \qquad f(\overline{P}) = \int_{\mathbb{R}} f(p)\, dE_P(p), \tag{8.2.1}$$

where

$$\overline{Q} = \int_{\mathbb{R}} q\, dE_Q(q) \qquad \text{and} \qquad \overline{P} = \int_{\mathbb{R}} p\, dE_P(p), \tag{8.2.2}$$

are the spectral representations in question[1]. On physical grounds, then, the associations

$$f(p) \mapsto f(\overline{P}) \qquad \text{and} \qquad f(q) \mapsto f(\overline{Q}) \tag{8.2.3}$$

should arise from the quantization formalism for an appropriate class of functions f.

These one-variable associations can be cast into a two-variable form by introducing the identity function

$$\mathrm{i}(x) = 1. \tag{8.2.4}$$

Using tensor product notation,

$$[\mathrm{i} \otimes f](p,q) = f(q) \qquad \text{and} \qquad [f \otimes \mathrm{i}](p,q) = f(p), \tag{8.2.5}$$

and so one of the demands placed on quantization is that the construction must lead to the results

$$\mathbf{\Delta}[\mathrm{i} \otimes f] = f(\overline{Q}) \qquad \text{and} \qquad \mathbf{\Delta}[f \otimes \mathrm{i}] = f(\overline{P}). \tag{8.2.6}$$

for suitable functions f.

Remark In the theory of probability and statistics, it may happen that a study involves the simultaneous observation of two different random variables X and Y. The resulting data is said to be *bivariate*, and to define a bivariate, or joint, distribution. The quality represented by either one of the two random variables X or Y alone is said to be a *marginal property*. If the joint density function of the bivariate distribution is $\rho(x,y)$ (where it is supposed that x, $y \in \mathbb{R}$ for the sake of definiteness), so that the probability of recording a measurement for the pair of random variables (X,Y) in the Borel subset V of the plane $\mathbb{R}^2$ is

$$\iint_V \rho(x,y)\,dx\,dy,$$

then the two functions of one variable,

$$\rho_X(x) = \int_{\mathbb{R}} \rho(x,y)\,dy \qquad \text{and} \qquad \rho_Y(y) = \int_{\mathbb{R}} \rho(x,y)\,dx$$

[1]Whether $f(\overline{Q})$ and $f(\overline{P})$ are bounded or smooth observables, or neither, will depend on regularity properties of the function f.

define two density functions, known as the marginal distributions of the bivariate distribution ρ, and represent the separate probability density functions for the random variables X and Y respectively. So, whereas the bivariate expectation of an arbitrary function $F(X,Y)$ of the two variables is given by

$$\mathsf{Exp}_\rho[F] = \iint_{\mathbb{R}^2} F(x,y)\rho(x,y)\,dx\,dy,$$

the two marginal distributions determine the restricted expectations

$$\mathsf{Exp}_{\rho_X}[f] = \int_{\mathbb{R}} f(x)\rho_X(x)\,dx \quad \text{and} \quad \mathsf{Exp}_{\rho_Y}[f] = \int_{\mathbb{R}} f(y)\rho_Y(y)\,dy,$$

for the expectations of functions $f(X)$ and $f(Y)$ of the random variables X and Y separately. Using tensor notation, these identities can be written as

$$\mathsf{Exp}_{\rho_X}[f] = \mathsf{Exp}_\rho[f \otimes \mathrm{i}] \qquad \text{and} \qquad \mathsf{Exp}_{\rho_Y}[f] = \mathsf{Exp}_\rho[\mathrm{i} \otimes f].$$

There is a rough analogy between this and what is being demanded of $\mathbf{\Delta}[f \otimes \mathrm{i}]$ and $\mathbf{\Delta}[\mathrm{i} \otimes f]$, and even though no question of a joint observable encompassing $\overline{P}$ and $\overline{Q}$ arises, that $\mathbf{\Delta}$ is required to satisfy equation (8.2.6) will be referred to by saying that $\mathbf{\Delta}$ has $\overline{P}$ and $\overline{Q}$ as marginals. This is a nontrivial requirement for, as will be seen in Chapter 14, there are "ordered" quantization procedures which do not have $\overline{P}$ and $\overline{Q}$ as marginals. ■

However, once true functions of both variables p and q are considered, there is no unique prescription for quantization. For example, consider the function

$$T(p,q) = p^2 q\,. \tag{8.2.7}$$

While the phase space functions p^2q, pqp and qp^2 are all the same, the observables $\overline{P}^2\overline{Q}$, $\overline{P}\,\overline{Q}\,\overline{P}$ and $\overline{Q}\,\overline{P}^2$ are each different operators on $L^2(\mathbb{R})$. Since an argument could be mounted which would propose any affine combination of these three operators as the "quantization" of T, the problems faced are evident.

What is therefore needed is a specific and consistent rule which can be used to provide a unique choice for the quantization of a phase space observable. To some extent, the actual choice of this rule is arbitrary, and there are many possibilities. One scheme which could be adopted would be to assign the operator $f(\overline{P})g(\overline{Q})$ to the phase space function $f \otimes g$, and another would assign the operator $g(\overline{Q})f(\overline{P})$ to the same function. These are valid quantization schemes, denoted *Q-ordered* and *P-ordered* quantization respectively, and will be discussed in detail in Chapter 14. It is clear from their definitions, however, that these quantization schemes choose to regard either $\overline{Q}$ or $\overline{P}$ in a preferential light. Weyl quantization, in contrast, treats $\overline{Q}$ and $\overline{P}$ on an equal footing — a compelling approach.

8.2.2 *Introducing Weyl Quantization*

The Fourier transform will be used in conjunction with test functions and distributions in much of what follows; this was discussed starting with equations (5.3.1.a) and (5.3.1.b).

Weyl's presentation of his quantization proposal [235, 236] was straightforward but, insofar as he offered any background, it was rather geometrical and based on the projective representations of groups. We prefer to offer an alternative heuristic derivation, which should provide some insight into our eventual rigorous development of the theory.

Suppose A is a bounded self-adjoint operator on $L^2(\mathbb{R})$, and that the function f is regular enough that $f(A)$ exists and is bounded as well. The direct expression for $f(A)$ in terms of the spectral representation of A,

$$f(A) \;=\; \int_{\sigma(A)} f(\lambda)\, dE_A(\lambda), \tag{8.2.8}$$

is not very helpful for our purposes. However, we can use the spectral theorem to calculate the one parameter unitary group $\mathfrak{U}$ which is generated by A,

$$\mathfrak{U}_\nu \;=\; e^{i\nu A}, \qquad\qquad \nu \in \mathbb{R}, \tag{8.2.9.a}$$

and then make use of the Fourier transform to obtain what is essentially an operator-valued Fourier inversion theorem,

$$f(A) \;=\; \frac{1}{\sqrt{2\pi}} \int_{\mathbb{R}} [\mathcal{F}f](\nu)\mathfrak{U}_\nu \, d\nu. \tag{8.2.9.b}$$

The prescriptions of Q-ordered and P-ordered quantization would then be seen as identifying with the phase space observable $f \otimes g$ the operators

$$f(\overline{P})g(\overline{Q}) = \frac{1}{2\pi} \iint_{\mathbb{R}^2} [\mathcal{F}f](a)[\mathcal{F}g](b)U(a)V(b)\, da\, db, \tag{8.2.10.a}$$

and

$$g(\overline{Q})f(\overline{P}) = \frac{1}{2\pi} \iint_{\mathbb{R}^2} [\mathcal{F}f](a)[\mathcal{F}g](b)V(b)U(a)\, da\, db, \tag{8.2.10.b}$$

respectively, where

$$U(a) = W(a,0) = e^{ia\overline{P}} \qquad \text{and} \qquad V(b) = W(0,b) = e^{ib\overline{Q}},$$

are the one-parameter unitary subgroups of the representation W of the Weyl group (in the Schrödinger representation) — see equation (4.6.4) — and these identities would lead to the associations

$$T(p,q) \mapsto \frac{1}{2\pi} \iint_{\mathbb{R}^2} [\mathcal{F}T](a,b)\, U(a)V(b)\, da\, db, \tag{8.2.11.a}$$

in the case of Q-ordered quantization, and the association

$$T(p,q) \mapsto \frac{1}{2\pi} \iint_{\mathbb{R}^2} [\mathcal{F}T](a,b)\, V(b)U(a)\, da\, db, \tag{8.2.11.b}$$

in the case of P-ordered quantization, for a phase space observable T.

The difference between these quantization schemes lies in the order in which the unitary groups U and V appear. However, a more symmetric scheme is possible, using the representation W of the Weyl group directly (and not just its unitary subgroups), which makes the association

$$T(p,q) \mapsto \frac{1}{2\pi} \iint_{\mathbb{R}^2} [\mathcal{F}T](a,b)W(a,b)\, da\, db \tag{8.2.12}$$

to a phase space observable T. Recalling the smearing formula (3.3.8) introduced in Section 3.3.3, this association can be written in the form

$$T(p,q) \mapsto \frac{1}{2\pi} W[\![\mathcal{F}T]\!], \tag{8.2.13}$$

and it makes sense so long as $T \in L^2(\Pi)$ is such that $\mathcal{F}T$ belongs to $L^1(\mathbb{R}^2)$. This is the association of Weyl quantization. The justification for working with it is that it treats position and momentum equally and it provides the correct marginal distributions for them. Moreover, since it is based upon the properties of the Weyl group W, it can be analyzed in a natural manner in both the bounded and the smooth models. It will be seen in Chapter 14 that it satisfies more desirable properties than do other quantization candidates. As Wigner [241] observed, it is essentially the simplest choice. Finally, using Weyl quantization takes advantage of Weyl's deep mathematical insight.

Axiom 8.8 ***(Weyl Quantization)*** *To every suitable function T on phase space there corresponds the operator*

$$\boldsymbol{\Delta}[T] \;=\; \frac{1}{2\pi}\, W[\![\mathcal{F}T]\!]. \tag{8.2.14}$$

However, at this stage, this is not so much an Axiom as the setting-forth of a programme. First and foremost, it is crucial to extend the definition of Weyl quantization in such a manner that it can be applied successfully to a much larger class of phase space observables than the class for which equation (8.2.14) is automatically valid — functions $T \in L^2(\Pi)$ such that $\mathcal{F}T \in L^1(\mathbb{R}^2)$. This will certainly require a weak interpretation of (8.2.14). Furthermore, it would be much more convenient to have a representation for $\boldsymbol{\Delta}[T]$ which depends explicitly upon T rather than its Fourier transform. Adopting the ansatz that it is possible to write

$$\boldsymbol{\Delta}[T] \;=\; \frac{1}{2\pi} \iint_{\Pi} T(p,q)\boldsymbol{\Delta}[p,q]\, dp\, dq, \tag{8.2.15}$$

where the integral is over all of phase space, the posited operator-valued phase space function $\boldsymbol{\Delta}[p,q]$ is determined by comparing equations (8.2.14) and (8.2.15). This yields

$$\boldsymbol{\Delta}[p,q] \;=\; \frac{1}{2\pi} \iint_{\mathbb{R}^2} e^{-i(ap+bq)} W(a,b)\, da\, db, \tag{8.2.16}$$

so that, in a formal sense at least, $\boldsymbol{\Delta}$ is the Fourier transform of W. By considering equation (8.2.16) weakly, the two-variable function $\langle \phi\,,\, \boldsymbol{\Delta}[p,q]\psi\rangle$ can be seen as the Fourier transform of the function $\langle \phi\,,\, W(a,b)\psi\rangle$ for any $\phi,\ \psi \in L^2(\mathbb{R})$, and as a result an explicit formula can be derived for the action of the desired operator $\boldsymbol{\Delta}[p,q]$.

Definition 8.1 *For each $p,\ q \in \mathbb{R}$, $\boldsymbol{\Delta}[p,q]$ is the bounded operator on $L^2(\mathbb{R})$ whose rule is*

$$\big[\boldsymbol{\Delta}[p,q]\,\phi\big](x) = 2e^{2ip(x-q)}\phi(2q-x), \qquad \phi \in L^2(\mathbb{R}). \tag{8.2.17}$$

By the formal Weyl quantization of a function $T : \Pi \to \mathbb{C}$ on phase space is meant the operator $\boldsymbol{\Delta}[T]$ defined in equation (8.2.15). The operator $\boldsymbol{\Delta}[p,q]$ is thus the formal Weyl quantization of the phase space Dirac measure $\delta_p \otimes \delta_q \in \mathcal{S}'(\Pi)$ concentrated at the point (p,q), and will be referred to as the point quantization operator.

This is an important definition, but is still strictly formal until equation (8.2.15) can be interpreted weakly in such a manner as to permit identification of a useful class of phase space observables T for which it is valid.

8.2.3 *Terminology*

In general usage, the term *quantization* is any consistent procedure for associating a phase space function with an operator on $L^2(\mathbb{R})$. As noted previously, unless specified otherwise, in this book quantization will refer to the procedure developed by Weyl. The operator $\boldsymbol{\Delta}[T]$ is then the quantization of the phase space function T.

Conversely, any consistent procedure which associates an operator B on $L^2(\mathbb{R})$ with a phase space function is called a *dequantization* scheme. Given a dequantization scheme, the function T will be called, variously, the dequantization of B or the symbol of B — the latter terminology deriving from partial differential equation theory.

The special usage in this book is that dequantization means the inverse process to Weyl quantization, so that it will become reasonable to write $T = \boldsymbol{\Delta}^{-1}[B]$ as the dequantization of B — again, when the appropriate class of quantum mechanical observables B which can be dequantized has been established.

When, in Chapter 14, other quantization schemes are considered, this particular quantization/dequantization scheme will be referred to as Weyl quantization/dequantization — this more or less transparent terminology is not standard, but we have adopted it because it is easy to remember.

8.3 The Wigner Transform Method

Having established a heuristic formulation of quantization, it is necessary to provide a rigorous implementation of that formalism which is valid for as large a class of phase space observables as possible. To do this requires the machinery of the smooth model, and the end result will be that equation

(8.2.15) can be extended in such a way that $\mathbf{\Delta}[T]$ can be defined for any tempered distribution $T \in \mathcal{S}'(\Pi)$, with the resulting operator $\mathbf{\Delta}[T]$ then being a continuous linear map from $\mathcal{S}(\mathbb{R})$ to $\mathcal{S}'(\mathbb{R})$. The space of all such linear mappings is denoted $\mathcal{L}(\mathcal{S}(\mathbb{R}), \mathcal{S}'(\mathbb{R}))$, and quantization $\mathbf{\Delta}$ is a linear bijection from $\mathcal{S}'(\Pi)$ to $\mathcal{L}(\mathcal{S}(\mathbb{R}), \mathcal{S}'(\mathbb{R}))$.

There is evidently something of a mismatch between the spaces used here to implement the quantization scheme and the algebras of observables, both classical and quantum. For example, the algebra $\mathcal{C}^\infty(\Pi)$ of classical observables is not a subset of $\mathcal{S}'(\Pi)$, nor is $\mathcal{S}'(\Pi)$ a subset of $\mathcal{C}^\infty(\Pi)$. Moreover, $\mathcal{S}'(\Pi)$ is not even an algebra. Similarly, while $\mathcal{L}(\mathcal{S}(\mathbb{R}), \mathcal{S}'(\mathbb{R}))$ contains both $\mathbb{B}(L^2(\mathbb{R}))$ and $\mathcal{L}^+(\mathcal{S}(\mathbb{R}))$ as proper subspaces, it is not equal to either of them, and is not itself an algebra.

However, it can be seen that the intersection $\mathcal{C}^\infty(\Pi) \cap \mathcal{S}'(\Pi)$ is a dense linear subspace of both $\mathcal{C}^\infty(\Pi)$ and of $\mathcal{S}'(\Pi)$, and that both $\mathbb{B}(L^2(\mathbb{R}))$ and $\mathcal{L}^+(\mathcal{S}(\mathbb{R}))$ are dense linear subspaces of $\mathcal{L}(\mathcal{S}(\mathbb{R}), \mathcal{S}'(\mathbb{R}))$. Thus, there are classical observables that cannot be quantized, and there are classical observables whose quantizations are not quantum observables (either bounded or smooth). However any classical observable can be approximated arbitrarily closely by another observable which can be quantized, and whose quantization is a quantum observable. Similarly, any quantum observable can be approximated arbitrarily closely by another observable whose dequantization is a classical observable. Thus, provided that such approximations are acceptable, the incompleteness of the correspondence between classical and quantum observables can be dealt with[2].

8.3.1 *Boundedness Of* $\mathbf{\Delta}[p,q]$

To get started, we show that $\mathbf{\Delta}[p,q]$ is a bounded operator for each p and q, but is not trace class.

Lemma 8.2 *The point quantization operator* $\mathbf{\Delta}[p,q]$ *is bounded, having norm 2, but is never trace class.*

---

[2]In view of the previously stated fact that there were, for example, quantum mechanical observables that had no natural classical analogue — electron spin, for example — the incompleteness of the correspondence to be provided by Weyl quantization is perhaps not that surprising.

Proof: Since equation (8.2.17) implies that

$$\mathbf{\Delta}[p,q] = 2e^{-2ipq}\,V(2p)\,U(-2q)\,\mathcal{P}, \tag{8.3.1}$$

where U and V are the standard unitary subgroups of the representation W of the Weyl group, and $\mathcal{P}$ is the parity operator on $L^2(\mathbb{R})$,

$$[\mathcal{P}\phi](x) = \phi(-x), \qquad \phi \in L^2(\mathbb{R}), \tag{8.3.2}$$

it follows that $\frac{1}{2}\mathbf{\Delta}[p,q]$ is a composition of four unitary operators on $L^2(\mathbb{R})$, so is itself unitary. Hence it is clear that $\mathbf{\Delta}[p,q]$ is bounded, with

$$\|\,\mathbf{\Delta}[p,q]\,\| = 2. \tag{8.3.3}$$

Since $L^2(\mathbb{R})$ is infinite dimensional, no unitary map on $L^2(\mathbb{R})$ can be trace class. Thus $\mathbf{\Delta}[p,q]$ is not trace class. ■

The point quantization operator $\mathbf{\Delta}[p,q]$ also has the following property.

Lemma 8.3 *The map* $(p,q) \mapsto \mathbf{\Delta}[p,q]$ *from* Π *to* $\mathbb{B}(L^2(\mathbb{R}))$ *is strongly (but not uniformly) continuous.*

Proof: Since $\mathbf{\Delta}[p,q]$ has the decomposition afforded by equation (8.3.1), the continuity properties of $\mathbf{\Delta}[p,q]$ now follow from the strong, but not uniform, continuity of the one parameter unitary subgroups U and V of the Weyl group W. ■

It is important to note from the outset that $\mathbf{\Delta}[p,q]$ is not a trace class operator. This is because the standard expression given for the dequantization symbol of an operator A in many physics books is [44]

$$T(p,q) = \mathrm{Tr}\,(\,\mathbf{\Delta}[p,q]\,A\,)\,.$$

However, since the point quantization operator $\mathbf{\Delta}[p,q]$ is not trace class, this formula can be undefined even when A is bounded[3]. Consequently, just as equation (8.2.15) will prove to be a starting point for a broader exposition of quantization, the above formula needs substantial extension before it can be seen as a means for defining dequantization. This will be discussed in greater detail in Chapter 12.

[3]However this formula makes sense, and gives the correct answer, when the operator A is trace class, as will be shown further on in this Chapter.

8.3.2 *The Wigner Transform*

As indicated previously, in order to develop a theory of quantization which will permit the quantization of a sufficiently large class of classical observables, it is necessary to interpret the heuristic formulæ of the previous Section weakly. Doing so makes a great difference, since the function

$$(p,q) \mapsto \langle g, \mathbf{\Delta}[p,q]f \rangle, \tag{8.3.4}$$

is simply a phase space function for each pair f and g. Clearly, what sort of function it is depends on what class f and g are. The model of quantization presented here depends on the key fact that the above function is a test function on phase space whenever $f,\ g \in \mathcal{S}(\mathbb{R})$. The resulting test function can then be evaluated against any tempered distribution $T \in \mathcal{S}'(\Pi)$, and the resulting quantity can be interpreted as the matrix coefficient of the quantization $\mathbf{\Delta}[T]$ with respect to the functions f and g.

Analysis of equation (8.2.17) enables us to express the function introduced in equation (8.3.4) in terms of an integral operator acting on the function $\overline{g} \otimes f$ on $\mathbb{R}^2$. It is convenient, as well as instructive, to note that this integral operator can be extended to act on all functions in $\mathcal{S}(\mathbb{R}^2)$ — making this extension renders subsequent analysis of mixed states much simpler. In a sense, the following definition and its consequences are key to the rest of quantization theory.

Definition 8.4 *By the Wigner transform we mean the map $\mathcal{G}$ from $\mathcal{S}(\mathbb{R}^2)$ to $\mathcal{S}(\Pi)$ given by*

$$\mathcal{G}(F)(p,q) = \frac{1}{2\pi}\int_{\mathbb{R}} F\left(q+\tfrac{1}{2}u, q-\tfrac{1}{2}u\right) e^{ipu}\,du, \qquad F \in \mathcal{S}(\mathbb{R}^2). \tag{8.3.5}$$

Proposition 8.5 *The Wigner transform $\mathcal{G}$ is bicontinuous. Its inverse is the map $\mathcal{G}^{-1} : \mathcal{S}(\Pi) \to \mathcal{S}(\mathbb{R}^2)$ whose defining formula is*

$$\mathcal{G}^{-1}(H)(x,y) = \int_{\mathbb{R}} H\left(v, \tfrac{1}{2}(x+y)\right) e^{-iv(x-y)}\,dv, \qquad H \in \mathcal{S}(\Pi). \tag{8.3.6}$$

Proof: We have to show that equation (8.3.5) is well defined on the domain $\mathcal{S}(\mathbb{R}^2)$, with $\mathcal{G}(F)$ being an element of $\mathcal{S}(\Pi)$ for every test function F, that the mapping $\mathcal{G}$ is continuous, and that it has a continuous inverse.

We observe that we can break up the integral transform $\mathcal{G}$ as a composition of three operations:

$$\mathcal{G} = \frac{1}{\sqrt{2\pi}} (M \otimes I) \mathcal{F}_1^{-1} \tau . \tag{8.3.7}$$

Here M is the endomorphism of $\mathcal{S}(\mathbb{R})$ such that $[Mf](x) = 2f(2x)$ for $f \in \mathcal{S}(\mathbb{R})$, so that

$$[(M \otimes I)F](x,y) = 2\,F(2x,y), \qquad\qquad F \in \mathcal{S}\big(\mathbb{R}^2\big) .$$

The operator $\mathcal{F}_1^{-1}$, as usual, represents taking the inverse Fourier transform of an element of $\mathcal{S}\big(\mathbb{R}^2\big)$ with respect to the first variable only, and τ is the endomorphism of $\mathcal{S}\big(\mathbb{R}^2\big)$ given by the formula

$$[\tau F](x,y) = F(y+x, y-x), \qquad\qquad F \in \mathcal{S}\big(\mathbb{R}^2\big) . \tag{8.3.8}$$

Now it is easy to show that M and $\mathcal{F}^{-1}$ are bicontinuous endomorphisms of $\mathcal{S}(\mathbb{R})$, and hence $M \otimes I$ and $\mathcal{F}_1^{-1}$ are bicontinuous endomorphisms of $\mathcal{S}\big(\mathbb{R}^2\big)$. Moreover the endomorphism τ, which represents a scaled rotation of coordinates, is also bicontinuous. Combining these results, we see that $\mathcal{G}$ is indeed bicontinuous. From the above results, it is seen that

$$\mathcal{G}^{-1} = \sqrt{2\pi}\,\tau^{-1}\,\mathcal{F}_1\,(M^{-1} \otimes I),$$

and this statement is equivalent to the integral transform equation (8.3.6) given in the statement of the Proposition. ∎

The following extremely important result is a consequence of comparing equations (8.3.4) and (8.3.5).

Proposition 8.6 *The identity*

$$\langle g\,,\,\mathbf{\Delta}\,[p,q]\,f\rangle = 2\pi\,\big[\mathcal{G}(\overline{g} \otimes f)\big](p,q) \tag{8.3.9}$$

holds for any f, $g \in \mathcal{S}(\mathbb{R})$.

Given the above, it is now possible to make a formal definition of the (Weyl) quantization of a tempered distribution $T \in \mathcal{S}'(\Pi)$.

Definition 8.7 ***(Weyl Quantization)*** *The Weyl quantization* $\mathbf{\Delta}[T]$ *of any tempered distribution* $T \in \mathcal{S}'(\Pi)$ *on phase space* Π *is the element of* $\mathcal{L}(\mathcal{S}(\mathbb{R}), \mathcal{S}'(\mathbb{R}))$ *defined by the equation*

$$[\![\mathbf{\Delta}[T] f , g]\!] = [\![T , \mathcal{G}(g \otimes f)]\!], \qquad f, g \in \mathcal{S}(\mathbb{R}). \qquad (8.3.10)$$

Important to the analysis of elements of $\mathcal{L}(\mathcal{S}(\mathbb{R}), \mathcal{S}'(\mathbb{R}))$ is the fact that any of its elements can be described in terms of an *integral kernel.* This description is important in the proof of a number of important results, but care should be taken, since the integral kernel of a general element of $\mathcal{L}(\mathcal{S}(\mathbb{R}), \mathcal{S}'(\mathbb{R}))$ is not a function, but rather a tempered distribution — for example, the integral kernel of the identity map is a delta distribution. This characterization of the elements of $\mathcal{L}(\mathcal{S}(\mathbb{R}), \mathcal{S}'(\mathbb{R}))$ is a result of the standard theory of tensor products of locally convex spaces, and we omit the proof.

Proposition 8.8 ***(Kernels And Distributions)*** *Every element* B *of* $\mathcal{L}(\mathcal{S}(\mathbb{R}), \mathcal{S}'(\mathbb{R}))$ *defines a unique distribution* $K_B \in \mathcal{S}'(\mathbb{R}^2)$ *such that*

$$[\![K_B , g \otimes f]\!] = [\![Bf , g]\!], \qquad f, g \in \mathcal{S}(\mathbb{R}). \qquad (8.3.11.a)$$

This distribution K_B *is called the* integral kernel *of* B. *Conversely, any distribution* T *in* $\mathcal{S}'(\mathbb{R}^2)$ *defines an element* $T^\sharp \in \mathcal{L}(\mathcal{S}(\mathbb{R}), \mathcal{S}'(\mathbb{R}))$ *with*

$$[\![T^\sharp f , g]\!] = [\![T , g \otimes f]\!], \qquad f, g \in \mathcal{S}(\mathbb{R}). \qquad (8.3.11.b)$$

The maps $K : \mathcal{L}(\mathcal{S}(\mathbb{R}), \mathcal{S}'(\mathbb{R})) \to \mathcal{S}'(\mathbb{R}^2)$ *and* $^\sharp : \mathcal{S}'(\mathbb{R}^2) \to \mathcal{L}(\mathcal{S}(\mathbb{R}), \mathcal{S}'(\mathbb{R}))$ *are mutually inverse.*

In general, K_B is a tempered distribution. However, if $B \in \mathcal{L}(\mathcal{S}(\mathbb{R}), \mathcal{S}'(\mathbb{R}))$ is such that $K_B \in \mathcal{S}(\mathbb{R}^2)$, then it follows that $Bf \in \mathcal{S}'(\mathbb{R})$ is a function for all $f \in \mathcal{S}(\mathbb{R})$, with

$$[Bf](x) = \int_{\mathbb{R}} K_B(x,y)\, f(y)\, dy\,; \qquad (8.3.12)$$

this is why K_B is called the integral kernel of B.

The following result expresses the integral kernel $K_{\mathbf{\Delta}[T]}$ of the quantization $\mathbf{\Delta}[T]$ of a tempered distribution $T \in \mathcal{S}'(\Pi)$ in terms of the Wigner transform.

Proposition 8.9 *For any $T \in \mathcal{S}'(\Pi)$, the integral kernel $K_{\boldsymbol{\Delta}[T]}$ of $\boldsymbol{\Delta}[T]$ is given by the formula*

$$K_{\boldsymbol{\Delta}[T]} = \mathcal{G}^{\mathrm{tr}}(T). \tag{8.3.13}$$

Proof: Since $\mathcal{G}$ is a continuous linear map from $\mathcal{S}(\mathbb{R}^2)$ to $\mathcal{S}(\Pi)$, it induces a continuous homomorphism $\mathcal{G}^{\mathrm{tr}}$ from $\mathcal{S}'(\Pi)$ to $\mathcal{S}'(\mathbb{R}^2)$, and it is clear that

$$[\![\mathcal{G}^{\mathrm{tr}}T \,,\, g \otimes f]\!] = [\![\boldsymbol{\Delta}[T] f \,,\, g]\!] = [\![K_{\boldsymbol{\Delta}[T]} \,,\, g \otimes f]\!]$$

for all $f,\, g \in \mathcal{S}(\mathbb{R})$. The density of $\mathcal{S}(\mathbb{R}) \otimes \mathcal{S}(\mathbb{R})$ in $\mathcal{S}(\mathbb{R}^2)$ then implies the identity (8.3.13), as required. ■

This result has the following crucial Corollary, legitimizing quantization of tempered distributions, which is a consequence of the bicontinuity of the Wigner transform $\mathcal{G}$, and hence also of its transpose $\mathcal{G}^{\mathrm{tr}}$.

Corollary 8.10 *The Weyl quantization map $\boldsymbol{\Delta}$ is a linear bijection from $\mathcal{S}'(\Pi)$ to $\mathcal{L}(\mathcal{S}(\mathbb{R}), \mathcal{S}'(\mathbb{R}))$.*

When, in Chapter 12, dequantization methods are studied in detail, an alternative approach to this result will be discussed.

Before proceeding further, it is essential to see how the above definitions implement the heuristics of the previous Section. If $T \in \mathcal{S}'(\Pi)$ is such that $\boldsymbol{\Delta}[T]$ is a linear map from $\mathcal{S}(\mathbb{R})$ to $L^2(\mathbb{R})$, then it is clear that $[\![\boldsymbol{\Delta}[T] f \,,\, \overline{g}]\!] = \langle g \,,\, \boldsymbol{\Delta}[T] f \rangle$ for any $f,\, g \in \mathcal{S}(\mathbb{R})$. Thus if $T \in L^1(\Pi)$ it is clear that

$$|\, [\![T \,,\, \mathcal{G}(\overline{g} \otimes f)]\!] \,| \leqslant \frac{1}{\pi} \| T \|_1 \, \| f \|_2 \, \| g \|_2$$

for any $f,\, g \in \mathcal{S}(\mathbb{R})$, and hence that $\boldsymbol{\Delta}[T] \in \mathbb{B}(L^2(\mathbb{R}))$, with

$$\| \boldsymbol{\Delta}[T] \| \leqslant \frac{1}{\pi} \| T \|_1 \,,$$

and

$$\langle g \,,\, \boldsymbol{\Delta}[T] f \rangle = \iint_{\Pi} T(p,q)\, \mathcal{G}(\overline{g} \otimes f)\, dp\, dq \,,$$

which establishes the validity of equation (8.2.15). Similar arguments show that $2\pi \boldsymbol{\Delta}[T] = W[\![\mathcal{F}T]\!]$, with $2\pi \| \boldsymbol{\Delta}[T] \| \leqslant \| \mathcal{F}T \|_1$, for any $T \in L^2(\Pi)$ for which $\mathcal{F}T \in L^1(\mathbb{R}^2)$, justifying equation (8.2.14).

As far as we are aware, the Wigner transform was first introduced [181] as a scaled unitary operator on $L^2(\mathbb{R}^2)$, rather than in the smooth form given above. The following result completes the connection between the smooth and unitary formalisms.

Proposition 8.11 *(Extension To Hilbert Space) The Wigner transform $\mathcal{G}$ can be extended to the whole of $L^2(\mathbb{R}^2)$. Denoting this extension by the same symbol $\mathcal{G}$, $\sqrt{2\pi}\mathcal{G}$ is unitary,*

$$\langle \mathcal{G}(\Phi), \mathcal{G}(\Psi) \rangle = \frac{1}{2\pi} \langle \Phi, \Psi \rangle, \qquad \Phi, \Psi \in L^2(\mathbb{R}^2). \tag{8.3.14}$$

Proof: Direct calculation shows that $2^{-\frac{1}{2}} M$ extends to a unitary map on $L^2(\mathbb{R})$. Moreover the inverse Fourier transform $\mathcal{F}^{-1}$ is a unitary map on $L^2(\mathbb{R})$. It can be shown easily that the endomorphism τ of $\mathcal{S}(\mathbb{R}^2)$ is such that $2^{\frac{1}{2}}\tau$ extends to a unitary map of $L^2(\mathbb{R}^2)$. Consequently we deduce that $\sqrt{2\pi}\,\mathcal{G} = (M \otimes I)\,\mathcal{F}_1^{-1}\,\tau$ can be extended to yield a unitary operator on $L^2(\mathbb{R}^2)$. ■

In summary, then, the Wigner transform $\mathcal{G}$ is a bicontinuous linear map from $\mathcal{S}(\mathbb{R}^2)$ to $\mathcal{S}(\Pi)$ such that $\sqrt{2\pi}\mathcal{G}$ extends to a unitary map from $L^2(\mathbb{R}^2)$ to $L^2(\Pi)$. Moreover, the Wigner transform can be used to define Weyl quantization as a bijective linear map from $\mathcal{S}'(\Pi)$ to $\mathcal{L}(\mathcal{S}(\mathbb{R}), \mathcal{S}'(\mathbb{R}))$ which implements the heuristic statements of the previous Section.

8.3.3 *Some Useful Identities Involving $\mathcal{G}$*

In order to be able to obtain information about the quantization procedure outlined above, it is necessary to have detailed technical knowledge about the properties of the Wigner transform. In this Section will be presented basic details of the behaviour of the Wigner transform with respect to certain elementary operations — multiplication by and partial differentiation with respect to the coordinate functions, Fourier transformation and complex conjugation. For simplicity of exposition, results will be presented concerning the action of these operations on functions of the form $\mathcal{G}(\overline{g} \otimes f)$, where f, $g \in \mathcal{S}(\mathbb{R})$ — similar results can be obtained for the action of these operations on more general functions of the form $\mathcal{G}(F)$, where $F \in \mathcal{S}(\mathbb{R}^2)$, but to do so makes the resulting formulæ less clear. To begin with, therefore, the next result establishes the behaviour of the Wigner transform with respect to elementary operations involving the coordinate functions[4].

Proposition 8.12 *The following identities*

$$p\,\big[\mathcal{G}(\overline{g}\otimes f)\big](p,q) = \tfrac{1}{2}\big[\mathcal{G}(\overline{Pg}\otimes f)+\mathcal{G}(\overline{g}\otimes Pf)\big](p,q), \quad (8.3.15.a)$$

$$q\,\big[\mathcal{G}(\overline{g}\otimes f)\big](p,q) = \tfrac{1}{2}\big[\mathcal{G}(\overline{Qg}\otimes f)+\mathcal{G}(\overline{g}\otimes Qf)\big](p,q), \quad (8.3.15.b)$$

$$\frac{\partial}{\partial p}\,\big[\mathcal{G}(\overline{g}\otimes f)\big](p,q) = i\,\big[\mathcal{G}(\overline{Qg}\otimes f)-\mathcal{G}(\overline{g}\otimes Qf)\big](p,q), \quad (8.3.15.c)$$

$$\frac{\partial}{\partial q}\,\big[\mathcal{G}(\overline{g}\otimes f)\big](p,q) = i\,\big[-\mathcal{G}(\overline{Pg}\otimes f)+\mathcal{G}(\overline{g}\otimes Pf)\big](p,q), \quad (8.3.15.d)$$

hold for $\mathcal{G}$, where Q and P are the usual Schrödinger operators of position and momentum.

Proof: All four identities can be established by similar methods, of which the following proof of the second result is representative. For

$$\begin{aligned} &q\,\big[\mathcal{G}(\overline{g}\otimes f)\big](p,q) \\ &= \frac{1}{2\pi}\int_{\mathbb{R}} \frac{1}{2}\big\{(q-\tfrac{1}{2}u)+(q+\tfrac{1}{2}u)\big\}\,\overline{g(q+\tfrac{1}{2}u)}\,f(q-\tfrac{1}{2}u)\,e^{ipu}\,du \\ &= \tfrac{1}{2}\big[\mathcal{G}(\overline{Qg}\otimes f)\big](p,q) + \tfrac{1}{2}\big[\mathcal{G}(\overline{g}\otimes Qf)\big](p,q), \end{aligned}$$

for any $f,\,g\in\mathcal{S}(\mathbb{R})$, as required. ∎

The next result outlines the properties of the Wigner transform with respect to the operation of Fourier transformation.

Proposition 8.13 *For any $f,\,g\in\mathcal{S}(\mathbb{R})$, the following identities hold:*

$$\big[\mathcal{G}(\overline{g}\otimes f)\big](p,q) = \big[\mathcal{G}(\overline{\mathcal{F}g}\otimes\mathcal{F}f)\big](-q,p), \quad (8.3.16.a)$$

$$\big[\mathcal{F}^{-1}\mathcal{G}(\overline{g}\otimes f)\big](a,b) = \frac{1}{2\pi}\,\langle g\,,\,W(a,b)f\rangle\,. \quad (8.3.16.b)$$

---

[4]It should perhaps be noted that the results of this Lemma can be used to obtain detailed information concerning the nature of the continuity of the Wigner transform as a map from $\mathcal{S}(\mathbb{R}^2)$ to $\mathcal{S}(\Pi)$.

Proof: Using the Fourier inversion formula and the properties of the Weyl group,

$$\begin{aligned}
\big[\mathcal{G}(\overline{g}\otimes f)\big](p,q) &= \frac{1}{2\pi}\,\langle g\,,\,\boldsymbol{\Delta}\,[\,p,q\,]\,f\rangle \\
&= \frac{1}{\pi}\,e^{-2ipq}\,\langle \mathcal{F}g\,,\,\mathcal{F}V(2p)U(-2q)\mathcal{P}f\rangle \\
&= \frac{1}{\pi}\,e^{2ipq}\,\langle \mathcal{F}g\,,\,V(-2q)U(-2p)\mathcal{P}\mathcal{F}f\rangle \\
&= \frac{1}{2\pi}\,\langle \mathcal{F}g\,,\,\boldsymbol{\Delta}\,[\,-q,p\,]\,\mathcal{F}f\rangle \\
&= \big[\mathcal{G}(\overline{\mathcal{F}g}\otimes\mathcal{F}f)\big](-q,p),
\end{aligned}$$

as required. For the second result, note that $\sqrt{2\pi}\,\mathcal{G}(F) = \mathcal{F}_1^{-1}\hat{F}$ for any $F \in \mathcal{S}(\mathbb{R}^2)$, where $\hat{F}(p,q) = (\tau F)(\frac{1}{2}p,q)$. From this it follows that

$$\sqrt{2\pi}\big[\mathcal{F}^{-1}\mathcal{G}F\big](p,q) = \big[(\mathcal{F}^{-2}\otimes\mathcal{F}^{-1})\hat{F}\big](p,q) = \big[\mathcal{F}_2^{-1}\hat{F}\big](-p,q)$$

for all $F \in \mathcal{S}(\mathbb{R}^2)$, and the desired equality for $\mathcal{F}^{-1}\mathcal{G}(\overline{g}\otimes f)$ is then immediate. ■

The first result shows that combined with the Wigner transform, the Fourier transform on $\mathcal{S}(\mathbb{R})$ implements a rotation of phase space[5] through an angle of 90°. The second result shows that, in the weak sense, the operator $\boldsymbol{\Delta}\,[\,p,q\,]$ is the Fourier transform of the Weyl group, a result that was used heuristically in the previous Section (see equation (8.2.16)) when formulating quantization.

To determine the marginal distributions for Weyl quantization rigorously, which will be done in Section 8.5.1, the following identities are necessary.

Corollary 8.14 *For any f, $g \in \mathcal{S}(\mathbb{R})$, we have*

$$\begin{aligned}
\int_{\mathbb{R}} \big[\mathcal{G}(\overline{g}\otimes f)\big](p,q)\,dp &= \overline{g(q)}f(q), \\
\int_{\mathbb{R}} \big[\mathcal{G}(\overline{g}\otimes f)\big](p,q)\,dq &= \overline{(\mathcal{F}g)(p)}(\mathcal{F}f)(p).
\end{aligned} \qquad (8.3.17)$$

[5]This observation will be useful in the discussion of the metaplectic representation at the end of this Chapter.

Proof: Since

$$\int_{\mathbb{R}} (Mf)(x)\,dx \;=\; \int_{\mathbb{R}} f(x)\,dx, \qquad \frac{1}{\sqrt{2\pi}} \int_{\mathbb{R}} (\mathcal{F}^{-1}f)(x)\,dx \;=\; f(0),$$

for any $f \in \mathcal{S}(\mathbb{R})$, it follows that

$$\int_{\mathbb{R}} [\mathcal{G}(\overline{g} \otimes f)](p,q)\,dp \;=\; \big[\tau(\overline{g} \otimes f)\big](0,q) \;=\; \overline{g(q)}f(q)\,,$$

which is the first result. The second identity is a consequence of this and equation (8.3.16.a), which enables us to switch the integration from the second to the first variable. ∎

Finally, the next result shows how $\mathcal{G}$ behaves with respect to complex conjugation:

Proposition 8.15 *For any f, $g \in \mathcal{S}(\mathbb{R})$ we have*

$$\overline{[\mathcal{G}(\overline{g} \otimes f)](p,q)} \;=\; [\mathcal{G}(\overline{f} \otimes g)](p,q) \tag{8.3.18}$$

Indeed, as elements of $L^2(\Pi)$, the functions $\overline{\mathcal{G}(\overline{\psi} \otimes \phi)}$ and $\mathcal{G}(\overline{\phi} \otimes \psi)$ are equal for any ϕ, $\psi \in L^2(\mathbb{R})$.

Although $\mathcal{L}(\mathcal{S}(\mathbb{R}), \mathcal{S}'(\mathbb{R}))$ is not an algebra of operators on Hilbert space, it is possible to equip it with an involution which naturally extends the involutive operations of taking the adjoint in either $\mathbb{B}\big(L^2(\mathbb{R})\big)$ or $\mathcal{L}^+(\mathcal{S}(\mathbb{R}))$. This is done by defining $X^+ \in \mathcal{L}(\mathcal{S}(\mathbb{R}), \mathcal{S}'(\mathbb{R}))$ for any $X \in \mathcal{L}(\mathcal{S}(\mathbb{R}), \mathcal{S}'(\mathbb{R}))$ by the formula

$$[\![\, X^+ g\,,\, \overline{f}\,]\!] \;=\; \overline{[\![\, Xf\,,\, \overline{g}\,]\!]}\,, \qquad f,\, g \in \mathcal{S}(\mathbb{R})\,. \tag{8.3.19}$$

This involution in $\mathcal{L}(\mathcal{S}(\mathbb{R}), \mathcal{S}'(\mathbb{R}))$ is connected, via quantization, to the standard involution in $\mathcal{S}'(\Pi)$ defined by the formula

$$[\![\, \overline{T}\,,\, F\,]\!] \;=\; \overline{[\![\, T\,,\, \overline{F}\,]\!]}, \qquad F \in \mathcal{S}(\Pi)\,, T \in \mathcal{S}'(\Pi)\,, \tag{8.3.20}$$

since equation (8.3.18) shows that

$$[\![\, \mathbf{\Delta}\,[T]^+ f\,,\, g\,]\!] \;=\; [\![\, \mathbf{\Delta}\,\big[\overline{T}\big]\, f\,,\, g\,]\!], \qquad f, g \in \mathcal{S}(\mathbb{R}),\, T \in \mathcal{S}'(\Pi), \tag{8.3.21.a}$$

and hence that

$$\mathbf{\Delta}\,[T]^+ \;=\; \mathbf{\Delta}\,\big[\overline{T}\big]\,. \tag{8.3.21.b}$$

In other words, quantization intertwines the involutive structures[6] of $\mathcal{S}'(\Pi)$ and $\mathcal{L}(\mathcal{S}(\mathbb{R}), \mathcal{S}'(\mathbb{R}))$.

8.4 Classes Of Bounded Observables

One of the associated problems of any quantization theory, once one has been established, is to correlate the class of the observable T with the class of its quantization $\mathbf{\Delta}[T]$. For instance, what properties must T satisfy in order that $\mathbf{\Delta}[T]$ be a smooth observable? Conversely, if B is a bounded operator, what can be said about the properties of its dequantization $\mathbf{\Delta}^{-1}[B]$? As with the cognate problem of the relation between an operator and its integral kernel [98], there are many such questions that can be asked about (Weyl) quantization, and only a few of them have definite answers. In this and the next Section some of these questions will be addressed.

In this Section, various classes of quantum mechanical observables which are bounded operators are considered — to be specific, questions will be asked as to what properties of a tempered distribution T ensure that its quantization $\mathbf{\Delta}[T]$ belongs to any of the following classes:

- $\mathcal{T}_0(L^2(\mathbb{R}))$, the finite-rank operators on $L^2(\mathbb{R})$,
- $\mathcal{T}_\infty(L^2(\mathbb{R}))$, the compact operators on $L^2(\mathbb{R})$,
- $\mathcal{T}_1(L^2(\mathbb{R}))$, the trace class operators on $L^2(\mathbb{R})$,
- $\mathcal{T}_2(L^2(\mathbb{R}))$, the Hilbert-Schmidt operators on $L^2(\mathbb{R})$,
- $\mathbb{B}(L^2(\mathbb{R}))$, the bounded operators on $L^2(\mathbb{R})$.

In general, if $\mathfrak{C} \subset \mathcal{L}(\mathcal{S}(\mathbb{R}), \mathcal{S}'(\mathbb{R}))$ is a chosen class of "generalized observables", the aim is to seek the class $\mathcal{O}[\mathfrak{C}] \subset \mathcal{S}'(\Pi)$ of phase space distributions for which quantization $\mathbf{\Delta} : \mathcal{O}[\mathfrak{C}] \to \mathfrak{C}$ is one to one and onto.

For the classes of bounded observables listed above,

$$\mathcal{T}_0(L^2(\mathbb{R})) \subseteq \mathcal{T}_1(L^2(\mathbb{R})) \subseteq \mathcal{T}_2(L^2(\mathbb{R})) \subseteq \mathcal{T}_\infty(L^2(\mathbb{R})) \subseteq \mathbb{B}(L^2(\mathbb{R}))\,;$$

[6]It should be noted that equation (8.3.20) could be applied to $\mathcal{S}'(\mathbb{R}^2)$ to obtain an involution formula which would then apply to the integral kernels of elements of $\mathcal{L}(\mathcal{S}(\mathbb{R}), \mathcal{S}'(\mathbb{R}))$. However doing so defines an involution on $\mathcal{S}'(\mathbb{R}^2)$ which is not consistent with the identification made between elements of $\mathcal{L}(\mathcal{S}(\mathbb{R}), \mathcal{S}'(\mathbb{R}))$ and their integral kernels in $\mathcal{S}'(\mathbb{R}^2)$ — another involution would have to be defined on $\mathcal{S}'(\mathbb{R}^2)$ in order to maintain consistency in this regard.

thus the corresponding distribution spaces $\mathcal{O}\left[\mathcal{T}_0\big(L^2(\mathbb{R})\big)\right]$, $\mathcal{O}\left[\mathcal{T}_1\big(L^2(\mathbb{R})\big)\right]$, $\mathcal{O}\left[\mathcal{T}_2\big(L^2(\mathbb{R})\big)\right]$, $\mathcal{O}\left[\mathcal{T}_\infty\big(L^2(\mathbb{R})\big)\right]$ and $\mathcal{O}\left[\mathbb{B}\big(L^2(\mathbb{R})\big)\right]$ are nested similarly.

Each of these spaces $\mathcal{O}[\mathfrak{C}]$ exist, but may not have a known description as a function space of a standard type. But it may be that an adequate implicit characterization of a space $\mathcal{O}[\mathfrak{C}]$ can be found. Even this may not be possible, and only partial conditions (necessary or sufficient but not both) may be all that is known.

8.4.1 *Finite-Rank Operators*

The space $\mathcal{T}_0\big(L^2(\mathbb{R})\big)$ is precisely the (finite) linear span of operators of the type $|\,\phi\,\rangle\,\langle\,\psi\,|$ for $\phi,\ \psi \in L^2(\mathbb{R})$. For this class a complete characterization of $\mathcal{O}\left[\mathcal{T}_0\big(L^2(\mathbb{R})\big)\right]$ is possible.

Proposition 8.16 *For any $\phi,\ \psi \in L^2(\mathbb{R})$, the function $\Xi_{\phi,\psi}$ given by*

$$\Xi_{\phi,\psi}(p,q) \;=\; 2\pi\left[\mathcal{G}(\overline{\psi}\otimes\phi)\right](p,q) \;=\; \langle\psi\,,\,\mathbf{\Delta}\,[\,p,q\,]\,\phi\rangle \tag{8.4.1}$$

belongs to $L^2(\Pi)$ and has the quantization

$$\mathbf{\Delta}\,[\,\Xi_{\phi,\psi}\,] \;=\; |\,\phi\,\rangle\,\langle\,\psi\,|. \tag{8.4.2}$$

Therefore $\mathcal{O}\left[\mathcal{T}_0\big(L^2(\mathbb{R})\big)\right]$ is equal to the subspace of $L^2(\Pi)$ spanned by all functions of the form $\Xi_{\phi,\psi}$ for $\phi,\ \psi \in L^2(\mathbb{R})$.

Proof: Simple calculations show that

$$\begin{aligned}
[\![\,\mathbf{\Delta}\,[\,\Xi_{\phi,\psi}\,]\,f\,,\,\overline{g}\,]\!] &= [\![\,\Xi_{\phi,\psi}\,,\,\mathcal{G}(\overline{g}\otimes f)\,]\!] \\
&= 2\pi\,\big\langle\mathcal{G}(\overline{\phi}\otimes\psi)\,,\,\mathcal{G}(\overline{g}\otimes f)\big\rangle \\
&= \langle g\,,\,\phi\rangle\,\langle\psi\,,\,f\rangle
\end{aligned}$$

for all $f,\ g \in \mathcal{S}(\mathbb{R})$, establishing the result. ■

8.4.2 *Compact Operators*

Compact operators form the basis of any discussion of many other classes of operators, including the trace and Hilbert-Schmidt classes. The following material is standard fare in operator theory, and details and proofs can be found in the references at the end of the Chapter.

A bounded operator A on a separable Hilbert space $\mathcal{H}$ is said to be *compact* if it maps bounded subsets of $\mathcal{H}$ to sets whose closures are compact. Equivalently, A is compact if from any bounded sequence $(\phi_n)_{n\geqslant 1}$ in $\mathcal{H}$ (meaning that $\sup_{n\geqslant 1} \| \phi_n \| < \infty$) a subsequence $\big(\phi_{n(k)}\big)_{k\geqslant 1}$ can be extracted such that the sequence $\big(A\phi_{n(k)}\big)_{k\geqslant 1}$ converges in $\mathcal{H}$. Then $\mathcal{T}_\infty(\mathcal{H})$ is a proper closed *-invariant ideal in $\mathbb{B}(\mathcal{H})$ — indeed it is the only such ideal in $\mathbb{B}(\mathcal{H})$ — and the space of finite rank operators $\mathcal{T}_0(\mathcal{H})$ is norm-dense in $\mathcal{T}_\infty(\mathcal{H})$. On a finite dimensional Hilbert space, all bounded operators are of finite rank, and hence compact. Consequently the theory of compact operators is only of interest with regard to infinite dimensional Hilbert spaces, and so it will now be assumed that the Hilbert space $\mathcal{H}$ in question is infinite dimensional (and separable).

The spectral theory of compact operators is particularly simple.

- The spectrum $\sigma(A)$ of a compact operator A consists of a countable set of nonzero eigenvalues, each of finite (geometric) multiplicity, together with the point 0. The spectral value 0 is the only possible limit point of the spectrum $\sigma(A)$.
- If A is a self-adjoint compact operator, there exists an orthonormal basis $(\phi_n)_{n\geqslant 1}$ of $\mathcal{H}$ consisting of eigenvectors of A. If, for each $n \in \mathbb{N}$, $\lambda_n \in \mathbb{R}$ is the eigenvalue of the eigenvector ϕ_n, then it is possible to ensure that $\lim_{n\to\infty} \lambda_n = 0$.
- Given a compact operator A, if $(\phi_n)_{n\geqslant 1}$ is an orthonormal basis of eigenvectors of the compact self-adjoint operator $|A| = \sqrt{A^*A}$, where $|A|\,\phi_n = \mu_n(A)\phi_n$, then defining $\psi_n = \mu_n(A)^{-1} A\phi_n$ for all $n \geqslant 1$ yields an orthonormal (but not necessarily complete) sequence $(\psi_n)_{n\geqslant 1}$ of vectors in $\mathcal{H}$ such that

$$A = \sum_{n\geqslant 1} \mu_n(A) |\, \psi_n \rangle \langle \phi_n \,| . \tag{8.4.3}$$

  This is known as the canonical form of the compact operator A, and the $\mu_n(A)$ are known as the singular values of A.

Since 0 must lie in the spectrum of a compact operator on $\mathcal{H}$, no unitary operator on $\mathcal{H}$ can be compact.

To our knowledge, no complete characterization of the observable space $\mathcal{O}\left[\mathcal{T}_\infty\big(L^2(\mathbb{R})\big)\right]$ is known. However, we can present two conditions on a distribution $T \in \mathcal{S}'(\Pi)$ which are sufficient to ensure that $\mathbf{\Delta}\,[\,T\,]$ is compact.

Proposition 8.17 ***(Quantizing To Compact Operators)*** *Either if T is in $L^1(\Pi)$, or else if T is in $L^2(\Pi)$ and its Fourier transform $\mathcal{F}T$ belongs to $L^1(\mathbb{R}^2)$, then $\boldsymbol{\Delta}[T]$ is compact.*

Proof: If $T \in L^1(\Pi)$, it has already been shown that $\boldsymbol{\Delta}[T] \in \mathbb{B}(L^2(\mathbb{R}))$ with $\pi \| \boldsymbol{\Delta}[T] \| \leqslant \| T \|_1$. Since $\mathcal{S}(\Pi)$ is dense in $L^1(\Pi)$, it is possible to find a sequence $(T_n)_{n\geqslant 1}$ of functions in $\mathcal{S}(\Pi)$ which converges to T with respect to the norm in $L^1(\Pi)$. Consequently the sequence of bounded operators $(\boldsymbol{\Delta}[T_n])_{n\geqslant 1}$ converges to $\boldsymbol{\Delta}[T]$ in the operator norm in $\mathbb{B}(L^2(\mathbb{R}))$. Later in this Section it will be shown that each operator $\boldsymbol{\Delta}[T_n]$ is Hilbert-Schmidt, and hence compact. Thus it follows that $\boldsymbol{\Delta}[T]$ is compact as well. On the other hand, if $T \in L^2(\Pi)$ is such that $\mathcal{F}T \in L^1(\mathbb{R}^2)$, it has already been noted that $\boldsymbol{\Delta}[T] = \frac{1}{2\pi} W[\![\mathcal{F}T]\!]$ belongs to $\mathbb{B}(L^2(\mathbb{R}))$, with $\| \boldsymbol{\Delta}[T] \| \leqslant \frac{1}{2\pi} \| \mathcal{F}T \|_1$. If $(S_n)_{n\geqslant 1}$ is a sequence in $\mathcal{S}(\mathbb{R}^2)$ which converges to $\mathcal{F}T$ in $L^1(\mathbb{R}^2)$, then the sequence of bounded operators $(\boldsymbol{\Delta}[\mathcal{F}^{-1}S_n])_{n\geqslant 1}$ converges to $\boldsymbol{\Delta}[T]$ in the operator norm in $\mathbb{B}(L^2(\mathbb{R}))$. Since each operator $\boldsymbol{\Delta}[\mathcal{F}^{-1}S_n]$ is Hilbert-Schmidt, and hence compact, the operator $\boldsymbol{\Delta}[T]$ is compact. ■

8.4.3 *Trace Class Operators*

The theory of trace class operators is delicate, and there are a number of pitfalls available for the unwary to fall into — these usually relate to a failure to distinguish between the trace of an operator and its trace norm. Positive trace class operators were discussed in Chapter 3 in connection with density matrices. However, in that discussion we did not give a precise definition of a trace class operator, and we must now rectify that omission. The general definition of a trace class operator is designed to avoid working with conditionally convergent sums, and so requires a slightly roundabout approach. As was the case with compact operators, our attention will be restricted to infinite dimensional Hilbert spaces.

If $B \in \mathbb{B}(\mathcal{H})$ is positive, the (finite or infinite) expression

$$\tau(B) = \sum_{n=1}^{\infty} \langle \xi_n , B\xi_n \rangle \tag{8.4.4.a}$$

is independent of the choice of orthonormal basis $(\xi_n)_{n\geqslant 1}$ for $\mathcal{H}$. A bounded operator A is said to be *trace class* if the expression $\tau(|A|)$ is finite, in which case the expression

$$\text{Tr}\,(A) = \sum_{n=1}^{\infty} \langle \xi_n\,,\, A\xi_n \rangle \tag{8.4.4.b}$$

is well-defined, finite, and independent of basis. The expression Tr (A) is then called the *trace* of A.

The collection $\mathcal{T}_1(\mathcal{H})$ of all trace class operators is a linear subspace of $\mathcal{T}_\infty(\mathcal{H})$ which is a Banach space with respect to the *trace norm*

$$\| A \|_1 = \text{Tr}\,(|A|)\,, \qquad A \in \mathcal{T}_1(\mathcal{H})\,, \tag{8.4.5.a}$$

which definition coincides with that in equation (3.4.8), with

$$\| A \|_1 \geqslant \| A \|\,, \qquad A \in \mathcal{T}_1(\mathcal{H})\,, \tag{8.4.5.b}$$

and the finite rank operators $\mathcal{T}_0(\mathcal{H})$ form a dense linear subspace of the Banach space $\mathcal{T}_1(\mathcal{H})$. It is important to note that Tr (A) and $\| A \|_1$ differ when A is not positive.

The space $\mathcal{T}_1(\mathcal{H})$ is a *-ideal in $\mathbb{B}(\mathcal{H})$ (but not a closed *-ideal) such that

$$\| AB \|_1 \leqslant \| A \|_1 \, \| B \|\,, \qquad A \in \mathcal{T}_1(\mathcal{H}),\, B \in \mathbb{B}(\mathcal{H})\,, \tag{8.4.6.a}$$

and

$$\| A^* \|_1 = \| A \|_1\,, \qquad A \in \mathcal{T}_1(\mathcal{H})\,, \tag{8.4.6.b}$$

and the trace is a continuous positive linear functional on the Banach space $\mathcal{T}_1(\mathcal{H})$ such that

$$\text{Tr}\,(AB) = \text{Tr}\,(BA)\,, \qquad A \in \mathcal{T}_1(\mathcal{H}),\, B \in \mathbb{B}(\mathcal{H})\,, \tag{8.4.7.a}$$

and

$$\text{Tr}\,(A^*) = \text{Tr}\,(A)\,, \qquad A \in \mathcal{T}_1(\mathcal{H})\,. \tag{8.4.7.b}$$

Remark Suppose that a bounded operator A is given and it is calculated that the series

$$\sum_{n=1}^{\infty} \langle \xi_n\,,\, A\xi_n \rangle$$

converges for some orthonormal basis $(\xi_n)_{n\geqslant 1}$. It is a tempting fallacy to conclude that A is trace class and that the sum is its basis-independent trace. From the information given, that may or may not be true, since whether A is trace class or not depends on the convergence of the series

$$\mathrm{Tr}\,(\,|A|\,) \;=\; \sum_{n=1}^{\infty} \langle \xi_n\,,\,|\,A\,|\,\xi_n\rangle\,,$$

about which (in general) nothing has been said. ■

Matrix theory has accustomed us to believe that the trace of an operator is the sum of its eigenvalues. The above observations show that this is still the case for positive trace class operators B, for then

$$\mathrm{Tr}\,(\,B\,) \;=\; \|\,B\,\|_1 \;=\; \sum_{n=1}^{\infty} \mu_n(B)\,,$$

and the singular values $\mu_n(B)$ are the eigenvalues of the operator $A \;=\; |\,B\,|$. Happily, this result is capable of further generalization. Every self-adjoint trace class operator A is compact, and hence its spectrum consists of a sequence $\big(\lambda_n(A)\big)_{n\geqslant 1}$ of real eigenvalues which converges to zero. A theorem of Weyl ensures that this sequence of eigenvalues belongs to ℓ^1, while the result of Lidskii establishes the trace formula

$$\mathrm{Tr}\,(\,A\,) \;=\; \sum_{n=1}^{\infty} \lambda_n(A), \tag{8.4.8}$$

(see equation (3.4.6)). For details and further references, see Simon [211].

Part of the folklore of physics is that an operator A is trace class if the diagonal of its integral kernel, $|\,K_A(x,x)\,|$ is integrable. This is not quite true, and must be replaced by the following.

Lemma 8.18 *Let K be a continuous function on $\mathbb{R}^2$ which satisfies the positivity condition,*

$$\sum_{j,k=1}^{n} \overline{z_j}\,K(x_j,x_k)\,z_k \;\geqslant\; 0 \tag{8.4.9.a}$$

for all $n \in \mathbb{N}$, $x_1, \ldots, x_n \in \mathbb{R}$ *and* $z_1, \ldots, z_n \in \mathbb{C}$. *In particular, this implies that* $K(x,x) \geqslant 0$ *for all* $x \in \mathbb{R}$. *Then if*

$$\int_{\mathbb{R}} K(x,x)\,dx \;<\; \infty, \tag{8.4.9.b}$$

there exists a positive trace class operator A *on* $L^2(\mathbb{R})$ *for which* $K = K_A$ *is its integral kernel, and then*

$$\mathrm{Tr}\,(\,A\,) \;=\; \|\,A\,\|_1 \;=\; \int_{\mathbb{R}} K(x,x)\,dx. \tag{8.4.9.c}$$

However the converse to this Lemma is not true. Given a bounded operator A on $L^2(\mathbb{R})$ determined by an integral kernel K, then convergence of the integral

$$\int_{\mathbb{R}} |\,K(x,x)\,|\;dx$$

is not sufficient to guarantee that A be trace class. Consequently, any attempt at a proof which makes such an assumption must be treated with extreme caution. On the positive side, however, Simon (*ibid*) tells us in his book on trace ideals that

> *However, the counter-examples that prevent nice theorems holding are generally rather contrived so that I have found the following to be true: If an operator with integral kernel occurs in some 'natural' way and* $\int |\,K(x,x)\,|\;dx < \infty$, *then the operator can (almost always) be proven to be trace class (although sometimes only after some considerable effort).*

However this observation, while reassuring, is not a charter allowing carefree calculations!

Just as was the case for compact operators, there is no known full characterization of the classical observable space $\mathcal{O}\left[\mathcal{T}_1\big(L^2(\mathbb{R})\big)\right]$. The partial results that are known are more delicate than those for compact operators. The first gives a sufficient condition on a distribution T for its quantization $\boldsymbol{\Delta}\,[\,T\,]$ to belong to $\mathcal{T}_1\big(L^2(\mathbb{R})\big)$. As it makes three requirements on the distribution T, it is somewhat impractical for everyday use.

Proposition 8.19 *If* $T \in L^1(\Pi)$ *is such that* $\boldsymbol{\Delta}\,[\,T\,]$ *is a positive bounded operator on* $L^2(\mathbb{R})$ *whose integral kernel* $K_{\boldsymbol{\Delta}[T]} = \mathcal{G}^{\mathrm{tr}}T$ *is a continuous* *bounded function on* $\mathbb{R}^2$*, then* $\mathbf{\Delta}[T]$ *is trace class, and*

$$\operatorname{Tr}(\mathbf{\Delta}[T]) = \frac{1}{2\pi}\iint_{\Pi} T(p,q)\,dp\,dq. \qquad (8.4.10)$$

Proof: This result essentially restates the results of Lemma 8.18. That $\mathbf{\Delta}[T]$ is a positive operator implies that its integral kernel $K_{\mathbf{\Delta}[T]}$ satisfies the required positivity condition, and since $T \in L^1(\Pi)$ the integral kernel can be written

$$K_{\mathbf{\Delta}[T]}(x,y) = \frac{1}{2\pi}\int_{\mathbb{R}} T\big(u, \tfrac{1}{2}(x+y)\big)\,e^{iu(x-y)}\,du,$$

and hence

$$K_{\mathbf{\Delta}[T]}(x,x) = \frac{1}{2\pi}\int_{\mathbb{R}} T(u,x)\,du.$$

Thus $\mathbf{\Delta}[T]$ is indeed trace class, with

$$\operatorname{Tr}(\mathbf{\Delta}[T]) = \int_{\mathbb{R}} K_{\mathbf{\Delta}[T]}(x,x)\,dx = \frac{1}{2\pi}\iint_{\Pi} T(u,v)\,du\,dv,$$

as required. ∎

It is clear from the above discussion that determining whether or not a particular operator $\mathbf{\Delta}[T]$ is trace class is difficult. However it is important since, as has been mentioned previously, the equality (hereinafter referred to as the *familiar formula*)

$$T(p,q) = \operatorname{Tr}(\mathbf{\Delta}[p,q]\,\mathbf{\Delta}[T]), \qquad (8.4.11)$$

is frequently cited as the dequantization formula.

But this formula is problematic as it stands, since it does not make sense for all classical observables T. Indeed the right hand side only has an obvious interpretation when $\mathbf{\Delta}[T]$ is trace class. By establishing a series of results pertaining to $\mathbf{\Delta}[p,q]$, it will be shown that the familiar formula (8.4.11) is valid precisely for observables T belonging to $\mathcal{O}\left[\mathcal{T}_1\big(L^2(\mathbb{R})\big)\right]$. From this analysis will arise some necessary conditions for the quantization of a classical observable to be trace class.

Given an arbitrary trace class operator A on $L^2(\mathbb{R})$, define the function

$$\alpha_A(p,q) = \operatorname{Tr}(\mathbf{\Delta}[p,q]\,A), \qquad p,\, q \in \mathbb{R}, \qquad (8.4.12)$$

It is clear that the function α_A is uniformly bounded on Π, with

$$|\alpha_A(p,q)| \leqslant 2\,\|A\|_1, \qquad (p,q) \in \Pi.$$

Moreover, it is continuous.

Theorem 8.20 *If A is a trace class operator, then the function α_A is continuous on Π.*

Proof: If A is the finite-rank operator $A = |\phi\rangle\langle\psi|$ for some $\phi, \psi \in L^2(\mathbb{R})$, it is clear that $\alpha_A(p,q) = \langle\psi\,,\, \mathbf{\Delta}[p,q]\,\phi\rangle$. By Lemma 8.3, $\mathbf{\Delta}[p,q]$ is certainly weakly continuous, and hence α_A is continuous in this case. By linearity it follows that α_A is continuous on Π for any finite-rank operator A.

In the general case, let A be trace class, and let p, $q \in \mathbb{R}$. Then in the canonical form, equation (8.4.3), the sequence of singular values $(\mu_n(A))_{n\geqslant 1}$ belongs to ℓ^1. Then, given $\varepsilon > 0$, there is an $M \in \mathbb{N}$ such that

$$\sum_{n=M+1}^{\infty} \mu_n(A) < \frac{1}{6}\varepsilon.$$

Setting

$$B = \sum_{n=1}^{M} \mu_n(A)\,|\,\psi_n\,\rangle\,\langle\,\phi_n\,|,$$

B is seen to be a finite rank operator on $L^2(\mathbb{R})$, and

$$\begin{aligned} \left|\alpha_{A-B}(u,v)\right| &\leqslant \sum_{n=M+1}^{\infty} \mu_n(A)\,|\langle\phi_n, \mathbf{\Delta}[u,v]\,\psi_n\rangle| \\ &\leqslant 2\sum_{n=M+1}^{\infty} \mu_n(A) < \frac{1}{3}\varepsilon \end{aligned}$$

for all u, $v \in \mathbb{R}$. Thus

$$\begin{aligned} &\left|\alpha_A(u,v) - \alpha_A(p,q)\right| \\ &\qquad \leqslant \left|\alpha_B(u,v) - \alpha_B(p,q)\right| + \left|\alpha_{A-B}(p,q)\right| + \left|\alpha_{A-B}(u,v)\right| \\ &\qquad < \left|\alpha_B(u,v) - \alpha_B(p,q)\right| + \frac{2}{3}\varepsilon \end{aligned}$$

for all $u, v \in \mathbb{R}$. Since α_B is continuous, we can find $\delta > 0$ such that

$$\begin{aligned} |(u,v)-(p,q)| < \delta \quad &\Rightarrow \quad |\alpha_B(u,v)-\alpha_B(p,q)| < \frac{1}{3}\varepsilon \\ &\Rightarrow \quad |\alpha_A(u,v)-\alpha_A(p,q)| < \varepsilon, \end{aligned}$$

so α_A is continuous at (p,q). Thus α_A is continuous on Π. ■

Note that the function α_A certainly belongs to $\mathcal{S}'(\Pi)$, since it is bounded and continuous. The following result determines the quantization $\mathbf{\Delta}[\alpha_A]$, thereby establishing the validity of the familiar formula (8.4.11) for trace class operators.

Theorem 8.21 ***(The "Familiar Formula")*** *For $A \in \mathfrak{T}_1(L^2(\mathbb{R}))$, its dequantization symbol $\mathbf{\Delta}^{-1}[A]$ is a bounded continuous function on Π, given by the formula*

$$\mathbf{\Delta}^{-1}[A](p,q) = \alpha_A(p,q) = \mathrm{Tr}\,(\mathbf{\Delta}[p,q]\,A), \qquad p,\, q \in \mathbb{R}. \tag{8.4.13}$$

Proof: For any $f,\, g \in \mathcal{S}(\mathbb{R})$, define the linear functional Φ_{fg} on $\mathfrak{T}_1(L^2(\mathbb{R}))$ by the formula

$$\Phi_{fg}(A) = [\![\mathbf{\Delta}[\alpha_A]f\,,\,\overline{g}]\!], \qquad A \in \mathfrak{T}_1(L^2(\mathbb{R})).$$

Since

$$\begin{aligned} \big|[\![\mathbf{\Delta}[\alpha_A]f\,,\,\overline{g}]\!]\big| &= \big|[\![\alpha_A\,,\,\mathcal{G}(\overline{g}\otimes f)]\!]\big| \leqslant \|\alpha_A\|_\infty \|\mathcal{G}(\overline{g}\otimes f)\|_1 \\ &\leqslant 2\|A\|_1 \|\mathcal{G}(\overline{g}\otimes f)\|_1 \end{aligned}$$

for all trace class operators A, Φ_{fg} is a continuous linear functional on $\mathfrak{T}_1(L^2(\mathbb{R}))$. Consequently there exists a bounded operator $Y_{fg} \in \mathbb{B}(L^2(\mathbb{R}))$ such that

$$\Phi_{fg}(A) = \mathrm{Tr}\,(A\,Y_{fg}), \qquad A \in \mathfrak{T}_1(L^2(\mathbb{R})).$$

But then

$$\begin{aligned} \langle\psi\,,\,Y_{fg}\phi\rangle &= \Phi_{fg}(|\phi\rangle\langle\psi|) \\ &= [\![\mathbf{\Delta}[\Xi_{\phi,\psi}]f\,,\,\overline{g}]\!] \\ &= \frac{1}{2\pi}\langle\mathcal{G}(\overline{f}\otimes g)\,,\,\mathcal{G}(\overline{\psi}\otimes\phi)\rangle \\ &= \langle\overline{f}\otimes g\,,\,\overline{\psi}\otimes\phi\rangle = \langle\psi\,,\,f\rangle\langle g\,,\,\phi\rangle \end{aligned}$$

for all ϕ, $\psi \in L^2(\mathbb{R})$, which implies that $Y_{fg} = |\, f\,\rangle\,\langle\, g\,|$, and hence

$$[\![\, \boldsymbol{\Delta}\,[\,\alpha_A\,]\,f\,,\,\overline{g}\,]\!] \;=\; \Phi_{fg}(A) \;=\; \langle g\,,\, Af\rangle\,, \qquad A \in \mathcal{T}_1\big(L^2(\mathbb{R})\big)\,.$$

Since this identity is true for all $f,\, g \in \mathcal{S}(\mathbb{R})$, it follows that $\boldsymbol{\Delta}\,[\,\alpha_A\,] = A$ for any trace class operator. This establishes the desired result. ■

Not only has it been shown that the familiar formula for dequantization makes sense for trace class operators, but it has also been shown that the resulting phase space symbol must be equal almost everywhere to a continuous function.

Corollary 8.22 *If $T \in L^1(\Pi)$ is not equal almost everywhere to a continuous function on Π, then $\boldsymbol{\Delta}\,[\,T\,]$ is compact but not trace class.*

Remark While the familiar formula (8.4.11) technically cannot be used to dequantize operators which are not trace class, it frequently is. For example, subtle formal manipulations of the innards of the familiar formula can be used to "show" that $\boldsymbol{\Delta}^{-1}\,[\,P\,] = p$, as should be. But calculations of this type crucially involve operations like an illegitimate interchange of limits, such as an infinite sum and an integration. In a sense, therefore, these "derivations" are *ad hoc* calculations which have been formulated in just such a way as to obtain results which can be derived by more complex methods, and are justified *ex post facto* by the fact that they yield the correct results.

But we cannot dismiss these results out of hand, for to do that is to overlook the fact that they (usually) do succeed in obtaining the correct dequantization of a quantum observable. We speculate that this is evidence that the trace formula could be extended to cover a wider class of observables than the trace class, perhaps by a summability field method such as Borel summability, or perhaps by a partial resummation method as in Fejér's Theorem in Fourier series. In the latter case, the method would single out a particular orthonormal basis for resummation, the Hermite-Gauss ones being the obvious first choice. The problem is then to identify the subset of $\mathcal{L}(\mathcal{S}(\mathbb{R}), \mathcal{S}'(\mathbb{R}))$ in a useful way for which this resummation formula will work. That such a class exists is clear, and it will certainly contain the trace class operators, and probably other observables such as polynomials in P and Q.

It would be interesting to attempt this analysis, but we are not aware of anyone having done so. Until such a theory is created, we must hold to the principle that a result obtained by the familiar formula where A is either not trace class or has not been shown to be, must be treated with some caution. ■

8.4.4 *Hilbert-Schmidt Operators*

The next important class of bounded operators on $L^2(\mathbb{R})$ to consider are the Hilbert-Schmidt operators. These operators can now be defined readily.

A bounded operator A on a separable Hilbert space $\mathcal{H}$ is said to be *Hilbert-Schmidt* if the positive operator A^*A is trace class. The collection of all Hilbert-Schmidt operators on $\mathcal{H}$ is denoted $\mathcal{T}_2(\mathcal{H})$, and is a linear subspace of $\mathcal{T}_\infty(\mathcal{H})$, which is a Hilbert space with respect to the inner product

$$\langle A\,,\,B\rangle_2 \;=\; \mathrm{Tr}\,(\,A^*\,B\,)\,, \qquad A,\,B \in \mathcal{T}_2(\mathcal{H}). \tag{8.4.14}$$

It is clear that a compact operator A is Hilbert-Schmidt if and only if its sequence $\big(\mu_n(A)\big)_{n\geqslant 1}$ of singular values belongs to ℓ^2. Thus every trace class operator on $\mathcal{H}$ is Hilbert-Schmidt. The subspace $\mathcal{T}_0(\mathcal{H})$ of finite rank operators is dense in the Hilbert space $\mathcal{T}_2(\mathcal{H})$.

The space $\mathcal{T}_2(\mathcal{H})$ of Hilbert-Schmidt operators on $\mathcal{H}$ is a *-ideal (but not a closed *-ideal) in $\mathbb{B}\big(L^2(\mathbb{R})\big)$ such that

$$\|\,AB\,\|_2 \;\leqslant\; \|\,A\,\|_2\;\|\,B\,\|\,, \qquad A \in \mathcal{T}_2(\mathcal{H}),\; B \in \mathbb{B}\big(L^2(\mathbb{R})\big)\,, \tag{8.4.15.a}$$

and

$$\|\,A^*\,\|_2 \;=\; \|\,A\,\|\,, \qquad A \in \mathcal{T}_2(\mathcal{H})\,, \tag{8.4.15.b}$$

where $\|\cdot\|_2$ is the norm on $\mathcal{T}_2(\mathcal{H})$ derived from the inner product $\langle\,\cdot\,,\,\cdot\,\rangle_2$.

A bounded operator A is Hilbert-Schmidt if and only if its integral kernel K_A belongs to $L^2\big(\mathbb{R}^2\big)$, in which case the equality

$$\langle A\,,\,B\rangle_2 \;=\; \langle K_A\,,\,K_B\rangle\,, \qquad A,\,B \in \mathcal{T}_2\big(L^2(\mathbb{R})\big)\,, \tag{8.4.16}$$

sets up a unitary isomorphism between $\mathcal{T}_2\big(L^2(\mathbb{R})\big)$ and $L^2\big(\mathbb{R}^2\big)$. This isomorphism can be used to advantage to characterize the phase space functions whose quantizations are Hilbert-Schmidt. Recalling that the Wigner transform can be extended from $\mathcal{S}(\Pi)$ to $L^2(\Pi)$, it can be shown that

$$K_{\mathbf{\Delta}[T]} = \mathcal{G}^{\rm tr} T = \frac{1}{2\pi}\overline{\mathcal{G}^{-1}\overline{T}} = \frac{1}{2\pi}\mathcal{R}\mathcal{G}^{-1}\,T = \frac{1}{\sqrt{2\pi}}\,\sigma\,\mathcal{F}_1^{-1}\,T \qquad (8.4.17.a)$$

for any $T \in L^2(\Pi)$, where $\mathcal{R}$ is the coordinate reversal operator,

$$\big[\mathcal{R}F\big](x,y) = F(y,x)\,, \qquad F \in L^2\big(\mathbb{R}^2\big)\,, \qquad (8.4.17.b)$$

and σ is the unitary mapping of $L^2\big(\mathbb{R}^2\big)$ into itself given by the formula

$$\big[\sigma F\big](x,y) = F\big(x-y, \tfrac{1}{2}(x+y)\big), \qquad F \in L^2(\Pi). \qquad (8.4.17.c)$$

Thus the mapping $T \mapsto \sqrt{2\pi}\,K_{\mathbf{\Delta}[T]}$ is a unitary mapping from $L^2(\Pi)$ to $L^2\big(\mathbb{R}^2\big)$, and so quantization provides a linear bijection from $L^2(\Pi)$ to $\mathcal{T}_2\big(L^2(\mathbb{R})\big)$. These results constitute Pool's Theorem [181]:

Theorem 8.23 ***(Hilbert-Schmidt Operators)*** *The map which sends T to $\sqrt{2\pi}\,\mathbf{\Delta}\,[\,T\,]$ is a unitary transformation from $L^2(\Pi)$ to $\mathcal{T}_2\big(L^2(\mathbb{R})\big)$. Thus $\mathcal{O}\,\big[\mathcal{T}_2\big(L^2(\mathbb{R})\big)\big] = L^2(\Pi)$.*

Pool's result thus provides a complete characterization of $\mathcal{O}\,\big[\mathcal{T}_2\big(L^2(\mathbb{R})\big)\big]$.

8.4.5 *Bounded Operators*

Any of the results obtained which are sufficient to ensure that $\mathbf{\Delta}\,[\,T\,]$ is compact, trace class or Hilbert-Schmidt will necessarily ensure that $\mathbf{\Delta}\,[\,T\,]$ is bounded. But as these are rather restrictive conditions, weaker sufficiency conditions would be useful.

Unfortunately, a complete characterization of the space $\mathcal{O}\,\big[\mathbb{B}\big(L^2(\mathbb{R})\big)\big]$ is not known. It is not necessary for a function T to vanish at infinity in phase space for $\mathbf{\Delta}\,[\,T\,]$ to be bounded, since the function $\mathrm{i}(p,q) = 1$ has the identity operator as its quantization. It is not sufficient for T to be bounded, since the function $T(p,q) = e^{2ipq}$ has quantization

$$[\![\, \mathbf{\Delta}\,[\,T\,]\,f\,,\,g\,]\!] = \frac{1}{2}\,g(0)\int_{\mathbb{R}} f(x)\,dx\,,$$

which is not a bounded operator. Nor is it necessary for T to be bounded, since any unbounded element of $L^2(\Pi)$ has a Hilbert-Schmidt, so bounded, quantization. For that matter, it is not even necessary for T to be a function, since simple calculations show us that $\pi\,\mathbf{\Delta}\left[\,\delta\otimes\delta\,\right] \;=\; \mathcal{P}$, where $\mathcal{P}$ is the parity operator on $L^2(\mathbb{R})$.

In the absence of a full characterization of elements of $\mathcal{O}\left[\mathbb{B}\big(L^2(\mathbb{R})\big)\right]$, establishing the boundedness of any particular quantization $\mathbf{\Delta}\left[\,T\,\right]$ may have to be done by hand. There are some interesting, if rather restrictive, results available. The first result assumes boundedness not only of T but of its derivatives up to order 3 [63]:

Proposition 8.24 ***(Calderón-Vaillancourt)*** *Suppose that $T \in \mathcal{C}^3(\Pi)$ and*

$$\sum_{\substack{m,n\\ m+n=3}} \sup_{p,q\in\Pi} \left| \frac{\partial^m}{\partial p^m}\frac{\partial^n}{\partial q^n}\, T(p,q) \right| \;<\; \infty. \tag{8.4.18}$$

Then $\mathbf{\Delta}\left[\,T\,\right]$ is bounded.

In a different direction Daubechies has obtained boundedness results involving the following Sobolev-type spaces [37]. Consider the Hilbertian norms

$$\mathfrak{p}_r(f)^2 \;=\; \left\langle f, \left[p^2+q^2-\frac{1}{4}\frac{\partial^2}{\partial p^2}-\frac{1}{4}\frac{\partial^2}{\partial q^2}+1\right]^r f\right\rangle \tag{8.4.19}$$

on $\mathcal{S}(\Pi)$. The space W_r is then defined to be the completion of $\mathcal{S}(\Pi)$ with respect to the seminorm $\mathfrak{p}_r$.

Proposition 8.25 ***(Daubechies)*** *If the quantization $\mathbf{\Delta}\left[\,T\,\right]$ of a tempered distribution $T \in \mathcal{S}'(\Pi)$ is bounded, then T belongs to W_{-1-t} for all $t > 0$ and there exists a positive constant C such that for all $t > 0$,*

$$\mathfrak{p}_{-1-t}(T) \;\leqslant\; C\, e^{-t/2}\left(\frac{3+t}{2}\right)^{(3+t)/2} (t+1)^{-1/2}. \tag{8.4.20}$$

For $r > 1$, W_r is a subspace of both $L^1(\Pi)$ and $L^2(\Pi)$, and this observation leads to a related result.

Proposition 8.26 ***(Daubechies)*** *If T belongs to W_{1+s} for any $s > 0$,* *then* $\mathbf{\Delta}[T]$ *is trace class, with the following bound on its trace norm:*

$$\| \mathbf{\Delta}[T] \|_1 \leqslant e^{-s/2} \left(\frac{3+s}{2}\right)^{(3+s)/2} (s+1)^{-1/2} \mathfrak{p}_{1+s}(T). \qquad (8.4.21)$$

8.5 Smooth Observables

In the previous Section, sufficient (and occasionally necessary) conditions on distributions $T \in \mathcal{S}'(\Pi)$ were found which ensure that $\mathbf{\Delta}[T]$ belong to one of the standard classes of bounded operators on $L^2(\mathbb{R})$. That discussion was particularly appropriate to the bounded model.

In this Section various conditions on distributions T which are sufficient to ensure that $\mathbf{\Delta}[T]$ belongs to $\mathcal{L}^+(\mathcal{S}(\mathbb{R}))$ will be considered. A complete characterization of the class $\mathfrak{O}[\mathcal{L}^+(\mathcal{S}(\mathbb{R}))]$ is not known, but there are some useful sufficiency conditions.

Before proceeding in detail, it is useful to expand on some of the discussion of previous Sections in a manner which is more specific and relevant to the smooth model.

It has been noted above that every element $B \in \mathcal{L}(\mathcal{S}(\mathbb{R}), \mathcal{S}'(\mathbb{R}))$ has an integral kernel K_B. Inspection of equation (8.3.11.a) makes it clear that B is a continuous endomorphism of $\mathcal{S}(\mathbb{R})$ whenever $B \in \mathcal{S}(\mathbb{R}^2)$. However, it is possible to establish more than this, as the following very important Proposition (due originally to G.A.Lassner) shows.

Proposition 8.27 ***(Smooth States And Kernels)*** *A necessary and sufficient condition for the operator* $B \in \mathcal{L}(\mathcal{S}(\mathbb{R}), \mathcal{S}'(\mathbb{R}))$ *to belong to* $\mathfrak{A}_*$ *for the smooth model is that its integral kernel* K_B *should belong to* $\mathcal{S}(\mathbb{R}^2)$.

Hence a density matrix $\rho \in \mathfrak{A}_*$ *for the smooth model can be regarded as a smooth observable* $\rho \in \mathcal{L}^+(\mathcal{S}(\mathbb{R}))$ *whose integral kernel* $K_\rho \in \mathcal{S}(\mathbb{R}^2)$ *is a test function in two variables ([147], [52]).*

Regarding density matrices in $\mathfrak{A}_*$ as special types of smooth observables has another useful consequence, in that it yields a useful formula for calculating generalized expectations.

Proposition 8.28 ***(Generalized Expectations)*** *If* $\rho \in \mathfrak{A}_*$ *is a density matrix for the smooth model and* $\mathbf{\Delta}[T] \in \mathcal{L}^+(\mathcal{S}(\mathbb{R}))$*, then* $\rho\,\mathbf{\Delta}[T]$ *is trace* *class, and*

$$\mathrm{Tr}\,(\rho\,\boldsymbol{\Delta}[T]) = [\![T\,,\,\mathcal{G}\,(\mathcal{R}K_\rho)\,]\!], \tag{8.5.1}$$

where the coordinate reversal map $\mathcal{R}$ *defined in equation (8.4.17.b) is now viewed as a continuous endomorphism of* $\mathcal{S}(\mathbb{R}^2)$.

Thus, if $\boldsymbol{\Delta}[T] \in \mathcal{L}(\mathcal{S}(\mathbb{R}), \mathcal{S}'(\mathbb{R}))$, we may take the well defined quantity $[\![T\,,\,\mathcal{G}\,(\mathcal{R}K_\rho)\,]\!]$ to be its "expectation value" in the smooth state determined by $\rho \in \mathfrak{A}_*$. Note that $\boldsymbol{\Delta}[T]$ need not even be an operator.

8.5.1 *Polynomials And Polynomial Bounds*

One of the key motivations of the smooth model was to be able to express the quantizations of the basic position and momentum coordinates q and p in terms of the standard observables Q and P of the Schrödinger representation. Moreover, Weyl quantization was designed to have Q and P marginals. Consequently we should hope to find that all polynomials in the coordinate functions q and p both belong to $\mathcal{O}\,[\mathcal{L}^+(\mathcal{S}(\mathbb{R}))]$.

However, polynomials are better-behaved than many distributions, since they are certainly well-defined functions at every point of Π. Indeed, polynomials are continuous, but for certain purposes there is an advantage to grouping them with other functions, not necessarily continuous, which share the same growth properties. The distributions of this class will prove important in a number of applications, including the quantization of radial functions. These distributions will be required with both one and two variables, so they will be defined in k-dimensional form, where k is a fixed positive integer.

Definition 8.29 *For each integer* $n \geqslant 0$, *define the sets*

$$\mathcal{O}^n(\mathbb{R}^k) = \Big\{ F : \mathbb{R}^k \to \mathbb{C} \,:\, \prod_{j=1}^{k} (1 + |\,x_j\,|)^{-n}\, F(x_1, \dots, x_k) \in L^2(\mathbb{R}^k) \Big\}, \tag{8.5.2.a}$$

and

$$\mathcal{O}^\infty(\mathbb{R}^k) = \bigcup_{n=0}^{\infty} \mathcal{O}^n(\mathbb{R}^k)\,. \tag{8.5.2.b}$$

It is clear that $\mathcal{S}(\mathbb{R}^k) \subseteq \mathcal{O}^n(\mathbb{R}^k)$, and that $\mathcal{O}^\infty(\mathbb{R}^k)$ contains all polynomials in the coordinate functions. More important is the fact that $\mathcal{O}^\infty(\mathbb{R}^k) \subseteq \mathcal{S}'(\mathbb{R}^k)$, so that the quantization of any function in $\mathcal{O}^\infty(\Pi)$ is defined.

Proposition 8.30 $\mathcal{O}^\infty(\mathbb{R}^k) \subseteq \mathcal{S}'(\mathbb{R}^k)$.

Proof: Let $F \in \mathcal{O}^\infty(\mathbb{R}^k)$, and consider any $f \in \mathcal{S}(\mathbb{R}^k)$. We can find $n \in \mathbb{N}$ such that the function

$$G(x_1, \ldots, x_k) = \Big(\prod_{j=1}^k (1 + |x_j|)^{-n}\Big) F(x_1, \ldots, x_k)$$

belongs to $L^2(\mathbb{R}^k)$. Because f is a test function, the function

$$g_{f,n}(x_1, \ldots, x_k) = \Big(\prod_{j=1}^k (1 + |x_j|)^{n}\Big) f(x_1, \ldots, x_k)$$

is in $L^2(\mathbb{R}^k)$ as well. But then

$$\int_{\mathbb{R}^k} F(x) f(x)\, d^k x = \int_{\mathbb{R}^k} G(x) g_{f,n}(x)\, d^k x = \langle \overline{G}\,,\, g_{f,n} \rangle$$

for all $f \in \mathcal{S}(\mathbb{R}^k)$, so that $|[\![F\,,\, f]\!]| \leqslant \|G\|\, \|g_{f,n}\|$ for all f in $\mathcal{S}(\mathbb{R}^k)$. Since the map $f \mapsto \|g_{f,n}\|$ is a continuous seminorm on $\mathcal{S}(\mathbb{R}^k)$, we deduce that F does indeed belong to $\mathcal{S}'(\mathbb{R}^k)$. ■

Inspection of the details of the above proof shows that, given a function F in $\mathcal{O}^\infty(\mathbb{R})$, the function $x \mapsto F(x)f(x)$ belongs to $L^2(\mathbb{R})$ for any $f \in \mathcal{S}(\mathbb{R})$, so that the unbounded operator $F(Q)$ contains $\mathcal{S}(\mathbb{R})$ in its domain. Similarly, $F(P)f = \mathcal{F}^{-1}F(Q)\mathcal{F}f$ belongs to $L^2(\mathbb{R})$ for any $f \in \mathcal{S}(\mathbb{R})$, and hence the operator $F(P)$ also contains $\mathcal{S}(\mathbb{R})$ in its domain. The space $\mathcal{O}^\infty(\mathbb{R})$ can therefore be used to generate a large number of classical observables relevant to the study of the marginals of quantization.

Proposition 8.31 ***(Quantization Of Marginals)*** *For any $F \in \mathcal{O}^\infty(\mathbb{R})$ and $f, g \in \mathcal{S}(\mathbb{R})$,*

$$\langle g\,,\, \mathbf{\Delta}[\mathrm{i} \otimes F]\, f \rangle = \langle g\,,\, F(Q) f \rangle\,, \qquad \langle g\,,\, \mathbf{\Delta}[F \otimes \mathrm{i}]\, f \rangle = \langle g\,,\, F(P) f \rangle\,, \tag{8.5.3}$$

where, as before, $\mathrm{i}(x) = 1$.

Proof: Recalling the results of Lemma 8.14, observe that

$$\begin{aligned}[\![\,\mathbf{\Delta}\,[\,\mathrm{i}\otimes F\,]\,f\,,\,\overline{g}\,]\!] &= \iint_\Pi F(q)\big[\mathcal{G}(\overline{g}\otimes f)\big](p,q)\,dp\,dq \\ &= \int_{\mathbb{R}} \overline{g(q)}F(q)f(q)\,dq \;=\; \langle g\,,\,F(Q)f\rangle\,,\end{aligned}$$

and, similarly,

$$\begin{aligned}[\![\,\mathbf{\Delta}\,[\,F\otimes\mathrm{i}\,]\,f\,,\,\overline{g}\,]\!] &= \iint_\Pi F(q)\big[\mathcal{G}(\overline{g}\otimes f)\big](p,q)\,dp\,dq \\ &= \int_{\mathbb{R}} \overline{(\mathcal{F}g)(p)}F(p)(\mathcal{F}f)(p)\,dp \\ &= \langle \mathcal{F}g\,,\,F(Q)\mathcal{F}f\rangle\,,\end{aligned}$$

for all f, $g\in\mathcal{S}(\mathbb{R})$ and $F\in\mathcal{O}^\infty(\mathbb{R})$, so that $\mathbf{\Delta}\,[\,\mathrm{i}\otimes F\,] = F(Q)$ and $\mathbf{\Delta}\,[\,F\otimes\mathrm{i}\,] = \mathcal{F}^{-1}F(Q)\mathcal{F} = F(P)$ (when restricted to $\mathcal{S}(\mathbb{R})$), as required. ∎

In determining which operator corresponds to a given phase space polynomial in p and q, the following generating function is useful.

Proposition 8.32 ***(Polynomial Generating Function)*** *The function*

$$E_{a,b}(p,q) \;=\; e^{i(ap+bq)} \tag{8.5.4}$$

belongs to $\mathcal{O}^\infty(\Pi)$*, and*

$$\mathbf{\Delta}\,[\,E_{a,b}\,] \;=\; W(a,b) \tag{8.5.5}$$

for any a, $b\in\mathbb{R}$*. Moreover, the function* $(a,b)\mapsto\langle g,W(a,b)f\rangle$ *is infinitely differentiable for any* f, g *in* $\mathcal{S}(\mathbb{R})$*, and*

$$\frac{\partial^m}{\partial a^m}\frac{\partial^n}{\partial b^n}\langle g\,,\,W(a,b)f\rangle\bigg|_{a=b=0} \;=\; i^{m+n}\,\langle g\,,\,\mathbf{\Delta}\,[\,p^m q^n\,]\,f\rangle \tag{8.5.6}$$

for any f, $g\in\mathcal{S}(\mathbb{R})$ *and* m, $n\geqslant 0$.

Proof: It is clear that $E_{a,b}\in\mathcal{O}^2(\Pi)$ for each a, $b\in\mathbb{R}$. Equation (8.3.16.b) in Lemma 8.13 then implies that

$$\begin{aligned}\langle g\,,\,\mathbf{\Delta}\,[\,E_{a,b}\,]\,f\rangle &= [\![\,E_{a,b}\,,\,\mathcal{G}(\overline{g}\otimes f)\,]\!] \\ &= 2\pi\big[\mathcal{F}^{-1}\mathcal{G}(\overline{g}\otimes f)\big](a,b) \;=\; \langle g\,,\,W(a,b)f\rangle\end{aligned}$$

for all $f,\, g \in \mathcal{S}(\mathbb{R})$, giving $\boldsymbol{\Delta}\left[\, E_{a,b} \,\right] = W(a,b)$, as required. The result concerning the derivatives of $\langle g\,,\, W(a,b)f\rangle$ is essentially a standard one of functional analysis, and its proof is of no particular interest, so it will be omitted. ■

Returning to polynomials, this result implies that

$$\boldsymbol{\Delta}\left[\,(ap+bq)^n\,\right] = (aP+bQ)^n \tag{8.5.7}$$

for every n. As a consequence of this it can be shown that $\boldsymbol{\Delta}\left[\,\mathrm{i}\,\right] = I$ and

$$\boldsymbol{\Delta}\left[pq\right] = \frac{1}{2}\,(PQ+QP)\,, \quad \boldsymbol{\Delta}\left[p^2q\right] = \frac{1}{4}\,\left(P^2Q+2\,PQP+QP^2\right), \tag{8.5.8}$$

for example. It can be argued that equation (8.5.5) is what distinguishes Weyl quantization from quantizations based on other orderings.

The next result is sufficient to show that the collection of polynomials in p and q forms a subspace of $\mathcal{O}\left[\mathcal{L}^+(\mathcal{S}(\mathbb{R}))\right]$. It is a straightforward result, and needs no proof.

Proposition 8.33 *$\boldsymbol{\Delta}\left[\,T\,\right]$ is a differential operator of order n if and only if T is a polynomial of degree n in p, and is a differential operator with polynomial coefficients if and only if T is a polynomial in p and q.*

8.5.2 *General Smooth Observables*

Finally, a discussion of more general conditions on distributions T which ensure that $\boldsymbol{\Delta}\left[\,T\,\right]$ is a smooth observable is in order. The first result is fairly elementary, but important, in that it shows that the fundamental phase space test functions behave well with respect to quantization in the smooth model.

Proposition 8.34 *If $T \in \mathcal{S}(\Pi)$, then $\boldsymbol{\Delta}\left[\,T\,\right] \in \mathcal{L}^+(\mathcal{S}(\mathbb{R}))$.*

Proof: By the same arguments used for Hilbert-Schmidt operators, it can be shown that

$$K_{\boldsymbol{\Delta}[T]} \;=\; \frac{1}{2\pi}\overline{\mathcal{G}^{-1}\overline{T}} \;=\; \frac{1}{2\pi}\mathcal{R}\mathcal{G}^{-1}T.$$

However, it is clear in this case that this implies that $K_{\boldsymbol{\Delta}[T]}$ belongs to $\mathcal{S}\left(\mathbb{R}^2\right)$. From Proposition 8.27, smooth states and kernels, this means that the operator $\boldsymbol{\Delta}\left[\,T\,\right]$ is a density matrix, and so is certainly a continuous endomorphism of $\mathcal{S}(\mathbb{R})$. Moreover, since $\overline{T} \in \mathcal{S}(\Pi)$, and $\mathbf{\Delta}[T]^+ = \mathbf{\Delta}[\overline{T}]$, it is clear that $\mathbf{\Delta}[T] \in \mathcal{L}^+(\mathcal{S}(\mathbb{R}))$, as required. ■

In the theory of partial differential equations, a variant of quantization is used, and symbol classes rather different from those discussed so far are commonly used. The way quantization is relevant to the study of partial differential equations can be understood as follows. Take, for simplicity, a differential operator in one variable x, with nonconstant coefficients, say $a(x) + b(x)\frac{d^2}{dx^2}$. Now if it were the case that a and b were constants, we could use Fourier transforms to construct solutions, Green's functions and so forth. This is possible, since conjugating the above operator with the Fourier transform yields the simple multiplication operator $a - bk^2$. However, for nonconstant a and b, Fourier analysis would yield an extremely complicated differential operator, and we would not be any better off. Now, if we could have our cake and eat it too, we would take a sort of partial Fourier transform which operated on the differential operator $\frac{d^2}{dx^2}$ while leaving the functions a and b alone, thereby obtaining the two-variable function $F(x,k) = a(x) - b(x)k^2$. The method that actually works is not as simple as this — a twist is needed — but the idea of replacing a differential operator in one variable by a function of two variables begins to take shape. Dequantization is, of course, a similar operation.

What can quantization theory then offer the study of partial differential equations aside from a neat trick? Anyone who has come across the familiar second order partial differential equations of mathematical physics knows that the nature of the solution (elliptic, parabolic, hyperbolic) can be discovered precisely from the properties of the analogue of $F(x,k)$ above. We may expect, therefore, that this method could tell us something about the nature of the solutions, about positivity, about continuity, and so on. Because quantization can deal with tempered distributions, it can handle formal differential operators whose coefficients are too singular for classical methods. It can even deal with infinite order operators to a certain extent. We call these nonclassical objects pseudo-differential operators[7]. The problems considered in the modern theory are formidable, and require very fine controls on the order of growth of $T(p,q)$. Consequently there has been a large quantity of work done studying a variety of carefully constructed

[7]This is not the technical definition!

function spaces, and suitable interpretation of some of the fruits of that study are of interest to us. The remainder of this Section will discuss one of the results from this theory, suitably interpreted in the context of Weyl quantization.

If r, s and t are real numbers, the space $\mathcal{S}^t_{r,s}$ consists of those functions T on Π which are infinitely differentiable, and which satisfy the following growth bounds on their derivatives:

$$\sup_{0\leqslant m+n\leqslant N} \sup_{p,q\in\Pi} \left| \left(\frac{\partial}{\partial p}\right)^m \left(\frac{\partial}{\partial q}\right)^n T(p,q) \right| (1+p^2+q^2)^{t+rm-sn} < \infty \tag{8.5.9}$$

for all positive integers N. Often good behaviour is obtained by restricting the parameters r and s to satisfy the condition

$$0 \leqslant s \leqslant r \leqslant 1, \qquad s < 1,$$

but this is not always necessary, and the following result is a case in point. A proof can be found in Folland's book (*ibid*).

Proposition 8.35 *If $T \in \mathcal{S}^t_{-s,s}$ for some $t \in \mathbb{R}$ and $0 < s < 1$, then $\mathbf{\Delta}[T] \in \mathcal{L}^+(\mathcal{S}(\mathbb{R}))$. Moreover, if (T_n) is a sequence of phase space functions in $\mathcal{S}^t_{-s,s}$ for some $0 < s < 1$, and if T_n converges to a phase space function T in the topology of $C^\infty(\Pi)$, then $\mathbf{\Delta}[T_n]f \to \mathbf{\Delta}[T]f$ for all $f \in \mathcal{S}(\mathbb{R})$.*

8.6 Positivity

Positivity is a property that is rather hard to pin down, for positivity of $\mathbf{\Delta}[T]$ does not guarantee that its phase space symbol T is a function taking positive values, or more generally is a positive distribution. This can be seen by considering the function ν, where $\nu + \frac{1}{2}$ is the harmonic oscillator Hamiltonian function, so that

$$\nu(p,q) = \frac{1}{2}\left(r^2 - 1\right) = \frac{1}{2}\left(p^2 + q^2 - 1\right). \tag{8.6.1}$$

This is important, as we shall see in the next Chapter that $\mathbf{\Delta}[\nu] = N$ is the number operator associated with our representation of the canonical commutation relation. Of course, the number operator N is a positive operator, but ν is not a positive function, since it is negative at the origin.

Conversely, positivity of T does not guarantee that $\mathbf{\Delta}[T]$ is a positive operator, as can be seen by considering the positive function $T = (pq)^2$. By adapting an observation of Daubechies [38], we observe that the property of positivity can be lost even for very regular phase space functions. For example, let h_1 be (as usual) the order 1 Hermite-Gauss function (for the Schrödinger representation). Direct calculation gives

$$\left[\mathcal{G}(\overline{h_1} \otimes h_1)\right](p,q) = \frac{1}{\pi}(2p^2 + 2q^2 - 1)e^{-p^2-q^2}.$$

Suppose now that $T \in \mathcal{C}^\infty(\Pi)$ is nonzero and positive, with support lying inside the disc of radius $2^{-\frac{1}{2}}$ centered at the origin. Since $T \in \mathcal{S}(\Pi)$, the operator $\mathbf{\Delta}[T]$ is a (smooth) density matrix, and hence is trace class. However it is clear that $\langle h_1, \mathbf{\Delta}[T]h_1\rangle < 0$, showing that $\mathbf{\Delta}[T]$ is not positive as an operator. This example also shows us that the function $\Xi_{h_1,h_1} = 2\pi\,\mathcal{G}(\overline{h_1} \otimes h_1)$ is a nonpositive phase space function whose quantization $\mathbf{\Delta}[\Xi_{h_1,h_1}] = |\,h_1\,\rangle\langle\,h_1\,|$ is a bounded positive operator on $L^2(\mathbb{R})$.

It is therefore perhaps surprising that any connection can be made at all between the classical and quantum mechanical concepts of positivity, other than for the marginals. The strongest result known seems to be due, essentially, to Gårding (Folland, *ibid*).

Proposition 8.36 *If $0 \leqslant s < r < 1$, let T be a positive valued function in $\mathcal{S}^t_{r,s}$. Then T can be decomposed as $T = P + C$, with $\mathbf{\Delta}[P] \geqslant 0$ and $C \in \mathcal{S}^{t-r+s}_{r,s}$.*

The moral of this story is similar to the one concerning boundedness: positivity of phase space distributions and positivity of their quantizations are two very different concepts.

8.7 The Heisenberg Group And Quantization

8.7.1 *Representations Of The Heisenberg Group*

Recall the composition law for the projective representation W of the Weyl group in vector form,

$$W(\xi)W(\zeta) = W(\xi+\zeta)\,e^{\frac{1}{2}i\Omega(\xi,\zeta)}, \qquad \xi,\,\zeta \in \mathbb{R}^2, \qquad (3.3.2.a)$$

where Ω is the classical mechanical symplectic form

$$\Omega(\xi,\zeta) = \xi_1\zeta_2 - \xi_2\zeta_1, \qquad \xi,\,\zeta \in \mathbb{R}^2. \tag{3.3.2.b}$$

There is a standard mathematical procedure for extending this sort of projective representation of $\mathbb{R}^2$ to a true unitary representation of a larger group. In this case the resulting larger group, denoted $\mathfrak{H}$, is called the Heisenberg group. As a set, $\mathfrak{H}$ is equal to $\mathbb{R}^3$. Notationally, we shall regard $\mathfrak{H}$ as the Cartesian product of $\mathbb{R}^2$ and $\mathbb{R}$, writing elements of $\mathfrak{H}$ in the form (ξ, t), where $\xi \in \mathbb{R}^2$ and $t \in \mathbb{R}$. The group structure of $\mathfrak{H}$ is given by the formula

$$(\xi,s)\cdot(\zeta,t) = \big(\xi+\zeta, s+t+\tfrac{1}{2}\Omega(\xi,\zeta)\big), \qquad \xi,\,\zeta \in \mathbb{R}^2,\ s,\,t \in \mathbb{R}, \tag{8.7.1}$$

with $(\xi,s)^{-1} = (-\xi,-s)$, and identity $e = (0,0)$.

The centre of $\mathfrak{H}$ is

$$\mathfrak{Z} = \{\,(0,t) : t \in \mathbb{R}\,\},$$

which is also its derived group. The quotient group $\mathfrak{H}/\mathfrak{Z}$ is isomorphic to $\mathbb{R}^2$, which is sometimes referred to in this context as the reduced Heisenberg group.

The formula

$$\big[\widehat{W}(\xi,t)\phi\big](x) = e^{it}\,\big[W(\xi)\phi\big](x), \qquad \xi \in \mathbb{R}^2,\, t \in \mathbb{R},\, \phi \in L^2(\mathbb{R}), \tag{8.7.2}$$

defines a unitary representation $\widehat{W}$ of $\mathfrak{H}$ on $L^2(\mathbb{R})$, which is such that

$$W(\xi) = \widehat{W}(\xi,0), \qquad \xi \in \mathbb{R}^2$$

retrieves the projective representation W of $\mathbb{R}^2$, which is the whole point of this construction.

Since $\mathfrak{H}$ is a Lie group, it has a three dimensional Lie algebra $\mathfrak{h}$ with basis $X(1)$, $X(2)$ and $X(3)$, where

$$e^{tX(1)} = (t,0,0), \qquad e^{tX(2)} = (0,t,0), \qquad e^{tX(3)} = (0,0,t),$$

for $t \in \mathbb{R}$. From these formulæ it is immediate that the brackets defining the Lie algebra structure of $\mathfrak{h}$ are

$$\big[X(1), X(2)\big] = X(3), \qquad \big[X(1), X(3)\big] = \big[X(2), X(3)\big] = 0. \tag{8.7.3.a}$$

Identifying $\mathfrak{h}$ with $\mathbb{R}^3$ by $X(1) = (1,0,0)$, $X(2) = (0,1,0)$ and $X(3) = (0,0,1)$, an element of $\mathfrak{h}$ can be written in $\mathbb{R}^3$ form as (ξ, s), with $\xi \in \mathbb{R}^2$ and $s \in \mathbb{R}$. The $\mathfrak{h}$-bracket is then

$$\big[\,(\xi,s)\,,\,(\zeta,t)\,\big] \;=\; \big(0,\Omega(\xi,\zeta)\big), \qquad \xi,\,\zeta \in \mathbb{R}^2,\ s,\,t \in \mathbb{R}. \qquad (8.7.3.b)$$

The fundamental connection of this construction with physics, and (ultimately) the reason why the Heisenberg group is of such key importance to quantum mechanics is that the map

$$X(\xi,s) \;=\; i(\xi_1 P + \xi_2 Q + sI), \qquad \xi \in \mathbb{R}^2,\ s \in \mathbb{R}, \qquad (8.7.4)$$

provides a faithful Lie algebra homomorphism X from $\mathfrak{h}$ to $\mathcal{L}^+(\mathcal{S}(\mathbb{R}))$.

Harmonic analysis gives detailed information about the irreducible unitary representations of the Heisenberg group, and their relations to physics. The extension from $\mathbb{R}^2$ to $\mathfrak{H}$ makes it possible to create a one-parameter family of unitary representations of $\mathfrak{H}$, allowing Planck's constant to be reinterpreted as a representation parameter.

This goes as follows. For any $u \in \mathbb{R}\backslash\{0\}$, define a unitary representation $\widehat{W}_u$ of $\mathfrak{H}$ on $L^2(\mathbb{R})$ by the formula

$$\widehat{W}_u(\xi,s) \;=\; \widehat{W}(u\xi_1,\xi_2,us), \qquad \xi \in \mathbb{R}^2,\ s \in \mathbb{R}. \qquad (8.7.5)$$

The representation with $u = \hbar$ is then the one realized in nature.

The original uniqueness theorem of von Neumann did not consider the possibility of different values of $\hbar$. The following result (due to Stone and von Neumann) does so, and should therefore be considered as an addendum to the von Neumann uniqueness theorem.

Theorem 8.37 ***(Stone-von Neumann)*** *The representations $\widehat{W}_u$ of $\mathfrak{H}$ on $L^2(\mathbb{R})$ are unitarily inequivalent for distinct $u \in \mathbb{R} \setminus \{0\}$. Every irreducible unitary representation of $\mathfrak{H}$ is unitarily equivalent either to $\widehat{W}_u$ on $L^2(\mathbb{R})$ (for some $u \neq 0$) or to an element of the two-parameter family of one-dimensional representations*

$$\widehat{W}_{r,s}(\xi,s)\,\zeta \;=\; e^{i(r\xi_1+s\xi_2)}\,\zeta, \qquad \xi \in \mathbb{R}^2,\ s \in \mathbb{R},\ z \in \mathbb{C}, \qquad (8.7.6)$$

acting on the one dimensional space $\mathbb{C}$.

An arbitrary unitary representation of $\mathfrak{H}$ on a separable Hilbert space reduces completely to a direct sum of a countable collection of irreducible unitary representations of $\mathfrak{H}$, each of which will be of one of the above forms.

This theorem can be given a classical limit gloss. Think of a form of the classical limit as a sequence of these representations as u tends to 0 from the value $\hbar$. For all $u > 0$, no matter how small, the representation is a scaled ($\hbar \to u$) version of quantum mechanics. But at the limit $u = 0$, the algebraic structure of the theory discontinuously changes to that of classical mechanics. These observations are related to the interpretation of quantum mechanics from the point of view of deformations of algebras.

8.7.2 *The Metaplectic Representation*

In the discussion of classical mechanics, consideration of the symplectic group $\mathrm{Sp}\,(2;\mathbb{R})$ and the symplectic form Ω arose from the study of canonical transformations. It is reasonable, therefore, to ask how symplectic transformations of phase space will pull through quantization. The answer is that there is a natural unitary action of the symplectic group $\mathrm{Sp}\,(2;\mathbb{R})$ on $L^2(\Pi)$ given by the formula

$$\big[A \cdot F\big](p,q) = F(p_A, q_A), \qquad A \in \mathrm{Sp}\,(2;\mathbb{R}),\ F \in L^2(\Pi), \tag{8.7.7.a}$$

where

$$\begin{pmatrix} p_A \\ q_A \end{pmatrix} = A^{-1} \begin{pmatrix} p \\ q \end{pmatrix}. \tag{8.7.7.b}$$

Note that $A \cdot F \in \mathcal{S}(\Pi)$ whenever $A \in \mathrm{Sp}\,(2;\mathbb{R})$ and $F \in \mathcal{S}(\Pi)$, and any element $A \in \mathrm{Sp}\,(2;\mathbb{R})$ acts continuously on $\mathcal{S}(\Pi)$. This action can be adjointed to obtain an action of $\mathrm{Sp}\,(2;\mathbb{R})$ on $\mathcal{S}'(\Pi)$ via the formula

$$[\![\, A \circ T\,,\, F \,]\!] = [\![\, T\,,\, A^{-1} \cdot F \,]\!], \qquad A \in \mathrm{Sp}\,(2;\mathbb{R}),\ T \in \mathcal{S}'(\Pi),\ F \in \mathcal{S}(\Pi). \tag{8.7.8}$$

The main result which connects the symplectic group $\mathrm{Sp}\,(2;\mathbb{R})$ with quantization procedure is the following,

Theorem 8.38 *There exists a unitary group representation π of $\mathrm{Sp}\,(2;\mathbb{R})$ on $L^2(\mathbb{R})$, known as the metaplectic representation, such that*

$$A \cdot \mathcal{G}(\overline{g} \otimes f) = \mathcal{G}\big(\overline{\pi(A)g} \otimes \pi(A)f\big), \qquad A \in \mathrm{Sp}\,(2;\mathbb{R}),\ f,\, g \in L^2(\mathbb{R}). \tag{8.7.9}$$

Moreover, each map $\pi(A)$ maps $\mathcal{S}(\mathbb{R})$ continuously into itself.

A detailed proof of the existence of the metaplectic representation on abstract grounds can be found in Folland (*ibid*), but it is sufficient here simply to exhibit it. Just as Lorentz transformations are best understood by decomposing them into combinations of boosts, reflections and rotations, a general symplectic matrix can be written as a product of matrices of the following types,

$$M(c) = \begin{pmatrix} 1 & c \\ 0 & 1 \end{pmatrix}, \qquad D(a) = \begin{pmatrix} a & 0 \\ 0 & a^{-1} \end{pmatrix}, \tag{8.7.10.a}$$

for $c \in \mathbb{R}$ and $a \in \mathbb{R} \setminus \{0\}$, and the familiar

$$J = \begin{pmatrix} 0 & -1 \\ 1 & 0 \end{pmatrix}. \tag{8.7.10.b}$$

These matrices satisfy the relations

$$M(c)D(a) = D(a)M(a^{-2}c), \qquad D(a)J = JD(a^{-1}), \tag{8.7.10.c}$$

for all $c \in \mathbb{R}$ and $a \in \mathbb{R} \setminus \{0\}$. Therefore, the metaplectic representation π will have been defined when the unitary maps $\pi\big(M(c)\big)$, $\pi\big(D(a)\big)$ and $\pi(J)$ have been given for all appropriate values of the constants c and a, provided that the operator identities

$$\begin{aligned} \pi\big(M(c)\big)\pi\big(D(a)\big) &= \pi\big(D(a)\big)\pi\big(M(a^{-2}c)\big), \\ \pi\big(D(a)\big)\pi(J) &= \pi(J)\pi\big(D(a^{-1})\big) \end{aligned}$$

are satisfied for all appropriate a and c. This can be done by introducing the family of unitary maps $\{ E_a : a \neq 0 \}$, where

$$\big[E_a\phi\big](x) = \sqrt{a}\,\phi(ax), \qquad \phi \in L^2(\mathbb{R}),\ a \in \mathbb{R} \setminus \{0\}. \tag{8.7.11}$$

Direct calculation shows that:

$$\begin{aligned} \left[\mathcal{G}\left(\overline{e^{-\frac{1}{2}icQ^2}g} \otimes e^{-\frac{1}{2}icQ^2}f\right)\right](p,q) &= \big[\mathcal{G}(\overline{g} \otimes f)\big](p - cq, q) \\ &= \big[M(c) \cdot \mathcal{G}(\overline{g} \otimes f)\big](p,q), \\ \big[\mathcal{G}(\overline{E_a g} \otimes E_a f)\big](p,q) &= \big[\mathcal{G}(\overline{g} \otimes f)\big](a^{-1}p, aq) \\ &= \big[D(a) \cdot \mathcal{G}(\overline{g} \otimes f)\big](p,q), \end{aligned}$$

for any f, $g \in \mathcal{S}(\mathbb{R})$, and equation (8.3.16.a) gives

$$\big[\mathcal{G}(\overline{\mathcal{F}g} \otimes \mathcal{F}f)\big](p,q) = \big[\mathcal{G}(\overline{g} \otimes f)\big](q,-p) = \big[J \cdot \mathcal{G}(\overline{g} \otimes f)\big](p,q),$$

for any $f,\, g \in \mathcal{S}(\mathbb{R})$. To check the required relationships between these operators is elementary. This establishes the following result, which gives a concrete form of the metaplectic representation.

Proposition 8.39 *We have the following identities,*

$$\pi\big(M(c)\big) \;=\; e^{-\frac{1}{2}icQ^2}, \qquad \pi\big(D(a)\big) \;=\; E_a, \qquad \pi(J) \;=\; \mathcal{F}, \tag{8.7.12}$$

for any $c \in \mathbb{R}$ and $a \in \mathbb{R} \setminus \{0\}$.

Applying the metaplectic representation to Weyl quantization, it is clear that

$$[\![\, \mathbf{\Delta}[\, A \circ T\,]\, f \,,\, \overline{g} \,]\!] \;=\; [\![\, \mathbf{\Delta}[\,T\,]\, \pi(A)^{-1} f \,,\, \overline{\pi(A)^{-1} g} \,]\!], \tag{8.7.13}$$

for all $A \in \mathrm{Sp}\,(2;\mathbb{R})$, $T \in \mathcal{S}'(\Pi)$ and $f,\, g \in \mathcal{S}(\mathbb{R})$. When $\mathbf{\Delta}\,[\,T\,]$ is bounded, it follows that $\mathbf{\Delta}\,[\,A \circ T\,]$ is bounded for all $A \in \mathrm{Sp}\,(2;\mathbb{R})$, and that

$$\mathbf{\Delta}[\,A \circ T\,] \;=\; \pi(A)\, \mathbf{\Delta}[\,T\,]\, \pi(A)^{-1}. \tag{8.7.14}$$

It is worth noting that the matrices $M(c)^{\mathrm{T}}$, for $c \in \mathbb{R}$, also belong to $\mathrm{Sp}\,(2;\mathbb{R})$, since $M(c)^{\mathrm{T}} \,=\, J^{-1}M(-c)J$ for any $c \in \mathbb{R}$. Moreover, it is often convenient to decompose elements of $\mathrm{Sp}\,(2;\mathbb{R})$ in a way which explicitly involves these matrices. For example,

$$A \;=\; M(a^{-1}c)^{\mathrm{T}}\, D(a)\, M(a^{-1}b), \qquad\qquad A \;=\; \begin{pmatrix} a & b \\ c & d \end{pmatrix} \in \mathrm{Sp}\,(2;\mathbb{R}),$$

provided that $a \neq 0$. Consequently it is useful to have an explicit expression for $\pi\big(M(c)^{\mathrm{T}}\big)$, and it is clear that

$$\pi\big(M(c)^{\mathrm{T}}\big) \;=\; \mathcal{F}^{-1}\, e^{\frac{1}{2}icQ^2}\, \mathcal{F} \;=\; e^{\frac{1}{2}icP^2}, \qquad c \in \mathbb{R}, \tag{8.7.15}$$

which, curiously, is the (time reversed) dynamical one parameter unitary group describing the time evolution of the one dimensional free particle.

The representation π can be differentiated, leading to the metaplectic representation of the Lie algebra $\mathrm{sp}(2;\mathbb{R})$. The type of operator that can be obtained through this procedure is indicated by the following example, since

$$\frac{\partial}{\partial a}\,\big[\pi\big(D(a)\big)f\big](x)\bigg|_{a=1} \;=\; \frac{1}{2}f(x) \,+\, xf'(x), \qquad f \in \mathcal{S}(\mathbb{R}),$$

and this gives an interesting new interpretation to the operator $\frac{1}{2}I \,+\, iQP$.

8.8 Additional Reading

For treatments of pseudo-differential operator theory touching on quantization, see Folland [63], Trèves [224] or Hörmander [120], for example. Integral kernels first appeared in the work of Schmidt [202] and Hilbert [117] on integral equations.

A few references for basic Hilbert space operator theory are Akhiezer [3]; Dunford & Schwartz [55]; Gohberg, Goldberg & Kaashoek [81]; Reed & Simon [186].

Part II

Quantization And Phase

Chapter 9

Quantization In Polar Coordinates

It has long been an axiom of mine that the little things are infinitely the most important.

– Sir Arthur Conan Doyle, *A Case Of Identity.*

9.1 Introduction

This Chapter investigates the theory of (Weyl) quantization in polar coordinates, in that it studies the properties of quantizations of distributions that depend on either the polar radius or the polar angle alone.

This problem of polar quantization has its roots in certain physical questions. For example, Klauder has used radial quantization in a problem concerning radar [139]. In the context of this book, the quantization of the phase plane angle function itself provides a proposal for a phase operator. More generally, in quantum mechanics the number operator N is certainly one of the more important observables, and $N = \mathbf{\Delta}[\nu]$, where

$$\nu = \frac{1}{2}(p^2 + q^2 - 1)$$

is a radial distribution. In any event, as a mathematical enterprise, polar quantization provides a number of interesting challenges. Some of these have been answered; others would undoubtedly yield if pushed.

Radial and angular quantization are quite different in nature and so will be considered separately. Radial quantization is much the easier of the two, and it is generally a straightforward matter to recognize when a Hilbert space operator has a radial phase space symbol.[1] Either as a direct study or in the course of a larger study, radial quantization has been considered in a number of works, and references to these will be found in Folland [63]. The basic technical material about angular quantization is

[1] Recall that T is the symbol, or dequantization, of $\mathbf{\Delta}[T]$.

due to the authors [212, 213, 53]. The idea of quantizing the angle function was independently suggested by Royer, who also considered this operator in non-Weyl variants of quantization [197, 198].

Before proceeding, it is important to establish conventions concerning phase space coordinate systems. The angle function on phase space can only be defined on a cut plane, and so a particular cut for Π must be chosen. The convention in this book is that the polar coordinates in the plane Π are the variables r and β defined by

$$p = r\cos\beta, \qquad q = r\sin\beta, \tag{9.1.1.a}$$

(note that the p-axis is the abscissa) together with the assumptions $r \geqslant 0$ and $-\pi < \beta \leqslant \pi$. In other words, Π is to be cut along the negative p-axis. In complex form,

$$p + iq = re^{i\beta}. \tag{9.1.1.b}$$

The angle function is just that, a function on phase space (which will be interpreted as a tempered distribution) which assigns to each point (p, q) in Π its associated polar angle β. The symbol φ is reserved for this function, so that

$$\varphi(p, q) = \beta, \tag{9.1.2}$$

on the cut plane. Note that φ is discontinuous across the cut at the negative p-axis.

If the plane were cut at a different inclination, the various formulæ change. But the physics is essentially the same, since the resultant quantized angle operator differs from the one described here by a (unitary) gauge transformation plus an additive constant [54].

9.2 The Hermite-Gauss Functions

9.2.1 *Generating Functions*

Many calculations here and later involve the Hermite-Gauss functions h_n of the Schrödinger representation. Often, the simplest way to do them is to use the generating function given in Proposition 5.1 of Chapter 5,

$$G_t(x) = \sum_{k=0}^{\infty} \frac{t^k}{\sqrt{2^k\, k!}}\, h_k(x) = \pi^{-\frac{1}{4}} \exp\left(-\frac{1}{4}t^2 + xt - \frac{1}{2}x^2\right), \tag{5.2.8}$$

where t is a real parameter. A particular advantage of working with G_t is that it is a Gaussian function, so the evaluation of integrals and Fourier transforms involving it are straightforward — and will therefore be performed without comment.

More importantly, the above series for G_t converges in the locally convex topology of $\mathcal{S}(\mathbb{R})$, and the associated doubly infinite series for $\overline{G_s} \otimes G_t$ in terms of the functions $h_m \otimes h_n$ converges in the topology of $\mathcal{S}(\mathbb{R}^2)$. Thus the identity

$$\begin{aligned} [\![\mathbf{\Delta}[T] G_t \,,\, \overline{G_s}]\!] &= [\![T \,,\, \mathcal{G}(\overline{G_s} \otimes G_t)]\!] \\ &= \sum_{m,n \geqslant 0} \frac{s^m t^n}{\sqrt{2^{m+n} m! n!}} [\![\mathbf{\Delta}[T] h_n \,,\, h_m]\!] \end{aligned} \tag{9.2.1}$$

is valid for any $T \in \mathcal{S}'(\Pi)$, and hence the matrix coefficients of the quantization $\mathbf{\Delta}[T]$ of any $T \in \mathcal{S}'(\Pi)$ can be determined from the Taylor series expansion of the function $[\![T \,,\, \mathcal{G}(\overline{G_s} \otimes G_t)]\!]$. Consequently, evaluation of the function $\mathcal{G}(\overline{G_s} \otimes G_t)$ is the first step in any analysis.

Proposition 9.1 *The Wigner transform of the function $\overline{G_s} \otimes G_t$ is given by the formula*

$$\big[\mathcal{G}(\overline{G_s} \otimes G_t)\big](p,q) = \frac{1}{\pi} \exp\Big[-(p^2+q^2)+(q+ip)s+(q-ip)t-\frac{1}{2}st\Big] \tag{9.2.2}$$

for any s, $t \in \mathbb{R}$.

Proof: From the definition of the Wigner transform it is clear that

$$\big[\mathcal{G}(\overline{G_s} \otimes G_t)\big](p,q) = \frac{1}{2}\pi^{-\frac{3}{2}} \int_{\mathbb{R}} e^{\Phi(s,t;p,q;u)}\, du,$$

where Φ is the function

$$\Phi(s,t;p,q;u) = -\frac{1}{4}(s^2+t^2) + (s+t)q + \frac{1}{2}(s-t+2ip)u - q^2 - \frac{1}{4}u^2.$$

Completing the square in u, this integral can be calculated, leading to equation (9.2.2). ∎

9.2.2 *Partial Polar Integrals*

For any $T \in \mathcal{S}'(\Pi)$ and f, $g \in \mathcal{S}(\mathbb{R})$, to calculate the matrix coefficient $[\![\mathbf{\Delta}[T] f \,,\, \overline{g}]\!]$ essentially requires integrating the function $\mathcal{G}(\overline{g} \otimes f)$ against the distribution T over all phase space. If the distribution T depends upon one of the polar variables only, then the integration with respect to the other polar variable will take place independently of the value of the distribution T. Consequently it is useful to know the integral of the function $\mathcal{G}(\overline{G_s} \otimes G_t)$ with respect to each of the polar variables. The first of these calculations is easy.

Lemma 9.2 *For any $s,\, t \in \mathbb{R}$ and $r > 0$, we have*

$$\begin{aligned}\int_{-\pi}^{\pi} \big[\mathcal{G}(\overline{G_s} \otimes G_t)\big](r\cos\beta, r\sin\beta)\, d\beta \\ &= 2e^{-\frac{1}{2}st} e^{-r^2} I_0(2r\sqrt{st}) \\ &= 2e^{-r^2} \sum_{n \geqslant 0} \frac{(-st)^n}{2^n\, n!} L_n(2r^2), \qquad (9.2.3)\end{aligned}$$

where I_0 is the modified Bessel function (of the first kind) of order 0, *and L_n is the nth Laguerre polynomial.*

Proof: Writing (9.2.2) in polar coordinates and expanding, the first of these two identities is elementary, since the desired integral is equal to

$$\begin{aligned}&\frac{1}{\pi} e^{-\frac{1}{2}st} e^{-r^2} \int_{-\pi}^{\pi} e^{ir(se^{-i\beta} - te^{i\beta})}\, d\beta \\ &= \frac{1}{\pi} e^{-\frac{1}{2}st} e^{-r^2} \sum_{n \geqslant 0} \frac{i^n r^n}{n!} \int_{-\pi}^{\pi} \left(se^{-i\beta} - te^{i\beta}\right)^n d\beta \\ &= 2e^{-\frac{1}{2}st} e^{-r^2} \sum_{n \geqslant 0} \frac{r^{2n}}{(n!)^2} s^n t^n \;=\; 2e^{-\frac{1}{2}st} e^{-r^2} I_0(2r\sqrt{st}).\end{aligned}$$

The second identity now follows by expanding $e^{-\frac{1}{2}st}$ as a power series and reordering terms. ■

Performing the integral with respect to the radial coordinate is significantly more complex, and requires a lengthy proof. The coefficients $g_{m,n}$ introduced in the next definition characterize angular quantization. The symbols $g_{m,n}$ and $s(m,n)$ will retain this fixed significance throughout the book.

Definition 9.3 *For any $m, n \geqslant 0$ the coefficient $g_{m,n}$ is defined by the formula*

$$g_{m,n} = \sqrt{\frac{\max(m,n)!}{\min(m,n)!}}\, 2^{-\frac{1}{2}|m-n|}\, \frac{\Gamma\big(\frac{1}{2}\min(m,n) + s(m,n)\big)}{\Gamma\big(\frac{1}{2}\max(m,n) + s(m,n)\big)}, \qquad (9.2.4.a)$$

where Γ is the usual Gamma function, and the coefficients $s(m,n)$ are

$$s(m,n) = \begin{cases} \frac{1}{2} & \min(m,n) \text{ even}, \\ 1 & \min(m,n) \text{ odd}, \end{cases} \qquad (9.2.4.b)$$

for any m, $n \geqslant 0$.

The $g_{m,n}$ are quite complicated, and not easy to handle analytically. Note, however, that they are symmetric, $g_{m,n} = g_{n,m}$, and are equal to unity on the main diagonal, $g_{n,n} = 1$. The complexity of the $g_{m,n}$ is due to the presence of the coefficients $s(m,n)$, which introduce different (asymptotic) behaviour in $g_{m,n}$ according as $\min(m,n)$ is even or odd. Those who enjoy computer mathematics are invited to investigate this behaviour.

For the present, it is sufficient to have comparatively simple controls on the behaviour of these coefficients, and Stirling's formula provides the fundamental inequality.

Lemma 9.4 *A constant $C > 1$ can be found such that*

$$C^{-1}\Big(\frac{\min(m,n)}{\max(m,n)}\Big)^{\frac{1}{4}} \leqslant g_{m,n} \leqslant C\Big(\frac{\max(m,n)}{\min(m,n)}\Big)^{\frac{1}{4}}, \qquad m,\, n \geqslant 0. \quad (9.2.5)$$

Proof: Introducing the coefficients $\xi_{n;j}$ defined by

$$\xi_{n;j} = 2^{\frac{1}{2}n}\, \Gamma\big(\tfrac{1}{2}n + j\big)\, \frac{1}{\sqrt{n!}},$$

where $n \geqslant 0$ and j is equal to either $\frac{1}{2}$ or 1, $g_{m,n}$ can be written in the form

$$g_{m,n} = \frac{\xi_{\min(m,n);s(m,n)}}{\xi_{\max(m,n);s(m,n)}}, \qquad m,\, n \geqslant 0.$$

From Stirling's formula,

$$\xi_{n;j} \sim (n+1)^{j-\frac{3}{4}}, \qquad n \to \infty,$$

for any value of j. Hence there are constants $0 < A < 1 < B$ such that

$$A(n+1)^{-\frac{1}{4}} \leqslant \xi_{n;\frac{1}{2}} \leqslant B(n+1)^{-\frac{1}{4}}$$
$$A(n+1)^{\frac{1}{4}} \leqslant \xi_{n;1} \leqslant B(n+1)^{\frac{1}{4}}$$

for all $n \geqslant 0$. The result now follows by putting $C = BA^{-1}$. ■

After a bit of work involving Gamma functions, the integral over the radial variable of $\mathcal{G}(\overline{G_s} \otimes G_t)$ can be expressed as a power series in s and t.

Lemma 9.5 *The identity*

$$\int_0^\infty \left[\mathcal{G}(\overline{G_s} \otimes G_t)\right](r\cos\beta, r\sin\beta)\, r\, dr$$
$$= \frac{1}{2\pi} \sum_{m,n\geqslant 0} \frac{i^{m-n}}{\sqrt{2^{m+n}\, m!\, n!}}\, g_{m,n}\, s^m t^n\, e^{i(n-m)\beta} \quad (9.2.6)$$

holds for any $s,\, t \in \mathbb{R}$ *and* $-\pi < \beta \leqslant \pi$.

Proof: Writing $\mathcal{G}(\overline{G_s} \otimes G_t)$ out in polar coordinates, expanding one of the constituent exponentials as a power series, and bringing this power series summation out through the integral (this procedure can be justified analytically), the desired integral is

$$\frac{1}{\pi} e^{-\frac{1}{2}st} \sum_{n\geqslant 0} \frac{i^n}{n!} \left(se^{-i\beta} - te^{i\beta}\right)^n \int_0^\infty e^{-r^2}\, r^{n+1}\, dr$$
$$= \frac{1}{2\pi} e^{-\frac{1}{2}st} \sum_{n\geqslant 0} \frac{i^n}{n!} \Gamma\left(\tfrac{1}{2}n+1\right) \left(se^{-i\beta} - te^{i\beta}\right)^n,$$

which is the same as

$$\frac{1}{2\pi} \sum_{j,k,m\geqslant 0} \left(-\tfrac{1}{2}\right)^m \frac{i^{j-k}}{j!\,k!\,m!} \Gamma\left(\tfrac{1}{2}j + \tfrac{1}{2}k + 1\right) s^{j+m} t^{k+m} e^{i(k-j)\beta},$$

after expanding all expressions involving s and t. Reordering this into a power series in s and t gives

$$\frac{1}{2\pi} \sum_{m,n\geqslant 0} i^{m-n} A_{m,n} s^m t^n e^{i(n-m)\beta},$$

where the $A_{m,n}$ are given for any $m,\ n \geqslant 0$ by the formula

$$A_{m,n} = \sum_{j=0}^{\min(m,n)} \frac{\Gamma\left(\frac{1}{2}m + \frac{1}{2}n + 1 - j\right)}{j!\,(m-j)!\,(n-j)!}\left(-\tfrac{1}{2}\right)^j.$$

The task now is to establish that $\sqrt{2^{m+n}\,m!\,n!}\,A_{m,n} = g_{m,n}$ for all $m,\ n \geqslant 0$. Inspection shows that this is true when $m = n$, and since $A_{m,n} = A_{n,m}$ for all $m,\ n \geqslant 0$, it remains to prove this identity for $m < n$. We shall need to make use of the Beta function

$$B(x,y) = 2\int_0^{\frac{1}{2}\pi} \sin^{2x-1}\beta\,\cos^{2y-1}\beta\,d\beta, \qquad x,\ y > 0,$$

which is related to the Gamma function by the identity

$$\Gamma(x)\,\Gamma(y) = \Gamma(x+y)\,B(x,y), \qquad x,\ y > 0.$$

Then, applying the Binomial Theorem and the change of variable $\gamma = 2\beta$,

$$\begin{aligned}
&m!\,\Gamma\left(\tfrac{1}{2}n - \tfrac{1}{2}m\right)A_{m,n}\\
&\quad= \sum_{j=0}^{m}\binom{m}{j}\left(-\tfrac{1}{2}\right)^j B\left(\tfrac{1}{2}n - \tfrac{1}{2}m,\ \tfrac{1}{2}n + \tfrac{1}{2}m + 1 - j\right)\\
&\quad= 2\sum_{j=0}^{m}\binom{m}{j}\left(-\tfrac{1}{2}\right)^j \int_0^{\frac{1}{2}\pi} \sin^{n-m-1}\beta\,\cos^{n+m+1-2j}\beta\,d\beta\\
&\quad= 2^{1-m}\int_0^{\frac{1}{2}\pi} \sin^{n-m-1}\beta\,\cos^{n-m+1}\beta\,\cos^m 2\beta\,d\beta\\
&\quad= 2^{-n}\int_0^{\pi} \sin^{n-m-1}\gamma\left[\cos^m\gamma + \cos^{m+1}\gamma\right]d\gamma\,.
\end{aligned}$$

Using the fact that the cosine function is an odd function about $\frac{1}{2}\pi$, this expression is equal to

$$\begin{aligned}
&2^{-n}\int_0^{\pi} \sin^{n-m-1}\gamma\,\cos^{m+2s(m,n)-1}\gamma\,d\gamma\\
&\qquad= 2^{-n}B\left(\tfrac{1}{2}n - \tfrac{1}{2}m,\ \tfrac{1}{2}m + s(m,n)\right)\\
&\qquad= 2^{-n}\,\Gamma\left(\tfrac{1}{2}n - \tfrac{1}{2}m\right)\frac{\Gamma\left(\frac{1}{2}m + s(m,n)\right)}{\Gamma\left(\frac{1}{2}n + s(m,n)\right)}
\end{aligned}$$

$$= \quad 2^{-\frac{1}{2}(m+n)}\sqrt{\frac{m!}{n!}}\,\Gamma\left(\tfrac{1}{2}n - \tfrac{1}{2}m\right) g_{m,n},$$

which shows that $\sqrt{2^{m+n}\,m!\,n!}\,A_{m,n} = g_{m,n}$, as required. ■

9.3 Radial Quantization

After the preliminary work of Lemma 9.2, the next step in radial quantization using the Weyl scheme is to decide what is meant by a distribution which is a "function of the radius". Because distributions are defined weakly, the notion of a radial distribution must be defined in a similarly weak manner.

There is a natural action of the rotation group $SO(2)$ on $\mathcal{S}(\Pi)$, and an action of $SO(2)$ on $\mathcal{S}'(\Pi)$. It is natural to consider elements of $\mathcal{S}'(\Pi)$ which are invariant under this action as radial distributions. The collection of radial distributions forms a closed linear subspace of $\mathcal{S}'(\Pi)$, being the image of a continuous projection on $\mathcal{S}'(\Pi)$ obtained by averaging over the group action of $SO(2)$.

Having defined radial distributions in this manner, it will be possible to find practical characterizations of such quantities, and analyze their quantizations.

9.3.1 *Radial Distributions*

As mentioned above, the two-dimensional rotation group $\mathrm{SO}(2)$ is a subgroup of the symplectic group $\mathrm{Sp}\,(2;\mathbb{R})$, and hence acts on both $\mathcal{S}(\Pi)$ and $\mathcal{S}'(\Pi)$ in the manner described in Chapter 8. Consequently, we make the following definition.

Definition 9.6 *A* radial distribution *is a distribution $T \in \mathcal{S}'(\Pi)$ such that $A \circ T = T$ for any $A \in SO(2)$. The space of radial distributions will be denoted $\mathcal{S}'_{\mathrm{rad}}(\Pi)$.*

Averaging over the action of $SO(2)$ on $\mathcal{S}(\Pi)$ yields the continuous linear endomorphism Σ of $\mathcal{S}(\Pi)$ defined on $F \in \mathcal{S}(\Pi)$ by

$$[\Sigma\, F](p,q) \;=\; \frac{1}{2\pi}\int_{-\pi}^{\pi} F(p\cos\beta + q\sin\beta, -p\sin\beta + q\cos\beta)\,d\beta\,. \quad (9.3.1.a)$$

Since it is clear that $\Sigma(A \cdot F) = A \cdot \Sigma F = \Sigma F$ for any $A \in SO(2)$ and $F \in \mathcal{S}(\Pi)$, it follows that the map Σ is a projection. Its image $\mathcal{S}_{\mathrm{rad}}(\Pi)$ is a closed linear subspace of $\mathcal{S}(\Pi)$, consisting of those Schwartz functions which are functions of the radius alone, in that

$$(\Sigma F)(r\cos\beta, r\sin\beta) = (\Sigma F)(r, 0), \qquad F \in \mathcal{S}(\Pi), \qquad (9.3.1.b)$$

for all $r > 0$ and $-\pi < \beta \leqslant \pi$. Transposing Σ leads to the space of distributions promised above. The proof of the following result is elementary, and will be omitted.

Proposition 9.7 *The space of radial distributions $\mathcal{S}'_{\mathrm{rad}}(\Pi)$ is the image of the continuous projection Σ^{tr} of $\mathcal{S}'(\Pi)$.*

Having defined the spaces $\mathcal{S}_{\mathrm{rad}}(\Pi)$ and $\mathcal{S}'_{\mathrm{rad}}(\Pi)$, it is necessary to study their structure and properties. To begin with, the space $\mathcal{S}_{\mathrm{rad}}(\Pi)$ of radial test functions possesses an orthogonal Schauder basis, consisting of the functions $\{\Phi_{m,m} : m \geqslant 0\}$ defined by the formula

$$\Phi_{m,m}(p,q) = 2(-1)^m e^{-r^2} L_m(2r^2), \qquad m \geqslant 0. \qquad (9.3.2)$$

This double index notation looks rather cumbrous, but it is there for a reason. The functions $\Phi_{m,m}$ comprise a subcollection of the so-called special Hermite functions studied in detail in Chapter 12, and presage a deep connection between Hermite and Laguerre functions. This connection has been known for a long time in terms of special functions, but finds its natural expression in terms of the Heisenberg group, mediated by Weyl quantization. Leaving the analysis to Chapter 12, the next Proposition gives the results needed here.

Proposition 9.8 *The collection $\{\Phi_{m,m} : m \geqslant 0\}$ is a Schauder basis for the closed linear subspace $\mathcal{S}_{\mathrm{rad}}(\Pi)$ of $\mathcal{S}(\Pi)$, and the series*

$$\sum_{m \geqslant 0} \xi_m \, \Phi_{m,m}$$

converges in $\mathcal{S}_{\mathrm{rad}}(\Pi)$ if and only if the sequence of coefficients $(\xi_m)_{m \geqslant 0}$ is rapidly decreasing (belongs to $\mathfrak{s}$). This identification between $\mathcal{S}_{\mathrm{rad}}(\Pi)$ and $\mathfrak{s}$ is a topological isomorphism. Moreover, the functions $\Phi_{m,m}$ are orthogonal, with

$$\langle \Phi_{m,m}, \Phi_{n,n} \rangle = 2\pi \, \delta_{mn}, \qquad m, n \geqslant 0. \qquad (9.3.3)$$

Standard results from topological vector space theory show that the topological dual of the subspace $\mathcal{S}_{\text{rad}}(\Pi)$ of $\mathcal{S}(\Pi)$ can be identified[2] with the subspace $\mathcal{S}'_{\text{rad}}(\Pi) = \Sigma^{\text{tr}}\mathcal{S}'(\Pi)$ of $\mathcal{S}'(\Pi)$. This observation has the following consequence.

Corollary 9.9 *The collection* $\{\phi_{m,m} : m \geqslant 0\}$ *is a Schauder basis for* $\mathcal{S}'_{\text{rad}}(\Pi)$*, and the series*

$$\sum_{m\geqslant 0} t_m\, \phi_{m,m}$$

converges in $\mathcal{S}'_{\text{rad}}(\Pi)$ *if and only if the sequence of coefficients* $(t_m)_{m\geqslant 0}$ *is of slow increase (belongs to* $\mathfrak{s}'$*).*

Another characterization of $\mathcal{S}'_{\text{rad}}(\Pi)$ of a more functional nature can be obtained in the following manner. The functions $\{\mathcal{L}_m : m \geqslant 0\}$ form an orthonormal basis for $L^2[0,\infty)$, where

$$\mathcal{L}_m(u) = \sqrt{2}L_m(2u)e^{-u}, \qquad m \geqslant 0. \tag{9.3.4}$$

Denote the finite linear span of these basis elements by $\mathcal{D}$, and consider the symmetric unbounded linear operator $H : \mathcal{D} \to L^2[0,\infty)$ given by

$$(Hf)(u) = -uf''(u) - f'(u) + uf(u), \qquad f \in \mathcal{D}. \tag{9.3.5.a}$$

From the standard properties of the Laguerre functions it can be shown that

$$H\mathcal{L}_m = (2m+1)\,\mathcal{L}_m, \qquad m \geqslant 0. \tag{9.3.5.b}$$

Thus H has a complete orthonormal set of eigenvectors, and can be shown to be essentially self-adjoint on $\mathcal{D}$. Denoting its closure by $\overline{H}$, this operator can be used to construct a space in the same way that $\mathcal{S}(\mathbb{R})$ was constructed from the number operator, (4.2.3),

$$\mathcal{S}[0,\infty) = \mathcal{D}^{\infty}(\overline{H}) = \bigcap_{n\geqslant 0} \mathcal{D}(\overline{H}^n). \tag{9.3.6}$$

[2] This identification is achieved by reinterpreting the continuous projection Σ from $\mathcal{S}(\Pi)$ to $\mathcal{S}(\Pi)$ as a continuous linear surjection $\widetilde{\Sigma} : \mathcal{S}(\Pi) \to \mathcal{S}_{\text{rad}}(\Pi)$, for then the transpose $\widetilde{\Sigma}^{\text{tr}}$ is a continuous linear injection from the dual of $\mathcal{S}_{\text{rad}}(\Pi)$ into $\mathcal{S}'(\Pi)$ whose image is $\mathcal{S}'_{\text{rad}}(\Pi)$.

This space is a dense linear subspace of $L^2[0,\infty)$ which contains $\mathcal{D}$ and consists precisely of those functions f on $[0,\infty)$ having an expansion of the form

$$f = \sum_{m \geqslant 0} a_m \mathcal{L}_m, \tag{9.3.7}$$

with $(a_m)_{m\geqslant 0} \in \mathfrak{s}$. Hence it is isomorphic as a nuclear Fréchet locally convex space to $\mathfrak{s}$. Elements of $\mathcal{S}[0,\infty)$ are polynomially bounded smooth functions on $[0,\infty)$. Elements of $\mathcal{S}'[0,\infty)$, the space of continuous linear functionals on $\mathcal{S}[0,\infty)$, can be represented through series of the form

$$R = \sum_{m \geqslant 0} r_m \mathcal{L}_m, \qquad (r_m)_{m\geqslant 0} \in \mathfrak{s}'. \tag{9.3.8}$$

Then the map $\mathcal{K} : \mathcal{S}_{\text{rad}}(\Pi) \to \mathcal{S}[0,\infty)$ defined by the formula

$$[\mathcal{K}F](u) = F(\sqrt{u}, 0), \qquad F \in \mathcal{S}_{\text{rad}}(\Pi), \tag{9.3.9.a}$$

is a bicontinuous linear bijection such that

$$\mathcal{K}\phi_{m,m} = (-1)^m \sqrt{2}\,\mathcal{L}_m, \qquad m \geqslant 0. \tag{9.3.9.b}$$

Consequently the map $\mathcal{K}^{\text{tr}}$ is a linear bijection from $\mathcal{S}'[0,\infty)$ to the topological dual of $\mathcal{S}_{\text{rad}}(\Pi)$, and hence defines a linear bijection $\widetilde{\Sigma}^{\text{tr}}\mathcal{K}^{\text{tr}}$ from $\mathcal{S}'[0,\infty)$ to the space $\mathcal{S}'_{\text{rad}}(\Pi)$ of radial distributions.

In other words, any radial distribution can be obtained by choosing a distribution $R \in \mathcal{S}'[0,\infty)$. Given a test function $F \in \mathcal{S}(\Pi)$, apply the projection Σ, obtaining what is essentially a function of one variable (the radius). Applying R to this function of one variable gives the value that the radial distribution $\widetilde{\Sigma}^{\text{tr}}\mathcal{K}^{\text{tr}}R$ derived from R takes[3] on the test function F.

---

[3]It should be noted that it is not necessary to introduce the square root in the definition of the operator $\mathcal{K}$ — we have chosen to do so in order to simplify the definition of the space $\mathcal{S}[0,\infty)$ somewhat, since the need for an element of $\mathcal{S}_{\text{rad}}(\Pi)$ to be differentiable at the origin forces it to be a smooth function of the square of the radius, rather than just of the radius.

9.3.2 *Quantizing Radial Distributions*

Using the special Hermite functions introduced above, Lemma 9.2 now states that

$$\Sigma\big(\mathcal{G}(\overline{G_s} \otimes G_t)\big) \;=\; \frac{1}{2\pi}\sum_{n\geqslant 0}\frac{s^n t^n}{2^n\, n!}\,\phi_{n,n}, \qquad s,\, t \in \mathbb{R}, \tag{9.3.10}$$

and so, for a radial distribution $T \in \mathcal{S}'_{\mathrm{rad}}(\Pi)$,

$$[\![T\,,\, \mathcal{G}(\overline{G_s} \otimes G_t)]\!] \;=\; [\![T\,,\, \Sigma\big(\mathcal{G}(\overline{G_s} \otimes G_t)\big)]\!] \;=\; \frac{1}{2\pi}\sum_{n\geqslant 0}\frac{s^n t^n}{2^n\, n!}[\![T\,,\, \phi_{n,n}]\!]$$

for all $s,\, t \in \mathbb{R}$. This in turn implies that

$$[\![\mathbf{\Delta}\,[T]\, h_n\,,\, h_m]\!] \;=\; \frac{1}{2\pi}[\![T\,,\, \phi_{n,n}]\!]\, \delta_{mn}, \qquad m,\, n \geqslant 0.$$

In other words, $\mathbf{\Delta}\,[T] \in \mathcal{L}(\mathcal{S}(\mathbb{R}), \mathcal{S}'(\mathbb{R}))$ is the diagonal continuous linear map such that

$$\mathbf{\Delta}\,[T]\; h_n \;=\; \frac{1}{2\pi}\,[\![T\,,\, \phi_{n,n}]\!]\, h_n, \qquad n \geqslant 0. \tag{9.3.11}$$

But the sequence $\big([\![T\,,\, \phi_{n,n}]\!]\big)_{n\geqslant 0}$ belongs to $\mathfrak{s}'$, and so $\mathbf{\Delta}\,[T]\,f$ belongs to $\mathcal{S}(\mathbb{R})$ for all $f \in \mathcal{S}(\mathbb{R})$. Thus $\mathbf{\Delta}\,[T] : \mathcal{S}(\mathbb{R}) \to \mathcal{S}(\mathbb{R})$ is both an unbounded linear operator on $L^2(\mathbb{R})$ and a continuous endomorphism of $\mathcal{S}(\mathbb{R})$.

The subspace $\mathcal{S}'_{\mathrm{rad}}(\Pi)$ is invariant under the involution of $\mathcal{S}'(\Pi)$ defined in equation (8.3.20), so

$$\langle \mathbf{\Delta}\,[T]\,f\,,\, g\rangle \;=\; \big\langle f\,,\, \mathbf{\Delta}\,[\,\overline{T}\,]\, g\big\rangle\,, \qquad T \in \mathcal{S}'_{\mathrm{rad}}(\Pi),\; f,\, g \in \mathcal{S}(\mathbb{R}), \tag{9.3.12}$$

and hence $\mathbf{\Delta}\,[T] \in \mathcal{L}^+(\mathcal{S}(\mathbb{R}))$ belongs to the algebra of smooth observables for any radial distribution T.

Moreover, $\mathbf{\Delta}\,[T]$ can be written in terms of the number operator, as follows. Given the radial distribution T, consider the function F_T on $\mathbb{N}\cup\{0\}$ defined by the formula

$$F_T(n) \;=\; \frac{1}{2\pi}[\![T\,,\, \phi_{n,n}]\!], \qquad n \geqslant 0\,. \tag{9.3.13.a}$$

Then it is clear from (9.3.11) that

$$\mathbf{\Delta}\,[T] \;=\; F_T(\overline{N}), \tag{9.3.13.b}$$

at least when both of these functions are restricted to $\mathcal{S}(\mathbb{R})$. Summarizing these results,

Theorem 9.10 *For any radial distribution $T \in \mathcal{S}'_{\mathrm{rad}}(\Pi)$, its quantization $\mathbf{\Delta}[T]$ belongs to $\mathcal{L}^+(\mathcal{S}(\mathbb{R}))$, with $\mathbf{\Delta}[T] = F_T(\overline{N})$, so that*

$$\mathbf{\Delta}[T]\,h_n \;=\; F_T(\overline{N})h_n \;=\; \frac{1}{2\pi}[\![T\,,\,\phi_{n,n}]\!]\,h_n, \qquad n \geqslant 0. \tag{9.3.14}$$

Equivalently, on $\mathcal{S}(\mathbb{R})$,

$$\mathbf{\Delta}[T] = \frac{1}{2\pi}\sum_{n\geqslant 0}[\![T\,,\,\phi_{n,n}]\!]\,P_n, \tag{9.3.15}$$

where P_n is the projection operator along h_n.

Thus the quantization of any radial distribution is an operator of a particularly simple type, and the analysis of such operators presents no particular difficulties.

Example 9.11 (Polynomially Bounded Distributions) For one variable, the radial test functions have been identified as polynomially bounded and continuous functions. It follows that amongst the radial distributions are those obtainable from functions $\mathfrak{f} \in \mathcal{O}^\infty(\mathbb{R})$ in the following way. Given $\mathfrak{f}$, form the function $\mathfrak{f}_{\mathrm{rad}} \in \mathcal{O}^\infty(\Pi)$ by setting

$$\mathfrak{f}_{\mathrm{rad}}(p,q) \;=\; \mathfrak{f}(\sqrt{p^2+q^2}), \qquad p,\,q \in \mathbb{R}. \tag{9.3.16}$$

Writing $\mathfrak{f}_{\mathrm{rad}} \;=\; \widetilde{\Sigma}^{\mathrm{tr}}\mathcal{K}^{\mathrm{tr}}\mathfrak{g}$, where $\mathfrak{g} \in \mathcal{S}'[0,\infty)$ is the function

$$\mathfrak{g}(u) \;=\; \pi\,\mathfrak{f}(\sqrt{u}), \qquad u \geqslant 0, \tag{9.3.17}$$

it is clear that $\mathfrak{f}_{\mathrm{rad}} \in \mathcal{S}'_{\mathrm{rad}}(\Pi)$ is a radial distribution.

From equation (9.3.14) it follows that $\mathbf{\Delta}[\mathfrak{f}_{\mathrm{rad}}]$ is diagonal with respect to the Hermite-Gauss functions, with

$$\mathbf{\Delta}[\mathfrak{f}_{\mathrm{rad}}]\,h_n \;=\; \rho_n(\mathfrak{f})\,h_n, \qquad n \geqslant 0, \tag{9.3.18.a}$$

where the eigenvalues $\rho_n(\mathfrak{f})$ are given by the integrals

$$\rho_n(\mathfrak{f}) \;=\; (-1)^n\int_0^\infty \mathfrak{f}(\sqrt{u})\,e^{-u}\,L_n(2u)\,du, \qquad n \geqslant 0. \tag{9.3.18.b}$$

Example 9.12 (Powers Of The Radius) Consider the functions

$$\mathfrak{f}^{(k)}(x) \;=\; |x|^k, \qquad k \geqslant 0. \tag{9.3.19.a}$$

These functions belong to $\mathcal{O}^\infty(\mathbb{R})$, and so determine radial distributions $\mathfrak{f}^{(k)}_{\mathrm{rad}}$, for which

$$\mathfrak{f}^{(k)}_{\mathrm{rad}}(p,q) \;=\; r^k, \qquad\qquad k \geqslant 0. \tag{9.3.19.b}$$

The eigenvalues $\rho_n(\mathfrak{f}^{(k)})$ are expressible in terms of the hypergeometric function ${}_2F_1$,

$$\rho_n(\mathfrak{f}^{(k)}) \;=\; (-1)^n\Gamma\big(\tfrac{1}{2}k+1\big)\,{}_2F_1\big(-n,\tfrac{1}{2}k+1;1;2\big), \qquad n \geqslant 0, \tag{9.3.20.a}$$

and have the generating function

$$\sum_{n\geqslant 0} \rho_n(\mathfrak{f}^{(k)})\,t^n \;=\; \Gamma\big(\tfrac{1}{2}k+1\big)\,(1-t)^{-\frac{1}{2}k-1}\,(1+t)^{\frac{1}{2}k}, \qquad |t|<1, \tag{9.3.20.b}$$

for any $k \geqslant 0$.

If $k \in \mathbb{N} \cup \{0\}$ is a nonnegative integer, the expression for $\rho_n(\mathfrak{f}^{(k)})$ simplifies somewhat. This is because the constants $g_{m,m+k}$ have generating function given by the formula

$$2^{\frac{1}{2}k}\Gamma\big(\tfrac{1}{2}k+\tfrac{1}{2}\big)\sum_{m\geqslant 0}\sqrt{\frac{(m+k)!}{m!}}\,g_{m,m+k}\,t^m \;=\; k!\sqrt{\pi}\,(1-t)^{-\frac{1}{2}k-1}\,(1+t)^{-\frac{1}{2}k} \tag{9.3.21.a}$$

for any $|t|<1$, and hence the coefficients $\rho_n(\mathfrak{f}^{(k)})$ can be written in terms of the $g_{m,n}$,

$$\rho_n(\mathfrak{f}^{(k)}) = \frac{\Gamma\big(\tfrac{1}{2}k+1\big)\Gamma\big(\tfrac{1}{2}k+\tfrac{1}{2}\big)}{k!\,\sqrt{\pi}}2^{\frac{1}{2}k}\sum_{j=0}^{\min(n,k)}\binom{k}{j}\sqrt{\frac{(n+k-j)!}{(n-j)!}}\,g_{n-j,n+k-j}, \tag{9.3.21.b}$$

for any $n \geqslant 0$. In particular,

$$\rho_n(\mathfrak{f}^{(1)}) \;=\; \begin{cases} (n+1)\,\dfrac{\Gamma\big(\tfrac{1}{2}n+1\big)}{\Gamma\big(\tfrac{1}{2}n+\tfrac{3}{2}\big)}, & n \text{ odd},\\[2ex] \big(n+\tfrac{1}{2}\big)\,\dfrac{\Gamma\big(\tfrac{1}{2}n+\tfrac{1}{2}\big)}{\Gamma\big(\tfrac{1}{2}n+1\big)}, & n \text{ even}, \end{cases} \tag{9.3.21.c}$$

which are the eigenvalues for the quantized phase space radius!

Even greater simplifications arise when k is an even nonnegative integer, for then $\rho_n(\mathfrak{f}^{(k)})$ is a polynomial expression in n, which implies that $\mathbf{\Delta}\big[\mathfrak{f}^{(k)}_{\mathrm{rad}}\big]$

is a polynomial function of the number operator N. For example,

$$\begin{aligned} \rho_n(\mathfrak{f}^{(2)}) &= 2n+1, \\ \rho_n(\mathfrak{f}^{(4)}) &= 4n^2+4n+2, \end{aligned} \qquad n \geqslant 0, \tag{9.3.22.a}$$

so

$$\begin{aligned} \mathbf{\Delta}\left[\mathfrak{f}^{(2)}_{\text{rad}}\right] &= 2N+1, \\ \mathbf{\Delta}\left[\mathfrak{f}^{(4)}_{\text{rad}}\right] &= 4N^2+4N+2I \;=\; (2N+I)^2+I. \end{aligned} \tag{9.3.22.b}$$

The first of these results is as expected, since $\mathfrak{f}^{(2)} = 2\nu+1$, and it has already been observed that $\mathbf{\Delta}[\nu] = N$. But the result for $\mathbf{\Delta}\left[\mathfrak{f}^{(4)}_{\text{rad}}\right]$ shows that $\mathbf{\Delta}\left[\mathfrak{f}^{(4)}_{\text{rad}}\right] \neq \mathbf{\Delta}\left[\mathfrak{f}^{(2)}_{\text{rad}}\right]^2$, even though $\mathfrak{f}^{(4)}$ is equal to $\left(\mathfrak{f}^{(2)}\right)^2$. For that matter, $\mathbf{\Delta}\left[\mathfrak{f}^{(2)}\right]$ is not equal to the square of $\mathbf{\Delta}\left[\mathfrak{f}^{(1)}\right]$, either. Thus, while (Weyl) quantization maps functions of the radius to functions of the number operator, its restriction to such functions is not an algebra homomorphism. Thus Weyl quantization does not provide marginals for the radius, as it does for position and momentum.

9.4 Angular Quantization

In the same way that it was necessary to understand what a radial distribution is, now it is necessary to define what is meant by a distribution which is a "function of the angle" as a preliminary step for angular quantization.

9.4.1 *Angular Distributions*

The multiplicative group of the positive reals, $\mathbb{R}^+_*$, is the group consisting of the strictly positive real numbers $(0,\infty)$ equipped with the usual multiplication as its group operation. This group acts on phase space test functions through the continuous endomorphism $\mathcal{E}_\alpha$ of $\mathcal{S}(\Pi)$ given by

$$[\mathcal{E}_\alpha F](p,q) \;=\; \alpha\, F(\sqrt{\alpha}\,p, \sqrt{\alpha}\,q), \qquad \alpha > 0,\; F \in \mathcal{S}(\Pi)\,. \tag{9.4.1}$$

Each map $\mathcal{E}_\alpha$ extends to a unitary map on $L^2(\Pi)$, so the collection of these maps determines a unitary representation of $\mathbb{R}^+_*$ on phase space.

Geometrically, the dilation $p \to \sqrt{\alpha}\,p$ and $q \to \sqrt{\alpha}\,q$, is a radial scaling which preserves the angle. If a phase space function does not change under such a transformation it cannot depend on the radial variable, and so it depends only on the angle variable. This idea can be extended naturally to distributions.

Definition 9.13 *Any distribution $T \in \mathcal{S}'(\Pi)$ such that $\mathcal{E}_\alpha^{\mathrm{tr}} T = T$ for all $\alpha > 0$ is called an* angular distribution. *The collection of angular distributions is written $\mathcal{S}'_{\mathrm{ang}}(\Pi)$.*

When considering true distributions, and not functions, some care needs to be taken in determining whether or not distributions are angular ones, since appearances can be deceiving. For example, the distribution $T_0 \in \mathcal{S}'(\Pi)$ defined by the formula

$$[\![T_0 , F]\!] = \int_0^\infty p\, F(-p,0)\, dp. \qquad F \in \mathcal{S}(\Pi), \tag{9.4.2}$$

is angular, although it may not look it, since it is (essentially) a delta distribution on the negative p-axis.

Having defined the space $\mathcal{S}'_{\mathrm{ang}}(\Pi)$ of angular distributions, the next objective is to characterize these angular distributions explicitly in terms of distributions of some angular variable — in other words, to show that $\mathcal{S}'_{\mathrm{ang}}(\Pi)$ is isomorphic to some space of distributions over the circle $\mathbb{T}$.

Consider the space $\mathcal{C}^\infty(\mathbb{T})$ of infinitely differentiable functions on the circle $\mathbb{T}$. As was observed in Section 5.7.4 of Chapter 5, $\mathcal{C}^\infty(\mathbb{T})$ is the space consisting of those functions $\omega \in L^2(\mathbb{T})$ whose Fourier coefficients $(\hat{\omega}_k)_{k\in\mathbb{Z}}$ with respect to the standard functions

$$\chi_k(e^{i\beta}) = e^{ik\beta}, \qquad k \in \mathbb{Z}, \tag{9.4.3}$$

is rapidly decreasing in both directions. It is clear, then, that $\mathcal{C}^\infty(\mathbb{T})$ can be equipped with a nuclear Fréchet topology defined by the family of seminorms

$$\mathfrak{q}_n(\omega) = \sum_{k\in\mathbb{Z}} (|k|+1)^n \, |\hat{\omega}_k|, \qquad \omega \in \mathcal{C}^\infty(\mathbb{T}),\ n \geqslant 0, \tag{9.4.4}$$

and that the collection $\{\chi_k : k \in \mathbb{Z}\}$ is a Schauder basis for $\mathcal{C}^\infty(\mathbb{T})$ with respect to this topology. Additionally, it can be shown that a sequence of functions in $\mathcal{C}^\infty(\mathbb{T})$ converges with respect to this topology if and only if the sequence, and all of its derivatives, converges uniformly on $\mathbb{T}$. In this sense, therefore, this topology on $\mathcal{C}^\infty(\mathbb{T})$ is a very natural one.

A concrete representation for the angular distributions is obtained by considering the continuous linear map $\mathcal{A} : \mathcal{S}(\Pi) \to \mathcal{C}^\infty(\mathbb{T})$ obtained by integrating test functions over the radial variable,

$$[\mathcal{A}F](e^{i\beta}) = \int_0^\infty F(r\cos\beta, r\sin\beta)\, r\, dr, \qquad F \in \mathcal{S}(\Pi). \quad (9.4.5.a)$$

In particular, the image of the Wigner transform of $\overline{G_s} \otimes G_t$, equation (9.2.6), is given by the formula

$$\mathcal{A}\mathcal{G}(\overline{G_s} \otimes G_t) = \frac{1}{2\pi} \sum_{m,n \geqslant 0} \frac{i^{m-n}}{\sqrt{2^{m+n}\, m!\, n!}}\, g_{m,n}\, s^m t^n\, \chi_{n-m}. \quad (9.4.5.b)$$

The family of all distributions on $\mathbb{T}$, the topological dual of $\mathcal{C}^\infty(\mathbb{T})$, will be denoted by $\mathcal{D}(\mathbb{T})$. By transposition of $\mathcal{A}$, any distribution $S \in \mathcal{D}(\mathbb{T})$ defines a distribution $S_{\text{ang}} = \mathcal{A}^{\text{tr}} S$ in $\mathcal{S}'(\Pi)$,

$$[\![\, S_{\text{ang}}\,,\, F\,]\!] = [\![\, S\,,\, \mathcal{A}F\,]\!], \qquad F \in \mathcal{S}(\Pi). \quad (9.4.6)$$

Since direct calculation shows that

$$\mathcal{A}\mathcal{E}_\alpha F = \mathcal{A}F$$

for all $\alpha > 0$ and $F \in \mathcal{S}(\Pi)$, it follows that $S_{\text{ang}} \in \mathcal{S}'_{\text{ang}}(\Pi)$ is an angular distribution. These are the only angular distributions, since it is possible to show that the map $S \mapsto S_{\text{ang}}$ from $\mathcal{D}(\mathbb{T})$ to $\mathcal{S}'_{\text{ang}}(\Pi)$ is a linear bijection. The proof of this result requires detailed knowledge of properties of the special Hermite functions, however, and is deferred until Chapter 12.

Note, in passing, how the angular distribution T_0 defined in equation (9.4.2) can be described in this new terminology, since $T_0 = \delta^{(-1)}_{\text{ang}}$, where $\delta^{(-1)} \in \mathcal{D}(\mathbb{T})$ is the delta distribution concentrated at the point -1, so that

$$[\![\, \delta^{(-1)}\,,\, \omega\,]\!] = \omega(-1), \qquad \omega \in \mathcal{C}^\infty(\mathbb{T}).$$

9.4.2 *Quantizing Angular Distributions*

It is clear from equation (9.4.5.b) what the matrix coefficients of the quantization of an angular distribution are.

Proposition 9.14 *For a distribution $S \in \mathcal{D}(\mathbb{T})$, the quantization $\boldsymbol{\Delta}[\,S_{\text{ang}}\,]$*

of the angular distribution S_{ang} is given through its matrix elements

$$[\![\mathbf{\Delta} [S_{\text{ang}}] h_n , h_m]\!] = \frac{1}{2\pi} i^{m-n} g_{m,n} [\![S , \chi_{n-m}]\!], \qquad m, n \geqslant 0. \quad (9.4.7)$$

As $L^2(\mathbb{T}) \subseteq \mathcal{D}(\mathbb{T})$, elements of $L^2(\mathbb{T})$ can be used to define angular distributions in $\mathcal{S}'(\Pi)$ — most of the examples that are considered in applications are constructed in this way. Any function $\omega \in L^2(\mathbb{T})$ thus defines the angular distribution[4]

$$\omega_{\text{ang}}(r \cos \beta, r \sin \beta) = \omega(e^{i\beta}) . \quad (9.4.8)$$

Equation (9.4.7) can be used to obtain the matrix coefficients of $\mathbf{\Delta} [\omega_{\text{ang}}]$, yielding

$$[\![\mathbf{\Delta} [\omega_{\text{ang}}] h_n , h_m]\!] = i^{m-n} g_{m,n} \hat{\omega}_{m-n}, \qquad m, n \geqslant 0 , \quad (9.4.9)$$

where, as usual, the $\hat{\omega}_k$ are the Fourier coefficients of ω.

Clearly, the functions χ_k (which form a Schauder basis for $\mathcal{D}(\mathbb{T})$ as well as for $\mathcal{C}^\infty(\mathbb{T})$) are of key importance here, so it is important to understand their quantizations. These turn out to be shifts of the Hermite-Gauss functions, weighted by the characteristic coefficients $g_{m,n}$:

Proposition 9.15 *The quantization $U_k = \mathbf{\Delta} [(\chi_k)_{\text{ang}}]$ of the exponential distribution χ_k, (9.4.3), is a bounded operator on $L^2(\mathbb{R})$ and a continuous endomorphism of $\mathcal{S}(\mathbb{R})$, for any $k \in \mathbb{Z}$. For any $k \geqslant 0$, the operator U_k is the weighted shift operator defined by the formula*

$$U_k h_n = i^k g_{n,n+k} h_{n+k}, \qquad n \geqslant 0, \quad (9.4.10.a)$$

and it satisfies the commutation relation

$$[U_k , N] f = -k U_k f, \qquad f \in \mathcal{S}(\mathbb{R}) . \quad (9.4.10.b)$$

with the number operator N. The adjoint of U_k is the operator U_{-k}, which is also a continuous endomorphism of $\mathcal{S}(\mathbb{R})$. The spectrum of the map U_1 is the unit disc $\{z : |z| \leqslant 1\}$, and contains no eigenvalues. Its adjoint U_{-1} also has the unit disc as its spectrum, and its eigenvalues are the complex numbers of modulus less than 1. All of these eigenvalues are nondegenerate.

---

[4]Note that the resultant distribution ω_{ang} is a function which belongs to $\mathcal{O}^\infty(\Pi)$ if $\omega \in L^\infty(\mathbb{T})$.

Proof: For any $k \geqslant 0$, the action of U_k on the Hermite-Gauss functions is evident from equation (9.4.9). Considering the sequence of coefficients $(g_{m,m+k})_{m\geqslant 0}$, the subsequence $(g_{2m,2m+k})_{m\geqslant 0}$ is monotonically decreasing, while the subsequence $(g_{2m+1,2m+1+k})_{m\geqslant 0}$ is monotonically increasing, with $g_{2m+1,2m+1+k} \leqslant g_{2m,2m+k}$ for all $m \geqslant 0$. Indeed we can find a constant $\mu^{(k)}$ such that

$$g_{2m+1,2m+1+k} \leqslant \mu^{(k)} \leqslant g_{2m,2m+k}, \qquad m \geqslant 0,$$

with $g_{m,m+k} \to \mu^{(k)}$ as $m \to \infty$. An application of Stirling's formula then implies that, in fact, $\mu^{(k)} = 1$. Hence U_k is a bounded operator, and an endomorphism of $\mathcal{S}(\mathbb{R})$, with $\| U_k \| = g_{0,k}$. Since χ_{-k} is the complex conjugate of χ_k, it follows that U_{-k} is the adjoint of U_k, and so is also bounded, mapping $\mathcal{S}(\mathbb{R})$ into itself. Having identified U_k as a weighted shift operator, the indicated commutation relation with the number operator N is immediate. Spectral results for U_1 and U_{-1} result from more detailed analysis of the asymptotic properties of the sequence $(g_{m,m+k})_{m\geqslant 0}$. Details can be found in [53][5]. ■

It is important to note that the operators U_k are not isometries.

9.4.3 *Representing Angular Functions And Distributions*

It is frequently the case that functions on $\mathbb{T}$ are described in terms of functions on some subinterval of the real line. Since this type of identification involves an arbitrary choice of the subinterval of the real line (except in that it should have length 2π), it is important to establish the notational conventions that will be used in this book. The convention to be adopted is the one that is consistent with the fact that we have chosen the radial angle β to lie in the interval $(-\pi, \pi]$. Thus any function $\mathfrak{f} \in L^2(-\pi,\pi]$ will be understood to define a function $\tilde{\mathfrak{f}} \in L^2(\mathbb{T})$ via the formula

$$\tilde{\mathfrak{f}}(e^{i\beta}) = \mathfrak{f}(\beta), \qquad \beta \in (-\pi, \pi]. \tag{9.4.11.a}$$

In other words, $\mathfrak{f}$ and $\tilde{\mathfrak{f}}$ are related by the identity

$$\tilde{\mathfrak{f}} = \mathfrak{f} \circ \mathfrak{p} \tag{9.4.11.b}$$

[5]Spectral properties for the operators U_k for different values of k can also be determined using the techniques found there.

where $\mathfrak{p} \in L^\infty(\mathbb{T})$ is the function

$$\mathfrak{p}(e^{i\beta}) = \beta, \qquad -\pi < \beta \leqslant \pi. \tag{9.4.11.c}$$

The function $\mathfrak{p}$ determines an angular distribution $\mathfrak{p}_{\mathrm{ang}}$ which is precisely the distribution whose quantization will be proposed as a quantum phase operator. As noted previously, the symbol φ is reserved for this distribution, so

Definition 9.16 *The angular distribution φ realizing the angle variable in phase space is*

$$\varphi = \mathfrak{p}_{\mathrm{ang}}, \tag{9.4.12.a}$$

so that

$$\varphi(r\cos\beta, r\sin\beta) = \beta, \qquad -\pi < \beta \leqslant \pi. \tag{9.4.12.b}$$

It follows that the angular distribution $\tilde{\mathfrak{f}}_{\mathrm{ang}}$ corresponding to a function $\mathfrak{f} \in L^2(\mathbb{T})$ is related to φ by

$$\tilde{\mathfrak{f}}_{\mathrm{ang}} = \mathfrak{f} \circ \varphi. \tag{9.4.13}$$

Since the identification between the functions $\mathfrak{f}$ and $\tilde{\mathfrak{f}}$ is so natural, we shall usually make this identification complete by omitting the tilde, referring to $\tilde{\mathfrak{f}}_{\mathrm{ang}}$ simply as $\mathfrak{f}_{\mathrm{ang}}$. This abuse of notation is rather convenient. It will always be clear, when discussing angular distributions, whether this distribution is being defined in terms of a function on $(-\pi, \pi]$ or a function on $\mathbb{T}$, and only in exceptional circumstances will this make any difference.

The detailed properties of $\mathbf{\Delta}[\varphi]$, U_k and U_k^* are of central importance to phase theory, and will be analyzed at length in future Chapters.

9.4.4 *Classes Of Operators*

Now that angular distributions have been defined and their matrix coefficients (with respect to the Hermite-Gauss functions) evaluated, it is sensible to look for conditions on angular distributions which ensure that their quantizations are either bounded or smooth observables. For notational simplicity, attention will be restricted to those angular distributions of the form $\mathfrak{f}_{\mathrm{ang}}$, where $\mathfrak{f} \in L^2(\mathbb{T})$. Various conditions will be given in terms of the growth properties of the Fourier coefficients $\hat{\mathfrak{f}}_n$ of the function $\mathfrak{f}$. Central to these results will be the sequence[6]

$$\mathfrak{f}^\sharp_n \;=\; (1+|n|)^{\frac{1}{4}} \left|\widehat{\mathfrak{f}}_n\right|, \qquad\qquad n \in \mathbb{Z}. \tag{9.4.14}$$

Since $\mathfrak{f} \in L^2(\mathbb{T})$, the sequence $\left(\hat{\mathfrak{f}}_n\right)$ belongs[7] to $\ell^2(\mathbb{Z})$, but of course the sequence $\mathfrak{f}^\sharp = \left(\mathfrak{f}^\sharp_n\right)$ may not.

The utility of the sequence $\mathfrak{f}^\sharp$ can be seen in that it provides a sufficient condition for the boundedness of $\mathbf{\Delta}\left[\,\mathfrak{f}_{\text{ang}}\,\right]$.

Proposition 9.17 *If $\mathfrak{f} \in L^2(\mathbb{T})$ is such that $\mathfrak{f}^\sharp$ belongs to $\ell^1(\mathbb{Z})$, then $\mathbf{\Delta}\left[\,\mathfrak{f}_{\text{ang}}\,\right]$ is a bounded operator.*

Proof: From Lemma 9.4, there exists a $C > 1$ such that

$$\left|\left[\!\left[\, \mathbf{\Delta}\left[\,\mathfrak{f}_{\text{ang}}\,\right] h_n\,,\, h_m \,\right]\!\right]\right| \;\leqslant\; C\,\mathfrak{f}^\sharp_{m-n}, \qquad\qquad m,\, n \geqslant 0.$$

From this it is clear that $\mathbf{\Delta}\left[\,\mathfrak{f}_{\text{ang}}\,\right] h_n \in L^2(\mathbb{R})$, with

$$\|\, \mathbf{\Delta}\left[\,\mathfrak{f}_{\text{ang}}\,\right] h_n \,\|^2 \;\leqslant\; C^2 \sum_{m\geqslant 0} (\mathfrak{f}^\sharp_{m-n})^2 \;\leqslant\; C^2 \,\|\, \mathfrak{f}^\sharp \,\|_2^2,$$

for all $n \geqslant 0$ (since the sequence $\mathfrak{f}^\sharp$ belongs to $\ell^1(\mathbb{Z})$, it certainly belongs to $\ell^2(\mathbb{Z})$). Consequently $\mathbf{\Delta}\left[\,\mathfrak{f}_{\text{ang}}\,\right] g \in L^2(\mathbb{R})$ whenever $g \in \mathcal{S}(\mathbb{R})$, and $\mathbf{\Delta}\left[\,\mathfrak{f}_{\text{ang}}\,\right]$ maps $\mathcal{S}(\mathbb{R})$ continuously into $L^2(\mathbb{R})$.
Since $\mathfrak{f}^\sharp \in \ell^1(\mathbb{Z})$, define the function $\mathfrak{f}^\dagger \in \mathcal{C}(\mathbb{T})$ by the absolutely and uniformly convergent series

$$\mathfrak{f}^\dagger(e^{i\beta}) \;=\; \sum_{m\in\mathbb{Z}} \mathfrak{f}^\sharp_m\, e^{im\beta}.$$

Using the material on Toeplitz operators in Chapter 5, $\mathfrak{f}^\dagger$ defines a bounded Toeplitz operator $\mathcal{M}(\mathfrak{f}^\dagger)$ on $L^2(\mathbb{R})$, where the norm

---

[6] The multiplying factor $(1+|\,n\,|)^{1/4}$ has not just been pulled from a hat, but results from the need to accommodate the complicated asymptotic behaviour of the coefficients $g_{m,n}$. That this multiplying factor is adequate for this purpose is suggested by the result of Lemma 9.4.

[7] For any $p > 1$, $\ell^p(\mathbb{Z})$ is the space of two-sided complex sequences $(a_n)_{n\in\mathbb{Z}}$ for which $\sum_{-\infty}^{\infty} |\, a_n \,|^p$ converges.

$\| \mathcal{M}(\mathfrak{f}^\dagger) \| \leqslant \| \mathfrak{f}^\dagger \|_\infty \leqslant \| \mathfrak{f}^\sharp \|_1$, such that $\langle h_m , \mathcal{M}(\mathfrak{f}^\dagger)h_n \rangle = \mathfrak{f}^\sharp_{m-n}$ for all m, $n \geqslant 0$. Thus

$$\begin{aligned} \big| \langle h_m , \boldsymbol{\Delta}[\mathfrak{f}_{\text{ang}}] f \rangle \big| &\leqslant C \sum_{n \geqslant 0} |f_n| \, \langle h_m , \mathcal{M}(\mathfrak{f}^\dagger)h_n \rangle \\ &= C \langle h_m , \mathcal{M}(\mathfrak{f}^\dagger)\tilde{f} \rangle \end{aligned}$$

for all $m \geqslant 0$ and $f \in \mathcal{S}(\mathbb{R})$, where $\tilde{f} \in \mathcal{S}(\mathbb{R})$ is given by the formula

$$\tilde{f} = \sum_{n \geqslant 0} \big| \langle h_n , f \rangle \big| \, h_n \, .$$

Therefore

$$\| \boldsymbol{\Delta}[\mathfrak{f}_{\text{ang}}] f \| \leqslant C \| \mathcal{M}(\mathfrak{f}^\dagger)\tilde{f} \| \leqslant C \, \| \mathfrak{f}^\sharp \|_1 \, \| \tilde{f} \| = C \, \| \mathfrak{f}^\sharp \|_1 \, \| f \|$$

for all $f \in \mathcal{S}(\mathbb{R})$, so that $\boldsymbol{\Delta}[\mathfrak{f}_{\text{ang}}]$ is bounded. ■

This condition is not necessary condition for boundedness. For example, $\boldsymbol{\Delta}[\varphi]$ will be shown to be bounded in the next Chapter, but the Fourier coefficients of $\mathfrak{p}$ do not satisfy this condition.

The sequence $\mathfrak{f}^\sharp$ can also be used to provide a complete characterization of the space $\mathcal{L}^+\big(\mathcal{S}(\mathbb{R}), L^2(\mathbb{R})\big)$ of closable unbounded linear maps A from $\mathcal{S}(\mathbb{R})$ to $L^2(\mathbb{R})$ for which $\mathcal{S}(\mathbb{R})$ is contained in the domain of the adjoint A^* of A. The Hellinger-Toeplitz Theorem then states that every such map A must map $\mathcal{S}(\mathbb{R})$ continuously into $L^2(\mathbb{R})$, and so

$$\mathcal{L}^+(\mathcal{S}(\mathbb{R})) \subseteq \mathcal{L}^+\big(\mathcal{S}(\mathbb{R}), L^2(\mathbb{R})\big) \subseteq \mathcal{L}(\mathcal{S}(\mathbb{R}), \mathcal{S}'(\mathbb{R})) \, . \tag{9.4.15}$$

If we restricted our discussion of quantum mechanics to consider only mappings which were operators (possibly unbounded), then $\mathcal{L}^+\big(\mathcal{S}(\mathbb{R}), L^2(\mathbb{R})\big)$ would be the largest subspace of $\mathcal{L}(\mathcal{S}(\mathbb{R}), \mathcal{S}'(\mathbb{R}))$ which could be considered. The reason that $\mathcal{L}^+\big(\mathcal{S}(\mathbb{R}), L^2(\mathbb{R})\big)$ was not chosen as the set of observables is that it is not an algebra. Moreover, it is futile to try and avoid the appearance of distributions, since they will intrude into the theory sooner or later. But $\mathcal{L}^+\big(\mathcal{S}(\mathbb{R}), L^2(\mathbb{R})\big)$ is clearly an interesting space to consider, and its mathematical place in the theory can be fixed precisely.

Proposition 9.18 *If* $\mathfrak{f} \in L^2(\mathbb{T})$, *then* $\boldsymbol{\Delta}[\mathfrak{f}_{\text{ang}}]$ *belongs to* $\mathcal{L}^+\big(\mathcal{S}(\mathbb{R}), L^2(\mathbb{R})\big)$ *if and only if* $\mathfrak{f}^\sharp$ *belongs to* $\ell^2(\mathbb{Z})$.

Proof: Suppose first that $\mathfrak{f}^\sharp \in \ell^2(\mathbb{Z})$. An inspection of the arguments in the proof of the previous Proposition shows that $\mathbf{\Delta}[\,\mathfrak{f}_{\mathrm{ang}}\,]$ is a continuous linear map from $\mathcal{S}(\mathbb{R})$ to $L^2(\mathbb{R})$. Since $(\bar{\mathfrak{f}})^\sharp_n = \mathfrak{f}^\sharp_{-n}$ for all $n \in \mathbb{Z}$, it is clear that $\mathbf{\Delta}\left[\,\bar{\mathfrak{f}}_{\mathrm{ang}}\,\right]$ is also a continuous linear map from $\mathcal{S}(\mathbb{R})$ to $L^2(\mathbb{R})$, and is the restriction of the adjoint of $\mathbf{\Delta}[\,\mathfrak{f}_{\mathrm{ang}}\,]$ to $\mathcal{S}(\mathbb{R})$. In other words, $\mathbf{\Delta}[\,\mathfrak{f}_{\mathrm{ang}}\,]$ belongs to $\mathcal{L}^+\big(\mathcal{S}(\mathbb{R}), L^2(\mathbb{R})\big)$. Conversely, if $\mathbf{\Delta}[\,\mathfrak{f}_{\mathrm{ang}}\,]$ belongs to $\mathcal{L}^+\big(\mathcal{S}(\mathbb{R}), L^2(\mathbb{R})\big)$, then

$$\begin{aligned} \left|\,\langle h_m\,,\,\mathbf{\Delta}[\,\mathfrak{f}_{\mathrm{ang}}\,]\,h_0\rangle\,\right| &= g_{m,0}\left|\widehat{\mathfrak{f}}_m\right| \\ &\geqslant B^{-1}\,\xi_{0;\frac{1}{2}}\,(m+1)^{\frac{1}{4}}\left|\widehat{\mathfrak{f}}_m\right| \;=\; B^{-1}\xi_{0;\frac{1}{2}}\,\mathfrak{f}^\sharp_m, \end{aligned}$$

and, similarly,

$$\left|\big\langle h_m\,,\,\mathbf{\Delta}[\,\mathfrak{f}_{\mathrm{ang}}\,]^*\,h_0\big\rangle\right| \geqslant B^{-1}\,\xi_{0;\frac{1}{2}}\,\mathfrak{f}^\sharp_{-m},$$

for all $m \geqslant 0$, using the notation of Lemma 9.4. Thus $\mathfrak{f}^\sharp \in \ell^2(\mathbb{Z})$, with

$$\xi^2_{0;\frac{1}{2}}\,\left\|\,\mathfrak{f}^\sharp\,\right\|_2^2 \leqslant B^2\left\{\left\|\,\mathbf{\Delta}[\,\mathfrak{f}_{\mathrm{ang}}\,]\,h_0\,\right\|^2 + \left\|\,\mathbf{\Delta}[\,\mathfrak{f}_{\mathrm{ang}}\,]^*\,h_0\,\right\|^2\right\},$$

as required. ■

A necessary and sufficient condition in order for $\mathbf{\Delta}[\,\mathfrak{f}_{\mathrm{ang}}\,] \in \mathcal{L}^+(\mathcal{S}(\mathbb{R}))$ to be a smooth observable can be given in terms of the Fourier coefficients of $\mathfrak{f}$ directly (the nature of the condition is such as to make the use of the associated sequence $\mathfrak{f}^\sharp$ unnecessary). The proof is very similar to the two given above, and so we shall omit it.

Proposition 9.19 *If $\mathfrak{f} \in L^2(\mathbb{T})$, then $\mathbf{\Delta}[\,\mathfrak{f}_{\mathrm{ang}}\,] \in \mathcal{L}^+(\mathcal{S}(\mathbb{R}))$ if and only if for any $j \geqslant 0$ we can find $k \geqslant 0$ and $K > 0$ such that*

$$\left|\widehat{\mathfrak{f}}_{m-n}\right|,\ \left|\widehat{\mathfrak{f}}_{n-m}\right| \leqslant K\,\frac{(n+1)^k}{(m+1)^j}, \qquad m,\,n \geqslant 0. \tag{9.4.16}$$

9.4.5 *The Method Of Wedges*

Notwithstanding the considerable work already done in this book on angular quantization, only one analytical technique has been employed — the use of Stirling's approximation to obtain bounds on the coefficients $g_{m,n}$. Further results require new methods, and the method of wedges is one such.

The idea behind it is quite simple. Amongst the angular distributions is the function which is 1 when β lies in the wedge $(\alpha_1, \alpha_2]$ and zero for all other angles. Knowledge about the quantization of this distribution can be transferred to any other wedge by action of the metaplectic group. Moreover, a reasonably well behaved angular distribution can be approximated by a linear combination of wedge functions, in the same way as an integral is approximated by a Riemann sum. This is the idea, but we shall see that its realization is technically rather difficult, and there are still many questions about wedge quantization left to answer.

The quantized wedge functions involve certain integral operators, and so the analysis must begin with some definitions.

Definition 9.20 *For any bounded function $h \in L^\infty[0,\infty)$, let the kernel function $\kappa_h : (0,\infty) \times (0,\infty) \to \mathbb{C}$ be given by*

$$\kappa_h(x,y) \;=\; \begin{cases} \dfrac{h(x+y)}{x+y}, & 0 < x \leqslant y, \\ 0, & 0 < y < x\,, \end{cases} \tag{9.4.17.a}$$

and let $K_h : L^2[0,\infty) \to L^2[0,\infty)$ be the operator with integral kernel κ_h:

$$[K_h\, u](x) \;=\; \int_0^\infty \kappa_h(x,y) u(y)\, dy \;=\; \int_x^\infty \frac{h(x+y)}{x+y} u(y)\, dy\,. \tag{9.4.17.b}$$

Introducing the continuous linear maps $\mathcal{P}_\pm : L^2(\mathbb{R}) \to L^2[0,\infty)$,

$$[\mathcal{P}_\pm g](x) \;=\; g(\pm x), \qquad g \in L^2(\mathbb{R}),\; x > 0, \tag{9.4.17.c}$$

it is possible to define a continuous linear map $\mathcal{K}_h : L^2(\mathbb{R}) \to L^2(\mathbb{R})$ by the formula

$$[\mathcal{K}_h g](x) \;=\; \begin{cases} [K_h \mathcal{P}_- g](x), & x > 0, \\ 0, & x = 0, \\ [K_h \mathcal{P}_+ g](-x), & x < 0, \end{cases} \qquad g \in L^2(\mathbb{R}), \tag{9.4.17.d}$$

Before all else, the well-definedness of these operators must be established.

Proposition 9.21 *K_h is a bounded linear operator on $L^2[0,\infty)$ with*

$$\|\, K_h \,\| \;\leqslant\; \tfrac{1}{2}\pi \;\|\, h \,\|_\infty\,, \tag{9.4.18.a}$$

and $\mathcal{K}_h$ is a bounded linear operator on $L^2(\mathbb{R})$ with the same bound:

$$\| \mathcal{K}_h \| \leqslant \frac{1}{2}\pi \, \| h \|_\infty . \tag{9.4.18.b}$$

Proof: Direct calculation establishes the inequalities,

$$\int_0^\infty |\kappa_h(s,t)| \frac{dt}{\sqrt{t}} \leqslant \frac{\pi}{2\sqrt{s}} \| h \|_\infty ,$$
$$\int_0^\infty |\kappa_h(t,s)| \frac{dt}{\sqrt{t}} \leqslant \frac{\pi}{2\sqrt{s}} \| h \|_\infty ,$$

for all $s > 0$. The Schur test [98] then shows that K_h is bounded with the indicated bound on its norm. Since the maps $\mathcal{P}_\pm$ are norm-decreasing, the results concerning $\mathcal{K}_h$ follow. ■

Although more general forms of these operators will be necessary in later Chapters, for the present only the operators $K \equiv K_{\mathfrak{i}}$ and $\mathcal{K} \equiv \mathcal{K}_{\mathfrak{i}}$ are needed, where $\mathfrak{i} \in L^\infty[0,\infty)$ is the constant function $\mathfrak{i}(x) = 1$.

The method of wedges begins with a study of the phase space function which is equal to $+1$ in the first and third quadrants, and -1 in the second and fourth quadrants. This function may be written as $\operatorname{sgn} \otimes \operatorname{sgn}$, where

$$\operatorname{sgn}(t) = \begin{cases} t/|t|, & \text{if } t \in \mathbb{R} \setminus \{0\}, \\ 0, & \text{if } t = 0. \end{cases} \tag{9.4.19}$$

Study of $\mathbf{\Delta}[\operatorname{sgn} \otimes \operatorname{sgn}]$ is made complicated by the fact that its integral kernel is singular, so that working with it involves the evaluation of various improper Riemann integrals. In order to be able to establish results about this operator, it has proved necessary to introduce an explicit formalism for the limiting process that will be used in evaluating these integrals — the necessary calculations can then be handled by Lebesgue integration theory. The presence of the "cut-off" function $g_{I(L)}$, which implements this process, should therefore be noted carefully.

Proposition 9.22 *The operator $\mathbf{\Delta}[\operatorname{sgn} \otimes \operatorname{sgn}]$ is given by*

$$\mathbf{\Delta}[\operatorname{sgn} \otimes \operatorname{sgn}] = i \operatorname{sgn}(Q) \circ \mathcal{H} - \frac{2i}{\pi} \mathcal{K} = \operatorname{sgn}(Q) \circ \operatorname{sgn}(P) - \frac{2i}{\pi} \mathcal{K}, \tag{9.4.20}$$

where $\mathcal{H}$ denotes the Hilbert transform. It is bounded, with

$$\| \mathbf{\Delta}[\operatorname{sgn} \otimes \operatorname{sgn}] \| \leqslant 2. \tag{9.4.21}$$

Proof: For any $f,\, g \in \mathcal{S}(\mathbb{R})$, let $F \in \mathcal{S}\big(\mathbb{R}^2\big)$ and $G \in \mathcal{S}(\mathbb{R})$ be the functions

$$\begin{aligned} F(x,y) &= \overline{g(y+\tfrac{1}{2}x)}\, f(y-\tfrac{1}{2}x), \\ G(x) &= \int_{\mathbb{R}} \operatorname{sgn}(y)\, F(x,y)\, dy. \end{aligned}$$

Then, after some elementary Fourier analysis,

$$\begin{aligned} [\![\, \boldsymbol{\Delta}\, [\operatorname{sgn} \otimes \operatorname{sgn}]\, f\,,\, \overline{g}\,]\!] &= \frac{1}{\sqrt{2\pi}} \int_{\mathbb{R}} \operatorname{sgn}(p)[\mathcal{F}^{-1}G](p)\, dp \\ &= \frac{i}{\pi} \lim_{L\to\infty} \big\langle g_{I(L)}\,,\, G\big\rangle \end{aligned}$$

where $g_{I(L)}$ is the function defined by

$$g_{I(L)}(x) = \begin{cases} x^{-1}, & L^{-1} \leqslant |x| \leqslant L, \\ 0, & \text{otherwise}, \end{cases} \tag{9.4.22}$$

for any $L > 0$.

The trick is to manipulate the expression for $\big\langle g_{I(L)}\,,\, G\big\rangle$ in such a way that the singularity of the limiting process and the discontinuity of the function sgn can be handled separately. To this end,

$$\begin{aligned} \big\langle g_{I(L)}\,,\, G\big\rangle &= \iint_{\mathbb{R}^2} g_{I(L)}(x-y)\operatorname{sgn}(x+y)\overline{g(x)}f(y)\, dx\, dy \\ &= \int_0^{\infty} \overline{g(x)} \int_{\mathbb{R}} g_{I(L)}(x-y) f(y)\, dy\, dx \\ &\quad - 2\int_0^{\infty} \overline{g(x)} \int_{-\infty}^{-x} g_{I(L)}(x-y) f(y)\, dy\, dx \\ &\quad - \int_{-\infty}^{0} \overline{g(x)} \int_{\mathbb{R}} g_{I(L)}(x-y) f(y)\, dy\, dx \\ &\quad + 2\int_{-\infty}^{0} \overline{g(x)} \int_{-x}^{\infty} g_{I(L)}(x-y) f(y)\, dy\, dx, \end{aligned}$$

and careful application of the Dominated Convergence Theorem yields

$$[\![\, \boldsymbol{\Delta}\, [\operatorname{sgn} \otimes \operatorname{sgn}]\, f\,,\, \overline{g}\,]\!] = \frac{i}{\pi} \lim_{L\to\infty} \big\langle \operatorname{sgn}(Q)g, g_{I(L)} * f\big\rangle - \frac{2i}{\pi} \langle g\,,\, \mathcal{K}f\rangle\,,$$

where $*$ denotes the standard operation of convolution. The above expression for $\boldsymbol{\Delta}[\,\mathrm{sgn}\otimes\mathrm{sgn}\,]$ is now immediate, as is the norm estimate. ∎

The operator $-2i\pi^{-1}\mathcal{K}$ measures the extent to which $\boldsymbol{\Delta}[\,\mathrm{sgn}\otimes\mathrm{sgn}\,]$ differs from $\mathrm{sgn}(Q)\circ\mathrm{sgn}(P)$, where, of course, $\mathrm{sgn}(Q) = \boldsymbol{\Delta}[\,\mathrm{i}\otimes\mathrm{sgn}\,]$ and $\mathrm{sgn}(P) = \boldsymbol{\Delta}[\,\mathrm{sgn}\otimes\mathrm{i}\,]$.

Remark The Hilbert transform $\mathcal{H}$ is the bounded operator on $L^2(\mathbb{R})$ defined in terms of the momentum operator P by

$$\mathcal{H} \;=\; -i\,\mathrm{sgn}(P)\,. \tag{9.4.23.a}$$

Taking the Fourier transform,

$$[\mathcal{F}\mathcal{H}\phi](x) \;=\; -i\,\mathrm{sgn}(x)\,[\mathcal{F}\phi](x)\,, \qquad \phi\in L^2(\mathbb{R})\,, \tag{9.4.23.b}$$

so that

$$\mathcal{F}\mathcal{H}\mathcal{F}^{-1} = -i\,\mathrm{sgn}(Q)\,. \tag{9.4.23.c}$$

Consequently, the Hilbert transform $\mathcal{H}$ is a unitary operator on $L^2(\mathbb{R})$. It may be verified that

$$\mathcal{H}\phi \;=\; \frac{1}{\pi}\lim_{L\to\infty} g_{I(L)} * \phi\,, \qquad \phi\in L^2(\mathbb{R})\,, \tag{9.4.24}$$

from which the classical representation of $\mathcal{H}$ as a singular integral operator follows,

$$[\mathcal{H}\phi](x) \;=\; \frac{1}{\pi}\,\mathrm{PV}\!\int_{\mathbb{R}} \frac{\phi(y)}{x-y}\,dy, \qquad \phi\in L^2(\mathbb{R})\,, \tag{9.4.25}$$

∎

To study general wedge functions, the following definitions are necessary.

Definition 9.23 *For any $0\leqslant\alpha\leqslant\pi$, the positive wedge function $\mathfrak{f}_+^{(\alpha)}$ in $L^\infty[-\pi,\pi]$ is defined by the formula*

$$[\mathfrak{f}_+^{(\alpha)}](\beta) \;=\; \begin{cases} 1, & \alpha\leqslant\beta\leqslant\pi, \\ 0, & -\pi\leqslant\beta<\alpha, \end{cases} \tag{9.4.26.a}$$

and the negative wedge function $\mathfrak{f}_-^{(\alpha)}$ *by*

$$[\mathfrak{f}_-^{(\alpha)}](\beta) = [\mathfrak{f}_+^{(\alpha)}](-\beta), \qquad -\pi \leqslant \beta \leqslant \pi. \tag{9.4.26.b}$$

The corresponding positive and negative wedge angular distributions are

$$D_\pm^{(\alpha)} = \big[\mathfrak{f}_\pm^{(\alpha)}\big]_{\mathrm{ang}}, \qquad 0 \leqslant \alpha \leqslant \pi. \tag{9.4.27}$$

The special cases

$$\begin{array}{lll} D_+^{(0)} = \mathrm{i} \otimes \chi_{[0,\infty)}, & D_+^{(\frac{1}{2}\pi)} = \chi_{(-\infty,0]} \otimes \chi_{[0,\infty)}, & D_+^{(\pi)} = 0, \\ D_-^{(0)} = \mathrm{i} \otimes \chi_{(-\infty,0]}, & D_-^{(\frac{1}{2}\pi)} = \chi_{(-\infty,0]} \otimes \chi_{(-\infty,0]}, & D_-^{(\pi)} = 0, \end{array} \tag{9.4.28}$$

of these distributions will be useful.

The quantizations of the various wedge distributions are interrelated by actions of the metaplectic group, and this fact enables us to prove the following result.

Proposition 9.24 *The operator* $\boldsymbol{\Delta}\big[D_\pm^{(\alpha)}\big]$ *is bounded, with*

$$\big\|\boldsymbol{\Delta}\big[D_\pm^{(\alpha)}\big]\big\| \leqslant \tfrac{5}{4}, \tag{9.4.29}$$

for all $0 \leqslant \alpha \leqslant \pi$.

Proof: The result is trivial if $\alpha = 0$ or π. Since it can be shown that

$$\begin{aligned} D_+^{(\frac{1}{2}\pi)} &= \chi_{(-\infty,0]} \otimes \chi_{[0,\infty)} \\ &= \frac{1}{4}\big\{\mathrm{i}\otimes\mathrm{i} - \mathrm{sgn}\otimes\mathrm{i} + \mathrm{i}\otimes\mathrm{sgn} - \mathrm{sgn}\otimes\mathrm{sgn}\big\}, \\ D_-^{(\frac{1}{2}\pi)} &= \chi_{(-\infty,0]} \otimes \chi_{(-\infty,0]} \\ &= \frac{1}{4}\big\{\mathrm{i}\otimes\mathrm{i} - \mathrm{sgn}\otimes\mathrm{i} - \mathrm{i}\otimes\mathrm{sgn} + \mathrm{sgn}\otimes\mathrm{sgn}\big\}, \end{aligned}$$

we deduce that

$$\boldsymbol{\Delta}\big[D_\pm^{(\frac{1}{2}\pi)}\big] = \frac{1}{4}\big\{I - \mathrm{sgn}(P) \pm \mathrm{sgn}(Q) \mp \boldsymbol{\Delta}\,[\mathrm{sgn}\otimes\mathrm{sgn}]\big\},$$

and so the result also holds when $\alpha = \frac{1}{2}\pi$.
For any $c \in \mathbb{R}$ we calculate that

$$\big[D_+^{(\frac{1}{2}\pi + \tan^{-1} c)}\big](p,q) = \big[D_+^{(\frac{1}{2}\pi)}\big](p + cq, q),$$

from which

$$D_+^{(\frac{1}{2}\pi+\tan^{-1}c)} = M(c)\circ D_+^{(\frac{1}{2}\pi)},$$

follows, and similarly,

$$D_-^{(\frac{1}{2}\pi-\tan^{-1}c)} = M(c)\circ D_-^{(\frac{1}{2}\pi)},$$

where $M(c)\in \mathrm{Sp}\,(2;\mathbb{R})$ is the matrix $\left(\begin{smallmatrix}1 & c\\ 0 & 1\end{smallmatrix}\right)$ introduced in Section 8.7.2 of Chapter 8. Thus

$$\mathbf{\Delta}\big[D_\pm^{(\frac{1}{2}\pi\,\pm\tan^{-1}c)}\big] = \pi\big(M(c)\big)\,\mathbf{\Delta}\big[D_\pm^{(\frac{1}{2}\pi)}\big]\,\pi\big(M(c)\big)^{-1},$$

where π is the representation of the metaplectic group previously discussed. Hence all the maps $\mathbf{\Delta}\big[D_\pm^{(\alpha)}\big]$ are bounded, having the desired upper bound on their norms. ■

Instead of dealing with the wedge distributions $D_+^{(\alpha)}$ and $D_-^{(\alpha)}$ separately as here, it is possible to obtain $D_-^{(\pi/2)}$ from $D_+^{(\pi/2)}$ by the action of the symplectic rotation matrix J, so $\mathbf{\Delta}\big[D_-^{(\pi/2)}\big]$ is obtained from $\mathbf{\Delta}\big[D_+^{(\pi/2)}\big]$ by conjugation with the Fourier transform. The boundedness properties for the negative wedge distributions could thus have been deduced from the boundedness properties for the positive ones.

The utility of these wedge distributions lies in the fact that, if the function $\mathfrak{f}$ is continuously differentiable, then the angular distribution $\mathfrak{f}_{\mathrm{ang}}$ can be written weakly as an integral over the wedge distributions, and hence $\mathbf{\Delta}\,[\,\mathfrak{f}_{\mathrm{ang}}\,]$ can be written as an integral of the quantizations of the wedge distributions. The precise formulation of these observations is as follows:

Theorem 9.25 *If the function $\mathfrak{f}\in\mathcal{C}^1[-\pi,\pi]$ is continuously differentiable, then $\mathbf{\Delta}\,[\,\mathfrak{f}_{\mathrm{ang}}\,]$ is bounded, with*

$$\|\,\mathbf{\Delta}\,[\,\mathfrak{f}_{\mathrm{ang}}\,]\,\| \leqslant |\,\mathfrak{f}(0)\,| + \frac{5}{4}\int_{-\pi}^{\pi}|\,\mathfrak{f}'(\alpha)\,|\,d\alpha\,, \tag{9.4.30.a}$$

and it has the integral decomposition

$$\mathbf{\Delta}\,[\,\mathfrak{f}_{\mathrm{ang}}\,] = \mathfrak{f}(0)\,I + \int_0^{\pi}\mathfrak{f}'(\alpha)\mathbf{\Delta}\big[D_+^{(\alpha)}\big]\,d\alpha - \int_0^{\pi}\mathfrak{f}'(-\alpha)\mathbf{\Delta}\big[D_-^{(\alpha)}\big]\,d\alpha\,. \tag{9.4.30.b}$$

Proof: For any $f,\, g \in \mathcal{S}(\mathbb{R})$, the functions

$$\alpha \mapsto \langle g\,,\, \mathbf{\Delta}\big[D_+^{(\alpha)}\big] f\rangle = \int_\alpha^\pi \big[\mathcal{AG}(\overline{g}\otimes f)\big](e^{i\beta})\, d\beta,$$

$$\alpha \mapsto \langle g\,,\, \mathbf{\Delta}\big[D_-^{(\alpha)}\big] f\rangle = \int_{-\pi}^{-\alpha} \big[\mathcal{AG}(\overline{g}\otimes f)\big](e^{i\beta})\, d\beta,$$

are continuous on $[0,\pi]$. After some simple manipulations of the integrals, the relations

$$\int_0^\pi \mathfrak{f}'(\alpha)\langle g\,,\, \mathbf{\Delta}\big[D_+^{(\alpha)}\big] f\rangle\, d\alpha = \int_0^\pi \big(\mathfrak{f}(\beta) - \mathfrak{f}(0)\big)\big[\mathcal{AG}(\overline{g}\otimes f)\big](e^{i\beta})\, d\beta,$$

$$\int_0^\pi \mathfrak{f}'(-\alpha)\langle g\,,\, \mathbf{\Delta}\big[D_-^{(\alpha)}\big] f\rangle\, d\alpha = \int_{-\pi}^0 \big(\mathfrak{f}(0) - \mathfrak{f}(\beta)\big)\big[\mathcal{AG}(\overline{g}\otimes f)\big](e^{i\beta})\, d\beta,$$

result, from which it may be deduced that

$$\int_0^\pi \mathfrak{f}'(\alpha)\langle g\,,\, \mathbf{\Delta}\big[D_+^{(\alpha)}\big] f\rangle\, d\alpha - \int_0^\pi \mathfrak{f}'(-\alpha)\langle g\,,\, \mathbf{\Delta}\big[D_-^{(\alpha)}\big] f\rangle\, d\alpha = [\![\, \mathbf{\Delta}\,[\,\mathfrak{f}_{\mathrm{ang}}\,]\, f\,,\, \overline{g}\,]\!] - \mathfrak{f}(0)\, \langle g, f\rangle\,.$$

The desired integral decomposition of $\mathbf{\Delta}\,[\,\mathfrak{f}_{\mathrm{ang}}\,]$ now follows, and so $\mathbf{\Delta}\,[\,\mathfrak{f}_{\mathrm{ang}}\,]$ is bounded with the given norm estimate. ∎

The condition that $\mathfrak{f}$ be continuously differentiable on $[-\pi,\pi]$ is not strictly necessary — only the facts that $\mathfrak{f}$ is continuously differentiable on the subintervals $[0,\pi]$ and $[-\pi,0]$ were used. Thus $\mathfrak{f}'$ may be permitted a jump discontinuity at 0.

This result is able to provide us with positive results in cases where the techniques of the previous Subsection could not. In particular, Proposition 9.17 is not applicable to angular distributions which are discontinuous across the cut in the plane Π, whereas the method of wedges is.

That Proposition 9.17 is limited in this way can be seen as follows. If $\mathfrak{f} \in \mathcal{C}^1[-\pi,\pi]$, then its Fourier coefficients are related to those of its derivative $\mathfrak{f}'$ by the formula

$$\widehat{f}_n = \frac{i}{n}(-1)^n[\mathfrak{f}(\pi) - \mathfrak{f}(-\pi)] - \frac{i}{n}\widehat{\mathfrak{f}'}_n$$

for any nonzero integer n, and the sequence $\left(\widehat{\mathfrak{f}'}_n\right)_{n\in\mathbb{Z}}$ belongs to $\ell^2(\mathbb{Z})$. Therefore $\tilde{f} = \mathfrak{f}\circ\mathfrak{p} \in L^\infty(\mathbb{T})$ satisfies the conditions of Proposition 9.17 (guaranteeing that $\mathbf{\Delta}[\,\mathfrak{f}_{\text{ang}}\,]$ is bounded) if and only if $\mathfrak{f}(\pi) = \mathfrak{f}(-\pi)$, so that the function $\tilde{\mathfrak{f}}$ is continuous on $\mathbb{T}$.

The result just obtained from the method of wedges is itself limited in that it does not give sharp bounds on $\mathbf{\Delta}[\,\mathfrak{f}_{\text{ang}}\,]$. Nor are we aware of any method that does, though in special cases the above bound can be improved.

To end the discussion of the method of wedges, a rather specialized class of distributions, for which a sharper bound can be found, will now be considered. The result below will be needed for the analysis of the phase operator $\mathbf{\Delta}[\,\varphi\,]$.

Given a function $\mathfrak{g}\in\mathcal{C}^1[0,\pi]$, define the quasi-periodic function $\mathfrak{f}(\mathfrak{g})$ on $[-\pi,\pi]$ by the formula

$$[\mathfrak{f}(\mathfrak{g})](\beta) = \begin{cases} \mathfrak{g}(\beta), & 0\leqslant\beta\leqslant\pi, \\ \mathfrak{g}(\pi+\beta)+\mathfrak{g}(0)-\mathfrak{g}(\pi), & -\pi\leqslant\beta\leqslant 0\,. \end{cases} \tag{9.4.31.a}$$

The function $\mathfrak{f}(\mathfrak{g})$ is continuous on $[-\pi,\pi]$ and continuously differentiable on both $[0,\pi]$ and $[-\pi,0]$, with derivatives

$$[\mathfrak{f}(\mathfrak{g})'](\beta-\pi) = [\mathfrak{f}(\mathfrak{g})'](\beta) = \mathfrak{g}'(\beta) \tag{9.4.31.b}$$

for all $0\leqslant\beta\leqslant\pi$.

Applying Theorem 9.25, $\mathbf{\Delta}[\,\mathfrak{f}(\mathfrak{g})_{\text{ang}}\,]$ is bounded, with

$$\|\,\mathbf{\Delta}[\,\mathfrak{f}(\mathfrak{g})_{\text{ang}}\,]\,\| \leqslant |\,\mathfrak{g}(0)\,| + \frac{5}{2}\int_0^\pi |\,\mathfrak{g}'(\beta)\,|\,d\beta. \tag{9.4.32}$$

However, this inequality can be bettered.

Proposition 9.26 *For any* $\mathfrak{g}\in\mathcal{C}^1[0,\pi]$, $\mathbf{\Delta}[\,\mathfrak{f}(\mathfrak{g})\,]$ *is bounded, with*

$$\|\,\mathbf{\Delta}[\,\mathfrak{f}(\mathfrak{g})_{\text{ang}}\,]\,\| \leqslant |\,\mathfrak{g}(0)\,| + \frac{3}{2}\int_0^\pi |\,\mathfrak{g}'(\beta)\,|\,d\beta\,. \tag{9.4.33}$$

Proof: For this class of distributions, the integral decomposition of Theorem 9.25 can be written as

$$\mathbf{\Delta}[\,\mathfrak{f}(\mathfrak{g})_{\text{ang}}\,] = \mathfrak{g}(0)\,I + \int_0^\pi \mathfrak{g}'(\alpha)\,\mathbf{\Delta}\big[\mathcal{D}^{(\alpha)}\big]\,d\alpha,$$

where $\mathcal{D}^{(\alpha)}$ is the composite wedge distribution

$$\mathcal{D}^{(\alpha)} = D_+^{(\alpha)} - D_-^{(\pi-\alpha)}, \qquad 0 \leqslant \alpha \leqslant \pi.$$

Previous calculations imply that

$$\mathcal{D}^{(\frac{1}{2}\pi+\tan^{-1}c)} = M(c) \circ \mathcal{D}^{(\frac{1}{2}\pi)}, \qquad c \in \mathbb{R},$$

where

$$\mathcal{D}^{(\frac{1}{2}\pi)} = D_+^{(\frac{1}{2}\pi)} - D_-^{(\frac{1}{2}\pi)} = \chi_{(-\infty,0]} \otimes \mathrm{sgn}.$$

Thus $2\mathcal{D}^{(\frac{1}{2}\pi)} = \mathfrak{i} \otimes \mathrm{sgn} - \mathrm{sgn} \otimes \mathrm{sgn}$, and so $\left\|\boldsymbol{\Delta}\left[\mathcal{D}^{(\frac{1}{2}\pi)}\right]\right\| \leqslant \frac{3}{2}$. Hence

$$\left\|\boldsymbol{\Delta}\left[\mathcal{D}^{(\alpha)}\right]\right\| \leqslant \frac{3}{2}, \qquad 0 \leqslant \alpha \leqslant \pi,$$

from which the desired result is immediate. ■

9.4.6 *Integral Kernels*

The comparatively simple nature of angular distributions on Π makes it possible to derive an explicit expression for the integral kernel of $\boldsymbol{\Delta}\left[\,\mathfrak{f}_{\mathrm{ang}}\,\right]$ for a large class of functions $\mathfrak{f}_{\mathrm{ang}}$. The construction starts with the introduction of some unlikely seeming integral transforms.

Definition 9.27 *For any continuous function $\mathfrak{f} \in C[-\pi,\pi]$, the functions $\mathcal{E}_{\pm 1}(\mathfrak{f})$, $\mathcal{C}_{\pm 1}(\mathfrak{f})$ and $\mathcal{S}_{\pm 1}(\mathfrak{f})$ are defined on $\mathbb{R}$ by the formulæ*

$$\left[\mathcal{E}_{\pm 1}(\mathfrak{f})\right](x) = \int_0^\pi \mathfrak{f}(\pm\vartheta)\, \exp i\left[x\cot^{-1}\vartheta\right] d\vartheta, \qquad (9.4.34.a)$$

$$\left[\mathcal{C}_{\pm 1}(\mathfrak{f})\right](x) = \int_0^\pi \mathfrak{f}(\pm\vartheta)\, \cos\left[x\cot^{-1}\vartheta\right] d\vartheta, \qquad (9.4.34.b)$$

$$\left[\mathcal{S}_{\pm 1}(\mathfrak{f})\right](x) = \int_0^\pi \mathfrak{f}(\pm\vartheta)\, \sin\left[x\cot^{-1}\vartheta\right] d\vartheta. \qquad (9.4.34.c)$$

Then $\mathcal{E}_{\pm 1}(\mathfrak{f}) = \mathcal{C}_{\pm 1}(\mathfrak{f}) + i\mathcal{S}_{\pm 1}(\mathfrak{f})$. These transforms have a natural interpretation when considered in terms of the phase space angle function $\mathfrak{f}_{\mathrm{ang}}$ associated with $\mathfrak{f}$.

Proposition 9.28 *The functions $\mathcal{E}_{\pm 1}(\mathfrak{f})$, $\mathcal{C}_{\pm 1}(\mathfrak{f})$, $\mathcal{S}_{\pm 1}(\mathfrak{f})$ are the Fourier, Fourier cosine and Fourier sine transforms of the functions*

$$p \mapsto \frac{\mathfrak{f}_{\mathrm{ang}}(p, \pm 1)}{p^2 + 1}$$

(which belong to $L^1(\mathbb{R})$), respectively, so that

$$\left[\mathcal{E}_{\pm 1}(\mathfrak{f})\right](x) = \int_{\mathbb{R}} \mathfrak{f}_{\mathrm{ang}}(p, \pm 1) \frac{e^{ipx}}{p^2+1}\, dp, \tag{9.4.35.a}$$

$$\left[\mathcal{C}_{\pm 1}(\mathfrak{f})\right](x) = \int_{\mathbb{R}} \mathfrak{f}_{\mathrm{ang}}(p, \pm 1) \frac{\cos px}{p^2+1}\, dp, \tag{9.4.35.b}$$

$$\left[\mathcal{S}_{\pm 1}(\mathfrak{f})\right](x) = \int_{\mathbb{R}} \mathfrak{f}_{\mathrm{ang}}(p, \pm 1) \frac{\sin px}{p^2+1}\, dp. \tag{9.4.35.c}$$

Thus $\mathcal{E}_{\pm 1}(\mathfrak{f})$, $\mathcal{C}_{\pm 1}(\mathfrak{f})$ and $\mathcal{S}_{\pm 1}(\mathfrak{f})$ are uniformly continuous on $\mathbb{R}$, and tend to zero at infinity.

It is clear that $\left[\mathcal{S}_{\pm 1}(\mathfrak{f})\right](0) = 0$. Moreover since, for any $0 < r < 1$,

$$\left| \left[\mathcal{S}_{\pm 1}(\mathfrak{f})\right](x) \right| \leqslant \|\mathfrak{f}\|_\infty \, |x|^r \int_{\mathbb{R}} \frac{|p|^r}{p^2+1}\, dp, \qquad x \in \mathbb{R},$$

it follows that $\left[\mathcal{S}_{\pm 1}(\mathfrak{f})\right](x) = O(|x|^r)$ as $x \to 0$ for any $0 < r < 1$.

Note that the functions $\mathcal{E}_{\pm 1}(\mathfrak{f})$, $\mathcal{C}_{\pm 1}(\mathfrak{f})$ and $\mathcal{S}_{\pm 1}(\mathfrak{f})$ are obtained by integrating along lines in Π which are parallel to, but disjoint from, the cut in Π along the negative p-axis. We need both the $+1$ and the -1 form of these integral transforms so as to be able to describe the effect of the behaviour of functions $\mathfrak{f}_{\mathrm{ang}}$ both above and below that cut.

The following technical result is presented in more detail in [53].

Lemma 9.29 *If $\mathfrak{f} \in C^1[-\pi, \pi]$ and $g \in \mathcal{S}(\mathbb{R})$, the identity*

$$\begin{aligned}
\int_{\mathbb{R}} \mathfrak{f}_{\mathrm{ang}}(p, q)(\mathcal{F}^{-1}g)(p)\, dp &= \sqrt{\frac{\pi}{2}} g(0)\{\mathfrak{f}(0) + \mathfrak{f}(\pi\, \mathrm{sgn}(q))\} \\
&\quad - \frac{i}{\sqrt{2\pi}} \mathrm{sgn}(q) \int_0^\infty \frac{g(x) - g(-x)}{x} \left[\mathcal{C}_{\mathrm{sgn}(q)} \mathfrak{f}'\right](qx)\, dx \\
&\quad + \frac{1}{\sqrt{2\pi}} \int_0^\infty \frac{g(x) + g(-x)}{x} \left[\mathcal{S}_{\mathrm{sgn}(q)} \mathfrak{f}'\right](qx)\, dx
\end{aligned} \tag{9.4.36}$$

holds for any nonzero q. The behaviour of $\mathcal{S}_{\pm 1}\mathfrak{f}'$ near the origin is sufficient to ensure that the last integral in this expression is well-defined.

Proof: For any nonzero q the function $p \mapsto \mathfrak{f}_{\mathrm{ang}}(p,q) = \mathfrak{f}(\varphi(p,q))$ is continuously differentiable on $\mathbb{R}$. Direct calculation, using integration by parts, leads to the identity

$$\frac{1}{\sqrt{2\pi}} \int_{-P}^{P} \mathfrak{f}_{\mathrm{ang}}(p,q) \left(\int_{|x| \geqslant \varepsilon} g(x) e^{ixp}\, dx \right) dp =$$

$$- \frac{i}{\sqrt{2\pi}} \mathfrak{f}_{\mathrm{ang}}(P,q) \int_{|x|\geqslant\varepsilon} \frac{g(x)}{x} e^{iPx}\, dx$$

$$+ \frac{i}{\sqrt{2\pi}} \mathfrak{f}_{\mathrm{ang}}(-P,q) \int_{|x|\geqslant\varepsilon} \frac{g(x)}{x} e^{-iPx}\, dx$$

$$- \frac{iq}{\sqrt{2\pi}} \int_{|x|\geqslant\varepsilon} \frac{g(x)}{x} \left(\int_{-P}^{P} \frac{\mathfrak{f}'(\varphi(p,q))}{p^2+q^2} e^{ipx}\, dp \right) dx.$$

Letting ε tend to zero yields the equality

$$\int_{-P}^{P} \mathfrak{f}_{\mathrm{ang}}(p,q)(\mathcal{F}^{-1}g)(p)\, dp$$

$$= -\frac{i}{\sqrt{2\pi}} \mathfrak{f}_{\mathrm{ang}}(P,q)\, I_P(g) + \frac{i}{\sqrt{2\pi}} \mathfrak{f}_{\mathrm{ang}}(-P,q)\, \overline{I_P(\overline{g})}$$

$$- \frac{iq}{\sqrt{2\pi}} \int_0^\infty \frac{g(x)-g(-x)}{x} \left(\int_{-P}^{P} \frac{\mathfrak{f}'(\varphi(p,q))}{p^2+q^2} \cos px\, dp \right) dx$$

$$+ \frac{q}{\sqrt{2\pi}} \int_0^\infty \frac{g(x)+g(-x)}{x} \left(\int_{-P}^{P} \frac{\mathfrak{f}'(\varphi(p,q))}{p^2+q^2} \sin px\, dp \right) dx,$$

where the expression

$$I_P(g) = \lim_{\varepsilon \to 0} \int_{|x|\geqslant\varepsilon} \frac{g(x)}{x} e^{iPx}\, dx\,, \tag{9.4.37.a}$$

is well-defined, and has the limit

$$\lim_{P\to\infty} I_P(g) = \pi i\, g(0). \tag{9.4.37.b}$$

The Dominated Convergence Theorem permits the limit $P \to \infty$ to be taken, giving the desired result. ∎

If $F \in \mathcal{S}(\mathbb{R}^2)$, recall (as was noted in Lemma 8.13) that $\sqrt{2\pi}\mathcal{G}F = \mathcal{F}_1^{-1}\hat{F}$, where $\hat{F}(p,q) = (\tau F)(\frac{1}{2}p, q)$. Thus

$$\begin{aligned}
[\![K_{\mathbf{\Delta}[\mathfrak{f}_{\mathrm{ang}}]}, F]\!] &= [\![\mathfrak{f}_{\mathrm{ang}}, \mathcal{G}F]\!] \\
&= \frac{1}{\sqrt{2\pi}} \int_{\mathbb{R}} \left(\int_{\mathbb{R}} \mathfrak{f}_{\mathrm{ang}}(p,q) \big[\mathcal{F}_1^{-1}\hat{F}\big](p,q)\, dp \right) dq \\
&= \frac{1}{2} \int_{\mathbb{R}} \big\{ \mathfrak{f}(0) + \mathfrak{f}(\pi\,\mathrm{sgn}(q)) \big\} \hat{F}(0,q)\, dq \\
&\quad + \frac{1}{2\pi} \iint_{\Pi} \big[\mathcal{S}_{\mathrm{sgn}(q)}\mathfrak{f}'\big](pq) \frac{\hat{F}(p,q)}{p}\, dp\, dq \\
&\quad - \frac{i}{2\pi} \int_{\mathbb{R}} \mathrm{sgn}(q) \left(\int_0^{\infty} \big[\mathcal{C}_{\mathrm{sgn}(q)}\mathfrak{f}'\big](pq) \frac{\hat{F}(p,q) - \hat{F}(-p,q)}{p}\, dp \right) dq.
\end{aligned}$$

Changing variables in the integral leads to an expression for the integral kernel. These calculations are summarized in this next Proposition.

Proposition 9.30 *If $\mathfrak{f} \in C^1[-\pi,\pi]$ is continuously differentiable, then the integral kernel $K_{\mathbf{\Delta}[\mathfrak{f}_{\mathrm{ang}}]}$ of $\mathbf{\Delta}[\mathfrak{f}_{\mathrm{ang}}]$ is given by*

$$\begin{aligned}
K_{\mathbf{\Delta}[\mathfrak{f}_{\mathrm{ang}}]}(p,q) &= \tfrac{1}{2}\big\{\mathfrak{f}(0) + \mathfrak{f}(\pi\,\mathrm{sgn}(q))\big\}\,\delta(p-q) \qquad (9.4.38) \\
&- \frac{i}{\sqrt{2\pi}(p-q)} \Big\{ \big[\mathcal{C}_{\mathrm{sgn}(p+q)}\mathfrak{f}'\big]\big(\tfrac{1}{2}(p^2-q^2)\big) + i\big[\mathcal{S}_{\mathrm{sgn}(p+q)}\mathfrak{f}'\big]\big(\tfrac{1}{2}(p^2-q^2)\big) \Big\},
\end{aligned}$$

where any integrals involving this expression must be evaluated in a principal value sense. More compactly,

$$\begin{aligned}
K_{\mathbf{\Delta}[\mathfrak{f}_{\mathrm{ang}}]}(p,q) &= \tfrac{1}{2}\big\{\mathfrak{f}(0) + \mathfrak{f}(\pi\,\mathrm{sgn}(q))\big\}\,\delta(p-q) \\
&- \frac{i}{\sqrt{2\pi}(p-q)} \big[\mathcal{E}_{\mathrm{sgn}(p+q)}\mathfrak{f}'\big]\big(\tfrac{1}{2}(p^2-q^2)\big). \qquad (9.4.39)
\end{aligned}$$

This integral kernel is, of course, singular. The singularity due to the presence of the delta distribution is comparatively easy to handle, since its effect is relatively simple. It is the singularity due to the presence of the factor $(p-q)^{-1}$ that is more complicated. However, it should be noted that the behaviour of $\mathcal{S}_{\pm 1}(\mathfrak{f}')$ near the origin is sufficient to ensure that the function

$$(p,q) \mapsto \big[\mathcal{S}_{\mathrm{sgn}(p+q)}\mathfrak{f}'\big]\big(\tfrac{1}{2}(p^2-q^2)\big)\,\frac{F(p,q)}{p-q}$$

belongs to $L^1(\mathbb{R}^2)$ for any $F \in \mathcal{S}(\mathbb{R}^2)$. Thus the only part of this integral kernel whose analysis needs the full machinery of principal value integrals is the term

$$-\frac{i}{\sqrt{2\pi}(p-q)}\left[\mathcal{C}_{\mathrm{sgn}(p+q)}\mathfrak{f}'\right]\left(\tfrac{1}{2}(p^2-q^2)\right).$$

Chapter 10

Phase Operators

Two wrongs don't make a right, but they make a good excuse.
– Thomas Szasz

Angular distribution of what?!
– John Klauder

10.1 Field Theory And Modes

In this Chapter various operators and families of states that are supposed to possess phaselike properties in some sense or another will be considered. Before doing so, two questions have to be addressed: how is the formalism of quantum phase related to that of the quantized electromagnetic field, and; what does the term *quantum phase* mean?

10.1.1 *The Free Quantized Electromagnetic Field*

The answer to the first question requires a brief description of the quantized electromagnetic field. A more complete discussion of this material can be found in the book of Mandel & Wolf [164]. As the interacting states are known only in renormalized perturbation theory, we shall suppose the field to be in the free state and confined to a cube V of edge L in space[1].

For this book, it is sufficient to begin with the classical electromagnetic field. In the Coulomb gauge, and in the absence of any charge distribution or current, the vector potential **A** is a real-valued vector field satisfying the

[1] Other spatial geometries can easily be accommodated by solving Helmholtz's equation for the appropriate boundary condition.

wave equation

$$\nabla^2 \mathbf{A} - \frac{1}{c^2}\frac{\partial^2}{\partial t^2}\mathbf{A} = \mathbf{0}, \tag{10.1.1.a}$$

subject to the Coulomb gauge condition

$$\nabla \cdot \mathbf{A} = 0. \tag{10.1.1.b}$$

The vector potential defines the electric and magnetic fields through the formulæ

$$\mathbf{E} = -\frac{\partial}{\partial t}\mathbf{A}, \qquad \mathbf{B} = \nabla \times \mathbf{A}. \tag{10.1.2}$$

Moreover, these equations are supplemented by choosing the spatial domain to be the "box"

$$V = \{(x_1, x_2, x_3) \in \mathbb{R}^3 : 0 \leqslant x_j \leqslant L, j = 1, 2, 3.\}, \tag{10.1.3}$$

and by imposing periodic boundary conditions on the surface of V.

It is clear that the vector potential $\mathbf{A}$ can be expressed as a Fourier series expansion

$$\mathbf{A}(\mathbf{r}, t) = \sum_{\mathbf{k}} \tilde{\boldsymbol{A}}_{\boldsymbol{k}}(t)\, e^{i\mathbf{k}\cdot\mathbf{r}}, \tag{10.1.4}$$

where the sum is taken over all vectors $\mathbf{k}$ of the form $\mathbf{k} = 2\pi L^{-1}\mathbf{n}$, where $\mathbf{n} \in \mathbb{Z}^3$. Solving the differential equation (10.1.1.a) and requiring that $\mathbf{A}$ be real-valued results in the solution

$$\mathbf{A}(\mathbf{r}, t) = \frac{1}{2\sqrt{\varepsilon_0 L^3}} \sum_{\mathbf{k}} \left[\mathbf{b}_{\mathbf{k}} e^{-i\omega(\mathbf{k})t} + \mathbf{b}^*_{-\mathbf{k}} e^{i\omega(\mathbf{k})t}\right] e^{i\mathbf{k}\cdot\mathbf{r}}, \tag{10.1.5.a}$$

where each vector $\mathbf{b}_{\mathbf{k}} \in \mathbb{C}^3$ is a constant, and the frequency $\omega(\mathbf{k})$ is defined to be

$$\omega(\mathbf{k}) = c|\mathbf{k}|. \tag{10.1.5.b}$$

Moreover, the gauge condition (10.1.1.b) implies that[2]

$$\mathbf{k} \cdot \mathbf{b}_{\mathbf{k}} = 0 \tag{10.1.5.c}$$

for each $\mathbf{k}$.

The electric and magnetic fields are then given by the formulæ

$$\mathbf{E}(\mathbf{r}, t) = \frac{i}{2\sqrt{\varepsilon_0 L^3}} \sum_{\mathbf{k}} \omega(\mathbf{k}) \left[\mathbf{b}_{\mathbf{k}} e^{-i\omega(\mathbf{k})t} - \mathbf{b}^*_{-\mathbf{k}} e^{i\omega(\mathbf{k})t} \right] e^{i\mathbf{k}\cdot\mathbf{r}}, \tag{10.1.6.a}$$

$$\mathbf{B}(\mathbf{r}, t) = \frac{i}{2\sqrt{\varepsilon_0 L^3}} \sum_{\mathbf{k}} \mathbf{k} \times \left[\mathbf{b}_{\mathbf{k}} e^{-i\omega(\mathbf{k})t} + \mathbf{b}^*_{-\mathbf{k}} e^{i\omega(\mathbf{k})t} \right] e^{i\mathbf{k}\cdot\mathbf{r}}. \tag{10.1.6.b}$$

The Poynting vector for this field is nonzero, but satisfies periodic boundary conditions. Consequently the integral of $\mathbf{E} \times \mathbf{B}$ over the boundary of V vanishes, and hence the cavity energy

$$H = \frac{1}{2} \iiint_V \left(\varepsilon_0 |\, \mathbf{E} \,|^2 + \frac{1}{\mu_0} |\, \mathbf{B} \,|^2 \right) dx_1 \, dx_2 \, dx_3$$

is constant, and direct calculation shows that

$$H = \frac{1}{2} \sum_{\mathbf{k}} \omega(\mathbf{k})^2 \,|\, \mathbf{b}_{\mathbf{k}} \,|^2 . \tag{10.1.7}$$

For physical reasons, for each (nonzero) $\mathbf{k}$ we introduce polarization vectors $\mathbf{e}_{\mathbf{k},1}$, $\mathbf{e}_{\mathbf{k},2} \in \mathbb{C}^3$ such that

$$\begin{aligned} \mathbf{k} \cdot \mathbf{e}_{\mathbf{k},i} &= 0\,, & i &= 1, 2\,, \\ \mathbf{e}^*_{\mathbf{k},i} \cdot \mathbf{e}_{\mathbf{k},j} &= \delta_{ij}\,, & i, j &= 1, 2. \\ \mathbf{e}_{\mathbf{k},1} \times \mathbf{e}_{\mathbf{k},2} &= \frac{1}{|\,\mathbf{k}\,|}\mathbf{k}\,. \end{aligned} \tag{10.1.8}$$

If the vectors $\mathbf{e}_{\mathbf{k},1}$ and $\mathbf{e}_{\mathbf{k},2}$ are real then they, together with the unit vector in the direction of $\mathbf{k}$, form a right-handed orthonormal triad in $\mathbb{R}^3$. Such a situation describes *linear polarization.* The general case describes *elliptic polarization.*

---

[2]These observations are not totally true, since they do not allow for the full range of possible solutions in the case $\mathbf{k} = \mathbf{0}$. However, the contributions made to $\mathbf{A}$ by the $\mathbf{k} = \mathbf{0}$ solutions have no physical importance, and so may be ignored. Indeed, physical considerations lead us to ignore modes of oscillation with wavelengths larger than the dimensions of the box V, which means that the Fourier series for $\mathbf{A}$ will be summed over those vectors $\mathbf{k}$ for which $|\,\mathbf{k}\,|$ is bounded away from zero. For the purposes of this mathematical discussion, however, it is sufficient simply to exclude the case $\mathbf{k} = \mathbf{0}$.

Each vector $\mathbf{b_k}$ can then be written in the form

$$\mathbf{b_k} = \xi_{\mathbf{k},1}\mathbf{e}_{\mathbf{k},1} + \xi_{\mathbf{k},2}\mathbf{e}_{\mathbf{k},2}\,, \tag{10.1.9}$$

where $\xi_{\mathbf{k},1}$ and $\xi_{\mathbf{k},2}$ are complex numbers. Any choice of $\mathbf{k}$ and $s \in \{1,2\}$ defines a single mode of oscillation (with momentum $\mathbf{k}$ and polarization vector $\mathbf{e}_{\mathbf{k},s}$). The vector potential $\mathbf{A}$ can then be written as a sum over modes as follows:

$$\mathbf{A}(\mathbf{r},t) = \frac{1}{2\sqrt{\varepsilon_0 L^3}} \sum_{\mathbf{k},s} \left\{ \xi_{\mathbf{k},s} e^{i(\mathbf{k}\cdot\mathbf{r}-\omega(\mathbf{k})t)} + \overline{\xi_{\mathbf{k},s}} e^{-i(\mathbf{k}\cdot\mathbf{r}-\omega(\mathbf{k})t)} \right\}; \tag{10.1.10.a}$$

similar expressions can be found for the electric and magnetic fields. In particular, the cavity energy is

$$H = \frac{1}{2} \sum_{\mathbf{k},s} \omega(\mathbf{k})^2 |\, \xi_{\mathbf{k},s} \,|^2\,, \tag{10.1.10.b}$$

a simple sum over all modes.

The canonical quantization of the electromagnetic field is then achieved by introducing an unbounded operator $a_{\mathbf{k},s}$ for each mode such that

$$\left[a_{\mathbf{k},i}\,,\, a^+_{\mathbf{l},j}\right] = \delta_{\mathbf{kl}}\,\delta_{ij} \tag{10.1.11}$$

for all $\mathbf{k}$, $\mathbf{l}$ and all i, j, and then defining the quantum mechanical Hamiltonian to be

$$H = \hbar \sum_{\mathbf{k},s} \omega(\mathbf{k})\, a^+_{\mathbf{k},s} a_{\mathbf{k},s}\,. \tag{10.1.12}$$

By doing so, $a_{\mathbf{k},s}$ can be regarded as the quantization of the classical quantity

$$\sqrt{\frac{\omega(\mathbf{k})}{2\hbar}}\,\xi_{\mathbf{k},s} = \frac{1}{\sqrt{2\hbar\omega(\mathbf{k})}} \left(\omega(\mathbf{k}) q_{\mathbf{k},s} + i p_{\mathbf{k},s}\right),$$

which expresses the coefficient $\xi_{\mathbf{k},s}$ in terms of the two real classical observables $q_{\mathbf{k},s}$ and $p_{\mathbf{k},s}$. In this way, the similarity of the quantization of each mode to that of the harmonic oscillator is clear.

At this point it is clearly appropriate to study the behaviour of a single mode of oscillation. For the remainder of this Chapter, therefore, it will be assumed that the electromagnetic field is described by a quantum mechanical Hamiltonian of the form $H = \hbar\omega A^+ A$, where A is an unbounded operator on $L^2(\mathbb{R})$ which provides a representation of the CCR. In other words, all the indices of the above discussion have been dropped, and the operator A has been capitalized for the sake of consistency with previous discussion.

This is not the whole story of the free quantized electromagnetic field of course, for there is the problem of the indefinite metric, which may be treated with the Gupta-Bleuler formalism [216], [217]. But that approach considers the character of the Fock space for the vector potential, and not the choice of CCR representation, and consequently is not an issue in the present discussion.

10.1.2 *Collective Excitations*

It is known that phenomena such as superfluidity or lattice vibrations have states which can be approximated by an independent excitation description. Put differently, it is expected that for these systems in these states, the exact dynamical solution could be interpreted as consisting mainly of an assemblage of independent oscillators describing the collective oscillations. As this exact solution is not available, to a high degree of accuracy it can be replaced by a phenomenological description by oscillators together with parameters supplied from observation. Higher approximations allow these parameters to vary, as in the notion of an energy dependent "mass", but such effects will not be considered in our description of quantum phase.

The operation of a laser is a collective phenomenon, so it is sensible to ask how appropriate the oscillator description is for laser light. After all, the (nonexistent) exact description of a laser would involve interacting electromagnetic fields, and the Fourier coefficients of these would not be oscillator operators. Indeed, all one knows about the commutation relations for the fields *in interaction* is that they vanish outside the appropriate light cones. Moreover, unlike free fields, an interacting field does not directly create and destroy particles. As Haag puts it [94], *The rôle of fields is to implement the principle of locality. The number and the nature of different basic fields needed in the theory is related to the charge structure, not to the empirical spectrum of particles. In the presently favoured gauge theories the basic fields are the carriers of charges called colour and flavour but are not directly associated to observed particles like protons.* Even worse, these are long range fields: photons are zero mass particles that can never be localized and are strictly relativistic objects.

Bearing these problems in mind, consider how laser light is created.

Essentially free and incoherent electromagnetic radiation is incident on the walls of a cavity. The effective interaction region for the field is within the cavity. Under the right circumstances, the incoming incoherent radiation is transformed by this interaction into a coherent state which is then leaked (emitted) from the laser device. At this point, coherent light is available to the experimentalist who does not have to be concerned with how it was created. A typical experiment would then have the coherent light confined to some (different) cavity, which includes half silvered mirrors, photodetectors and the like.

So there are two principal stages to a quantum optics experiment. In the first, laser light is created in the background. In the second, the created light is used for some purpose. The laser model described in Chapter 11 considers the first stage only, and results in a totally coherent state. In a more nearly complete model of the laser, an exact solution would presumably result in a state in which this coherence is the principal effect, but modified by higher order correlations (known as photon statistics). So far as we know, at present there is no way to obtain this sort of state without more or less putting it in by hand, and such models are not so much predictive as descriptive. Even so, there does not seem to us to be a clean connection between a model of coherent light creation, however phenomenological, and the quantum optics description that begins with some favourite vector state in $L^2(\mathbb{R})$.

Let us restate this. The degrees of freedom describing radiation in the creation cavity utilize smooth observables in interaction with a sink and with the cavity atoms. But these radiation observables are not the same as the smooth observables that describe the field in the apparatus in which the experiment is done. These latter operators describe a completely different system, wholly phenomenological, and if there is any dynamics in this description, it is not that of the interaction with the atoms in the creation cavity.

Within the formalism of a model of the creation of coherent radiation, we have been able to trace how the various phase operators behave in that process. As there are questions still open about the relation between the output radiation from the model and the radiation observed in experiments with lasers, so there must be questions open about what operator describes the light phase in any given experimental arrangement. This is clearly a matter of great importance for the theory, and deserves more attention paid to it than heretofore.

10.2 What Do We Mean By Quantum Phase?

Notwithstanding questions of principle, certain operators have been proposed as describing quantum phase for one degree of freedom (mode). In this Chapter, some of these operators will be examined as a problem in basic quantum theory. But in order eventually to judge how appropriate a proposal is physically, it is necessary to decide what the term *quantum phase observable* should mean.

As a general principle of quantum mechanics, any quantum observable carries its own meaning in that there is, in theory, an experimental arrangement that will respond to the physical property the operator represents[3]: the apparatus will register spectral values and emit corresponding eigenfunctions as output states. This is the end of the matter as far as the strict Copenhagen interpretation of quantum mechanics is concerned. But this is too austere a standpoint to satisfy most physicists, for unless we can gain some understanding of some identifiable aspect of the world around us, such experiments are pointless even if possible. It was Bohr who emphasized that understanding in this sense requires a connection to a classical concept in some sense or another[4]. Only by a careful and structurally complete analysis of an experiment can the precise connection between the quantum operator and some classical quality be validated. It is by not doing so, by substituting metaphor for analysis, that (supposed) paradoxes creep into the discussion.

An example of incomplete analysis is the tendency to identify the classical quality of a quantum operator from its spectrum alone, which is clearly wrong. This is exemplified by the momentum and position operators P and Q, which have the same spectral values but quite different physical meanings. At the end of this Chapter we shall consider what the minimum information in terms of (probability) distributions is necessary to identify an observable.

---

[3]We are aware that this assumes a strong correspondence between the mathematical theory and physical reality.

[4]Bohr also believed that the experimental arrangement required a macroscopic registration component which was describable only by classical mechanics, but we believe that it can always be described within the quantum formalism by using methods similar to those developed for statistical mechanics.

10.3 Some Candidate Phase Operators

Rather curiously, a good place to begin a discussion of phase operators is by considering what a generalized eigenstate of a canonical phase operator might be — if there were such a thing.

10.3.1 *Pure Phase States*

Consider how the canonically conjugate pair Q and P act on each other's (generalized) eigenfunctions. The nature of these actions can be seen most simply in exponentiated form. Recalling the generalized eigenfunctions δ_x ($x \in \mathbb{R}$) and T_k ($k \in \mathbb{R}$) of Q and P respectively introduced in Section 7.1 of Chapter 7, it is evident that

$$\left(e^{iaP}\right)^{\mathrm{tr}}\delta_x \;=\; \delta_{x+a}\,, \qquad \left(e^{iaQ}\right)^{\mathrm{tr}}T_k \;=\; T_{k-a}\,, \tag{10.3.1}$$

for all $a, x, k \in \mathbb{R}$. In other words, in both cases the action is simply that of translating the parameter.

Now, as we know, there is no operator which is canonically conjugate to N but, if there were one, it could be expected to have the same mutual eigendistribution translation properties with N as do P and Q. Consequently, it may well be interesting to look for distributions Ψ_β ($\beta \in \mathbb{R}$) which satisfy the identity[5]

$$\left(e^{i\theta N}\right)^{\mathrm{tr}}\Psi_\beta \;=\; \Psi_{\theta+\beta}\,, \qquad \theta,\,\beta \in \mathbb{R}\,. \tag{10.3.2.a}$$

Equivalently, this condition may be written

$$N^{\mathrm{tr}}\Psi_\beta \;=\; -i\,\frac{d\Psi_\beta}{d\beta}\,. \tag{10.3.2.b}$$

Since the Hermite-Gauss functions h_n form a topological basis for $\mathcal{S}'(\mathbb{R})$, there must be a unique slowly growing sequence $(c_n(\beta))_{n\geqslant 0}$ for each β such that

$$\Psi_\beta \;=\; \sum_{n=0}^{\infty} c_n(\beta) h_n.$$

The above conditions then imply that

$$c_n(\beta) \;=\; c_n(0)e^{in\beta}\,,$$

[5]Note that mathematical consistency requires Ψ_β to be periodic in β of period 2π.

for each n, so

$$\Psi_\beta = \sum_{n=0}^{\infty} c_n(0) e^{in\beta} h_n. \tag{10.3.3}$$

If the sequence $\big(c_n(0)\big)_n$ belongs to ℓ^2, then the distribution Ψ_β must belong to $L^2(\mathbb{R})$. However, such a state weights the various h_n differently, whereas the commonly-held view is that a phase operator should treat them all identically.

Accepting this reasoning would lead us to chose $c_n(0) = 1$ for all n, resulting in the tempered distributions

$$\Lambda_\beta = \sum_{n=0}^{\infty} e^{in\beta} h_n\,, \qquad \beta \in \mathbb{R}\,. \tag{10.3.4}$$

These distributions were first introduced as early as 1926 by Fritz London [151, 152]. Hence we call the Λ_β London distributions. (Some authors include a factor of $(2\pi)^{-1/2}$ in Λ_β for normalization purposes, but this is taken care of here by our using the measure $d\beta/2\pi$.) Notice that if $f = \sum b_n h_n$ is in $\mathcal{S}(\mathbb{R})$, then

$$[\![\Lambda_\beta\,, f]\!] = \sum_{n=0}^{\infty} b_n e^{in\beta} = \big(\mathfrak{U}_T f\big)(e^{i\beta}), \tag{10.3.5}$$

for any $\beta \in \mathbb{R}$, where $\mathfrak{U}_T$ is the unitary isomorphism between $L^2(\mathbb{R})$ and $H^2(\mathbb{T})$ given by

$$\mathfrak{U}_T h_n = \chi_n, \qquad n \geqslant 0, \tag{5.7.24}$$

in Chapter 5. Thus the integral kernel of $\mathfrak{U}_T$ is given by the symbolic expression

$$\mathfrak{U}_T(e^{i\beta}, x) = \Lambda_\beta(x)\,. \tag{10.3.6}$$

Thus, pairing with the London distribution implements the isomorphism $\mathfrak{U}_T$, which is why Hardy space recurs frequently in phase theory.

Another approach sets the first few coefficients $c_n(0)$ all equal, and then sets all the remaining coefficients to be zero. In other words, setting

$$c_n(0) = \begin{cases} \dfrac{1}{\sqrt{s+1}}\,, & 0 \leqslant n \leqslant s\,, \\ 0\,, & n > s\,, \end{cases}$$

leads to the unit vectors[6]

$$\zeta_s[\beta] = \frac{1}{\sqrt{s+1}} \sum_{n=0}^{s} e^{-in\beta}\, h_n\,, \qquad -\pi < \beta \leqslant \pi\,, \tag{10.3.7}$$

for any non-negative integer s. These states were used systematically by Lerner, Huang & Walters [149], and so will be termed *LHW states* — they define pure states in either the smooth or bounded model. We shall return to considering these states toward the end of this Chapter.

10.3.2 *Operators From The London Distributions*

There is an important family of phase-related operators which frequently appear in connection with the London distributions, since the London distributions can be used to transport bounded operators on Hardy space $H^2(\mathbb{T})$ to bounded operators on $L^2(\mathbb{R})$. In particular, this procedure can be applied to Toeplitz operators on $H^2(\mathbb{T})$. Recall that any function ω in $L^\infty(\mathbb{T})$ defines the bounded Toeplitz operator $M(\omega)$ on $H^2(\mathbb{T})$ via the formula

$$M(\omega)F = P_+(\omega F)\,, \qquad F \in H^2(\mathbb{T})\,, \tag{5.7.15}$$

where $P_+ : L^2(\mathbb{R}) \to H^2(\mathbb{T})$ is the Szëgo-Riesz projection. The unitary mapping $\mathfrak{U}_T : L^2(\mathbb{R}) \to H^2(\mathbb{T})$ can then be used to define the bounded operator $\mathcal{M}(\omega)$ on $L^2(\mathbb{R})$ by the formula

$$\mathcal{M}(\omega) = \mathfrak{U}_T^{-1} \circ M(\omega) \circ \mathfrak{U}_T\,; \tag{10.3.8}$$

the map $\mathcal{M}(\omega)$ is often (somewhat inaccurately) referred to as a Toeplitz operator. With respect to the standard Hermite-Gauss basis $\{h_n : n \geqslant 0\}$ for $L^2(\mathbb{R})$, the operator $\mathcal{M}(\omega)$ clearly has matrix coefficients

$$\begin{aligned} \langle h_m\,, \mathcal{M}(\omega)h_n\rangle &= \frac{1}{2\pi}\int_{-\pi}^{\pi} \omega(e^{i\beta})e^{i(n-m)\beta}\,d\beta \\ &= \hat{\omega}_{m-n}, \end{aligned} \tag{10.3.9}$$

for $m, n \geqslant 0$ — simple functions of the Fourier coefficients of ω.

[6]Note the change of sign in the exponential terms in $\zeta_s[\beta]$. As previously remarked, the natural embedding of $L^2(\mathbb{R})$ into $\mathcal{S}'(\mathbb{R})$ is antilinear, and so is frequently composed with the complex structure of complex conjugation — the change of sign in these exponentials represents the necessary complex conjugation.

However, the literature is not uniform concerning the definition of angle functions. For purposes of consistency with later discussion, we intend to adopt a slightly different definition for Toeplitz operators. We do this by defining the bounded operator $\overline{\mathcal{M}}(\omega)$ on $L^2(\mathbb{R})$ by the formula

$$\overline{\mathcal{M}}(\omega) = \mathcal{F}^{-1} \circ \mathcal{M}(\omega) \circ \mathcal{F}, \qquad (10.3.10.a)$$

for any $\omega \in L^\infty(\mathbb{T})$, where we recall that the Fourier transform $\mathcal{F}$ is a unitary automorphism of $L^2(\mathbb{R})$ such that

$$\mathcal{F} h_n = i^{-n} h_n, \qquad n \geqslant 0. \qquad (10.3.10.b)$$

When interpreted back in $H^2(\mathbb{T})$, the effect of this conjugation by $\mathcal{F}$ is to rotate the angle in Hardy space through 90°. The operator $\overline{\mathcal{M}}(\omega)$ then has matrix coefficients

$$\langle h_m, \overline{\mathcal{M}}(\omega) h_n \rangle = i^{m-n} \langle h_m, \mathcal{M}(\omega) h_n \rangle = i^{m-n} \hat{\omega}_{m-n}. \qquad (10.3.10.c)$$

Definition 10.1 *The phase-related operators associated with the London distributions are the bounded Toeplitz operators* $X = \overline{\mathcal{M}}(\mathfrak{p})$, $E = \overline{\mathcal{M}}(\chi_{-1})$, $E^* = \overline{\mathcal{M}}(\chi_1)$, $S = \overline{\mathcal{M}}(\sin)$ *and* $C = \overline{\mathcal{M}}(\cos)$, *where* $\mathfrak{p} \in L^\infty(\mathbb{T})$ *is the function* $\mathfrak{p}(e^{i\beta}) = \beta$ *(for* $-\pi < \beta \leqslant \pi$*) defined in formula (9.4.11.c) of the previous Chapter.*

It is elementary to establish the following facts concerning the operators E and E^*. The matrix coefficients of these operators with respect to the Hermite-Gauss functions are

$$\langle h_m, E h_n \rangle = -i\, \delta_{m+1,n}, \qquad \langle h_m, E^* h_n \rangle = i\, \delta_{m,n+1}. \qquad (10.3.11)$$

Consequently

$$E E^* = I, \qquad E^* E = I - P_0, \qquad (10.3.12)$$

where P_0 is the orthogonal projection onto the subspace of $L^2(\mathbb{R})$ spanned by the vector h_0. Consequently the map E^* is isometric, but E is not — it is only a partial isometry.

Since it is clear that

$$C = \frac{1}{2}(E^* + E), \qquad S = \frac{1}{2i}(E^* - E), \qquad (10.3.13)$$

the matrix coefficients of the self-adjoint operators C and S with respect to the Hermite-Gauss functions can be readily obtained from the above identities. It is important to note that since

$$CS - SC = \frac{1}{2i} P_0 , \tag{10.3.14.a}$$

the operators C and S do not commute with each other. Moreover, since

$$C^2 + S^2 = I - \frac{1}{2} P_0 , \tag{10.3.14.b}$$

it follows that $C^2 + S^2$ is not equal to I.

The matrix coefficients of X with respect to the Hermite-Gauss vectors can be calculated to be

$$\langle h_m , X h_n \rangle = i^{m-n} \hat{\mathfrak{p}}_{m-n} = \begin{cases} \dfrac{i^{n-m+1}}{m-n} , & m \neq n , \\ 0 , & m = n . \end{cases} \tag{10.3.15}$$

Although E, E^*, S and C all belong to $\mathcal{L}^+(\mathcal{S}(\mathbb{R}))$ (in addition to being bounded operators on $L^2(\mathbb{R})$), X does not. This can be seen by observing that the vector Xh_0 does not belong to $\mathcal{S}(\mathbb{R})$, since

$$X h_0 = \sum_{m \geqslant 0} \frac{i^{1-m}}{m} h_m .$$

Spectral theory for Toeplitz operators was briefly discussed in Section 5.7.2. The particular results of that theory, applied to the above five operators, are listed in Table 10.1 below. For the standard mathematical analysis of the properties of these operators, appropriate analysis texts should be consulted, cf, [97], [118], [196], [25], [219].

Table 10.1 Spectral Properties of Toeplitz Operators

Properties	E	E^*	C	S	X
Spectrum	$\overline{\mathbb{D}}$	$\overline{\mathbb{D}}$	$[-1,1]$	$[-1,1]$	$[-\pi,\pi]$
Eigenvalues	$\mathbb{D}$	$\emptyset$	$\emptyset$	$\emptyset$	$\emptyset$
Continuous Spectrum	$\mathbb{T}$	$\mathbb{T}$	$[-1,1]$	$[-1,1]$	$[-\pi,\pi]$
Residual Spectrum	$\emptyset$	$\mathbb{D}$	$\emptyset$	$\emptyset$	$\emptyset$

The eigenvalue $z \in \mathbb{D}$ of E is nondegenerate, with normalized eigenvector

$$\Psi_z = (1 - |z|^2)^{-\frac{1}{2}} \sum_{n \geqslant 0} i^n z^n h_n . \tag{10.3.16}$$

It is clear that the London distributions form a family of generalized eigenvectors for the operator E^*, since

$$[\![\Lambda_\beta , E^* f]\!] = e^{i(\beta + \frac{\pi}{2})} [\![\Lambda_\beta , f]\!] , \qquad f \in \mathcal{S}(\mathbb{R}), -\pi < \beta \leqslant \pi . \tag{10.3.17}$$

It is also possible to find families of generalized eigenvectors for S and C, since

$$[\![\Lambda^{(S)}_{\sin\beta} , Sf]\!] = \sin\beta \, [\![\Lambda^{(S)}_{\sin\beta} , f]\!], \quad f \in \mathcal{S}(\mathbb{R}), -\frac{1}{2}\pi < \beta < \frac{1}{2}\pi, \tag{10.3.18.a}$$

where

$$\Lambda^{(S)}_{\sin\beta} = \sum_{n \geqslant 0} \cos\left[\frac{1}{2} n\pi - (n+1)\beta\right] h_n , \qquad -\frac{1}{2}\pi < \beta < \frac{1}{2}\pi , \tag{10.3.18.b}$$

and

$$[\![\Lambda^{(C)}_{\cos\beta} , Cf]\!] = \cos\beta \, [\![\Lambda^{(C)}_{\cos\beta} , f]\!] , \qquad f \in \mathcal{S}(\mathbb{R}), 0 < \beta < \pi , \tag{10.3.19.a}$$

where

$$\Lambda^{(C)}_{\cos\beta} = \sum_{n \geqslant 0} i^{-n} \sin\left[(n+1)\beta\right] h_n . \tag{10.3.19.b}$$

The generalized eigendistributions for S and C provide a weak spectral representation in accordance with equations (7.1.11) to (7.1.15).

The subject of quantum phase began with Dirac's consideration of the operators E and E^*, which appear naturally in the polar decompositions of the raising and lowering operators for the harmonic oscillator:

$$\mathcal{F}^{-1} \circ A \circ \mathcal{F} = \sqrt{N}\, E, \qquad \mathcal{F}^{-1} \circ A^{+} \circ \mathcal{F} = \sqrt{N}\, E^*. \tag{10.3.20}$$

This was, essentially, Dirac's starting point in 1927 when he further supposed that E and E^* were unitary. Were this so, then writing $E = \exp(i\Phi)$ would yield an operator Φ canonically conjugate to N. But the No-Go Theorem assures us that this is not possible and, as is clear, neither E nor E^* are unitary.

Given the highly plausible nature of the derivation of the London distributions Λ_β and the associated operators E, E^*, S, C and X, it is natural to ask what the physical significance of the quantity β is. The strongest possible supposition would be that there is a self-adjoint observable which has the London distributions as its family of generalized eigendistributions. In other words, it might be supposed that there exists a self-adjoint bounded operator Ξ on $L^2(\mathbb{R})$ which satisfies the weak eigenvalue equation

$$[\![\Lambda_\beta , \Xi f]\!] = \beta [\![\Lambda_\beta , f]\!] , \qquad f \in \mathcal{S}(\mathbb{R}), -\pi < \beta \leqslant \pi . \tag{10.3.21}$$

However, this is impossible, since this would imply that $\mathfrak{U}_T \circ \Xi \circ \mathfrak{U}_T^{-1}$ was a bounded operator on $H^2(\mathbb{T})$ such that

$$\mathfrak{U}_T \circ \Xi \circ \mathfrak{U}_T^{-1} F = \mathfrak{p} F , \qquad F \in \mathfrak{U}_T \mathcal{S}(\mathbb{R}) , \tag{10.3.22}$$

and this would imply in particular that the function $\mathfrak{p}$ itself belonged to $H^2(\mathbb{T})$, which is not true. Hence we have a variant of the No-Go Theorem:

Proposition 10.2 ***(Another No-Go Theorem)*** *There is no bounded self adjoint operator Ξ on $L^2(\mathbb{R})$ whose spectrum is $\mathbb{T}$ and whose generalized eigenfunctions are the London distributions.*

Thus β has no physical meaning in this direct sense. Why, then, should β be accounted a phase angle at all? In principle, X, S and C are measurable[7]. As far as we know, it has not been shown convincingly how any particular experimental apparatus can measure these (approximating) operators, or that their eigenstates show significant coherence effects. But granting that this might be done, this leaves a rather curious phenomenon to account for. The operators X, S and C do not commute with each other, so an ideal measurement of one, which would prepare a state with a definite value for the associated angle variable, does not give a definite value to the angles appearing in the definition of the other two operators, notwithstanding that they are meant to be the same independent variable β on Hardy space. In particular, if S is diagonal C is not, and vice versa. Experiments to measure these operators will presumably involve breaking parity symmetry in some sense.

[7] As will be argued in Chapter 16, an experimental apparatus designed to measure one of these operators will in fact be measuring some approximation to these operators which has discrete spectrum.

However, it is possible to show that the operators E, E^*, S, C and X do bear some relation to the classical phase angle, although not directly. This can be seen by evaluating the symbols of these operators, namely their Weyl dequantizations. The following results will be derived in Chapter 12. Note that by $\mathbf{\Delta}^{-1}[B]$ is meant the phase space distribution T whose Weyl quantization $\mathbf{\Delta}[T]$ is equal to B.

Proposition 10.3 *The symbol of E is the function*

$$\begin{aligned} &\mathbf{\Delta}^{-1}[E](r\cos\beta, r\sin\beta) \\ &= e^{-i\beta}2^{3/2}re^{-r^2}\sum_{j\geqslant 0}(-1)^j\frac{1}{\sqrt{j+1}}L_j^{(1)}(2r^2) \\ &= \frac{re^{-i\beta}}{\sqrt{2\pi}}\int_0^\infty \exp\left[-r^2\tanh\left(\tfrac{1}{2}\xi\right)\right]\operatorname{sech}^2\left(\tfrac{1}{2}\xi\right)\frac{d\xi}{\sqrt{\xi}}. \end{aligned} \qquad (10.3.23)$$

The limit

$$\lim_{r\to\infty}\mathbf{\Delta}^{-1}[E](r\cos\beta, r\sin\beta) = e^{-i\beta} \qquad (10.3.24)$$

is true both pointwise and (appropriately defined) in the sense of distributions. Taking the complex conjugate of the above results yields the corresponding results for E^*.

In the case of the operator X, a more complicated formula for $\mathbf{\Delta}^{-1}[X]$ (which is also a function) can be found, and it can be shown that

$$\lim_{r\to\infty}\mathbf{\Delta}^{-1}[X](r\cos\beta, r\sin\beta) = \mathfrak{p}(e^{i\beta}) = \varphi(\beta) \qquad (10.3.25)$$

for all $-\pi < \beta \leqslant \pi$, in the same pointwise and distributional sense. Thus the *(Weyl) symbols of these operators are not functions of the classical phase angle alone, but more complicated functions which approximate the angular functions at large radii/energy.* It is our view that these results make these operators interesting, but imprecise, quantities when studying quantum phase effects.

10.3.3 *The Bargmann-Segal Phase Operator*

As has been seen above, the London distributions have been used as a means for introducing candidate phase observables. When considered from a mathematical point of view, the London distributions have simply implemented the unitary equivalence of $L^2(\mathbb{R})$ and $H^2(\mathbb{T})$. Since there is, equally well, a unitary equivalence between $L^2(\mathbb{R})$ and the Bargmann-Segal representation space $\mathcal{B}$, it is not surprising that there is a candidate phase observable which owes its provenance to the formalism of $\mathcal{B}$. In the construction of this operator, the so-called *coherent states* play the same rôle for the new operator that the London distributions did previously[8].

Firstly it is necessary to define a coherent state and to explain why it is so named. Our convention is that a coherent state is the pure state determined by a coherent vector, and a coherent vector is a normalized vector of the form

$$\Phi_\zeta = e^{-\frac{1}{2}|\zeta|^2} \sum_{n\geqslant 0} \frac{\overline{\zeta}^n}{\sqrt{n!}} h_n , \qquad (10.3.26.a)$$

for any $\zeta \in \mathbb{C}$. This series can be summed, giving

$$\Phi_\zeta(q) = \pi^{-1/4} \exp\left[-\frac{1}{2}(\overline{\zeta}^2 + |\zeta|^2) - \frac{1}{2}q^2 + \sqrt{2}\,\overline{\zeta} q\right] . \qquad (10.3.26.b)$$

We shall write ω_ζ for the coherent state determined by the vector Φ_ζ.

Coherent vectors are important not least because they are eigenvectors of the lowering operator,

$$A\Phi_\zeta = \overline{\zeta}\,\Phi_\zeta . \qquad (10.3.27)$$

It follows from this that coherent states factor over normal ordered operators, in that

$$\omega_\zeta\left([A^+]^r A^s\right) = \left[\omega_\zeta\left(A^+\right)\right]^r \left[\omega_\zeta\left(A\right)\right]^s = \zeta^r\,\overline{\zeta}^s , \qquad (10.3.28)$$

for any $r, s \in \mathbb{N} \cup \{0\}$. This result for monomials in A and A^* extends linearly to one for normal ordered polynomials[9]. However factorization does not hold for polynomials in the number operator, since (for example)

$$\omega_\zeta(N^2) = \omega_\zeta(N)^2 + \omega_\zeta(N) .$$

[8]Some discussions of this operator start with considering the expression $\log A - \log A^+$. Working with A is not strictly necessary, since doing so involves a spurious appearance of the number operator, but working instead with an expression like $\log E - \log E^*$, which does not involve the number operator, does not greatly simplify matters, since any approach of this nature would require considerable care, as the logarithm of the operator E is not well-defined. Fortunately, it is not necessary to proceed in this way.

[9]It can be extended further, but we shall have no need for that result.

When A and A^+ describe a mode of the electromagnetic field, this result can be interpreted as the factorization of normal ordered correlation functions of the field. Following the reasoning of Glauber [76, 77, 78], [164], this is a characteristic property of coherent radiation. It is usually said in this regard that these states show classical behaviour, since classical fields factor in this way. However, this is only an approximation to real coherent light, which shows higher order correlations.

An additional property of coherent states is that they are states of minimal uncertainty for position and momentum, since elementary calculations show that $\mathbf{Unc}_{\omega_\zeta}[P] = \mathbf{Unc}_{\omega_\zeta}[Q] = 1/\sqrt{2}$ — see [221], Volume III.

Whatever the exact description of coherent states of the interacting quantized electromagnetic field may turn out to be, the case for considering coherent states in the context of quantum phase theory is clear.

We recall the definition of the Bargmann-Segal representation, as described in Section 5.6. Direct calculation shows us that the unitary map $\mathfrak{U}_{BS} : L^2(\mathbb{R}) \to \mathcal{B}$ defined in that Section can be written

$$\big[\mathfrak{U}_{BS}\phi\big](z) \;=\; e^{\frac{1}{2}|z|^2}\big\langle \Phi_z\,,\,\phi\big\rangle\,, \qquad \phi \in L^2(\mathbb{R})\,. \tag{10.3.29}$$

Thus the coherent states implement the unitary map $\mathfrak{U}_{BS}$ between $L^2(\mathbb{R})$ and $\mathcal{B}$ in the same way the London distributions implement the unitary mapping $\mathfrak{U}_T$ between $L^2(\mathbb{R})$ and $H^2(\mathbb{T})$.

Any bounded measurable function F on $\mathbb{C}$ defines a bounded operator $\mathcal{M}_F$ on $\mathcal{H}$ by multiplication via the formula

$$\big[\mathcal{M}_F G\big](z) \;=\; F(z)G(z)\,, \qquad G \in \mathcal{H}\,, \tag{10.3.30}$$

and also defines a bounded operator $\mathfrak{M}_F$ on the Bargmann-Segal representation space $\mathcal{B}$ by the formula

$$\mathfrak{M}_F \;=\; \mathcal{P}_{BS} \circ \mathcal{M}_F\,, \tag{10.3.31}$$

where $\mathcal{P}_{BS} : \mathcal{H} \to \mathcal{B}$ is the orthogonal projection of the Hilbert space $\mathcal{H}$ onto its closed subspace $\mathcal{B}$. Finally, conjugation with the unitary map $\mathfrak{U}_{BS}$ leads to the bounded operator $\Xi(F)$ on $L^2(\mathbb{R})$ given by

$$\Xi(F) \;=\; \mathfrak{U}_{BS}^{-1} \circ \mathfrak{M}_F \circ \mathfrak{U}_{BS}\,, \tag{10.3.32}$$

and we note that $\|\,\Xi(F)\,\| \leqslant \|\,F\,\|_\infty$. There are clearly many similarities between this definition of these Bargmann-Segal operators and the definition of the Toeplitz operators given above. In particular, $\mathfrak{M}_F$ is not a multiplication operator on $\mathcal{B}$, due to the presence of the projection operator $\mathcal{P}_{BS}$.

Direct calculation shows that

$$\langle \phi \,,\, \Xi(F)\psi \rangle \;=\; \frac{1}{\pi} \int_{\mathbb{C}} \langle \phi \,,\, \Phi_z \rangle \, F(z) \, \langle \Phi_z \,,\, \psi \rangle \, dA(z) \,,$$

which leads to the following weak spectral representation[10] for $\Xi(F)$,

$$\Xi(F) \;=\; \frac{1}{\pi} \int_{\mathbb{C}} F(z) \,|\, \Phi_z \,\rangle \, \langle \, \Phi_z \,|\, dA(z) \,. \qquad (10.3.33)$$

As usual, any function $F \in L^\infty(\Pi)$ can be identified with a function (also to be denoted by F) in $L^\infty(\mathbb{C})$, simply by the usual identification of $(p, q) \in \Pi$ with the complex number $(q - ip)/\sqrt{2}$. Then

Definition 10.4 *The bounded operator* $\Xi(\varphi)$ *is called the* Bargmann-Segal phase operator.

An insight into the physical meaning of the angle/argument function in the Bargmann-Segal complex plane can be obtained by deriving the symbol (Weyl dequantization) of the observable $\Xi(\varphi)$, and by comparing that symbol with the angle function in phase space. To do this we note that

$$\Xi_{\Phi_\zeta, \Phi_\zeta}(p, q) \;=\; 2\pi \Big[\mathcal{G}\big(\overline{\Phi_\zeta} \otimes \Phi_\zeta \big) \Big](p, q) \;=\; 2 e^{-2|\, w - \zeta \,|^2} \,, \qquad (10.3.34)$$

where $\sqrt{2} w \;=\; (q - ip)$, recalling the phase space function $\Xi_{\phi,\psi}$ defined in Section 8.4.1 for any $\phi,\ \psi \in L^2(\mathbb{R})$. It is important to note the difference between functions of the form $\Xi_{\phi,\psi}$ and Bargmann-Segal operators like $\Xi(F)$.

Proposition 10.5 *Let* F *be a bounded measurable function on* $\mathbb{C}$. *Then the dequantization symbol of the operator* $\Xi(F)$ *is the phase space function*

$$D_F(p, q) \;=\; \frac{2}{\pi} \int_{\mathbb{C}} F(z) \, e^{-2|\, z - w \,|^2} \, dA(w), \qquad w = \frac{1}{\sqrt{2}}(q - ip). \quad (10.3.35)$$

Proof: Recalling from Section 8.4.1 the fact that $\Xi_{\phi,\psi}$ is the symbol of the operator $|\, \phi \,\rangle \, \langle \, \psi \,|$ for any $\phi,\ \psi \in L^2(\mathbb{R})$, from the weak spectral

[10] The use of coherent states to define operators via a weak spectral representation seems to have been suggested originally by Turski [225], and later developed by Paul [175].

representation of $\Xi(F)$ can be deduced the fact that the symbol of $\Xi(F)$ is the function

$$D_F(p,q) = \frac{1}{\pi}\int_{\mathbb{C}} F(z)\,\Xi_{\Phi_z,\Phi_z}(p,q)\,dA(z)\,, \tag{10.3.36}$$

from which the result is immediate. ■

In particular, the symbol of the Bargmann-Segal phase operator $\Xi(\varphi)$ is

$$D_\varphi(p,q) = \frac{1}{\pi}\iint_{\mathbb{R}^2} e^{-(p-u)^2-(q-v)^2}\,\varphi(u,v)\,dudv\,, \tag{10.3.37}$$

which is not equal to the phase space angle function $\varphi(p,q)$. Again, just as for the Toeplitz operator X considered above, the relationship of the Bargmann-Segal phase operator to the classical phase angle is at best indirect. In Chapter 15 it will be shown that

$$D_\varphi(r\cos\beta, r\sin\beta) - \varphi(r\cos\beta, r\sin\beta) \to 0\,, \qquad -\pi < \beta < \pi\,,$$

quite rapidly as $r \to \infty$, so D_φ is an approximation to φ which is increasingly accurate at large radii (large energies). In this limiting (and limited) sense, the Bargmann-Segal phase angle has a classical phase connection.

The above formulæ yield an integral expression for the matrix coefficients of an operator $\Xi(F)$ with respect to the Hermite-Gauss functions,

$$\langle h_m\,, \Xi(F)h_n\rangle = \frac{1}{\pi\sqrt{m!\,n!}}\int_{\mathbb{C}} F(w)\overline{w}^m w^n e^{-|w|^2}\,dA(w), \qquad m,n \geqslant 0, \tag{10.3.38}$$

and so, in the particular case where $F = \varphi$, the matrix coefficients of the Bargmann-Segal phase operator $\Xi(\varphi)$ are

$$\begin{aligned} \langle h_m, \Xi(\varphi)h_n\rangle &= \frac{i^{m-n}}{\sqrt{m!\,n!}}\Gamma\big(\tfrac{1}{2}(m+n)+1\big)\hat{\mathfrak{p}}_{m-n} \\ &= \begin{cases} \dfrac{i^{n-m+1}}{\sqrt{m!\,n!}}\Gamma\big(\tfrac{1}{2}(m+n)+1\big)\dfrac{1}{m-n}, & m \neq n, \\ 0, & m = n, \end{cases} \end{aligned} \tag{10.3.39}$$

which coefficients should be compared with those given in equation (10.3.15) for the Toeplitz operator X.

10.3.4 *The Barnett-Pegg Operators*

All of the considerations so far have been based on operators and states with an infinite dimensional character. In contrast, in a series of articles beginning with [176], Barnett & Pegg have proposed a theory based on the finite sum LHW states, equation (10.3.7), and corresponding finite rank operators. An infinite-dimensional limit is taken, but only after all algebraic manipulations have been performed. Leaving the discussion of the physical aspects of this theory to a later stage, the basic operators of their theory will be introduced here. (See the special issue *Quantum Phase and Phase Dependent Measurements* of Physica Scripta, **T48**, 1993, for articles and further references.)

10.3.4.1 *Weak And Strong Convergence*

Barnett & Pegg theory is particularly concerned with the convergence of (moments of) sequences of operators. If reliable results are to be obtained about such sequences, it is important to know the manner in which the sequences of operators converge. So let us recall the main types of operator convergence.

In the following, $\big(B_n\big)_n$ will be a sequence of linear operators from some dense subspace[11] $\mathcal{D}$ of $L^2(\mathbb{R})$ to $L^2(\mathbb{R})$, and B will be another linear map from $\mathcal{D}$ to $L^2(\mathbb{R})$.

The sequence $\big(B_n\big)_n$ is said to *converge weakly* to B if

$$\lim_{n\to\infty} \langle f\,,\, B_n g\rangle \;=\; \langle f\,,\, Bg\rangle\,, \qquad f, g \in \mathcal{D}\,, \tag{10.3.40.a}$$

and is said to *converge strongly* to B if

$$\lim_{n\to\infty} \|\, B_n f \;-\; Bf\,\| \;=\; 0\,, \qquad f \in \mathcal{D}\,. \tag{10.3.40.b}$$

Moreover, if $\big(B_n\big)_n$ is a sequence of bounded operators, and if B is also a bounded operator, then the sequence is said to *converge uniformly* to B if

$$\lim_{n\to\infty} \|\, B_n \;-\; B\,\| \;=\; 0\,. \tag{10.3.40.c}$$

It is clear that any strongly convergent sequence is also weakly convergent, and that any uniformly convergent sequence of bounded operators is

[11] The space $\mathcal{D}$ will, in general, be either $L^2(\mathbb{R})$ or $\mathcal{S}(\mathbb{R})$, according as the operators B_n and B are bounded or smooth observables.

strongly convergent. However the reverse implications, in general, are not true.

Strongly convergent sequences of operators are quite easy to manipulate — for example the functional calculus can be used to obtain other strongly convergent sequences [55]. In particular, if $\big(B_n\big)_n$ is a sequence of observables (either bounded or smooth) which converges strongly to the observable B, then the sequence $\big(B_n^2\big)_n$ converges strongly to B^2, and hence $\mathbf{Unc}_f\,[B_n]$ converges to $\mathbf{Unc}_f\,[B]$ as $n \to \infty$ for any vector state $f \in \mathcal{D}$. Results of this nature are not available to weakly convergent sequences. Consequently care must be taken with the theory of Barnett & Pegg, which (as a rule) deals in weak convergence.

10.3.4.2 *The Truncation Subspaces $\mathcal{H}^{(s)}$.*

Barnett & Pegg theory starts by choosing a nonnegative integer $s \geqslant 0$, and then subdividing the circle into $s+1$ equal wedges, defining the angles[12]

$$\theta_{s,j} \;=\; -\pi + \frac{2\pi j}{s+1}\,, \qquad 0 \leqslant j \leqslant s. \tag{10.3.41}$$

It is crucial to note that the angle $\theta_{s,j}$ depends upon both s and j — consequently omitting the subscript s (as do some authors) is misleading. The angles $\theta_{s,j}$ are then used to define the LHW states $\zeta_s[\theta_{s,j}]$. For reasons of consistency, we must again apply the unitary automorphism $\mathcal{F}^{-1}$, and define the transformed LHW states

$$\eta_s[\theta] \;=\; \mathcal{F}^{-1}\big(\zeta_s[\theta]\big) \tag{10.3.42.a}$$

for any $\theta \in \mathbb{R}$ and, in particular, the states

$$\eta_{s,j} \;=\; \eta_s[\theta_{s,j}] \;=\; \mathcal{F}^{-1}\big(\zeta_s[\theta_{s,j}]\big)\,, \qquad 0 \leqslant j \leqslant s\,, \tag{10.3.42.b}$$

noting that $\big\{\eta_{s,j} \,:\, 0 \leqslant j \leqslant s\big\}$ is an orthonormal basis for the subspace $\mathcal{H}^{(s)}$ spanned by the Hermite-Gauss vectors $\big\{h_n \,:\, 0 \leqslant n \leqslant s\big\}$. If $P^{(s)}$ is the orthogonal projection of $L^2(\mathbb{R})$ onto $\mathcal{H}^{(s)}$, it is clear that

$$P^{(s)} \;=\; \sum_{n=0}^{s} P_n \;=\; \sum_{j=0}^{s} P_{s,j}\,, \tag{10.3.43}$$

[12] The term $-\pi$ can be replaced by an arbitrary reference angle θ_0, but no essential physics results from doing so.

where, as usual, $P_n = |\, h_n \rangle \langle h_n \,|$ is the orthogonal projection onto the subspace of $L^2(\mathbb{R})$ spanned by h_n, and $P_{s,j}$ is the orthogonal projection $|\, \eta_{s,j} \rangle \langle \eta_{s,j} \,|$ of $L^2(\mathbb{R})$ onto the one-dimensional subspace of $L^2(\mathbb{R})$ spanned by $\eta_{s,j}$.

It is clear that $\mathcal{H}^{(s)}$ is a subspace of $\mathcal{S}(\mathbb{R})$, and so any observable B (whether bounded or smooth) has a projection onto $\mathcal{H}^{(s)}$ given by

$$B^{(s)} = P^{(s)} B P^{(s)}, \tag{10.3.44}$$

and elementary analysis can be used to show that the sequence of bounded operators $\big(B^{(s)}\big)_s$ converges strongly to B as $s \to \infty$.

10.3.4.3 *Barnett & Pegg Theory*

Barnett & Pegg, and other authors, propose what is described as a phase observable by introducing the family of self-adjoint operators

$$X_s = \sum_{j=0}^{s} \theta_{s,j}\, P_{s,j}, \qquad s \in \mathbb{N}. \tag{10.3.45}$$

These operators are dealt with according to the following ansatz: expectations are calculated for functions of the operators X_s. The limit of these quantities as $s \to \infty$ is then taken, and the resulting expressions are then seen as the corresponding expectations of these functions of their phase observable.

There are evident calculational advantages to this approach, since the operators X_s are clearly diagonal, and hence any application of the functional calculus is elementary. For example, any continuous function ω in $\mathcal{C}[-\pi, \pi]$ defines the bounded operator $\omega(X_s)$ by the formula

$$\omega(X_s) = \sum_{j=0}^{s} \omega(\theta_{s,j})\, P_{s,j} + \omega(0)\big(I - P^{(s)}\big), \tag{10.3.46.a}$$

and this operator can be shown to have matrix coefficients

$$\big\langle h_m\,, \omega(X_s) h_n \big\rangle = \begin{cases} \dfrac{i^{m-n}}{s+1} \displaystyle\sum_{j=0}^{s} \omega(\theta_{s,j}) e^{i(n-m)\theta_{s,j}}, & 0 \leqslant m, n \leqslant s, \\ \omega(0)\delta_{mn}, & \text{otherwise,} \end{cases} \tag{10.3.46.b}$$

and Riemann integration theory then implies that

$$\begin{aligned} \lim_{s\to\infty} \langle h_m \,,\, \omega(X_s)h_n \rangle &= \frac{i^{m-n}}{2\pi} \int_{-\pi}^{\pi} \omega(\beta) e^{i(n-m)\beta}\, d\beta \\ &= \langle h_m \,,\, \overline{\mathcal{M}}(\omega)h_n \rangle \,, \end{aligned} \qquad (10.3.46.c)$$

for all $m, n \geqslant 0$. Since the sequence $\big(\omega(X_s)\big)_s$ is uniformly bounded, it follows that it converges weakly to $\overline{\mathcal{M}}(\omega)$. Note that we are making a standard identification between ω as a function on $[-\pi, \pi]$ and ω as a function on $\mathbb{T}$.

In particular, introducing the functions $\mathfrak{p}_r \in \mathcal{C}[-\pi, \pi]$ given by

$$\mathfrak{p}_r(\beta) = \beta^r \,, \qquad -\pi \leqslant \beta \leqslant \pi \,, \qquad (10.3.47)$$

for any $r \in \mathbb{N}$ (so that $\mathfrak{p}_1 = \mathfrak{p}$), it is clear that the sequence $\big({X_s}^r\big)_s$ converges weakly to $\overline{\mathcal{M}}(\mathfrak{p}_r)$ as $s \to \infty$ for any $r \in \mathbb{N}$; in particular X_s converges weakly to $\overline{\mathcal{M}}(\mathfrak{p}) = X$. However the convergence of the sequence $\big(X_s\big)_s$ is not strong, as will now be shown.

Since

$$e^{iX_s} = \sum_{j=0}^{s} e^{i\theta_{s,j}} P_{s,j}. \qquad (10.3.48)$$

direct calculation shows that[13]

$$\begin{aligned} e^{iX_s} &= P^{(s)}\, E^*\, P^{(s)} - i^s |\, h_0 \,\rangle \langle\, h_s \,| + I - P^{(s)} \\ &= (E^*)^{(s)} - i^s |\, h_0 \,\rangle \langle\, h_s \,| + I - P^{(s)} \,. \end{aligned} \qquad (10.3.49)$$

Since, consequently,

$$\big\| e^{iX_s}\phi - E^*\phi \big\| \leqslant \Big\| (E^*)^{(s)}\phi - E^*\phi \Big\| + |\langle h_s \,,\, \phi\rangle| + \Big\| \phi - P^{(s)}\phi \Big\|$$

for any $\phi \in L^2(\mathbb{R})$, it follows that the sequence $\big(e^{iX_s}\big)_s$ converges strongly to E^*. However E^* is not unitary, and in particular is not equal to the unitary operator e^{iX}. Thus the sequence $\big(e^{iX_s}\big)_s$ does not converge strongly to the unitary map e^{iX}, which would be the case if $\big(X_s\big)_s$ converged strongly to X [55].

---

[13]Note the boundary term involving the indices 0 and s. This term appears as a consequence of the process of truncation to the spaces $\mathcal{H}^{(s)}$ that Barnett & Pegg theory employs.

The final topic in this preliminary discussion of Barnett & Pegg theory concerns the moments of the operators X_s. As implied above, the quantity

$$BP_k(\psi) = \lim_{s\to\infty} \langle \psi, X_s^k \psi \rangle, \qquad k \in \mathbb{N}, \ \psi \in L^2(\mathbb{R}), \tag{10.3.50}$$

is understood to represent the kth moment of the Barnett & Pegg phase observable in the vector state ψ. It has been shown that

$$BP_k(\psi) = \mathbf{Exp}_\psi \left[\overline{\mathcal{M}}(\mathfrak{p}_k) \right], \qquad k \in \mathbb{N}, \ \psi \in L^2(\mathbb{R}), \tag{10.3.51}$$

and, since $\overline{\mathcal{M}}(\mathfrak{p}_k) \neq X^k$ for $k \geqslant 2$, these quantities cannot be interpreted as the moments of any single observable (in the ordinary manner). Thus the quantities $BP_k(\psi)$ will have to serve on their own. Barnett & Pegg theory would then view the quantity

$$V_{BP}(\psi) = BP_2(\psi) - BP_1(\psi)^2, \qquad \psi \in L^2(\mathbb{R}), \tag{10.3.52}$$

as the variance of the Barnett & Pegg phase observable in the vector state $\psi \in L^2(\mathbb{R})$. Since

$$BP_2(\psi) = \| \mathfrak{p}\mathfrak{U}_T \mathcal{F}\psi \|^2,$$
$$\mathbf{Exp}_\psi \left[X^2 \right] = \| P_+ \mathfrak{p}\mathfrak{U}_T \mathcal{F}\psi \|^2,$$

for any $\psi \in L^2(\mathbb{R})$, it follows that $BP_2(\psi) \geqslant \mathbf{Exp}_\psi \left[X^2 \right]$ for any $\psi \in L^2(\mathbb{R})$, and hence that

$$V_{BP}(\psi) \geqslant \mathbf{Var}_\psi [X], \qquad \psi \in L^2(\mathbb{R}). \tag{10.3.53}$$

We shall return to a discussion as to why the Barnett & Pegg variance should be expected to be greater than the variance of X in Chapter 16.

For the Hermite-Gauss vectors, explicit results are obtainable, since

$$V_{BP}(h_n) = \frac{1}{2\pi} \int_{-\pi}^{\pi} \beta^2 \, d\beta = \frac{1}{3}\pi^2, \qquad n \geqslant 0. \tag{10.3.54}$$

This is accounted an important result by the proponents of Barnett & Pegg theory, as they expect the variance of a phase operator to be uniformly distributed over the Hermite-Gauss functions.Unfortunately, since $V_{BP}(h_n)$ is not the variance of any operator in the state ψ, it is hard to see what connection equation (10.3.54) has with this desideratum.

For comparison, the variance of the Toeplitz operator X in the state h_n can be calculated, since

$$\begin{aligned}
\mathsf{Var}_{h_n}[X] &= \| X h_n \|^2 = \sum_{m\geqslant 0, m\neq n} (n-m)^{-2} \\
&= 2\sum_{k=1}^{\infty} k^{-2} - \sum_{k=n+1}^{\infty} k^{-2} \\
&= \frac{1}{3}\pi^2 - \sum_{k=n+1}^{\infty} k^{-2}. \qquad (10.3.55)
\end{aligned}$$

Thus the sequence of variances $\big(\mathsf{Var}_{h_n}[X]\big)_n$ is monotonic increasing and convergent to $\frac{1}{3}\pi^2$. Moreover,

$$\mathsf{Var}_{h_n}[X] \sim \frac{1}{3}\pi^2 - \frac{1}{n}, \qquad n \to \infty. \qquad (10.3.56)$$

10.3.5 *The Quantized Angle Function*

In preceding Sections, we have considered three of the families of operators presumed by various advocates to be phase-related in one sense or another. Since these operators have constructed using some form of angle as independent variable, it is a tempting fallacy to assume them to represent the same thing physically. While each family arises in some mathematically natural way, none of them has an *a priori* relation to a classical angle as such. In all these proposals, the operator was constructed with no reference to a classical limit, which had to be determined subsequently.

These problems can be addressed by reversing the process, and deriving a quantum mechanical observable from what is indisputably a phase space angle function. Thus we propose to consider the (Weyl) quantizations[14] of the phase space angle function φ and various functions of it. In the following analysis, much of the work done in generality in Chapters 8 and 9 can now be applied in this specific context.

---

[14]Although we do not do so until Chapter 14, it would clearly be possible to consider other quantization schemes than that of Weyl.

10.3.5.1 *Elementary Properties*

Recalling the basic results of the preceding Chapter, any function $\omega \in L^2(\mathbb{T})$ defines an operator $\mathbf{\Delta}[\omega_{\text{ang}}]$ given by the formula

$$[\![\mathbf{\Delta}[\omega_{\text{ang}}] h_n , h_m]\!] = i^{m-n} g_{m,n} \hat{\omega}_{m-n}, \qquad m, n \geqslant 0, \qquad (9.4.9)$$

where the $g_{m,n}$ are defined in equation (9.3). In particular, we recall the operators $U_1 = \mathbf{\Delta}\left[e^{i\varphi}\right]$ and $U_{-1} = \mathbf{\Delta}\left[e^{-i\varphi}\right]$ introduced in Subsection 9.4.2, which have the following action on the Hermite-Gauss vectors:

$$U_1 h_n = i\, g_{n,n+1} h_{n+1}, \qquad n \geqslant 0, \qquad (10.3.57.a)$$

$$U_{-1} h_n = \begin{cases} -i\, g_{n-1,n} h_{n-1}, & n \geqslant 1, \\ 0, & n = 0. \end{cases} \qquad (10.3.57.b)$$

Most of the standard spectral properties of the operators U_1 and U_{-1} have already been summarized in Proposition 9.15, but the results bear repeating and amplifying. The reader is directed to the work of Lerner, Huang & Walters [149], as well as the paper [53] for details of the necessary proofs.

Both of the operators U_1 and U_{-1} are bounded, with

$$\| U_1 \| = \| U_{-1} \| = \sqrt{\frac{\pi}{2}}, \qquad (10.3.58)$$

and U_{-1} is the adjoint of U_1. Moreover, both U_1 and U_{-1} belong to $\mathcal{L}^+(\mathcal{S}(\mathbb{R}))$. The spectral properties of these operators, already announced, are summarized and extended in Table 10.2.

Table 10.2 Spectral Properties of Quantized Exponentials

Properties	$U_1 = \mathbf{\Delta}\left[e^{i\varphi}\right]$	$U_{-1} = \mathbf{\Delta}\left[e^{-i\varphi}\right]$
Spectrum	$\overline{\mathbb{D}}$	$\overline{\mathbb{D}}$
Eigenvalues	$\emptyset$	$\mathbb{D}$
Continuous Spectrum	$\mathbb{T}$	$\mathbb{T}$
Residual Spectrum	$\mathbb{D}$	$\emptyset$

Each eigenvalue $z \in \mathbb{D}$ of U_{-1} is nondegenerate, with eigenvector

$$e_z = h_0 + \sum_{n=1}^{\infty} \left(\prod_{k=1}^{n} \lambda_{k-1,k}^{-1} \right) (iz)^n h_n . \qquad (10.3.59)$$

The eigenvectors $\{e_z : z \in \mathbb{D}\}$ of U_{-1} are not mutually orthogonal, but they do form an overcomplete set, in the sense that

$$\phi = \int_{\mathbb{D}} \langle e_z , \phi \rangle \, e_z \, dA(z) , \qquad \phi \in L^2(\mathbb{R}) . \qquad (10.3.60.a)$$

In other words, we have a weak spectral resolution of the identity:

$$I = \int_{\mathbb{D}} | \, e_z \rangle \langle \, e_z \, | \, dA(z) . \qquad (10.3.60.b)$$

It is clear, however, that neither U_1 nor U_{-1} is unitary[15], since $U_1 \, U_{-1} \neq I$.

Evidently, the operators U_1 and U_{-1} are the Weyl quantization analogues of the Toeplitz operators E^* and E, of the Bargmann-Segal operators $\Xi(e^{i\varphi})$ and $\Xi(e^{-i\varphi})$ and of the calculations of Barnett & Pegg derived from considering the families of operators $\left(e^{iX_s}\right)_s$ and $\left(e^{-iX_s}\right)_s$ respectively. Corresponding operators for the different approaches are very similar, with closely matching (yet different) definitions and matrix coefficients. The distinctive nature of the observables U_1 and U_{-1} is the fact that they are directly related to an angle function in phase space which has a definite physical meaning.

The main observable of interest to us is $\mathbf{\Delta}\,[\,\varphi\,]$, the Weyl quantization of φ. This is the Weyl quantization analogue of the Toeplitz operator X, the Bargmann-Segal phase observable $\Xi(\varphi)$ and the phase observable of Barnett & Pegg. We can now summarize some of its basic properties. It is clear that $\mathbf{\Delta}\,[\,\varphi\,]$ has matrix coefficients

$$\begin{aligned} \left\langle h_m \, , \, \mathbf{\Delta}\,[\,\varphi\,]\, h_n \right\rangle &= i^{m-n} \, g_{m,n} \, \hat{\mathfrak{p}}_{m-n} \\ &= \begin{cases} \dfrac{i^{n-m+1}}{m-n} \, g_{m,n} \, , & m \neq n \, , \\ 0 \, , & m = n \, . \end{cases} \end{aligned} \qquad (10.3.61)$$

These matrix coefficients should be compared with those of the Toeplitz operator X and with those of the Bargmann-Segal phase operator $\Xi(\varphi)$.

[15]These matters will be discussed further in Chapter 13, where it will be observed that the Moyal product $e^{2i\varphi} \star e^{-2i\varphi}$ of the phase space observables $e^{2i\varphi}$ and $e^{-2i\varphi}$ is not equal to i.

In Section 9.4.4 something of the relation between the angular distribution $\mathfrak{f}_{\text{ang}}$ and the operator class of $\boldsymbol{\Delta}[\mathfrak{f}_{\text{ang}}]$ was discussed. When applied to the observable $\boldsymbol{\Delta}[\varphi]$, the result is as follows.

Proposition 10.6 *The phase operator $\boldsymbol{\Delta}[\varphi]$ belongs to $\mathcal{L}^+(\mathcal{S}(\mathbb{R}), L^2(\mathbb{R}))$, but not to $\mathcal{L}^+(\mathcal{S}(\mathbb{R}))$, and so is not a smooth observable.*

Proof: It is trivial to test the sequence determined by equation (9.4.14) for $\boldsymbol{\Delta}[\varphi]$ against the criteria in the three Propositions in Section 9.4.4. ■

The general results of Section 9.4.4 do not determine the boundedness of $\boldsymbol{\Delta}[\varphi]$, but Proposition 9.26 does this, and gives an upper bound on the norm.

Proposition 10.7 *$\boldsymbol{\Delta}[\varphi]$ is a bounded operator, with*

$$\|\boldsymbol{\Delta}[\varphi]\| \leqslant \frac{3}{2}\pi. \tag{10.3.62}$$

Comparatively little is known about the spectral properties of $\boldsymbol{\Delta}[\varphi]$. In particular, no spectral decomposition is known. However it is shown in [113] that

$$[-\pi, \pi] \subseteq \mathrm{Sp}(\boldsymbol{\Delta}[\varphi])$$

which in turn implies that

$$\pi \leqslant \|\boldsymbol{\Delta}[\varphi]\| \leqslant \frac{3}{2}\pi.$$

It is our belief, albeit an unproven one, that $\|\boldsymbol{\Delta}[\varphi]\| = \pi$ and that $\mathrm{Sp}(\boldsymbol{\Delta}[\varphi]) = [-\pi, \pi]$ — suppositions which are supported by various numerical calculations.

In Section 9.4.6 the integral kernels for operators of the form $\boldsymbol{\Delta}[\mathfrak{f}_{\text{ang}}]$ were determined. For the operators $\boldsymbol{\Delta}[\varphi]$, U_1 and U_{-1} being considered here, equation (9.4.39) of Proposition 9.30 takes the following specific form.

Proposition 10.8 *The kernel of $\boldsymbol{\Delta}[\varphi]$ may be derived from the formula*

$$\begin{aligned}(\boldsymbol{\Delta}[\varphi]g)(x) &= \frac{\pi}{2}\,\mathrm{sgn}(x)g(x) \\ &\quad - \frac{i}{2}\,\mathrm{PV}\int_{\mathbb{R}} \mathrm{sgn}(x+y)\,\frac{1}{x-y}\,e^{-\frac{1}{2}|x^2-y^2|}\,g(y)\,dy,\end{aligned} \tag{10.3.63}$$

where $g \in \mathcal{S}(\mathbb{R})$. The kernels for the quantized exponentials are complex conjugate distributions, with

$$\begin{aligned} &K_{\mathbf{\Delta}[e^{i\varphi}]}(x,y) \qquad\qquad (10.3.64)\\ &= \frac{i}{2\pi}(x+y)\Big[K_0\big(\tfrac{1}{2}\,|\,x^2-y^2\,|\big)+\pi^{1/2}\mathrm{sgn}(x^2-y^2)K_1\big(\tfrac{1}{2}\,|\,x^2-y^2\,|\big)\Big], \end{aligned}$$

where K_0 and K_1 are the modified Bessel functions of the second kind, of index 0 and 1, respectively.

The integral kernel for $\mathbf{\Delta}[\varphi]$ is clearly a difficult object to work with when attempting specific calculations, like variances. However, after substantial work, the integral kernel formula for $\mathbf{\Delta}[\varphi]$ can be manipulated to determine expressions for the variances of $\mathbf{\Delta}[\varphi]$ in states determined by the Hermite-Gauss vectors. Details of this proof can be found in [113].

Proposition 10.9 *The variance of $\mathbf{\Delta}[\varphi]$ in the vector state determined by the Hermite-Gauss vector h_n is given by the expression* *(see page 458)*

$$\begin{aligned} \mathsf{Var}_{h_n}[\mathbf{\Delta}[\varphi]] &= \frac{3}{8}\pi^2 + \\ &\sum_{\substack{0\leqslant j,k\leqslant \lfloor (n-2)/2\rfloor \\ j+k\leqslant \lfloor (n-2)/2\rfloor}} \frac{1}{(2j+1)(2k+1)} - \left(\sum_{j=0}^{\lfloor (n-1)/2\rfloor}\frac{1}{2j+1}\right)^2 \qquad (10.3.65.a) \end{aligned}$$

for any $n \geqslant 0$. Consequently

$$\mathsf{Var}_{h_n}[\mathbf{\Delta}[\varphi]] = \frac{1}{3}\pi^2 + \mathrm{O}\big(\frac{\log n}{n}\big), \qquad n\to\infty. \qquad (10.3.65.b)$$

It should be noticed that the convergence of the sequence $\big(\mathsf{Var}_{h_n}[\mathbf{\Delta}[\varphi]]\big)_n$ to $\frac{1}{3}\pi^2$ exhibits an oscillatory behaviour characteristic of all calculations concerning $\mathbf{\Delta}[\varphi]$ — the subsequence $\big(\mathsf{Var}_{h_{2n}}[\mathbf{\Delta}[\varphi]]\big)_n$ decreases monotonically to the limit, while the subsequence $\big(\mathsf{Var}_{h_{2n+1}}[\mathbf{\Delta}[\varphi]]\big)_n$ increases monotonically to the same limit. A similar oscillatory behaviour can be found in the sequence $\big(\mathsf{Var}_{h_n}[U_1]\big)_n$, which converges to 1. Although this mode of convergence is more complicated than in cases previously considered, the sequences $\big(\mathsf{Var}_{h_n}[\mathbf{\Delta}[\varphi]]\big)_n$ and $\big(\mathsf{Var}_{h_n}[U_1]\big)_n$ converge to the limits that are consistent with some notion of a uniform distribution of phase with respect to the Hermite-Gauss functions.

10.3.5.2 *Noncanonicity*

As has been stated previously, the study of quantum phase has been frequently directed towards trying to find a quantum mechanical observable which is canonically conjugate to the number operator, on the grounds that the symbol $\nu(p,q) = \frac{1}{2}(p^2 + q^2 - 1)$ of the number operator and the phase space angle function φ are (classically) canonically conjugate. The No-Go Theorem precludes the existence of any such operator on a physically meaningful domain. Consequently, it is not surprising that none of the operators X, $\Xi(\varphi)$ and $\mathbf{\Delta}\,[\varphi]$ are canonically conjugate to the number operator N.

This, however, should not be a cause for concern, for the basic premise which motivates this search for a canonically conjugate phase observable is flawed, since the classical observables ν and φ are not canonically conjugate, contrary to popular belief. The reason for this is subtle — the standard definition for the Poisson bracket of two observables is only valid for smooth observables. But φ is not a smooth observable, being discontinuous on Π, and hence any study of the Poisson bracket of ν and φ must extend the definition of the Poisson bracket to deal with (at least to some extent) distributions on phase space Π.

The formalism of classical mechanics developed in Chapter 2 was based on the observable space of $\mathcal{C}^\infty(\Pi)$ and state space of $\mathcal{E}'(\Pi)$. However, this pairing is not sufficiently flexible for us to be able to perform the calculations we intend, and so we start by restricting the definition of the Poisson bracket to the subspace $\mathcal{S}(\Pi)$ of $\mathcal{C}^\infty(\Pi)$. Doing so will enable us to extend the Poisson bracket to $\mathcal{S}'(\Pi)$ in the following manner. The Poisson bracket $\{\cdot\,,\,\cdot\} : \mathcal{S}(\Pi) \times \mathcal{S}(\Pi) \to \mathcal{S}(\Pi)$ is a bicontinuous bilinear mapping, and hence we can define a bicontinuous bilinear mapping $\{\!\{\,\cdot\,,\,\cdot\}\!\}$ from $\mathcal{S}'(\Pi) \times \mathcal{S}(\Pi)$ into $\mathcal{S}'(\Pi)$ by the formula

$$[\![\, \{\!\{\, T\,,\, f\}\!\}\,,\, g\,]\!] = [\![\, T\,,\, \{f,g\}\,]\!]\,, \qquad T \in \mathcal{S}'(\Pi)\,,\; f,g \in \mathcal{S}(\Pi)\,. \quad (10.3.66)$$

Under the natural linear embedding of $\mathcal{S}(\Pi)$ into $\mathcal{S}'(\Pi)$, the map $\{\!\{\,\cdot\,,\,\cdot\}\!\}$ is an extension of the Poisson bracket $\{\cdot\,,\,\cdot\}$.

Given two distributions $S, T \in \mathcal{S}'(\Pi)$, it may not always be possible to define their Poisson bracket. However it can be done in some circumstances. For example, if $T \in \mathcal{S}'(\Pi)$ is such that $\{\!\{\, T\,,\, f\}\!\} \in \mathcal{S}(\Pi)$ for all $f \in \mathcal{S}(\Pi)$, and if the map $f \mapsto \{\!\{\, T\,,\, f\}\!\}$ is a continuous endomorphism of $\mathcal{S}(\Pi)$, then the distribution $\{\!\{ S, T \}\!\} \in \mathcal{S}'(\Pi)$ can be defined by the formula

$$[\![\{\!\{ S, T \}\!\}, f]\!] = [\![S, \{\!\{ T, f \}\!\}]\!], \qquad f \in \mathcal{S}(\Pi). \qquad (10.3.67)$$

Again, this definition extends the previous two. Thus the Poisson bracket on $\mathcal{S}'(\Pi)$ defines what is known as a partial (Lie) algebra structure on $\mathcal{S}'(\Pi)$, since the Poisson bracket can only be defined for certain pairs of elements of $\mathcal{S}'(\Pi)$, but not for all.

Direct calculation shows that the distribution $\nu \in \mathcal{S}'(\Pi)$ is such that $\{\!\{ \nu, f \}\!\} \in \mathcal{S}(\Pi)$ for all $f \in \mathcal{S}(\Pi)$, with

$$\{\!\{ \nu, f \}\!\}(p,q) = p(\partial_2 f)(p,q) - q(\partial_1 f)(p,q)$$

for any $f \in \mathcal{S}(\Pi)$. Thus the Poisson bracket $\{\!\{ \varphi, \nu \}\!\}$ can be defined, yielding

$$[\![\{\!\{ \varphi, \nu \}\!\}, f]\!] = -\iint_\Pi f(p,q)\,dpdq - 2\pi \int_{-\infty}^{0} f(p,0)p\,dp \qquad (10.3.68.a)$$

for any $f \in \mathcal{S}(\Pi)$, so that

$$\{\!\{ \varphi, \nu \}\!\}(p,q) = -1 - 2\pi p \chi_{(-\infty,0)}(p)\delta(q). \qquad (10.3.68.b)$$

Details of this calculation can be found in [53]. Thus the Poisson bracket $\{\!\{ \varphi, \nu \}\!\}$ is not equal to $-\mathrm{i}$, and so φ and ν are not canonically conjugate — small wonder that $\mathbf{\Delta}[\varphi]$ and N are not, either!

What, then, is the commutator of $\mathbf{\Delta}[\varphi]$ and N? To answer this question requires an extension of the given quantum mechanical formalism, since although $\mathbf{\Delta}[\varphi]Nf$ is a well-defined element of $L^2(\mathbb{R})$ for any $f \in \mathcal{S}(\mathbb{R})$, $N\mathbf{\Delta}[\varphi]f$ is not. The generalized commutator between any pair of maps A, $B \in \mathcal{L}^+(\mathcal{S}(\mathbb{R}), L^2(\mathbb{R}))$ is defined to be the sesquilinear form $\tau_{A,B} : \mathcal{S}(\mathbb{R}) \times \mathcal{S}(\mathbb{R}) \to \mathbb{C}$ given by the formula

$$\tau_{A,B}(f,g) = \langle A^+ f, Bg \rangle - \langle B^+ f, Ag \rangle, \qquad f,g \in \mathcal{S}(\mathbb{R}). \qquad (10.3.69)$$

By a representation theorem given in [53], there exists a mapping $X_{A,B}$ in $\mathcal{L}(\mathcal{S}(\mathbb{R}), \mathcal{S}'(\mathbb{R}))$ implementing $\tau_{A,B}$, in that

$$\tau_{A,B}(f,g) = [\![X_{A,B}(g), \overline{f}]\!], \qquad f,g \in \mathcal{S}(\mathbb{R}). \qquad (10.3.70)$$

The theorem further asserts that when A and B are smooth observables, $X_{A,B}$ is a smooth observable and equal to their usual commutator.

Calculation establishes that the commutator of $\mathbf{\Delta}[\varphi]$ and N is given by the formula

$$\begin{aligned} [\![X_{\mathbf{\Delta}[\varphi],N} g\,,\, f]\!] &= i[\![g\,,\, f]\!] - \tfrac{1}{2}\pi\{g(0)f'(0) - g'(0)f(0)\} \\ &\quad + \tfrac{1}{2}i[\![Ug\,,\, f']\!] - \tfrac{1}{2}i[\![Ug'\,,\, f]\!] \end{aligned} \qquad (10.3.71.a)$$

for any $f, g \in \mathcal{S}(\mathbb{R})$, where $U \in \mathcal{L}(\mathcal{S}(\mathbb{R}), \mathcal{S}'(\mathbb{R}))$ is the operator

$$[\![Ug\,,\, f]\!] = \lim_{L\to\infty} \int_{\mathbb{R}} g_{I(L)}(x)\, g(x)\, f(-x)\, dx\,, \qquad f, g \in \mathcal{S}(\mathbb{R})\,; \qquad (10.3.71.b)$$

$g_{I(L)}$ being the cut-off function defined in equation (9.4.22). Thus

$$X_{\mathbf{\Delta}[\varphi],N}\, g = ig + \tfrac{1}{2}\pi\{g(0)\delta' + g'(0)\delta\} - \tfrac{1}{2}i\{(Ug)' + U(g')\} \qquad (10.3.71.c)$$

for any $g \in \mathcal{S}(\mathbb{R})$. In particular we observe that $X_{\mathbf{\Delta}[\varphi],N} \neq iI$, due to the presence of two additional noncanonical terms.

What is interesting to note is the fact that

$$i\, X_{\mathbf{\Delta}[\varphi],N} = \mathbf{\Delta}[\{\!\{ \varphi\,,\, \nu \}\!\}]\,, \qquad (10.3.72)$$

so that the quantization of the Poisson bracket of φ and ν corresponds correctly with the commutator of their quantizations. Consequently $\mathbf{\Delta}[\varphi]$ and N are canonically conjugate in the sense that they satisfy the basic condition of Dirac in his discussion of q-numbers, just as P and Q do. Thus $\mathbf{\Delta}[\varphi]$ is *as nearly canonical as a phase observable can be*[16]. In terms of the Moyal product for phase space observables, this result can be summarized as follows:

Proposition 10.10 ***(The Classical Bracket Theorem)*** *The Poisson bracket of φ and ν is equal to their Moyal bracket*[17] *$i(\varphi \star \nu - \nu \star \varphi)$.*

---

[16]In the review of [53] in Mathematical Reviews, it was stated that this result cleared up the mystery of the quantum phase operator. This is perhaps optimistic, but it is certainly a strong argument for considering $\mathbf{\Delta}[\varphi]$ as the most natural of the proposals for a phase operator.

[17]The Moyal product and bracket are discussed at length in Chapter 13.

10.4 Distribution Functions And Phase

Given a self-adjoint observable B on $L^2(\mathbb{R})$, any state ω of the system determines a quantum probability distribution, whose moments are the expectations $\omega(B^n)$ $(n \in \mathbb{N})$. If these probability distributions are known for all states, it is possible to reconstruct the observable B. If, however, the probability distributions are only known for some of the states, under what conditions is it still possible to perform this reconstruction and, if it is possible, is the resultant observable uniquely defined by the given distributions?

This is not an idle question, since a number of authors advocate the description of quantum phase primarily through such probability distributions. However, for this approach to provide a legitimate quantum mechanical description of quantum phase, any such definition must define an observable. This section outlines a necessary and sufficient condition for a collection of probability distributions to define a self-adjoint operator[18]. This condition will not meet all needs, and is certainly not the best possible. It requires rather detailed knowledge of properties of the various probability distributions, and moreover requires that such distributions are known for all vector states. Moreover, it does not address the question of the representation of (smooth) unbounded observables — fortunately, however, phase observables are generally bounded. It seems clear, though, that any result of this type will be of a similar form.

In order to proceed efficiently, these families of distributions need to be grouped. To see how this is done, suppose that the self-adjoint operator B on $L^2(\mathbb{R})$ is represented by a positive operator valued measure E, so that

$$B = \int_{\mathbb{R}} \theta \, dE(\theta) . \tag{10.4.1}$$

Any vector $\psi \in L^2(\mathbb{R})$ then defines a positive Borel measure μ_ψ by means of the formula

$$\mu_\psi(\Delta) = \langle \psi , E(\Delta)\psi \rangle , \qquad \Delta \in \mathrm{Bor}(\mathbb{R}) , \tag{10.4.2}$$

thereby yielding a function $\mu : L^2(\mathbb{R}) \to \mathrm{BM}(\mathbb{R})$.

[18] Approaches to quantum phase for which the distributions being considered are acknowledged not to come from a single observable, such as the theory of Barnett & Pegg, are clearly outside the scope of these results. A different analysis of the Barnett & Pegg theory will be considered in Chapter 16.

The question that we shall address is the following: given a function $\mu : L^2(\mathbb{R}) \to \mathrm{BM}(\mathbb{R})$, what conditions ensure that μ can be derived from some bounded self-adjoint observable B in the above manner?

Proposition 10.11 ***(Reconstruction Of Quantum Probabilities)*** *Let μ be a function from $L^2(\mathbb{R})$ to $\mathrm{BM}(\mathbb{R})$, the space of (positive) Borel measures on $\mathbb{R}$, satisfying the following four conditions*

1. $\mu_{z\psi} = |z|^2 \mu_\psi$ *for any* $z \in \mathbb{C}$ *and* $\psi \in L^2(\mathbb{R})$,
2. *There exists a constant* $K > 0$ *such that*

$$\mu_\psi(-\infty, \theta] = \|\psi\|^2, \qquad \mu_\psi(-\infty, -\theta] = 0$$

   for all $\theta > K$,
3. *For any* $\theta \in \mathbb{R}$ *the function*

$$\psi \mapsto \mu_\psi(-\infty, \theta]$$

   from $L^2(\mathbb{R})$ *to* $\mathbb{R}$ *is (norm-)continuous,*
4. *The parallelogram identity,*

$$\mu_{\phi+\psi} + \mu_{\phi-\psi} = 2\big[\mu_\phi + \mu_\psi\big]$$

   holds for any ϕ, $\psi \in L^2(\mathbb{R})$.

Then μ defines a unique bounded self-adjoint operator. Conversely, any bounded self-adjoint operator defines a function μ of the above type.

Proof: The converse is easy to establish, since it is clear that the function μ defined from a bounded self-adjoint operator B via equation (10.4.2) satisfies the above four conditions, with $K = \|B\|$.
Let us now suppose that $\mu : L^2(\mathbb{R}) \to \mathrm{BM}(\mathbb{R})$ is a function which satisfies the required conditions. For any $\theta \in \mathbb{R}$, the formula

$$\begin{aligned} E_\theta(\phi, \psi) &= \frac{1}{4}\left[\mu_{\phi+\psi}(-\infty, \theta] - \mu_{\phi-\psi}(-\infty, \theta]\right] \\ &\quad - \frac{i}{4}\left[\mu_{\phi+i\psi}(-\infty, \theta] - \mu_{\phi-i\psi}(-\infty, \theta]\right], \end{aligned} \tag{10.4.3}$$

where $\phi, \psi \in L^2(\mathbb{R})$, defines a jointly continuous sesquilinear hermitian form E_θ on $L^2(\mathbb{R})$. Thus there exists a bounded self-adjoint operator $E(\theta)$ on $L^2(\mathbb{R})$ such that

$$E_\theta(\phi, \psi) = \langle \phi, E(\theta)\psi \rangle, \qquad \phi, \psi \in L^2(\mathbb{R}). \tag{10.4.4}$$

Since

$$0 \leqslant \langle \phi\,,\, E(\theta)\phi \rangle \;=\; E_\theta(\phi,\phi) \;=\; \mu_\phi(-\infty,\theta] \;\leqslant\; \mu_\phi(\mathbb{R}) \;=\; \|\,\phi\,\|^2$$

for any $\phi \in L^2(\mathbb{R})$, it follows that $0 \leqslant E(\theta) \leqslant I$. Moreover, it is clear that $E(-\theta) = 0$ and $E(\theta) = I$ for all $\theta > K$.
If $\theta_1 < \theta_2$ then

$$0 \leqslant \langle \psi\,,\, E(\theta_2)\psi \rangle - \langle \psi\,,\, E(\theta_1)\psi \rangle \;=\; \mu_\psi(\theta_1,\theta_2]$$

for any $\psi \in L^2(\mathbb{R})$, and hence $E(\theta_1) \leqslant E(\theta_2)$. Moreover, since $(\theta_1,\theta_2] \to \varnothing$ as $\theta_2 \downarrow \theta_1$, the function $\theta \mapsto \langle \psi\,,\, E(\theta)\psi \rangle$ is right-continuous on $\mathbb{R}$ for any $\psi \in L^2(\mathbb{R})$. By standard polarization identities, the function $\theta \mapsto E(\theta)$ from $\mathbb{R}$ to $\mathbb{B}\big(L^2(\mathbb{R})\big)$ is weakly right-continuous. In other words, we have defined a positive operator valued measure E on $\mathbb{R}$, which defines a bounded self-adjoint operator X on $L^2(\mathbb{R})$, with $\|\,X\,\| \leqslant K$, via the weak formula

$$X \;=\; \int_{\mathbb{R}} \theta \, dE(\theta), \tag{10.4.5.a}$$

so that

$$\langle \psi\,,\, X\psi \rangle \;=\; \int_{\mathbb{R}} \theta \, d\mu_\psi(-\infty,\theta] \tag{10.4.5.b}$$

for any $\psi \in L^2(\mathbb{R})$. It is clear from the method of derivation that the operator X is uniquely determined by the function μ. ■

Thus any measure-valued function μ on $L^2(\mathbb{R})$ of the above sort defines a unique self-adjoint bounded operator, and every self-adjoint bounded operator comes from some such a measure-valued function. However, and this is a key point to remember, a given self-adjoint bounded operator is associated with a measure-valued function for every positive operator valued measure which describes it, and since it can be represented by more than one positive operator valued measure, it can be obtained from more than one measure-valued function of the above type. So a given self-adjoint bounded operator does not define a unique measure-valued function, even though it defines a unique projection valued measure.

As a consequence of this result, we consider what conditions permit the definition of some form of phase operator from a family of distributions labelled by an angular parameter $\theta \in \mathbb{T}$, in the same way that the Toeplitz operator X was derived from the London distributions.

Proposition 10.12 *Suppose that $\Upsilon_\theta \in \mathcal{S}'(\mathbb{R})$ is a tempered distribution for every $\theta \in [-\pi,\pi]$. If the function*

$$G_f(\theta) \;=\; [\![\, \Upsilon_\theta \,,\, f \,]\!]\,, \qquad \theta \in [-\pi,\pi]$$

belongs to $L^2[-\pi,\pi]$ for every $f \in \mathcal{S}(\mathbb{R})$, with

$$\frac{1}{2\pi}\int_{-\pi}^{\pi} |\, G_f(\theta)\,|^2 \, d\theta \;=\; \|\, f\,\|^2\,, \qquad f \in \mathcal{S}(\mathbb{R})\,,$$

then the family $\big\{\Upsilon_\theta \,:\, \theta \in [-\pi,\pi]\big\}$ can be used to define a bounded self-adjoint operator Y on $L^2(\mathbb{R})$, with $\|\, Y\,\| \leqslant \pi$.

Proof: The above conditions imply that the linear map G from $\mathcal{S}(\mathbb{R})$ into $L^2[-\pi,\pi]$ is such that $\|\, G_f\,\|_2 \;=\; \|\, f\,\|_2$ for any $f \in \mathcal{S}(\mathbb{R})$. Consequently G extends uniquely to an isometric linear map G from $L^2(\mathbb{R})$ to $L^2[-\pi,\pi]$. For any $\psi \in L^2(\mathbb{R})$ the formula

$$\mu_\psi(\Delta) \;=\; \frac{1}{2\pi}\int_{\Delta\cap[-\pi,\pi]} |\, G_\psi(\theta)\,|^2 \, d\theta\,, \qquad \Delta \in \mathrm{Bor}(\mathbb{R})\,,$$

defines a positive measure $\mu_\psi \in \mathrm{BM}(\mathbb{R})$, and it is clear that the resulting map $\mu : L^2(\mathbb{R}) \to \mathrm{BM}(\mathbb{R})$ satisfies the conditions of the preceding Proposition with $K \;=\; \pi$. Hence there exists a bounded self-adjoint map Y on $L^2(\mathbb{R})$ with $\|\, Y\,\| \;\leqslant\; \pi$ which has been defined by μ, so that

$$\langle f\,,\, Yg\rangle \;=\; \frac{1}{2\pi}\int_{-\pi}^{\pi} \theta\overline{G_f(\theta)}G_g(\theta)\, d\theta$$

for all $f, g \in \mathcal{S}(\mathbb{R})$. ■

Corollary 10.13 *If the London distributions Λ_θ are used to define Υ_θ via the formula*

$$\Upsilon_\theta \;=\; \mathcal{F}\Lambda_\theta\,, \qquad \theta \in [-\pi,\pi]\,,$$

then the bounded self-adjoint observable resulting from the previous result is the Toeplitz operator X.

Proof: It is clear that

$$G_f(\theta) \;=\; [\![\, \Upsilon_\theta \,,\, f \,]\!] \;=\; [\![\, \Lambda_\theta \,,\, \mathcal{F}f \,]\!] \;=\; \big(\mathfrak{U}_T\mathcal{F}f\big)(e^{i\theta})$$

for any $f \in \mathcal{S}(\mathbb{R})$ and $\theta \in [-\pi, \pi]$, from which it follows that the required conditions are satisfied by the distributions Υ_θ. We must check that the conditions required of the family of distributions are satisfied, and so the previous Proposition can be used to define a bounded self-adjoint observable Y. Direct calculation shows that Y has matrix coefficients

$$\langle h_m \,, Y h_n \rangle \;=\; \frac{i^{m-n}}{2\pi} \int_{-\pi}^{\pi} \theta e^{i(n-m)\theta} \, d\theta$$

for any $m, n \geqslant 0$. Thus Y is our old friend the Toeplitz operator X, as asserted. ■

Now the whole (presumed) point of this construction of phase observables by probability distributions is that, for any normalized vector f in $\mathcal{S}(\mathbb{R})$, the function $F_f = |\, G_f \,|^2$ has the interpretation in phase theory as a probability density function for some posited "phase" observable in the vector state f. As we have just seen, the correct observable would then be the operator Y constructed above, for then

$$\frac{1}{2\pi} \int_{-\pi}^{\pi} \theta F_f(\theta) \, d\theta \;=\; \langle f \,, Y f \rangle \;=\; \mathsf{Exp}_f\,[Y] \qquad (10.4.3.a)$$

for any $f \in \mathcal{S}(\mathbb{R})$. However, this interpretation falls foul of the same problems that have already been encountered with the theory of Toeplitz operators, of Bargmann-Segal phase observables and in the theory of Barnett & Pegg, because F_f cannot define a classical probability distribution since, for example, the second moment of this distribution is not the expectation of the observable Y^2 in the state f,

$$\frac{1}{2\pi} \int_{-\pi}^{\pi} \theta^2 F_f(\theta) \, d\theta \;\neq\; \langle f \,, Y^2 f \rangle \,. \qquad (10.4.3.b)$$

Our conclusion has to be that these methods do not give anything that could not be obtained from using operators in the standard quantum mechanical formalism of operators on Hilbert space, and they must produce a theory with the same sort of structure as those based on the London distributions and the coherent states, with their built-in lack of a product calculus. Thus, no (valid) distributional formalism can avoid the consequences of the No-Go Theorem.

It is worth noting that it was not strictly necessary to introduce the tempered distributions Υ_θ in the above Proposition — all that was needed was the isometric map G from $\mathcal{S}(\mathbb{R})$ to $L^2[-\pi, \pi]$. However, candidate phase observables usually considered in quantum phase theory are normally selected on the basis of some intuitive notions of how phase observables ought to behave, and the consequence of these assumptions is to involve a map G which has indeed been derived from some family of tempered distributions. Thus the inclusion of the distributions Υ_θ is not necessary mathematically, but rather forms a template for the standard theories in this area.

CHAPTER 11

THE LASER MODEL

Provide thyself with a Teacher, and eschew doubtful matters, and tithe not overmuch by guesswork.

– *The Wisdom of the Fathers*

11.1 Introduction

Accounts of the physics of lasers are readily available, from popular accounts to highly technical monographs on their design. Much of this is probably known to the reader, so there is no point in repeating such matters here. Knowledge of the rigorous approach to the thermodynamic limit and how that relates to the laser is perhaps less common. If this connection is to be appreciated and not dismissed as mere mathematical detail, it must be justified on physical grounds, and that is the purpose of the introductory sections of this Chapter. We then go on to treat the model itself. At the end of the Chapter, having described the mathematics, we shall discuss what conclusions can be drawn regarding phase operators. For those uninterested in the calculations, it should be possible to omit a detailed reading and still understand the conclusions. Thus, for reasons of space, and because it seems too far from the main theme of this book to expound the formalism of quantum statistical mechanics for systems with infinitely many degrees of freedom, the model calculations will be rather condensed.

11.1.1 *Background*

Lasers form a broad category of devices which can transform incoherent electromagnetic radiation into coherent radiation. As a group they operate over a wide range of frequencies from above the optical band to the microwave level and beyond.

This production of coherent radiation depends on three processes. In the first process, atoms are prepared in a particular excited state, which has to be relatively long-lived (or metastable) for the process to proceed. The result is known as *population inversion*, this term emphasizing the fact that there are more atoms in the excited metastable state than are in some lower state to which they can decay.

In the second process, the atoms decay back down to this lower-lying state by spontaneous emission, which occurs even when *there is no radiation present.* This does not mean that there is no interaction between the atoms and the radiation field, but rather that this field is in its (dressed) ground state. From the standpoint of quantum field theory, the ground state is a very complicated entity indeed, involving polarization of the vacuum, virtual pair production and the like. Hence, despite the absence of real photons in the radiation field, it is the atom-field interaction which causes spontaneous emission.

The atoms, together with the radiation they emit, are enclosed in a radiation feedback arrangement — the precise nature of which depends upon the type of device. For lasers this is a passive optical resonator, for instance a pair of reflecting mirrors whose common normal defines the coherence direction. Most of the spontaneously emitted radiation will be absorbed by the walls of the enclosure, or otherwise lost, but a certain fraction will have the correct momentum and polarization to be reflected back into the region where the atoms are situated. In the third process, these reflected photons induce some of the remaining atoms in the excited state to emit photons by stimulated emission. Moreover, this stimulation tends to result in the emission of photons in the same state as the stimulating photons. As this process continues, therefore, it is to be expected that the radiation that builds up and has properties of *coherence.*[1]

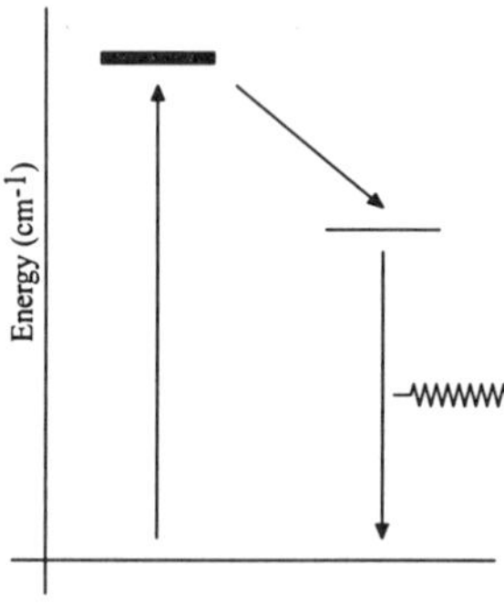

Fig. 11.1 Emission of Radiation

[1]The atomic levels and transitions operative for a laser device can be a good deal more complicated than the description given above conveys. Some possible complications will be discussed below in connection with the examples of a ruby laser and a He-Ne gas laser, but the above observations cover the basic issues. More detailed treatments can be found in specialist books on lasers, for example [164], [42].

11.1.2 *Coherence And Factorization*

In what sense is laser radiation coherent? A starting point for any such discussion must be the classical notion of *temporal*, or *longitudinal*, *coherence*[2]. Consider a quasi-monochromatic wave *in vacuo*, namely a signal composed of radiation of frequencies confined to a band of width much smaller than its mean value. A quasi-monochromatic wave, then, will comprise radiation of frequencies lying in some interval $[\nu_0 - \frac{1}{2}\Delta\nu, \nu_0 + \frac{1}{2}\Delta\nu]$, where $\Delta\nu \ll \nu_0$. Analysis shows that while the resulting signal has (on average) a frequency of ν_0, its amplitude and phase vary slowly with time. The signal is then a nearly periodic function which modulates slowly in amplitude and phase, with the frequency of modulation being smaller than $\frac{1}{2}\Delta\nu$. This modulation is therefore negligible over any time scale which is significantly smaller than $T_\ell = \Delta\nu^{-1}$, the so-called (*longitudinal*) *coherence time*. Associated to the coherence time is the (*longitudinal*) *coherence length* $L_\ell = cT_\ell$. That quasi-monochromatic radiation is described as coherent is reflected in the fact that, in experiments which investigate properties of the radiation on a time-scale smaller than T_ℓ, the radiation behaves to a large extent as if it were monochromatic of frequency ν_0 — thus, the smaller the value of $\Delta\nu$, the greater the degree of coherence. Such effects can be measured experimentally. For example, in a Michelson interferometry experiment, moving the interferometer arms a distance of more than L_ℓ apart destroys the fringe patterns, which were observable at smaller displacements.

Putting these ideas on a more mathematical footing, the (complex) signal describing the radiation is a function $V : \mathbb{R}^3 \times \mathbb{R} \to \mathbb{C}$ which has Fourier expansion

$$V(\mathbf{r}, t) = \int_0^\infty \widetilde{V}(\mathbf{r}, \nu)\, e^{-2\pi i \nu t}\, d\nu\,, \tag{11.1.1}$$

where the real-valued function $\widetilde{V}(\mathbf{r}, \nu)$ is presumed to be concentrated in some narrow interval about the frequency ν_0. In signal theory it is assumed that there are fluctuations in the signal due to uncontrollable, essentially

[2]The notion of *spatial coherence* can also be considered. Spatial coherence relates to the lateral spread of the radiation. This is usually not as important theoretically as longitudinal coherence, since it is strongly dependent upon the geometry of the optical resonator in the device, as well as other factors such as thermal and acoustic vibrations. However it is often useful to have a beam of light that approximates a plane wave closely and so does not spread much, and hence the design of effective resonators is a continuing development process.

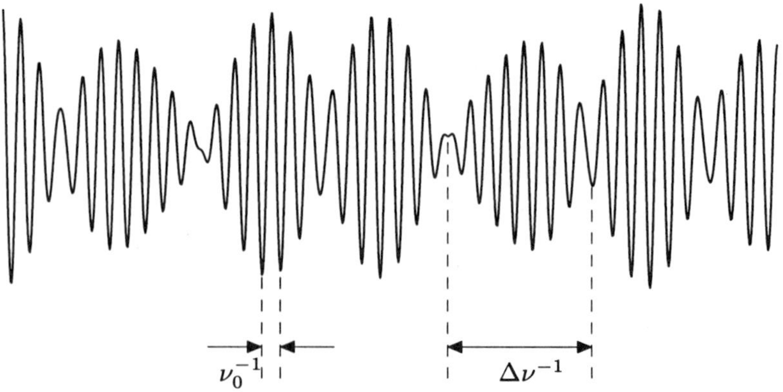

Fig. 11.2 A Quasi-Monochromatic Wave

random, factors such as thermal fluctuations and noisy circuits. These random factors need to be smoothed away when studying the signal, which involves taking the *ensemble average* over all possible instances of the signal. Indicating this ensemble average by double angular brackets, the key quantities[3] to consider are the *cross-correlation function* $\Gamma(\mathbf{r}_1, t_1; \mathbf{r}_2, t_2)$ and the *cross-spectral density function* $W(\mathbf{r}_1, \mathbf{r}_2, \nu)$ defined by the formulæ

$$\Gamma(\mathbf{r}_1, t_1; \mathbf{r}_2, t_2) = \langle\!\langle \overline{V(\mathbf{r}_1, t_1)}\, V(\mathbf{r}_2, t_2) \rangle\!\rangle\,, \qquad (11.1.2.a)$$

$$\langle\!\langle \overline{\widetilde{V}(\mathbf{r}_1, \nu_1)}\, \widetilde{V}(\mathbf{r}_2, \nu_2) \rangle\!\rangle = W(\mathbf{r}_1, \mathbf{r}_2, \nu_1)\, \delta(\nu_1 - \nu_2)\,. \qquad (11.1.2.b)$$

It should be noted that both of these definitions, and indeed many of the formulæ studied in signal theory, are inherently distributional in nature, so that all calculations concerning them must be performed in a weak sense. Under the reasonable technical assumptions of *stationarity* and *ergodicity* for the signal, the cross-correlation function $\Gamma(\mathbf{r}_1, t_1; \mathbf{r}_2, t_2)$ can be shown to depend on the difference $t_2 - t_1$ of the two times, and moreover to be equal to the time average $\Gamma(\mathbf{r}_1, \mathbf{r}_2, t_2 - t_1)$, where

$$\Gamma(\mathbf{r}_1, \mathbf{r}_2, \tau) = \lim_{T \to \infty} \frac{1}{2T} \int_{-T}^{T} \overline{V(\mathbf{r}_1, t)}\, V(\mathbf{r}_2, t + \tau)\, dt\,. \qquad (11.1.3)$$

The cross-correlation function and the cross-spectral density functions are

[3] There are higher correlation functions that can be studied, but we shall not need to mention them here.

related by the identity

$$\Gamma(\mathbf{r}_1,\mathbf{r}_2,\tau) \;=\; \int_0^\infty W(\mathbf{r}_1,\mathbf{r}_2,\nu)\, e^{-2\pi i\nu t}\, d\nu\,. \tag{11.1.4}$$

As special cases of these quantities, the functions

$$\Gamma(\mathbf{r},t) \;=\; \Gamma(\mathbf{r},\mathbf{r},t)\,, \tag{11.1.5.a}$$

$$S(\mathbf{r},\nu) \;=\; W(\mathbf{r},\mathbf{r},\nu)\,, \tag{11.1.5.b}$$

are called the *self-coherence function* and the *spectral density function* respectively, and are related by the identity

$$\Gamma(\mathbf{r},t) \;=\; \int_0^\infty S(\mathbf{r},\nu)\, e^{-2\pi i\nu t}\, d\nu\,. \tag{11.1.6}$$

Provided that these last two functions are sufficiently well-behaved that the following formulæ make sense, the equations

$$T_\ell(\mathbf{r})^2 \;=\; \frac{\displaystyle\int_{\mathbb{R}} t^2 |\,\Gamma(\mathbf{r},t)\,|^2\, dt}{\displaystyle\int_{\mathbb{R}} |\,\Gamma(\mathbf{r},t)\,|^2\, dt}\,, \tag{11.1.7.a}$$

$$\overline{\nu}(\mathbf{r}) \;=\; \frac{\displaystyle\int_0^\infty \nu S(\mathbf{r},\nu)^2\, d\nu}{\displaystyle\int_0^\infty S(\mathbf{r},\nu)^2\, d\nu}\,, \tag{11.1.7.b}$$

$$(\Delta\nu)(\mathbf{r})^2 \;=\; \frac{\displaystyle\int_0^\infty (\nu-\overline{\nu}(\mathbf{r}))^2\, S(\mathbf{r},\nu)^2\, d\nu}{\displaystyle\int_0^\infty S(\mathbf{r},\nu)^2\, d\nu}\,, \tag{11.1.7.c}$$

define the coherence length T_ℓ, average frequency $\overline{\nu}$ and bandwidth $\Delta\nu$ for the radiation. Given the Fourier transform relationship between the self-coherence function and the spectral density function, it is standard that

$$T_\ell(\mathbf{r})\,\cdot\,(\Delta\nu)(\mathbf{r}) \;\geqslant\; \frac{1}{4\pi}\,. \tag{11.1.8}$$

Under certain reasonable conditions on the nature of the radiation, the above inequality can be sharpened, so that $T_\ell(\mathbf{r})\,\cdot\,(\Delta\nu)(\mathbf{r}) \;\sim\; (4\pi)^{-1}$. This observation provides a relationship of the same nature between the coherence time and the bandwidth of the radiation as was introduced in the above heuristic discussion.

As has been mentioned, experiments exist which can be used to measure the coherence length of radiation. For example, sunlight over the frequency range 400 - 700 nanometers has a coherence length of roughly 10^{-3} millimeters. A low pressure mercury lamp has a coherence length of a few centimeters. But if an interferometry experiment is carried out on a typical He-Ne laser, results are obtained which would indicate a coherence length of many kilometers! No wonder that laser light is said to be coherent.

However, laser radiation is not a classical phenomenon, and so the above theory is not sufficient to explain the concept of coherence in lasers. To this end, Glauber [76, 77] has developed a theory of quantum correlation functions and their relation to quantum optical coherence. In effect, he replaces the complex signals by their quantum field counterparts, and ensemble averages by expectations in field states[4].

In particular, Glauber considers the N-point Wightman functions

$$\left[\mathcal{W}_\rho^{(N)}\right]^{\mu,\nu}(x_1,\ldots,x_N) = \mathrm{Tr}\left(\rho\prod_{j=1}^{N} F^{\mu,\nu}(x_j)\right), \quad 0 \leqslant \mu,\nu \leqslant 3, \tag{11.1.9}$$

where $F^{\mu,\nu}$ is the electromagnetic field operator and $x_1,\ldots,x_N \in \mathbb{R}^4$. Since $F^{\mu,\nu}$ satisfies the operator form of Maxwell's equations (weakly in the sense of Gupta & Bleuler), these tempered distributions $\mathcal{W}_\rho^{(N)}$ are interrelated by a hierarchy of partial differential equations involving the source 4-current. In particular, the 1-point functions satisfy the first order differential equation

$$\nabla_\mu \left[\mathcal{W}_\rho^{(N)}\right]^{\mu,\nu} = \mathrm{Tr}\left(\rho j^\nu\right), \tag{11.1.10}$$

which is a classical Maxwell's equation with source. The issue of *factorization* relates to whether, and for which values of N, the formula

$$\left[\mathcal{W}_\rho^{(N)}\right]^{\mu,\nu}(x_1,\ldots,x_N) = \prod_{j=1}^{N}\left[\mathcal{W}_\rho^{(1)}\right]^{\mu,\nu}(x_j) \tag{11.1.11}$$

holds, writing N-point functions as products of 1-point functions. If factorization occurs, there is correlation between the statistics at different spacetime points, and hence a degree of coherence in the field. Were equa-

---

[4]Other expectations are possible, for example ones which include factors due to the atoms in the system.

tion (11.1.11) satisfied for all integers N (this condition is termed *complete factorization*), the field would be strictly classical, would have infinite longitudinal coherence length and would in addition be spatially coherent. Such a situation is not possible physically, but represents an ideal situation to which coherent radiation should be a reasonable approximation.

Although the exact physical interpretation of the factorization of N-point functions (except in the case $N = 2$) is unclear[5], it is generally accepted that a quantum mechanical electromagnetic field exhibits a greater degree of coherence as equation (11.1.11) is satisfied for more and more values of N. It is worth noting that a free field does not exhibit such factorization into 1-point functions, which reflects the fact that the factorization property is dependent upon field-atom interactions.

Consequently, just what coherence properties a real laser beam has (in these terms) is not known with complete precision. It should be noted in advance that the solution of the laser model described in this Chapter yields complete factorization in the thermodynamic limit. In other words, the simplicity of the model suppresses higher order effects, and so the model is certainly not complete. Nonetheless it is a useful first step on the road to a fuller solution, which does not yet exist.

11.1.3 *The Phase Transition*

The above discussion describes which factors contribute to the generation of coherent radiation in a laser device. However these factors need to be set against others which operate to inhibit this process. For example depletion of the inverted population, or excessive photon absorption, will certainly reduce and (unless things are arranged just so) may even prevent the build-up of coherent radiation. Indeed, the absence of coherent radiation is the normal situation. An intrinsic part of the design of a laser device lies in ensuring that the factors which assist the generation of coherent radiation outweigh these negative factors.

In a model of a laser, this effect can be described by means of a real "pumping" parameter, chosen so that its increase is proportional to the buildup of coherent radiation. Eventually, if the system is well designed,

[5]It seems that fields exhibiting factorization for all k-point functions with $1 \leqslant k \leqslant N$ for some finite value of $N \geqslant 2$, but no further factorization, have not been observed experimentally.

the pumping parameter increases to some threshold value, and coherent output is then obtained from the device[6]. Since coherent electromagnetic radiation has quite different characteristics from incoherent radiation, it is seen that a sudden change of state of the field occurs at a threshold value of a characteristic parameter. But this is what is meant by a phase transition, so it ought to be possible to observe this in a proper mathematical model, and we shall. In anticipation of these results, we note that this is an *order-disorder phase transition* associated with a *spontaneous breakdown of gauge symmetry, far from thermal equilibrium.*

Experience from statistical mechanics tells us that the correct (idealized) description of a phase transition requires a thermodynamic limit. The proper limit is to let the number of atoms increase to infinity. This is intuitively sensible, for the build-up of coherent radiation is a collective effect of all the atoms acting in concert through their interaction with the field.

In summary, then, any model for the production of laser light must address the issues of coherence, factorization and phase transitions.

11.1.4 *The Ruby And He-Ne Lasers*

The idea of using stimulated emission from a feedback device to react with an inverted population in order to produce coherent amplified electromagnetic radiation is due, independently, to Gordon, Zeiger & Townes [83], [84] and Basov & Prokhorov [15]; these devices operated in the microwave frequency range, hence were given the acronym *masers.* It is now possible to construct a number of different types of laser — solid state, gas, semiconductor, amongst others — that can operate both as pulsed and continuous output devices. The first laser was the pulsed ruby laser of Maiman [162], and the first continuous output laser was the Helium-Neon device of Javan, Bennett & Herrott [132]. Because these were the first lasers in their class, it seems worthwhile examining how they work[7].

---

[6]In real lasers, there is a time delay between the pumping parameter's reaching this critical value and the onset of coherent radiation.

[7]Neither of these devices are used for the everyday lasers found in CD players and the like, which are semiconductor lasers, chosen because of ease of large scale manufacture, price and size, if not of efficiency.

11.1.4.1 *The Ruby Laser*

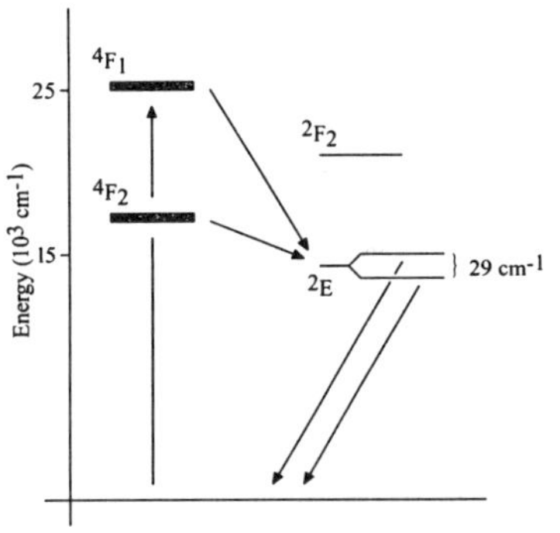

Fig. 11.3 Cr^{3+} Ions in Sapphire

The ruby laser population inversion system consists of a sapphire crystal[8] doped with Cr^{3+} (about one part in two thousand). There are a number of possible laser transitions possible for this atomic system, and which ones occur is controlled by the physical setup and the frequencies of radiation involved. The transition originally observed is a bit more complicated than the simple description associated with Figure 11.1. In Figure 11.3 we see two broadened excited states ${}^4F_1$ and ${}^4F_2$, which are filled as the result of excitation by an intense flashlamp pulse.

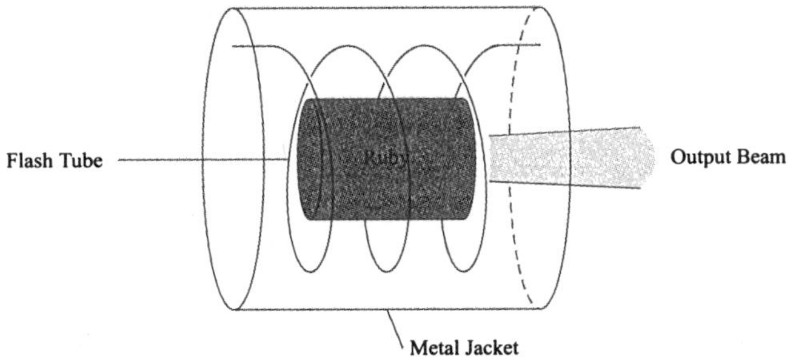

Fig. 11.4 Schematic of a Flashlamp

Spontaneous transitions then occur into the ^{2}E state just below. But this state is two-fold split, so the inverted population is distributed between these two levels, which are separated by no more than about 10^{11} cps. The lasing transitions are from these two states down to the ground state. By control of the excitation flash, the dominating laser emission is from the lower of the two split states, for preference. This transition is visible as a deep red flash. A rough schematic is shown in Figure 11.4. In the ruby laser the optical cavity is formed by polishing the ends of the crystals to be exactly perpendicular to the beam direction, and silvering them. One end is only partially silvered to allow a small fraction of the radiation to escape

[8]The ruby laser is thus an example of one in which the active atoms are held in a crystal or vitreous matrix, since the sapphire acts as a holding matrix for the chromium.

as the output beam.

11.1.4.2 *The He-Ne Laser*

The Helium-Neon gas laser, as its name implies, consists of a mixture of helium and neon gases, in the ratio of anywhere between five and ten to one, respectively. The helium atoms in their ground state are excited by a radio frequency generator, causing occupation of the 2^1S_0 and 2^3S_1 metastable states. Figure 11.5 gives a simplified energy diagram.

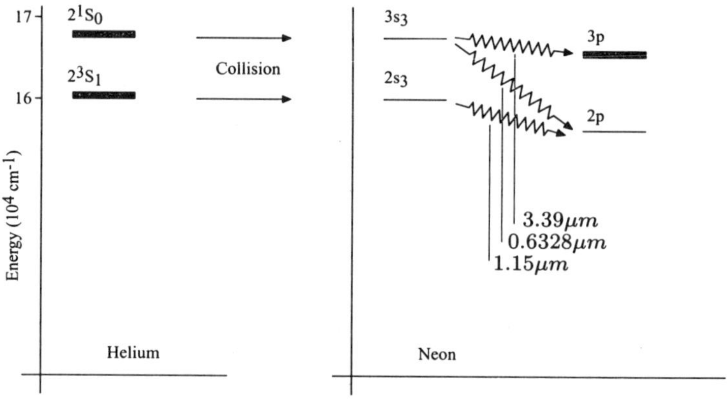

Fig. 11.5 Energy Levels for the He-Ne Laser

Amongst other things that might happen, these excited helium atoms can lose energy to the neon atoms in their ground state through collisions with them. The neon atoms are then excited to their $2s_3$ and $3s_3$ states, leaving them inverted relative to the neon 2p and 3p states. It is transition to these latter two states which are the principal laser transitions. The $3s_3$-2p transition is visible as red light.

The passive optical resonator in the original construction consisted of plane parallel mirrors, with the arrangement movable to a certain extent to allow alignment, output being very sensitive to this.

Modern He-Ne lasers are arranged somewhat differently and are less sensitive, but the essentials of the device remain the same.

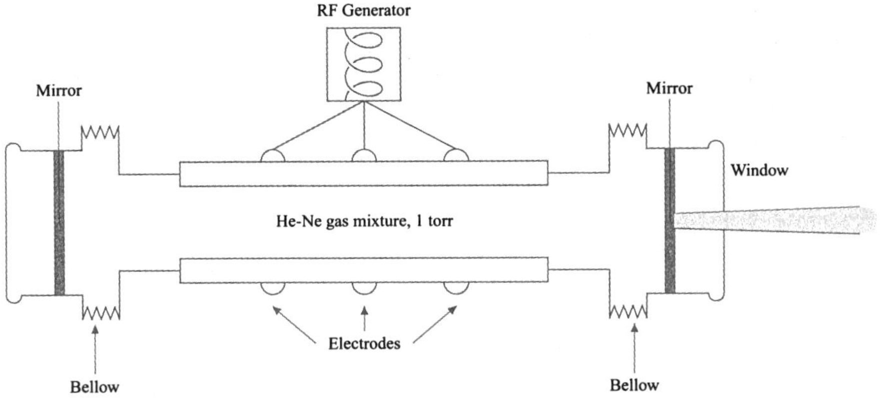

Fig. 11.6 A He-Ne Laser

11.1.5 *Laser Models*

Ideally, one would like to transcribe the mechanism of any given laser device into the language of quantum electrodynamics (it is a pretty safe assumption that quantum electrodynamics is the correct formalism for the description of laser energies), with exact initial and boundary conditions, and then solve the resulting model fully. That solution would be an exact description of the various states of radiation the device could create, and all its properties would be known by analysis. There would then be no question but that we would know exactly what laser light is. However, it is clear that such a state of knowledge will not be available to us for the near future, and perhaps longer than that. Of course, this is a fairly common situation in physics, and the standard procedure is to devise a model as near as possible to the exact case which can be solved, at least approximately.

The main business of this Chapter is to describe an analysis of a quantum model for the creation of laser light patterned after the initial work of Dicke [46], and extended by Graham & Haken [88], Haken [116, 96], Hepp & Lieb [115], Sewell [210] and Alli & Sewell [5], which we shall call the quantum laser model, or QL-model for short. This model is not completely realistic, nor does it provide the usual "coherent" or "squeezed state" description of laser radiation usually assumed by physicists in their analysis of optical experiments involving lasers. Hence we shall concentrate exclusively on the creation problem here, leaving the problem of the dynamical origin of coherent or squeezed states as an open question.

An early precursor of the QL-model was the semiclassical single mode laser model devised by Lamb [145]. Since only the lasing transition is of interest, Lamb chose to describe the atoms as two-level quantum systems, with no line broadening[9]. Lamb then chose to represent the coherent light as a classical monochromatic electromagnetic field, but in a self-consistent way. There are two crucial simplifications built into this assumption. First, since the field is monochromatic, it has infinite coherence length *ab initio*. Second, the field is to be commutative, and so the quantum correlation functions factorize completely[10].

The initial classical field produces a dipole moment in each atom, which is averaged over the collection for use in the interaction. In this way, the atoms act collectively to produce a macroscopic polarization field. The self-consistency of the electromagnetic field in the Lamb model can be seen in the assumption that it satisfies Maxwell's equations with this polarization field as source. Additionally, there has to be a cavity sink, so that there can be a balance between gain and loss, measured by a pumping parameter. In the Lamb model, this is represented by an enclosing conducting medium, which introduces a term proportional to the electric field[11]. The geometry of the cavity is accounted for in the form taken for the initial field. In other words, the electric field is written as

$$\mathbf{E}(\mathbf{r},t) \;=\; e^{-i\omega t}\mathbf{u}(\mathbf{r})\mathcal{E}(t) \,+\, e^{i\omega t}\overline{\mathbf{u}(\mathbf{r})}\,\overline{\mathcal{E}(t)}\,, \tag{11.1.12}$$

where $\mathbf{u}$ is a solution of the Helmholtz equation for frequency ω satisfying the relevant cavity boundary conditions, and the frequency ω is to be nearly equal to the frequency difference between the atomic levels.

The properties of the Lamb model are thus reflected in the behaviour of the unknown function $\mathcal{E}(t)$, and this turns out to satisfy the nonlinear differential equation

$$\frac{d\mathcal{E}(t)}{dt} \;=\; b\left[\frac{a-c}{b} - |\,\mathcal{E}(t)\,|^2\right]\mathcal{E}(t), \tag{11.1.13}$$

[9]This simplification carries over into the QL-model.

[10]These important simplifications render the model soluble, but are introduced in an *ad hoc* fashion.

[11]In a simple conducting medium, the current density is equal to the conductivity times the electric field.

where a, b and c are (positive) parameters of the model. It is simpler to work with the dimensionless quantity

$$\mathcal{F}(t) = \sqrt{\frac{b}{a}}\,\mathcal{E}(t)\,, \tag{11.1.14.a}$$

and introduce the pumping parameter

$$\eta = 1 - \frac{c}{a}\,, \tag{11.1.14.b}$$

for then the (dimensionless) intensity $\mathcal{I}(t) = |\mathcal{F}(t)|^2$ of the field satisfies the differential equation

$$\frac{d\mathcal{I}(t)}{dt} = 2a\big(\eta - \mathcal{I}(t)\big)\mathcal{I}(t)\,, \tag{11.1.14.c}$$

which equation is completely soluble. There is a unique dynamically stable steady-state solution[12] for each value of the pumping parameter η, namely

$$\mathcal{I}(t) = \begin{cases} \eta\,, & \eta > 0\,, \\ 0\,, & \eta \leqslant 0\,. \end{cases} \tag{11.1.15}$$

Moreover, for any value of the pumping parameter η, the intensity $\mathcal{I}(t)$ tends asymptotically to its appropriate steady-state value $\max(\eta, 0)$ in an overdamped manner, namely without any oscillation about that value[13]. It should also be noted that $\mathcal{F}(t)$ differs from $\sqrt{\mathcal{I}(t)}$ solely by some phase factor, which remains constant with time — thus this phase factor has no significant physical effect on the model.

As has been recognized by a number of people, this mathematical formalism is exactly the same as that for the magnetization M in the Curie-Weiss model of a ferromagnetic material. In this analogy, the constant a corresponds to the temperature T, while the constant c corresponds to the critical temperature T_c, and so the pumping parameter η corresponds to the familiar expression $1 - T_c/T$. Thus the phase transition observed above in the Lamb model corresponds to the phase transition which occurs at temperature T_c in the ferromagnetic model. As is well-known, this phase transition in the ferromagnetic model is an order-disorder transition, and $|M(t)|$ is the order parameter. As the pumping increases past the critical

[12]The steady-state solution $\mathcal{I}(t) = 0$ for $\eta > 0$ is not stable.

[13]Actual experiments may record some oscillation about the steady-state value, but this effect may well be due to random fluctuations which have been excluded from the Lamb model as currently enunciated.

value, there is a spontaneous increase in the order parameter, indicating the appearance of a collective action of the atoms. One might say that, when the pumping is sufficiently large, the atoms act coherently. It is important to realize that the steady state resulting from this phase transition is not one of thermal equilibrium — it is far from that. It is important to emphasize, however, that this is only an analogy — the parameter a in the Lamb model should not be interpreted as temperature, since the model is implicitly only valid at absolute zero.

A similar form of phase transition will be observed in the following rigorous treatment of the QL-model — it cannot be emphasized too strongly that the creation of coherent radiation is a collective effect, and there can be no proper description of it without a model that shows this clearly.

The completely classical nature of the electromagnetic field in the Lamb model is evidently a weakness. The model can be improved by modifying the equation for $\mathcal{E}(t)$ to read

$$\frac{d\mathcal{E}(t)}{dt} - b\left[\frac{a-c}{b} - |\,\mathcal{E}(t)\,|^2\right]\mathcal{E}(t) = \mathcal{W}(t), \tag{11.1.16}$$

where $\mathcal{W}(t)$ is a stochastic noise term modelled after the effect of the free Bose field sink on the oscillator lowering operator in the damped oscillator model, see equation (7.4.27). However this modification is evidently an empirical procedure, and it would be a decided improvement to include dynamical interactions which will result in the required statistical effects in the solution[14].

11.2 QL-Model Kinematics

11.2.1 *Preliminaries*

The quantum laser model adopted here is a refinement of the ideas of a number of authors, based on earlier work of Dicke, who studied the interaction of light with two level systems [46]. For an extensive list of references, see Mandel & Wolf [164].

In the early laser models, a dipolar interaction Hamiltonian is used to determine the equation of motion for the (pure) state of the system (using

[14] It should be noted that the full Lamb model is more complicated than the above description might indicate, and an interested reader is directed to the literature [164] for details.

the interaction picture). This Hamiltonian for a single mode of angular frequency ω is essentially that of the Lamb model, but with the classical electromagnetic field replaced by the operator

$$\mathbf{E}(\mathbf{r}) = i\lambda\sqrt{\frac{\omega}{V}}\left[\mathbf{u}(\mathbf{r})A - \overline{\mathbf{u}(\mathbf{r})}A^{+}\right], \tag{11.2.1}$$

where $\mathbf{u}$ is an appropriate mode function determined by the cavity geometry, V is the cavity volume, λ is a constant, and A, A^{+} are the lowering and raising operators for the mode. Hence it is implicit that the smooth model is being used.

At the second stage, the equation of motion for the state is modified by assuming that initially some atoms occupy the higher and some the lower energy level, and gain and loss rates are put in by hand, resulting in a master equation with A, B and C coefficients which are adjusted to satisfy the Einstein balance relations.

This system was later modified in accordance with the general principles of quantum statistical mechanics of infinite systems. In this treatment, the basic Hamiltonian of the earlier models is retained for N atoms, but now the gain and loss mechanism is controlled by sources and sinks. These reservoirs are constructed from free Bose and Fermi fields, [115], [210], and the system is treated as evolving according to a reservoir driven open dynamics, as discussed in Chapter 7. From the discussion there it follows that the Heisenberg equations of motion for the observables will be of Langevin type.

At this stage, Hepp & Lieb made a significant improvement in the treatment of the production of laser light as a collective process by requiring the cooperative behaviour of a large number N of atoms, and then allowing the number of atoms to increase without limit, considering the limit as $N \to \infty$. In this manner, the possibility of a phase transition associated with a spontaneous breakdown of symmetry becomes possible. As in all collective phenomena based on a microscopic dynamics, the macroscopic physics must be described by intensive variables, and the associated equations of motion will exhibit irreversible behaviour.

The implication is that it will be necessary to scale the observables of the quantum (microscopic) system with appropriate powers of the number of atoms N (which is proportional to the volume V), and at the same time-scale whatever initial state is chosen for the system so as to obtain a finite energy density in the limit. The (time dependent) macroscopic variables of the system will be the limits of the expectations of the time evolved scaled observables in the scaled state (using the Heisenberg picture). Macroscopic equations of motion are obtained by taking the limits of the expectations of the quantum equations of motion.

The macroscopic physics described by the model is encoded in these equations. The values of the control constants in the generator of the time translations determine a pumping parameter, and associated with this parameter are two critical values. For values of the pumping parameter less than both of these critical values, states of the system are normal radiation states (even those of pure phase). For values of the pumping parameter lying between these two critical values, states of the system exhibit properties of coherent laser light. Finally, for values of the pumping parameter greater than both of the critical values, monochromatic laser radiation gives way to chaotic behaviour described by a Lorentz strange attractor. It was emphasized that this must be the case by Haken [96, 95], and was first shown rigorously by Alli & Sewell [5]. Their work involved two further refinements of the Hepp & Lieb treatment. The first was to use a more general choice of parameters than did Hepp & Lieb, and the second was to include more than one mode of the electromagnetic field. The result is that if there are L modes, there are $L+1$ critical values for the pumping parameter. The parametric regions of the solutions are consequently more involved than for the one mode model. The reader is referred to their paper for a detailed description of the chaotic regions. Alli & Sewell also succeeded in constructing a mathematically sound description of the open system dynamics involving an unbounded generator of Lindblad type.

Most work on quantum phase makes no mention of a dynamical scheme to generate the coherent radiation. Indeed, the laser light seems to arrive like Athena, springing full-blown from Zeus' head, described by a one mode coherent state. We know of no rigorous model for generating such vectors dynamically in the sense of a quantum laser model, and so feel that justification of the accuracy of this description is inferential rather than direct. While not suggesting that these states do not represent coherent laser light, it must be pointed out that it is certainly neither obvious nor proved that they do so, and this situation cannot be considered satisfactory until a rigorous dynamical model exists. Further foundational work evidently remains to be done on this important problem. Notwithstanding its drawbacks, the QL-model is the only model of coherent light production (so far as we know) in which all the assumptions are clear from the start, the treatment is entirely rigorous (including a proper treatment of the infinite number of degrees of freedom inherent in the thermodynamic limit) and the answer is exact. Moreover, it has a respectable pedigree in terms of the more heuristic and phenomenological models that precede it. For all these reasons, the conclusions drawn from it must be taken seriously.

In the remainder of the Chapter, the QL-model will be considered in some detail, although a number of proofs have been omitted due to their length and technical difficulty. In particular this is so for the existence of the dynamics, as will be made clear at the relevant points in the argument.

11.2.2 *The Matter*

Each atom is taken to be a two level quantum system, and such a system was described in Section 7.5. The algebra of observables is then the set $M_2(\mathbb{C})$ of all 2×2 matrices. The states are given through density matrices, which are the positive matrices of unit trace.

The matter consists of an atom at each point of the one dimensional linear lattice $\mathbb{N}$. In this linear array, atom r is at site r, with $r = 1, 2, \ldots$. When a matrix B refers to this atom, it is denoted B_r. Correspondingly, the algebra of observables for atom r is denoted $[M_2(\mathbb{C})]_r$.

To combine the atoms into a single system, we distinguish the systems with one atom (at site 1), two atoms (at sites 1, 2), and so on. For N atoms, the "matter" system will be denoted $\Sigma^{(\mathsf{M};N)}$. These are compounded (without statistics) from the single atom systems. The algebra of observables for $\Sigma^{(\mathsf{M};N)}$ is the C^*-tensor product algebra,

$$\mathfrak{A}^{(\mathsf{M};N)} = \bigotimes_{r=1}^{N} [M_2(\mathbb{C})]_r \,. \qquad (11.2.2)$$

In the usual way, the operator B_r for atom r (where $1 \leqslant r \leqslant N$) is then identified with the observable

$$I_1 \otimes \cdots \otimes B_r \otimes \cdots \otimes I_N \,. \qquad (11.2.3)$$

in $\mathfrak{A}^{(\mathsf{M};N)}$. The states $\mathfrak{A}_*^{(\mathsf{M};N)}$ for $\Sigma^{(\mathsf{M};N)}$ are the positive and normalized linear functionals on $\mathfrak{A}^{(\mathsf{M};N)}$. Every such state is a linear combination of product states (tensor products of density matrices).

The complete matter system $\Sigma^{(\mathsf{M})}$ is defined to be the C^*-inductive limit of the $\mathfrak{A}^{(\mathsf{M};N)}$,

$$\mathfrak{A}^{(\mathsf{M})} = \liminf_{N\to\infty} \mathfrak{A}^{(\mathsf{M};N)} = \bigotimes_{r\geqslant 1} [M_2(\mathbb{C})]_r \,. \tag{11.2.4}$$

The identifications in equation (11.2.4) need explanation. Any of the N-atom algebras $\mathfrak{A}^{(\mathsf{M};N)}$ can be regarded as a subalgebra of the C^*-algebra $\bigotimes_{r\geqslant 1} \left[M_2(\mathbb{C})\right]_r$ by identifying $B \in \mathfrak{A}^{(\mathsf{M};N)}$ with the operator

$$B \otimes \Big(\bigotimes_{k>N} I_k \Big) \in \bigotimes_{r\geqslant 1} \left[M_2(\mathbb{C})\right]_r .$$

Under these identifications, it is clear that $\mathfrak{A}^{(\mathsf{M};N_1)} \subseteq \mathfrak{A}^{(\mathsf{M};N_2)}$ whenever $N_1 \leqslant N_2$. Thus the union of the subalgebras $\mathfrak{A}^{(\mathsf{M};N)}$ is itself a subalgebra of $\bigotimes_{r\geqslant 1} \left[M_2(\mathbb{C})\right]_r$ which, in a standard terminology of statistical mechanics, is called the local matter algebra,

$$\mathfrak{A}^{(\mathsf{M})}_{\mathrm{LOC}} = \bigcup_{N\in\mathbb{N}} \mathfrak{A}^{(\mathsf{M};N)} . \tag{11.2.5}$$

The connection between the local matter algebra $\mathfrak{A}^{(\mathsf{M})}_{\mathrm{LOC}}$ and the algebra $\mathfrak{A}^{(\mathsf{M})}$, which would be termed the quasi-local matter algebra in statistical mechanics, is that $\mathfrak{A}^{(\mathsf{M})}$ is the norm-closure of $\mathfrak{A}^{(\mathsf{M})}_{\mathrm{LOC}}$. The differences between local and quasi-local matter observables are of physical significance, and we refer the reader to the appropriate texts for a discussion of such issues, for example [28], [57], [94], [199], [210].

As usual, the matter states are the positive and normalized linear functionals on the quasi-local matter algebra $\mathfrak{A}^{(\mathsf{M})}$.

The system $\Sigma^{(\mathsf{M})}$ has infinitely many degrees of freedom, like the Bose field of the damped oscillator. However, unlike that field, every element of $\mathfrak{A}^{(\mathsf{M})}$ is bounded[15].

The Pauli matrices σ_ν (where the index ν can, as usual, take any of the values 1, 2, 3, + or −) are the basic operators in the description of an atom

[15] Hidden in the above formalism is our decision to use the same notation for B_r as an element of $[M_2(\mathbb{C})]_r$, $\mathfrak{A}^{(\mathsf{M};N)}$ or $\mathfrak{A}^{(\mathsf{M})}$, except when doing so will lead to a definite error. Otherwise we should have been burdened with the matrix B appearing as B_r, $B_r^{(\mathsf{M};N)}$ and $B_r^{(\mathsf{M})}$, and the appearance of symbols for the associated isomorphisms and identifications. In the same way, as far as possible we shall avoid labels on the states. The algebra $\mathfrak{A}^{(\mathsf{M})}$ does not depend on the particular choice of subalgebras $\mathfrak{A}^{(\mathsf{M};N)}$ used in the above description. For further details concerning this construction, see eg, [28], [51], [57], [125], [199].

in this model, and the Pauli matrices associated with the rth atom will be denoted $\sigma_\nu[r]$.

The spin density operators,

$$\mathfrak{s}_\nu^{(N)} = \frac{1}{N}\sum_{r=1}^{N}\sigma_\nu[r], \qquad N \in \mathbb{N}, \tag{11.2.6}$$

are important to the physics, since they are the macroscopic atomic variables. Since the operators $\mathfrak{s}_\nu^{(N)}$ concern the first N atoms, the term macroscopic becomes physically sensible for large values of N. Note also that the variables $\mathfrak{s}_\nu^{(N)}$ are intensive, in that both the numerators and the denominators in their definition increase linearly with N. When considering these intensive variables, the vector notation $\mathfrak{s}^{(N)} = \left(\mathfrak{s}_1^{(N)}, \mathfrak{s}_2^{(N)}, \mathfrak{s}_3^{(N)}\right)$ is often convenient.

In the same way that only certain states in quantum mechanics describe thermal equilibrium, only certain states of $\Sigma^{(\mathsf{M})}$ will describe coherent radiation. Although we do not know a necessary and sufficient condition which ensures that a state on $\Sigma^{(\mathsf{M})}$ describes coherent radiation, it is certainly sufficient that the state has the properties of *homogeneity* and *clustering.*

A state $\varpi \in \mathfrak{A}_*^{(\mathsf{M})}$ is *homogeneous* if the value that ϖ takes on the Pauli matrices $\sigma_\nu[r]$ is independent of r (although it will be dependent upon ν). Thus any state with the property of homogeneity is associated with a vector $\mathbf{s} \in \mathbb{R}^3$, in that

$$\varpi\left(\sigma_\nu[r]\right) = s_\nu\,, \qquad r \geqslant 1\,. \tag{11.2.7}$$

A state of $\Sigma^{(\mathsf{M})}$ which is homogeneous and which is associated with the vector $\mathbf{s} \in \mathbb{R}^3$ will be denoted $\varpi_{\mathbf{s}}$.

A state $\varpi \in \mathfrak{A}_*^{(\mathsf{M})}$ has the *clustering* property if

$$\lim_{N\to\infty}\left\{\varpi\left[\left(\mathbf{b}\cdot\mathfrak{s}^{(N)}\right)^2\right] - \left[\varpi\left(\mathbf{b}\cdot\mathfrak{s}^{(N)}\right)\right]^2\right\} = 0 \tag{11.2.8}$$

for any $\mathbf{b} \in \mathbb{R}^3$. In other words, the uncertainty of the operator $\mathbf{b}\cdot\mathfrak{s}^{(N)}$ in the state ϖ vanishes in the limit as $N \to \infty$ for any $\mathbf{b} \in \mathbb{R}^3$,

$$\lim_{N\to\infty}\mathfrak{Unc}_\varpi\left[\mathbf{b}\cdot\mathfrak{s}^{(N)}\right] = 0\,. \tag{11.2.9}$$

Homogeneity of a state $\varpi_{\mathbf{s}} \in \mathfrak{A}_*^{(\mathsf{M})}$ implies that

$$\varpi_{\mathbf{s}}\left(\mathbf{b}\cdot\mathfrak{s}^{(N)}\right) = \mathbf{b}\cdot\mathbf{s}, \tag{11.2.10}$$

for all $N \geqslant 1$ and all $\mathbf{b} \in \mathbb{R}^3$, and so the clustering requirement can be written as

$$\lim_{N\to\infty} \varpi_{\mathbf{s}}\big[\,(\mathbf{b}\cdot \mathfrak{s}^{(N)})^2\,\big] \;=\; (\mathbf{b}\cdot\mathbf{s})^2\,. \tag{11.2.11}$$

The conditions of homogeneity and clustering do not imply that a state ϖ is a product state, although it certainly must be primary[16]. However, any homogeneous product state $\varpi_{\mathbf{s}}$ is such that

$$\varpi_{\mathbf{s}}\big[\,(\mathbf{b}\cdot \mathfrak{s}^{(N)})^2\,\big] \;=\; (\mathbf{b}\cdot\mathbf{s})^2 + \frac{1}{N}\big[\mathbf{b}\cdot\mathbf{b} - (\mathbf{b}\cdot\mathbf{s})^2\,\big]\,, \tag{11.2.12}$$

for any $N \in \mathbb{N}$, and hence automatically satisfies the clustering property.

11.2.3 *The Radiation*

Alli & Sewell considered the QL-model with more than one radiation mode, but the essential features of the model are present in the single mode case. In view of this, the radiation will be taken to consist of only one monochromatic mode. That is, the electromagnetic field is to be given by equation (11.2.1) above.

Thus, the radiation system is taken to be a single copy of the smooth model for one oscillator, distinguished for foundational purposes by an appropriate label,

$$\Sigma^{(\mathrm{R})} \;=\; [\mathcal{H}^{(\mathrm{R})}, \mathcal{S}^{(\mathrm{R})}, \mathfrak{A}^{(\mathrm{R})}, \mathfrak{A}_*^{(\mathrm{R})}]\,. \tag{11.2.13}$$

Thus $\Sigma^{(\mathrm{R})}$ is irreducible and isomorphic to the Schrödinger representation, so the meaning of all these symbols has already been considered in detail.

All the smooth and bounded model operators previously encountered must therefore have an exact counterpart here. Thus $A^{(\mathrm{R})}$, $(A^{(\mathrm{R})})^+$ denote the radiation ladder operators and

$$W^{(\mathrm{R})}[z] \;=\; \exp\Big\{i\overline{[zA^{(\mathrm{R})} + \overline{z}(A^{(\mathrm{R})})^+]}\Big\}\,, \tag{11.2.14}$$

the radiation Weyl group (see equation (4.6.5)). Evidently Weyl quantization can be defined, with radiation observables being denoted $\mathbf{\Delta}^{(\mathrm{R})}[T]$ for $T \in \mathcal{S}'(\Pi)$, where $\mathbf{\Delta}^{(\mathrm{R})}[T]$ is an element of $\mathcal{L}(\mathcal{S}^{(\mathrm{R})}, [\mathcal{S}^{(\mathrm{R})}]')$.

[16] A state ω on a C^*-algebra $\mathcal{A}$ is primary if the weak closure $\pi_\omega(\mathcal{A})''$ of the GNS representation is a factor.

A radiation state on $\mathfrak{A}^{(\mathsf{R})}$ is given by a density matrix whose integral kernel belongs to the completed tensor product $\mathcal{S}^{(\mathsf{R})}\overline{\bigotimes}\mathcal{S}^{(\mathsf{R})}$, which is isomorphic to $\mathcal{S}(\mathbb{R}^2)$. No restriction is placed by the model on the choice of radiation state — the model only requires that the state be scaled appropriately, as will be discussed below.

11.2.4 *Combining Matter And Radiation*

The QL-model system $\Sigma^{(\mathsf{S})}$ is the combination of the matter and radiation subsystems, so that $\Sigma^{(\mathsf{S})} = \Sigma^{(\mathsf{M})} \otimes \Sigma^{(\mathsf{R})}$. However, for practical purposes, we shall only be interested in $\Sigma^{(\mathsf{S})}$ insofar as it contains the important N atom subsystems $\Sigma^{(\mathsf{S};N)} = \Sigma^{(\mathsf{M};N)} \otimes \Sigma^{(\mathsf{R})}$ (where $N \in \mathbb{N}$), with corresponding algebras of observables[17] $\mathfrak{A}^{(\mathsf{S};N)} = \mathfrak{A}^{(\mathsf{M};N)} \otimes \mathfrak{A}^{(\mathsf{R})}$ and collections of states $\mathfrak{S}^{(\mathsf{S};N)}$ contained in $\mathfrak{A}_*^{(\mathsf{S};N)} = \mathfrak{A}_*^{(\mathsf{M};N)} \otimes \mathfrak{A}_*^{(\mathsf{R})}$. The model will provide a separate dynamics for each subsystem $\Sigma^{(\mathsf{S};N)}$, and the dynamics of physical interest will arise from a limit being taken of this sequence of dynamics. Thus it would be mistaken to consider there being a well-defined dynamics on the full system $\Sigma^{(\mathsf{S})}$ — cf [210], §§2.4.4, 2.4.5.

Operators that belong to either of the matter and radiation subsystems can be represented in the combined system in a natural manner, so we have the spin densities

$$\mathfrak{s}_\nu^{(\mathsf{S};N)} = \mathfrak{s}_\nu^{(N)} \otimes I^{(\mathsf{R})}, \tag{11.2.15}$$

the radiation ladder operators

$$A^{(\mathsf{S};N)} = I^{(\mathsf{M};N)} \otimes A^{(\mathsf{R})} \quad \text{and} \quad (A^{(\mathsf{S};N)})^+ = I^{(\mathsf{M};N)} \otimes (A^{(\mathsf{R})})^+ , \tag{11.2.16}$$

the Weyl group

$$W^{(\mathsf{S};N)}[z] = I^{(\mathsf{M};N)} \otimes W^{(\mathsf{R})}[z] , \tag{11.2.17}$$

and the general radiation observables

$$\mathbf{\Delta}^{(\mathsf{S};N)}[T] = I^{(\mathsf{M};N)} \otimes \mathbf{\Delta}^{(\mathsf{R})}[T] , \qquad T \in \mathcal{S}'(\Pi) . \tag{11.2.18}$$

---

[17] Since $\mathfrak{A}^{(\mathsf{M};N)}$ is finite-dimensional, this tensor product of algebras is algebraic, and not topological.

11.2.5 *The Macroscopic Variables*

Perhaps the most basic point in the physics of collective phenomena is the necessity of scaling microscopic variables appropriately, to yield macroscopic variables which are suitable for consideration in the thermodynamic limit. We have already defined the basic macroscopic matter observables, namely the spin densities $\mathfrak{s}_{\nu}^{(\mathsf{S};N)}$ which were defined in equation (11.2.15). We must now determine the proper scaling for radiation observables in contact with matter.

The key to this scaling in the QL-model is energy. It will be clear how the extensive microscopic Hamiltonian for $\Sigma^{(\mathsf{S};N)}$ can be scaled to an intensive quantity. This will involve writing it in terms of the spin densities and other factors which are obtained by dividing the radiation ladder operators by certain factors of N. This procedure then prescribes how all matter and radiation observables must be scaled to yield appropriate macroscopic variables. There must also be a corresponding scaling of the state, but that will be discussed after we have dealt with the observables. The (single mode) free radiation Hamiltonian is proportional to the number operator, which must be divided by N to obtain a density. As this Hamiltonian is a product of the radiation raising and lowering operators, it follows that the macroscopic ladder observables for $\Sigma^{(\mathsf{S};N)}$ are

$$\mathfrak{a}^{(\mathsf{S};N)} = \frac{1}{\sqrt{N}}\big(I^{(\mathsf{M};N)} \otimes A^{(\mathsf{R})}\big), \qquad (\mathfrak{a}^{(\mathsf{S};N)})^+ = \frac{1}{\sqrt{N}}\big(I^{(\mathsf{M};N)} \otimes (A^{(\mathsf{R})})^+\big). \tag{11.2.19}$$

A feature of this model is that this scaling suffices to describe the full dynamics of the system $\Sigma^{(\mathsf{S};N)}$, since the relevant interaction Hamiltonian $H_{int}^{(\mathsf{S};N)}$ for this system is an extensive observable which can be written in terms of the macroscopic observables $\mathfrak{s}_{\pm}^{(N)}$, $\mathfrak{a}^{(\mathsf{S};N)}$ and $(\mathfrak{a}^{(\mathsf{S};N)})^+$, see equation (11.3.6). This also implies that $N^{-1}H_{int}^{(\mathsf{S};N)}$ is an intensive macroscopic variable, as it should be.

Because the Weyl group can be written in terms of the raising and lowering operators, by replacing these in the exponential by their scaled counterparts, we obtain the scaled Weyl group. Consulting equation (11.2.14), the scaled Weyl group is thus

$$\begin{aligned}\mathfrak{W}^{(\mathsf{S};N)}[z] &= I^{(\mathsf{M};N)} \otimes \exp\left\{i\overline{[z\mathfrak{a}^{(\mathsf{S};N)} + \overline{z}(\mathfrak{a}^{(\mathsf{S};N)})^+]}\right\} \\ &= I^{(\mathsf{M};N)} \otimes W^{(\mathsf{R})}[N^{-1/2}z] = W^{(\mathsf{S};N)}[N^{-1/2}z]\,, \end{aligned} \tag{11.2.20}$$

so that $\mathfrak{W}^{(\mathsf{S};N)}[z]$ is the appropriate macroscopic variable to be associated with $W^{(\mathsf{R})}[z]$.

Every other radiation variable can now be scaled correctly. The following heuristic argument is the crux of the connection between quantization and the QL-model. Informally viewing $\mathbf{\Delta}^{(\mathsf{R})}[T]$ as equal to $W^{(\mathsf{R})}[\![\mathcal{F}T]\!]$ divided by 2π, the necessary scaling of $\mathbf{\Delta}^{(\mathsf{R})}[T]$ can be obtained from the scaling of $W^{(\mathsf{R})}$. That is, the unscaled radiation observable $\mathbf{\Delta}^{(\mathsf{S};N)}[T]$ should be replaced by the macroscopic observable $\mathfrak{D}^{(\mathsf{S};N)}[T] = (2\pi)^{-1}\mathfrak{W}^{(\mathsf{S};N)}[\![\mathcal{F}T]\!]$.

Since operators of the form $\mathfrak{D}^{(\mathsf{S};N)}[T]$ are the ones that will be considered in the thermodynamic limit, it is important to establish a rigorous formulation of such operators. We do this by pushing the scaling through the quantization procedure onto the symbol T, and this necessitates scaling the independent variables of T. Since T is, in general, a distribution and not a function, doing this is not entirely elementary. Consider the continuous endomorphisms $\mathcal{E}_N$ and $\mathcal{E}_N^{\mathrm{tr}}$ of $\mathcal{S}(\mathbb{R}^2)$ defined by the formulæ:

$$\big[\mathcal{E}_N F\big](p,q) \;=\; N\,F(p\sqrt{N},q\sqrt{N})\,, \tag{11.2.21.a}$$

$$\big[\mathcal{E}_N^{\mathrm{tr}} F\big](p,q) \;=\; F\big(\tfrac{p}{\sqrt{N}},\tfrac{q}{\sqrt{N}}\big)\,, \tag{11.2.21.b}$$

for any $F \in \mathcal{S}(\mathbb{R}^2)$. Then

$$[\![F\,,\mathcal{E}_N G]\!] \;=\; [\![\mathcal{E}_N^{\mathrm{tr}} F\,, G]\!]\,, \tag{11.2.22.a}$$

$$[\![F\,,\mathcal{E}_N^{\mathrm{tr}} G]\!] \;=\; [\![\mathcal{E}_N F\,, G]\!]\,, \tag{11.2.22.b}$$

for any $F,G \in \mathcal{S}(\mathbb{R}^2)$. Thus it is clear that $\mathcal{E}_N^{\mathrm{tr}}$ is the restriction to $\mathcal{S}(\mathbb{R}^2)$ of the continuous endomorphism $(\mathcal{E}_N)^{tr}$ of $\mathcal{S}'(\mathbb{R}^2)$, and moreover that $\mathcal{E}_N$ is the restriction to $\mathcal{S}(\mathbb{R}^2)$ of the continuous endomorphism $(\mathcal{E}_N^{\mathrm{tr}})^{tr}$ of $\mathcal{S}'(\mathbb{R}^2)$. It is then convenient to denote the endomorphism $(\mathcal{E}_N^{\mathrm{tr}})^{tr}$ of $\mathcal{S}'(\mathbb{R}^2)$ by the symbol $\mathcal{E}_N$, and the endomorphism $(\mathcal{E}_N)^{tr}$ of $\mathcal{S}'(\mathbb{R}^2)$ by the symbol $\mathcal{E}_N^{\mathrm{tr}}$. This identification of $\mathcal{E}_N^{\mathrm{tr}}$ with $(\mathcal{E}_N)^{tr}$ justifies our choice of notation — the identification of $\mathcal{E}_N$ with $(\mathcal{E}_N^{\mathrm{tr}})^{tr}$ is serendipitous.

As well as restricting to the continuous endomorphisms of $\mathcal{S}(\mathbb{R}^2)$ defined above, the maps $\mathcal{E}_N$ and $\mathcal{E}_N^{\mathrm{tr}}$ also restrict to yield continuous endomorphisms of $L^2(\mathbb{R}^2)$ (and, in this context, $\mathcal{E}_N^{\mathrm{tr}}$ is the adjoint of $\mathcal{E}_N$, and $N^{-1/2}\mathcal{E}_N$ is unitary). Thus each of the symbols $\mathcal{E}_N$ and $\mathcal{E}_N^{\mathrm{tr}}$ can refer to three different maps, according as whether they are regarded as endomorphisms of $\mathcal{S}'(\mathbb{R}^2)$, $\mathcal{S}(\mathbb{R}^2)$, or $L^2(\mathbb{R}^2)$. Doubling the confusion, we shall use exactly the same symbols to refer to the analogously defined maps[18], seen as continuous endomorphisms of $\mathcal{S}(\Pi)$, $\mathcal{S}'(\Pi)$ and $L^2(\Pi)$. It should be clear from context, however, which form of any of these maps is intended at any time[19].

These scaling maps are intertwined by the Fourier transform, since

$$\mathcal{F} \circ \mathcal{E}_N = \mathcal{E}_N^{\mathrm{tr}} \circ \mathcal{F}\,, \tag{11.2.23.a}$$

$$\mathcal{F} \circ \mathcal{E}_N^{\mathrm{tr}} = \mathcal{E}_N \circ \mathcal{F}\,. \tag{11.2.23.b}$$

These formulæ are primarily true in the spaces $\mathcal{S}'(\mathbb{R}^2)$ and $\mathcal{S}'(\Pi)$, but by restriction are true on any of the other four spaces that might be considered.

The utility of these maps lies in the following identity concerning the scaled Weyl group,

$$\mathfrak{W}^{(\mathsf{S};N)}[\![F]\!] = W^{(\mathsf{S};N)}[\![\mathcal{E}_N F]\!]\,, \qquad F \in \mathcal{S}(\mathbb{R}^2)\,, \tag{11.2.24}$$

which in turn implies that

$$\begin{aligned} \mathfrak{W}^{(\mathsf{S};N)}[\![\mathcal{F}F]\!] &= W^{(\mathsf{S};N)}[\![\mathcal{E}_N \mathcal{F} F]\!] \\ &= W^{(\mathsf{S};N)}[\![\mathcal{F}\mathcal{E}_N^{\mathrm{tr}} F]\!] \\ &= 2\pi\,\Delta^{(\mathsf{S};N)}\big[\mathcal{E}_N^{\mathrm{tr}} F\big] \end{aligned} \tag{11.2.25}$$

for any $F \in \mathcal{S}(\Pi)$. Thus we are led to make the following definition.

Definition 11.1 *For any $T \in \mathcal{S}'(\Pi)$, the macroscopic observable $\mathfrak{D}^{(\mathsf{S};N)}[T]$ corresponding to the microscopic radiation observable $\boldsymbol{\Delta}^{(\mathsf{S};N)}[T]$ is*

$$\mathfrak{D}^{(\mathsf{S};N)}[T] = \boldsymbol{\Delta}^{(\mathsf{S};N)}\left[\mathcal{E}_N^{\mathrm{tr}} T\right]\,. \tag{11.2.26}$$

The key macroscopic observables for the system $\Sigma^{(\mathsf{S};N)}$ comprise the set

$$\mathfrak{J}^{(\mathsf{S};N)} = \left\{ I^{(\mathsf{S};N)}, \mathfrak{a}^{(\mathsf{S};N)}, (\mathfrak{a}^{(\mathsf{S};N)})^+, \mathfrak{s}_-^{(\mathsf{S};N)}, \mathfrak{s}_+^{(\mathsf{S};N)}, \mathfrak{s}_3^{(\mathsf{S};N)} \right\}. \tag{11.2.27}$$

Proposition 11.2 *The elements of $\mathfrak{J}^{(\mathsf{S};N)} \subset \mathfrak{A}^{(\mathsf{S};N)}$ obey the commutation relations*

$$\left[\mathfrak{a}^{(\mathsf{S};N)}, (\mathfrak{a}^{(\mathsf{S};N)})^+\right] = \frac{1}{N}\, I^{(\mathsf{S};N)}\,, \tag{11.2.28.a}$$

[18]This multiple use of the same symbol is analogous to the convention already adopted concerning the Fourier transform.

[19]Much of this may seem like pedantry, but it is necessary to be precise about the domains of action of these maps when analyzing the thermodynamic limit accurately.

$$\left[\mathfrak{a}^{(\mathrm{S};N)}, \mathfrak{s}_{\nu}^{(\mathrm{S};N)}\right] = 0, \tag{11.2.28.b}$$

$$\left[\mathfrak{s}_{+}^{(\mathrm{S};N)}, \mathfrak{s}_{-}^{(\mathrm{S};N)}\right] = \frac{1}{N}\,\mathfrak{s}_{3}^{(\mathrm{S};N)}, \tag{11.2.28.c}$$

$$\left[\mathfrak{s}_{+}^{(\mathrm{S};N)}, \mathfrak{s}_{3}^{(\mathrm{S};N)}\right] = -\frac{2}{N}\,\mathfrak{s}_{+}^{(\mathrm{S};N)}, \tag{11.2.28.d}$$

$$\left[\mathfrak{s}_{-}^{(\mathrm{S};N)}, \mathfrak{s}_{3}^{(\mathrm{S};N)}\right] = \frac{2}{N}\,\mathfrak{s}_{-}^{(\mathrm{S};N)}, \tag{11.2.28.e}$$

and hence generate a Lie algebra.

Why should it be that $\mathfrak{I}^{(\mathrm{S};N)}$ generates a Lie algebra? For each atom, of course, the spin operators generate a copy of su(2). In an N atom system, N such representations are tensored together, yielding a reducible representation of su(2). The ladder operators of $\Sigma^{(\mathrm{R})}$ generate a representation of the (restricted) Heisenberg Lie algebra. Together, the unscaled commutation relations determine a representation of the Lie algebra $\mathfrak{g}$ which is equal to the tensor product of su(2) with the (restricted) Heisenberg Lie algebra. Scaling can then be considered as an isomorphism from $\mathfrak{g}$ to the Lie algebra determined by the above commutation relations. So, mathematically speaking, $\mathfrak{I}^{(\mathrm{S};N)}$ is a representation of $\mathfrak{g}$ for each N. Each of these representations describes a different physical system of course, since they each contain a different number of cavity atoms.

Note that, in the limit as $N \to \infty$, the commutator of $\mathfrak{a}^{(\mathrm{S};N)}$ and $(\mathfrak{a}^{(\mathrm{S};N)})^{+}$ converges to zero. (Indeed, this is true of all commutators of elements of $\mathfrak{I}^{(\mathrm{S};N)}$.) This observation is a harbinger of the classical nature of the description of the radiation that will result from the thermodynamic limit.

11.2.6 *Scaling the Initial States*

The necessity of scaling the states has been noted several times. Choosing a matter state which satisfies the two conditions of homogeneity and clustering is sufficient, and no further scaling is required. However, things are different for the radiation state. This must be scaled in such a way that it gives an expectation of $(\mathfrak{a}^{(\mathrm{S};N)})^{+}\,\mathfrak{a}^{(\mathrm{S};N)}$ which is bounded by a constant independent of N, since otherwise the energy density would diverge in the limit as $N \to \infty$. However, even this requirement is not sufficient by itself to guarantee coherent radiation. The following approach describes a particular procedure for scaling radiation states which satisfies the requirement of finite energy density in the limit, and which also ensure the production of coherent radiation — these two results are sufficient justification for the procedure[20].

Proposition 11.3 *For each $\vartheta \in \mathbb{C}$, define the automorphism $\mathcal{T}_\vartheta^{(\mathsf{S};N)}$ of $\mathfrak{A}^{(\mathsf{S};N)}$ by the formula*

$$\mathcal{T}_\vartheta^{(\mathsf{S};N)}[\Xi] = \left(I^{(\mathsf{M};N)} \otimes W^{(\mathsf{R})}[-i\sqrt{N}\vartheta]\right) \Xi \left(I^{(\mathsf{M};N)} \otimes W^{(\mathsf{R})}[i\sqrt{N}\vartheta]\right) \quad (11.2.29)$$

for any observable $\Xi \in \mathfrak{A}^{(\mathsf{S};N)}$. This automorphism is the lifting through quantization of the phase space translation $z \mapsto z + \vartheta$ of $\mathbb{C}$, as can be seen from the formulæ

$$\mathcal{T}_\vartheta^{(\mathsf{S};N)}[\mathfrak{a}^{(\mathsf{S};N)}] = \mathfrak{a}^{(\mathsf{S};N)} + \overline{\vartheta}, \qquad \mathcal{T}_\vartheta^{(\mathsf{S};N)}[(\mathfrak{a}^{(\mathsf{S};N)})^+] = (\mathfrak{a}^{(\mathsf{S};N)})^+ + \vartheta \quad (11.2.30)$$

describing the action of $\mathcal{T}_\vartheta^{(\mathsf{S};N)}$ on the scaled lowering and raising operators. As a consequence, the action of $\mathcal{T}_\vartheta^{(\mathsf{S};N)}$ on the scaled Weyl group is given by the formula

$$\mathcal{T}_\vartheta^{(\mathsf{S};N)}\left(\mathfrak{W}^{(\mathsf{S};N)}[z]\right) = e^{i(z\overline{\vartheta} + \overline{z}\vartheta)}\, \mathfrak{W}^{(\mathsf{S};N)}[z]\,. \quad (11.2.31)$$

We can now calculate how the automorphism $\mathcal{T}_\vartheta^{(\mathsf{S};N)}$ acts on a general macroscopic radiation observable. If $\vartheta \in \mathbb{C}$ is parametrized as $\sqrt{2}\vartheta = \beta - i\alpha$, where α and β are real, let us define the continuous endomorphism τ_ϑ of $\mathcal{S}(\Pi)$ by the formula

$$\left[\tau_\vartheta F\right](p,q) = F(p+\alpha, q+\beta)\,, \quad (11.2.32.a)$$

where $F \in \mathcal{S}(\Pi)$. Making the usual identification between $\mathbb{C}$ and Π, this endomorphism could be written

$$\left[\tau_\vartheta F\right](w) = F(w+\vartheta)\,, \qquad F \in \mathcal{S}(\Pi)\,, w \in \mathbb{C}\,. \quad (11.2.32.b)$$

In other words, τ_ϑ translates the independent variables of elements of $\mathcal{S}(\Pi)$ by ϑ. The map τ_ϑ is related to the automorphism $\mathcal{T}_\vartheta^{(\mathsf{S};N)}$ through the formula

$$\mathcal{T}_\vartheta^{(\mathsf{S};N)}\left[\mathfrak{W}^{(\mathsf{S};N)}[\![\mathcal{F}F]\!]\right] = \mathfrak{W}^{(\mathsf{S};N)}[\![\mathcal{F}\tau_\vartheta F]\!]\,,$$

for any $F \in \mathcal{S}(\Pi)$, which leads to the following description of the action of $\mathcal{T}_\vartheta^{(\mathsf{S};N)}$ on macroscopic radiation observables:

---

[20]The authors are indebted to Professor Sewell for suggesting this scaling procedure.

Corollary 11.4 *For $T \in \mathcal{S}'(\Pi)$, the action of $\mathcal{T}_\vartheta^{(\mathsf{S};N)}$ on $\mathfrak{D}^{(\mathsf{S};N)}[T]$ is given by*

$$\mathcal{T}_\vartheta^{(\mathsf{S};N)}\left[\mathfrak{D}^{(\mathsf{S};N)}[T]\right] = \mathfrak{D}^{(\mathsf{S};N)}\left[(\tau_{-\vartheta})^{\mathrm{tr}}T\right] . \tag{11.2.33}$$

Note that $(\tau_{-\vartheta})^{tr}$ is a continuous endomorphism of $\mathcal{S}'(\Pi)$ which restricts to the continuous endomorphism τ_ϑ of $\mathcal{S}(\Pi)$.

Our prescription for scaling states simply requires us to compose them with the automorphism $\mathcal{T}_\vartheta^{(\mathsf{S};N)}$.

Definition 11.5 *Given an initial matter state $\varpi_{\mathbf{s}}$ satisfying the conditions of homogeneity and clustering, an initial radiation state $\omega^{(\mathsf{R})}$, and any ϑ in $\mathbb{C}$, the corresponding (s,ϑ)-scaled initial system state $\Psi_{\mathbf{s};\vartheta}^{(\mathsf{S};N)} \in \mathfrak{A}_*^{(\mathsf{S};N)}$ of $\Sigma^{(\mathsf{S};N)}$ is defined by the formula*

$$\Psi_{\mathbf{s};\vartheta}^{(\mathsf{S};N)}(\Xi) = \left(\varpi_{\mathbf{s}}^{(\mathsf{M};N)} \otimes \omega^{(\mathsf{R})}\right)\left[\mathcal{T}_\vartheta^{(\mathsf{S};N)}(\Xi)\right], \qquad \Xi \in \mathfrak{A}^{(\mathsf{S};N)} . \tag{11.2.34}$$

Hence, for any $T \in \mathcal{S}'(\Pi)$, the expectation of the macroscopic radiation observable $\mathfrak{D}^{(\mathsf{S};N)}[T]$ in the state $\Psi_{\mathbf{s};\vartheta}^{(\mathsf{S};N)}$ is given by the formula

$$\begin{aligned}\Psi_{\mathbf{s};\vartheta}^{(\mathsf{S};N)}\left(\mathfrak{D}^{(\mathsf{S};N)}[T]\right) &= \left(\varpi_{\mathbf{s}}^{(\mathsf{M};N)} \otimes \omega^{(\mathsf{R})}\right)\left(\mathfrak{D}^{(\mathsf{S};N)}\left[(\tau_{-\vartheta})^{\mathrm{tr}}T\right]\right) \\ &= \omega^{(\mathsf{R})}\left(\mathbf{\Delta}^{(\mathsf{R})}\left[\mathcal{E}_N^{\mathrm{tr}}(\tau_{-\vartheta})^{\mathrm{tr}}T\right]\right) . \end{aligned} \tag{11.2.35}$$

11.3 QL-Model Dynamics

The dynamics of the QL-model are based on the following considerations. Were there no interaction between the matter and the radiation, the dynamics for each atom would be a copy of the pumped spin model of Chapter 7, and the dynamics for the (single) mode of the field would be a copy of that of the singularly coupled damped oscillator of that Chapter. However, each atom interacts with the radiation mode through a dipole coupling, as previously noted.

Employing an interaction picture, this interaction can be treated perturbatively, and the resulting expansions are convergent in an appropriate sense. The number of atoms of the system, N, is a parameter in these calculations, and asymptotic expansions (as $N \to \infty$) can be obtained for each of the leading terms of these series. A number of convergence problems now arise, and must be dealt with. Firstly, an asymptotic expansion (as $N \to \infty$) must be obtained for the complete series, and not simply for its first few terms. Secondly, in order to understand the importance of this model for phase operator theory, the behaviour of general radiation observables must be expressed in terms of Weyl quantization so that, ultimately, questions of convergence for these observables can be cast as questions concerning distributions on phase space. Although these operations can be performed, their justification is lengthy, and so we shall only report the results. Details will be found in [5] and [111].

11.3.1 *Free Dynamics*

The dynamics for each atom is that of a pumped spin, so following the treatment in Subsection 7.5.2, the generator $Z^{(\mathrm{M})}[r]$ for the dynamics of atom r is defined by linearity from its action on the spins,

$$\begin{aligned} Z^{(\mathrm{M})}[r](\sigma_{\pm}[s]) &= (-u \pm i\varepsilon)\, \delta_{rs}\, \sigma_{\pm}[s], \\ Z^{(\mathrm{M})}[r](\sigma_3[s]) &= -v\, \delta_{rs}\, (\sigma_3[s] - \eta\, I^{(\mathrm{M})})\,, \\ Z^{(\mathrm{M})}[r](I^{(\mathrm{M})}) &= 0\,, \end{aligned} \tag{11.3.1}$$

see equations (7.5.8.a), (7.5.8.b), (7.5.8.c). Since the atoms do not interact with each other, the generator for the matter dynamics is the operator obtained by summing over the N atoms,

$$Z^{(\mathrm{M};N)} = \sum_{r=1}^{N} Z^{(\mathrm{M})}[r]\,. \tag{11.3.2}$$

The free dynamics of the radiation mode alone is that of a singularly coupled damped oscillator, which was discussed in Section 7.4. Rather than employing a Bose field as a reservoir, it is convenient to use the open system generator formalism. Thus

$$Z^{(\mathrm{R})}(B^{(\mathrm{R})}) = i\omega\, [N^{(\mathrm{R})}, B^{(\mathrm{R})}] + 2\kappa\, (A^{(\mathrm{R})})^{+} B^{(\mathrm{R})} A^{(\mathrm{R})} - \kappa\, \{N^{(\mathrm{R})}, B^{(\mathrm{R})}\}_{+}\,, \tag{11.3.3}$$

for any $B^{(\mathrm{R})} \in \mathfrak{A}^{(\mathrm{R})}$, where $N^{(\mathrm{R})}$ is the number operator $(A^{(\mathrm{R})})^{+} A^{(\mathrm{R})}$. Recall that g_0 is a standard parameter of the singularly coupled damped oscillator, and we have introduced the abbreviation

$$\kappa = \pi g_0^2\,. \tag{11.3.4}$$

Recall that the time translations generated by $Z^{(\mathsf{R})}$ are *not* algebra homomorphisms of $\mathfrak{A}^{(\mathsf{R})}$.

11.3.2 *The Microscopic Equations Of Motion*

In the Lamb model, the interaction term between an atom and the field was taken to be proportional to the dot product of the electric field and the electric dipole moment for transition of the atom. The physics behind this choice is the crucial step in understanding the mechanism of coherent light production, and remains the same in the fully quantum domain. The interaction Hamiltonian used in this model is the same as that of the earlier quantum models, the only difference being the explicit appearance of the number of atoms in a form that permits the taking of the thermodynamic limit. Using the quantum ansatz of taking the appropriate classical expression and (as nearly as possible) replacing c-numbers by q-numbers, so to speak, leads to the expression which, in a sense, defines the model.

Definition 11.6 *The interaction Hamiltonian for the N-atom system will be taken to be*

$$H_{int}^{(\mathsf{S};N)} = \frac{i\lambda}{\sqrt{N}} \sum_{j=1}^{N} \left\{ \sigma_-[j] \otimes (A^{(\mathsf{R})})^+ - \sigma_+[j] \otimes A^{(\mathsf{R})} \right\} \tag{11.3.5}$$

$$= i\lambda N \left\{ \mathfrak{s}_-^{(\mathsf{S};N)} \left(\mathfrak{a}^{(\mathsf{S};N)}\right)^+ - \mathfrak{s}_+^{(\mathsf{S};N)} \mathfrak{a}^{(\mathsf{S};N)} \right\} . \tag{11.3.6}$$

The electric field operator used in this construction was given in equation (11.2.1), and the electric dipole moment contribution now includes the spin raising and lowering operators. The N-independent coupling constant λ contains the mean electric dipole moment for the transition projected along the polarization direction, and the factor $N^{-1/2}$ comes from the volume factor in equation (11.2.1).

The full microscopic dynamics is described by the generator

$$Z^{(\mathsf{S};N)} = i\left[H_{int}^{(\mathsf{S};N)}, \cdot\right] + Z^{(\mathsf{M};N)} \otimes I^{(\mathsf{R})} + I^{(\mathsf{M};N)} \otimes Z^{(\mathsf{R})} , \tag{11.3.7}$$

in the following sense. The time evolution of the full matter-radiation system is described by a one-parameter family $\left\{\mathfrak{U}_t^{(\mathsf{S};N)} : t \geqslant 0\right\}$ of continuous endomorphisms of $\mathfrak{A}^{(\mathsf{S};N)} \equiv \mathfrak{A}^{(\mathsf{M};N)} \otimes \mathfrak{A}^{(\mathsf{R})}$ such that

$$\frac{d}{dt}\mathfrak{U}_t^{(\mathsf{S};N)}(\Xi) = \mathfrak{U}_t^{(\mathsf{S};N)}\left(Z^{(\mathsf{S};N)}(\Xi)\right), \qquad \Xi \in \mathfrak{A}^{(\mathsf{S};N)} . \tag{11.3.8}$$

Moreover, this family of automorphisms restricts to a one-parameter family of contractions of the subalgebra $\mathfrak{A}^{(\mathsf{M};N)} \otimes \mathbb{B}(\mathcal{H}^{(\mathsf{R})})$ of bounded operators in $\mathfrak{A}^{(\mathsf{S};N)}$. This result was established in the case of bounded observables in [5], while the full extension to the smooth observable space $\mathfrak{A}^{(\mathsf{S};N)}$ is dealt with in [111].

It will be sufficient for our purposes to know the action of $Z^{(\mathsf{S};N)}$ on the intensive variables in the collection $\mathfrak{J}^{(\mathsf{S};N)}$ detailed in equation (11.2.27). Combining the results of Chapter 7 with calculations involving the interaction Hamiltonian $H_{int}^{(\mathsf{S};N)}$ leads to the equations

$$Z^{(\mathsf{S};N)}\left(\mathfrak{a}^{(\mathsf{S};N)}\right) = -\zeta\mathfrak{a}^{(\mathsf{S};N)} + \lambda\mathfrak{s}_-^{(\mathsf{S};N)}, \tag{11.3.9.a}$$

$$Z^{(\mathsf{S};N)}\left(\mathfrak{s}_-^{(\mathsf{S};N)}\right) = -(u+i\varepsilon)\mathfrak{s}_-^{(\mathsf{S};N)} + \lambda\mathfrak{s}_3^{(\mathsf{S};N)}\mathfrak{a}^{(\mathsf{S};N)}, \tag{11.3.9.b}$$

$$\begin{aligned} Z^{(\mathsf{S};N)}\left(\mathfrak{s}_3^{(N)}\right) &= -v\left(\mathfrak{s}_3^{(\mathsf{S};N)} - \eta I^{(\mathsf{S};N)}\right) \\ &\quad - 2\lambda\left[\mathfrak{s}_-^{(\mathsf{S};N)}\left(\mathfrak{a}^{(\mathsf{S};N)}\right)^+ + \mathfrak{s}_+^{(\mathsf{S};N)}\mathfrak{a}^{(\mathsf{S};N)}\right]. \end{aligned} \tag{11.3.9.c}$$

Here ζ is the complex natural frequency for the singularly coupled damped oscillator, namely

$$\zeta = \pi g_0^2 + i\omega. \tag{7.4.36.b}$$

The actions of $Z^{(\mathsf{S};N)}$ on $(\mathfrak{a}^{(\mathsf{S};N)})^+$ and $\mathfrak{s}_+^{(\mathsf{S};N)}$ can be readily obtained from the above by taking adjoints. The nonlinear terms in these equations are crucial, since they supply the nontriviality of the model. One implication of the presence of these terms is that while, for example, the observables $\mathfrak{s}_3^{(\mathsf{S};N)}$ and $\mathfrak{a}^{(\mathsf{S};N)}$ commute, their time evolutes $\mathfrak{U}_t^{(\mathsf{S};N)}(\mathfrak{s}_3^{(\mathsf{S};N)})$ and $\mathfrak{U}_t^{(\mathsf{S};N)}(\mathfrak{a}^{(\mathsf{S};N)})$ no longer do so.

11.4 The Thermodynamic Limit

We now proceed to investigate the behaviour of the model in the thermodynamic limit $N \to \infty$. In order to understand the material that follows, it is useful to preview the results that we shall obtain. In what follows we shall assume that a radiation state $\omega^{(\mathsf{R})} \in \mathfrak{A}_*^{(\mathsf{R})}$, a matter state $\varpi_{\mathbf{s}} \in \mathfrak{A}_*^{(\mathsf{M})}$ and a scaling parameter $\vartheta \in \mathbb{C}$ have been chosen, leading to the initial state $\Psi_{\mathbf{s};\vartheta}^{(\mathsf{S};N)} \in \mathfrak{A}_*^{(\mathsf{S};N)}$ for all integers N, in accordance with Definition 11.5. Consider first a microscopic radiation observable $\mathbf{\Delta}^{(\mathsf{R})}[T]$ and a time $t \geqslant 0$. We scale the observable, obtaining the corresponding macroscopic observable

$\mathfrak{D}^{(\mathsf{S};N)}[T]$, calculate the expectation of $\mathfrak{D}^{(\mathsf{S};N)}[T]$ in the state $\Psi^{(\mathsf{S};N)}_{\mathbf{s};\vartheta}$ at time t, and then consider the limit as $N \to \infty$, which process corresponds to allowing the radiation field to interact with the entire assemblage of atoms. If this limit exists[21], we write the result as

$$\varPi[T;t]_{\mathbf{s};\vartheta} = \lim_{N\to\infty} \Psi^{(\mathsf{S};N)}_{\mathbf{s};\vartheta}\Big(\mathfrak{U}^{(\mathsf{S};N)}_t\big(\mathfrak{D}^{(\mathsf{S};N)}[T]\big)\Big). \tag{11.4.1}$$

What sort of a quantity is $\varPi[T;t]_{\mathbf{s};\vartheta}$? It evidently corresponds to a limiting description of the microscopic observable $\mathbf{\Delta}^{(\mathsf{R})}[T]$ at time t. Moreover, if we consider k radiation observables $\mathbf{\Delta}^{(\mathsf{R})}[T_1]$, ..., $\mathbf{\Delta}^{(\mathsf{R})}[T_k]$ and k corresponding times t_1, ..., $t_k \geqslant 0$, when we take the thermodynamic limit for their product, we obtain the identity

$$\lim_{N\to\infty} \Psi^{(\mathsf{S};N)}_{\mathbf{s};\vartheta}\Big(\mathfrak{U}^{(\mathsf{S};N)}_{t_1}\big(\mathfrak{D}^{(\mathsf{S};N)}[T_1]\big)\cdots\mathfrak{U}^{(\mathsf{S};N)}_{t_k}\big(\mathfrak{D}^{(\mathsf{S};N)}[t_k]\big)\Big)$$
$$= \varPi[T_1;t_1]_{\mathbf{s};\vartheta}\cdots\varPi[T_k;t_k]_{\mathbf{s};\vartheta}, \tag{11.4.2}$$

at least for certain well-behaved distributions T_1, ..., T_k. Since each quantity $\varPi[T_j;t_j]_{\mathbf{s};\vartheta}$ represents the limiting description of the corresponding microscopic radiation observable, the most striking feature of this result is the complete factorization. As the microscopic radiation observables do not, in general, commute with each other, this thermodynamic limit has destroyed the quantum correlations. One consequence of this is that the limiting descriptions of these variables, whatever they may be, are sharp with no uncertainty. Another is that the limiting descriptions of the microscopic radiation ladder operators no longer satisfy the canonical commutation relation. Tentatively, then, we find that a classical-like description of the radiation has emerged from this limit.

In further confirmation of the classical nature of this limiting description, it will be shown that

$$\varPi[\overline{z};0]_{\mathbf{s};\vartheta} = \lim_{N\to\infty} \Psi^{(\mathsf{S};N)}_{\mathbf{s};\vartheta}\big(\mathfrak{a}^{(\mathsf{S};N)}\big) = \overline{\vartheta}, \tag{11.4.3.a}$$

$$\varPi[z;0]_{\mathbf{s};\vartheta} = \lim_{N\to\infty} \Psi^{(\mathsf{S};N)}_{\mathbf{s};\vartheta}\big(\big(\mathfrak{a}^{(\mathsf{S};N)}\big)^+\big) = \vartheta. \tag{11.4.3.b}$$

We will obtain the result that the limiting description of the microscopic radiation observable $\mathbf{\Delta}^{(\mathsf{R})}[T]$ is the quantity $T(\vartheta)$, namely the value that the distribution T takes at the point ϑ in phase space (in complex form) — at least for certain types of distribution. Thus a classical description of the

[21] We defer the question of the existence of this limit to the following Subsections.

radiation field has indeed emerged from this limit and is, moreover, a phase space description. We will thus be able to recognize when the QL-model yields coherent radiation, for in that case the limiting description $\Pi[T;t]_{\mathbf{s};\vartheta}$ of a microscopic radiation observable will be of the form $T(\vartheta(t))$, where the complex phase space parameter $\vartheta(t)$ is undergoing simple harmonic motion.

It is of fundamental importance, therefore, to determine the class of microscopic radiation observables for which the limit in equation (11.4.1) exists, and to study the properties of such limits.

To begin with, we consider the case when $t = 0$, which is notationally and technically much simpler.

11.4.1 *Convergence At Time* 0

Using a generating function is the most convenient way to obtain the limiting descriptions of polynomials in the variables $\mathfrak{a}^{(\mathsf{S};N)}$, $(\mathfrak{a}^{(\mathsf{S};N)})^+$ and $\mathfrak{s}_\nu^{(\mathsf{S};N)}$ in a systematic manner. First consider spin densities in the absence of radiation observables. The basic convergence result is elementary.

Lemma 11.7 *The expectation value of the intensive variable* $e^{i\mathbf{b}\cdot\mathfrak{s}^{(\mathsf{S};\mathbf{N})}}$ *in the state* $\Psi_{\mathbf{s};\vartheta}^{(\mathsf{S};N)}$ *converges to*

$$\lim_{N\to\infty} \Psi_{\mathbf{s};\vartheta}^{(\mathsf{S};N)}\left(e^{i\mathbf{b}\cdot\mathfrak{s}^{(\mathsf{S};\mathbf{N})}}\right) = e^{i\mathbf{b}\cdot\mathbf{s}}. \tag{11.4.4}$$

Proof: The first step is to write the expectation value as

$$\Psi_{\mathbf{s};\vartheta}^{(\mathsf{S};N)}\left(e^{i\mathbf{b}\cdot\mathfrak{s}^{(\mathsf{S};\mathbf{N})}}\right) = e^{i\mathbf{b}\cdot\mathbf{s}}\,\Psi_{\mathbf{s};\vartheta}^{(\mathsf{S};N)}\left(e^{i\mathbf{b}\cdot[\mathfrak{s}^{(\mathsf{S};\mathbf{N})}-\mathbf{s}]}\right). \tag{11.4.5}$$

Simply using the Cauchy-Schwarz inequality gives

$$\begin{aligned}
&\left|\Psi_{\mathbf{s};\vartheta}^{(\mathsf{S};N)}\left(e^{i\mathbf{b}\cdot[\mathfrak{s}^{(\mathsf{S};\mathbf{N})}-\mathbf{s}]} - I^{(\mathsf{S};N)}\right)\right|^2 \\
&\quad\leqslant 2\left|\Psi_{\mathbf{s};\vartheta}^{(\mathsf{S};N)}\left(I^{(\mathsf{S};N)} - \cos\{\mathbf{b}\cdot[\mathfrak{s}^{(\mathsf{S};\mathbf{N})}-\mathbf{s}]\}\right)\right| \\
&\quad\leqslant \Psi_{\mathbf{s};\vartheta}^{(\mathsf{S};N)}\left(\{\mathbf{b}\cdot[\mathfrak{s}^{(\mathsf{S};\mathbf{N})}-\mathbf{s}]\}^2\right) \\
&\quad= \varpi_{\mathbf{s}}\left(\{\mathbf{b}\cdot[\mathfrak{s}^{(\mathsf{S};\mathbf{N})}-\mathbf{s}]\}^2\right),
\end{aligned}$$

and the clustering property implies that this last expression converges to zero. ∎

For the radiation alone the basic result is also available in closed form.

Lemma 11.8 *The expectation value of the macroscopic variable* $\mathfrak{W}^{(\mathsf{S};N)}[z]$ *in the state* $\Psi^{(\mathsf{S};N)}_{\mathbf{s};\vartheta}$ *converges to*

$$\lim_{N\to\infty} \Psi^{(\mathsf{S};N)}_{\mathbf{s};\vartheta}\left(\mathfrak{W}^{(\mathsf{S};N)}[z]\right) = e^{i\left(z\overline{\vartheta}+\overline{z}\vartheta\right)}. \tag{11.4.6}$$

Proof: Consulting Proposition 11.3,

$$\Psi^{(\mathsf{S};N)}_{\mathbf{s};\vartheta}\left(\mathfrak{W}^{(\mathsf{S};N)}[z]\right) = e^{i\left(z\overline{\vartheta}+\overline{z}\vartheta\right)}\,\omega^{(\mathsf{R})}\left(W^{(\mathsf{R})}[N^{-1/2}z]\right),$$

and as $W^{(\mathsf{R})}[N^{-1/2}z]$ converges strongly to I as $N\to\infty$, the result is immediate. ■

These two results can be combined into a single convergence result for the convergence of a sequence of characteristic functions of probability measures [180], out of which the classical structure of the solution will emerge. In order to make this clear, the parameters $\mathbf{s}$ and ϑ will be taken as constituting a point of a set that we shall call the *phase space* for the model. This terminology extends the original phase space Π to include the matter component.

Definition 11.9 *By the phase space* $\widetilde{\Pi}$ *for the QL-model is meant the set*

$$\widetilde{\Pi} = \mathbb{R}^3\times\Pi \equiv \mathbb{R}^5, \tag{11.4.7}$$

which is a real inner product space equipped with the standard inner product

$$\mathbf{a}\cdot\mathbf{b} = \sum_{j=1}^{5} a_j b_j, \qquad \mathbf{a},\mathbf{b}\in\widetilde{\Pi}. \tag{11.4.8}$$

There are natural complexifications for both the matter and radiation portions of $\widetilde{\Pi}$. We handle the complexification of the radiation portion by allowing any vector $\mathbf{a}\in\widetilde{\Pi}$ to be written in the form $(\mathbf{s},\vartheta)\in\mathbb{R}^3\times\mathbb{C}$, where

$$\mathbf{s} = (s_1,s_2,s_3) = (a_1,a_2,a_3), \qquad \vartheta = \frac{1}{\sqrt{2}}(a_5 - ia_4), \tag{11.4.9}$$

and the complexification of the matter portion is handled by allowing this vector $\mathbf{s}$ to be represented by the pair $(\gamma,p)\in\mathbb{C}\times\mathbb{R}$, where

$$\gamma = \frac{1}{2}(a_1 - ia_2), \qquad p = a_3. \tag{11.4.10}$$

With respect to these various notations, the inner product of the vectors $\mathbf{a}_1$, $\mathbf{a}_2$ in $\widetilde{\Pi}$ may be written

$$\begin{aligned} \mathbf{a}_1 \cdot \mathbf{a}_2 &= \mathbf{s}_1 \cdot \mathbf{s}_2 + \vartheta_1 \overline{\vartheta_2} + \overline{\vartheta_1} \vartheta_2 \\ &= 2(\gamma_1 \overline{\gamma_2} + \overline{\gamma_1} \gamma_2) + p_1 p_2 + \vartheta_1 \overline{\vartheta_2} + \overline{\vartheta_1} \vartheta_2 \,. \end{aligned}$$

Having introduced the space $\widetilde{\Pi}$, it is now natural to denote the system state $\Psi^{(\mathsf{S};N)}_{\mathbf{s};\vartheta}$ by the simpler symbol $\Psi^{(\mathsf{S};N)}_{\mathbf{a}}$, and to denote the limiting description $\mathit{\Pi}[T;t]_{\mathbf{s};\vartheta}$ of a microscopic radiation observable $\boldsymbol{\Delta}^{(\mathsf{R})}[T]$ in the sequence of states $\big(\Psi^{(\mathsf{S};N)}_{\mathbf{s};\vartheta}\big)_{N\geqslant 1}$ by the simpler symbol $\mathit{\Pi}[T;t]_{\mathbf{a}}$.

Introducing the space $\widetilde{\Pi}$ allows us to define a single generating function for all of the fundamental observables in $\mathfrak{J}^{(\mathsf{S};N)}$ in a notationally simple manner. Doing this provides us with a single formalism for calculating the thermodynamic limit of these fundamental observables, and permits us to calculate thermodynamic limits for a much wider class of observables. We therefore make the following Definition.

Definition 11.10 *For any* $\mathbf{b} = (\mathbf{b}, \xi) \in \widetilde{\Pi}$, *by the* generalized Weyl operator *we mean the unitary operator* $U^{(\mathsf{S};N)}[\mathbf{b}] \in \mathfrak{A}^{(\mathsf{S};N)}$ *given by the formula*

$$U^{(\mathsf{S};N)}[\mathbf{b}] = e^{i\mathbf{b}\cdot\mathfrak{s}^{(\mathsf{S};N)}} \,\mathfrak{W}^{(\mathsf{S};N)}[\xi] \,. \tag{11.4.11}$$

The next task is to consider the expectation of this generalized Weyl operator (which is a macroscopic observable), and to determine its limit as $N \to \infty$. Accordingly, we note that any $\mathbf{a} \in \widetilde{\Pi}$ determines a function $\mu^{(\mathsf{S};N)}_{\mathbf{a}} : \widetilde{\Pi} \to \mathbb{C}$ through the formula

$$\mu^{(\mathsf{S};N)}_{\mathbf{a}}[\mathbf{b}] = \Psi^{(\mathsf{S};N)}_{\mathbf{a}}\big(U^{(\mathsf{S};N)}[\mathbf{b}]\big) \,, \qquad \mathbf{b} \in \widetilde{\Pi} \,. \tag{11.4.12}$$

Combining the preceding two Lemmata yields the thermodynamic limit for the generalized Weyl operator.

Proposition 11.11 *For any* $\mathbf{a}, \mathbf{b} \in \widetilde{\Pi}$,

$$\lim_{N\to\infty} \mu^{(\mathsf{S};N)}_{\mathbf{a}}[\mathbf{b}] = e^{i\mathbf{a}\cdot\mathbf{b}} \,. \tag{11.4.13}$$

Now it can be shown that $\mu^{(\mathsf{S};N)}_{\mathbf{a}}$ is the characteristic function of a Borel probability measure on $\widetilde{\Pi}$, and the above result then states that $\mu^{(\mathsf{S};N)}_{\mathbf{a}}$ converges (as $N \to \infty$) to the characteristic function of the Dirac measure concentrated at the point $\mathbf{a} \in \widetilde{\Pi}$. This will imply that all correlation functions factor in the limit, as we anticipated. Thus complete Glauber factorization obtains in this model — there are no higher order correlations. Note also that all details of the initial state have been lost, except for those aspects which are reflected in the parameters represented by $\mathbf{a}$ — this is a "coarse graining" result which is typical of a thermodynamic limit.

For technical purposes, it is vital that we know more about the manner of the convergence in the above Proposition. To be specific, this convergence is not simply pointwise with respect to $\mathbf{a}$ and $\mathbf{b}$ but, for example, is uniform as $\mathbf{b}$ ranges over compact subsets of $\widetilde{\Pi}$. Even more than this is true, and it is possible to differentiate these generating functions to determine the thermodynamic limits of polynomials in the observables in $\mathfrak{J}^{(\mathsf{S};N)}$. We shall not go into details here, referring the reader to [5] and [111] for specifics, and shall simply state the results which concern us. Later on, we shall incorporate the time-evolution into this limiting procedure — see Proposition 11.15.

We are primarily concerned with the behaviour of radiation observables, and to this end we shall regard the vectors $\mathbf{s}, \mathbf{b} \in \mathbb{R}^3$ of the phase space points $\mathbf{a} = (\mathbf{s}, \vartheta)$ and $\mathbf{b} = (\mathbf{b}, \xi)$ in $\widetilde{\Pi}$ as constant vectors, and thus regard $\mu_{\mathbf{a}}^{(\mathsf{S};N)}[\mathbf{b}]$ as a function of the two complex variables ϑ and ξ. For any $\vartheta \in \mathbb{C}$, $\mu_{\mathbf{a}}^{(\mathsf{S};N)}[\mathbf{b}]$ is a Schwartz function of ξ, and we know that

$$\frac{1}{2\pi}[\![\, \mathcal{F}T\,,\, \mu_{\mathbf{a}}^{(\mathsf{S};N)}]\!] \;=\; e^{i\mathbf{s}\cdot\mathbf{b}}\, \Psi_{\mathbf{a}}^{(\mathsf{S};N)}\big(\mathfrak{D}^{(\mathsf{S};N)}[T]\big) \tag{11.4.14}$$

for any $T \in \mathcal{S}'(\Pi)$. Moreover, for any $F \in \mathcal{S}(\Pi)$, the function

$$\mu_{F,\mathbf{s}}^{(\mathsf{S};N)}[\mathbf{b}] \;=\; \int_{\mathbb{C}} F(\vartheta)\mu_{\mathbf{a}}^{(\mathsf{S};N)}[\mathbf{b}]\, dA(\vartheta) \tag{11.4.15}$$

is a Schwartz function of ξ, with

$$\frac{1}{2\pi}[\![\, \mathcal{F}T\,,\, \mu_{F;\mathbf{s}}^{(\mathsf{S};N)}]\!] \;=\; e^{i\mathbf{s}\cdot\mathbf{b}} \int_{\mathbb{C}} F(\vartheta)\Psi_{\mathbf{a}}^{(\mathsf{S};N)}\big(\mathfrak{D}^{(\mathsf{S};N)}[T]\big)\, dA(\vartheta) \tag{11.4.16}$$

for any $T \in \mathcal{S}'(\Pi)$.

The main aim of this discussion is to show that the limiting description $\mathit{\Pi}[T;0]_{\mathbf{a}}$ of the microscopic radiation observable $\mathbf{\Delta}^{(\mathsf{R})}[T]$ is $T(\vartheta)$ for any $T \in \mathcal{S}'(\Pi)$. However, since $T \in \mathcal{S}'(\Pi)$, the quantity $T(\vartheta)$ may not exist, so the thermodynamic limit will have to be performed weakly. That we can do this is a consequence of the good nature of the convergence in Proposition 11.11, which implies that

$$\lim_{N\to\infty} \mu_{F;\mathbf{s}}^{(\mathsf{S};N)}[\mathbf{b}] \;=\; 2\pi\, e^{i\mathbf{s}\cdot\mathbf{b}}(\mathcal{F}^{-1}F)(\xi) \tag{11.4.17}$$

for any $F \in \mathcal{S}(\Pi)$, where this convergence is with respect to the Fréchet topology in $\mathcal{S}(\mathbb{R}^2)$. This result will be discussed again in detail (and in a less complex notational setting!) in Chapter 12. Hence

$$\lim_{N\to\infty} \int_{\mathbb{C}} F(\vartheta)\Psi_{\mathbf{a}}^{(\mathsf{S};N)}\big(\mathfrak{D}^{(\mathsf{S};N)}[T]\big)\, dA(\vartheta) \;=\; [\![T\,,\, F]\!]$$

for any $F \in \mathcal{S}(\Pi)$ and any $T \in \mathcal{S}'(\Pi)$, the result we need. Summarizing these observations,

Proposition 11.12 *Regarded as tempered distributions in ϑ, the limit*

$$\lim_{N\to\infty} \Psi_{\mathbf{a}}^{(\mathsf{S};N)}\big(\mathfrak{D}^{(\mathsf{S};N)}[T]\big) \;=\; T(\vartheta) \;=\; T(a_4, a_5) \qquad (11.4.18.a)$$

is valid weakly in $\mathcal{S}'(\Pi)$ for any $T \in \mathcal{S}'(\Pi)$. In other words,

$$\text{Д}[T;0]_{\mathbf{a}} \;=\; T(\vartheta)\,, \qquad (11.4.18.b)$$

is the variable representing the microscopic radiation observable $\mathbf{\Delta}^{(\mathsf{R})}[T]$ at time $t=0$, with respect to the sequence of states $\big(\psi_{\mathbf{a}}^{(\mathsf{S};N)}\big)_{N\geqslant 1}$, in the classical description which emerges from the thermodynamic limit. Note that, since no spin densities are present, the vector component $\mathbf{s}$ of $\mathbf{a}$ does not appear in the limit.

In particular, this result implies that the classical descriptions corresponding to the phase observables $\mathbf{\Delta}^{(\mathsf{R})}[\varphi]$ and $\mathbf{\Delta}^{(\mathsf{R})}[e^{\pm i\varphi}]$ are given by the formulæ

$$\text{Д}[\varphi;0]_{\mathbf{a}} \;=\; \varphi(a_4, a_5)\,, \qquad (11.4.19.a)$$

$$\text{Д}[e^{\pm i\varphi};0]_{\mathbf{a}} \;=\; e^{\pm i\varphi(a_4,a_5)}\,, \qquad (11.4.19.b)$$

and it also justifies equations (11.4.3.a) and (11.4.3.b).

The fact that taking the thermodynamic limit yields the Weyl dequantization of microscopic radiation observables gives a remarkably simple connection between the classical description of the physics which emerges from the thermodynamic limit and the microscopic radiation observables. In addition, it emphasizes again the special status of Weyl quantization amongst the various quantization schemes that we will discuss in Chapter 14, and moreover confirms the particular importance of the operator $\mathbf{\Delta}\,[\,\varphi\,]$ amongst the various proposed phase operators. While the Weyl symbol φ of $\mathbf{\Delta}\,[\,\varphi\,]$ is a function of the angle in phase space alone, this is not true of any other phase observable, and so the classical description of these other phase observables that arises from the thermodynamic limit will not be one which related purely to the phase of emitted coherent radiation, since the square of the radius in phase space will shortly be seen to represent the intensity of that radiation.

That the thermodynamic limit results in complete factorization for radiation observables is what permits us to assign a classical interpretation to the description of the physics which arises in the thermodynamic limit. We shall now see how this factorization comes about. Recall that we remarked after Proposition 11.2 that the commutation relation between the macroscopic ladder operators converged to zero as $N \to \infty$, and that this fact was a precursor of this factorization. A more general indication of factorization can be obtained from the following observations. If $F, G \in \mathcal{S}(\Pi)$, it can be shown that

$$\mathfrak{D}^{(\mathsf{S};N)}[F]\,\mathfrak{D}^{(\mathsf{S};N)}[G] \;=\; \mathfrak{D}^{(\mathsf{S};N)}[F \star_{(1/N)} G]\,,$$

where $\star_{(1/N)}$ is the (parametrized) Moyal product to be discussed in Chapter 13. Since we shall find that

$$F \star_{(1/N)} G \;=\; F \cdot G + O(N^{-1})\,,$$

for any $F, G \in \mathcal{S}(\Pi)$, it follows that

$$\begin{aligned}\lim_{N\to\infty} \Psi_{\mathbf{a}}^{(\mathsf{S};N)}\big(\mathfrak{D}^{(\mathsf{S};N)}[F]\,\mathfrak{D}^{(\mathsf{S};N)}[G]\big) &= \lim_{N\to\infty} \Psi_{\mathbf{a}}^{(\mathsf{S};N)}\big(\mathfrak{D}^{(\mathsf{S};N)}[F \star_{(1/N)} G]\big)\\ &= \Pi[F;0]_{\mathbf{a}}\,\Pi[G;0]_{\mathbf{a}}\\ &= F(\vartheta)\,G(\vartheta)\,. \qquad (11.4.20)\end{aligned}$$

As equation (11.4.20) can be extended to include any finite number of factors, complete Glauber factorization at time zero can now be seen to be a result of the fact that the Moyal product $\star_{(1/N)}$ tends to the simple pointwise product as $N \to \infty$.

These results can be extended to encompass a larger class of observables in $\mathfrak{A}^{(\mathsf{R})}$ than just the Weyl quantizations of elements of $\mathcal{S}(\Pi)$ — they are also valid, for example, for observables which are polynomials in $A^{(\mathsf{R})}$ and $(A^{(\mathsf{R})})^{+}$. It is therefore appropriate to interpret every radiation observable $\mathbf{\Delta}^{(\mathsf{R})}[T] \in \mathfrak{A}^{(\mathsf{R})}$, in the thermodynamic limit at time zero, by the (distributional) classical quantity $\Pi[T;0]_{\mathbf{a}} = T(\vartheta)$.

Of the radiation observables, the number operator $N^{(\mathsf{R})} = (A^{(\mathsf{R})})^{+}A^{(\mathsf{R})}$ is particularly important, since it counts excitations of the electromagnetic field. Hence the thermodynamic limit of the expectation of that operator is proportional to the light intensity. This thermodynamic limit could, of course, be determined using the above result, but can also be calculated directly, since

$$\Psi_{\mathbf{a}}^{(\mathsf{S};N)}\left((\mathfrak{a}^{(\mathsf{S};N)})^{+}\mathfrak{a}^{(\mathsf{S};N)}\right) = \tag{11.4.21}$$
$$\frac{1}{N}\omega^{(\mathsf{R})}\left((A^{(\mathsf{R})})^{+}A^{(\mathsf{R})}\right) + \frac{\vartheta}{\sqrt{N}}\omega^{(\mathsf{R})}\left(A^{(\mathsf{R})}\right) + \frac{\overline{\vartheta}}{\sqrt{N}}\omega^{(\mathsf{R})}\left((A^{(\mathsf{R})})^{+}\right) + |\,\vartheta\,|^{2}.$$

Proposition 11.13 *Since the description of the radiation number operator $N^{(\mathsf{R})}$ in the thermodynamic limit is*

$$\varPi[|\,z\,|^{2};0]_{\mathbf{a}} = \lim_{N\to\infty}\Psi_{\mathbf{a}}^{(\mathsf{S};N)}\left((\mathfrak{a}^{(\mathsf{S};N)})^{+}\mathfrak{a}^{(\mathsf{S};N)}\right) = |\,\vartheta\,|^{2}, \tag{11.4.22}$$

the parameter ϑ is such that $|\,\vartheta\,|^{2}$ is the light intensity at time $t = 0$. Thus the value of ϑ in the state scaling determines the limiting radiation intensity.

To put these results into context it must be remembered that to some extent they are a consequence of the special nature of the initial state. It is not known to us how far a solution can be found which simply satisfies the minimal conditions required for convergence (nor do we know what those initial conditions are). Alli & Sewell [5] have considered a more general case, where the matter component of the state still satisfies the conditions of homogeneity and clustering, and where the radiation component of the state satisfies a condition which places an upper bound on the system energy. This condition is satisfied by the ϑ-scaled states that we have considered, but can also be satisfied by other states. They have then shown that the thermodynamic limit of the expectation of the operator $U^{(\mathsf{S};N)}[\mathbf{b}]$ in this state is equal to the characteristic function of some Borel probability measure on $\widetilde{\Pi}$, and moreover that the thermodynamic limit is uniform over compacta in $\widetilde{\Pi}$.

The ϑ-scaled states have the particular property of ensuring that this limiting measure is a Dirac measure. Obtaining a Dirac measure in this context yields coherent radiation with complete Glauber factorization, and we have already noted that such complete factorization is not physical (it being an idealization, rather than what would be observed physically).

Consequently this model only provides a limited description of coherent radiation, while being nonetheless very interesting. The possibility remains, however, of finding radiation states which satisfy the more general conditions of Alli & Sewell, but which also yield coherent radiation exhibiting some degree of higher order correlations.

11.4.2 *The Limiting Dynamics*

The problem of obtaining the thermodynamic limit at positive time t is much more complicated. This is due to the nonlinear coupling of the matter and radiation observables produced by the interaction Hamiltonian, and the nature of the dynamics for open systems.

Initially considering bounded operators only, $\{\mathfrak{U}_t^{(\mathsf{S};N)} : t \geqslant 0\}$ forms a one-parameter semigroup of contractions of the algebra $\mathfrak{A}^{(\mathsf{M};N)} \otimes \mathbb{B}(\mathcal{H}^{(\mathsf{R})})$ of bounded operators in $\mathfrak{A}^{(\mathsf{S};N)}$. However, these contractions are not unitarily implemented, in that there does not exist a unitary operator $W(t)$ on the Hilbert space $\otimes^N \mathbb{C}^2 \otimes \mathcal{H}^{(\mathsf{R})}$ such that

$$\mathfrak{U}_t^{(\mathsf{S};N)}(\Xi) = W(t)^{-1}\,\Xi\, W(t)\,, \qquad \Xi \in \mathfrak{A}^{(\mathsf{M};N)} \otimes \mathbb{B}(\mathcal{H}^{(\mathsf{R})})\,,$$

and, moreover, $\mathfrak{U}_t^{(\mathsf{S};N)}$ is not an algebra homomorphism of $\mathfrak{A}^{(\mathsf{M};N)} \otimes \mathbb{B}(\mathcal{H}^{(\mathsf{R})})$. As discussed in Chapter 7, this problem can be partially addressed by expanding the system explicitly to include both the matter and the radiation reservoirs. On this extended system it is then possible to define a unitarily implemented time-evolution which, when projected back to $\mathfrak{A}^{(\mathsf{M};N)} \otimes \mathbb{B}(\mathcal{H}^{(\mathsf{R})})$, yields $\mathfrak{U}_t^{(\mathsf{S};N)}$. Even this extended time-evolution is highly involved, being a composition of the non-interacting time-evolution (namely, the evolution that would occur if the matter and the radiation portions of the system did not interact, while still allowing the matter and radiation subsystems to interact with their respective reservoirs) with an interaction term which involves the interaction Hamiltonian $H_{int}^{(\mathsf{S};N)}$, this interaction being constructed via a (relatively) standard procedure using time-ordered integrals.

Notwithstanding these difficulties, the mathematics can be dealt with in this bounded case, and details are presented in [5]. More complicated still is the problem of extending this formalism to the smooth model. This can be done, however, with the previously stated result that $\{\mathfrak{U}_t^{(\mathsf{S};N)} : t \geqslant 0\}$ is a one-parameter family of continuous endomorphisms of $\mathfrak{A}^{(\mathsf{M};N)} \otimes \mathfrak{A}^{(\mathsf{R})}$. Details of this analysis can be found in [111].

The results of Alli & Sewell in [5] concerning the thermodynamic limit can be extended, at least to some degree, to the smooth case. The outcome of this extension enables us to take the thermodynamic limit of a large class of radiation observables including (to within reasonable approximations) physically interesting phase observables. Some technical difficulties remain which prevent a complete proof, which would permit our handling all radiation observables, but we believe strongly that such a proof can be found.

Having noted all this, we intend to avoid becoming embroiled in technical details, and shall limit ourselves to stating the relevant results. Fundamental to these is the following problem in differential equation theory.

Proposition 11.14 *There exists a uniquely defined smooth one-parameter family* $\{\tau_t : t \geqslant 0\}$ *of continuous endomorphisms of* $\widetilde{\Pi}$ *such that, writing* $\tau_t(\mathbf{a}) = (\mathbf{s}(t), \vartheta(t)) = (\gamma(t), p(t), \vartheta(t))$*, the differential equations*

$$\frac{d}{dt}\gamma(t) = -(u+i\varepsilon)\gamma(t) + \lambda p(t)\overline{\vartheta(t)}, \tag{11.4.23.a}$$

$$\frac{d}{dt}p(t) = v\eta - vp(t) - 2\lambda\big(\gamma(t)\vartheta(t) + \overline{\gamma(t)\vartheta(t)}\big), \tag{11.4.23.b}$$

$$\frac{d}{dt}\vartheta(t) = -\overline{\zeta}\vartheta(t) + \lambda\overline{\gamma(t)}, \tag{11.4.23.c}$$

are satisfied. Moreover, the coefficients of $\tau_t(\mathbf{a})$ *are smooth functions of the coefficients of* $\mathbf{a}$.

Extending the notation of the previous Section, we are led to define the function $\mu_{\mathbf{a},t}^{(\mathsf{S};N)} : \widetilde{\Pi} \to \mathbb{C}$ by the formula

$$\mu_{\mathbf{a},t}^{(\mathsf{S};N)}[\mathbf{b}] = \Psi_{\mathbf{a}}^{(\mathsf{S};N)}\big(\mathfrak{U}_t^{(\mathsf{S};N)}(U^{(\mathsf{S};N)}[\mathbf{b}])\big), \qquad \mathbf{b} \in \widetilde{\Pi}, \tag{11.4.24}$$

for any $\mathbf{a} \in \widetilde{\Pi}$ and $t \geqslant 0$. As before, we choose to consider $\mu_{\mathbf{a},t}^{(\mathsf{S};N)}[\mathbf{b}]$ as a function of ϑ and ξ, regarding $\mathbf{s}$ and $\mathbf{b}$ as parameters of the problem, where $\mathbf{a} = (\mathbf{s}, \vartheta)$ and $\mathbf{b} = (\mathbf{b}, \xi)$. Smearing this function[22] with respect to the parameter ϑ, we define the function

$$\mu_{\chi,t}^{(\mathsf{S};N)}[\mathbf{b}] = \int_{\mathbb{C}} \chi(\vartheta)\mu_{\mathbf{a},t}^{(\mathsf{S};N)}[\mathbf{b}]\, dA(\vartheta) \tag{11.4.25}$$

[22] Such smearings will allow us to approximate distributions in localized regions of phase space.

for any $\chi \in \mathcal{D}(\mathbb{C})$, the space of smooth functions on $\mathbb{C}$ of compact support[23]. The key results of [5] and [111] may then be summarized as follows:

Proposition 11.15 *For any* $\mathbf{a} \in \widetilde{\Pi}$, $t \geqslant 0$ *and* $N \in \mathbb{N}$, $\mu^{(\mathsf{S};N)}_{\mathbf{a},t}[\mathbf{b}]$ *is a Schwartz function of* ξ, *and*

$$\lim_{N\to\infty} \mu^{(\mathsf{S};N)}_{\mathbf{a},t}[\mathbf{b}] = \mu_{\mathbf{a},t}[\mathbf{b}] = e^{i\tau_t(\mathbf{a})\cdot\mathbf{b}}, \tag{11.4.26}$$

where this convergence is uniform as $\mathbf{b}$ *varies over compacta in* $\widetilde{\Pi}$.

Also, for any $\chi \in \mathcal{D}(\mathbb{C})$, $t \geqslant 0$ *and* $N \in \mathbb{N}$, $\mu^{(\mathsf{S};N)}_{\chi,t}[\mathbf{b}]$ *is a Schwartz function of* ξ, *and*

$$\lim_{N\to\infty} \mu^{(\mathsf{S};N)}_{\chi,t}[\mathbf{b}] = \mu_{\chi,t}[\mathbf{b}] = \int_{\mathbb{C}} \chi(\vartheta) e^{i\tau_t(\mathbf{a})\cdot\mathbf{b}}\, dA(\vartheta), \tag{11.4.27}$$

where this convergence is also uniform as $\mathbf{b}$ *varies over compacta in* $\widetilde{\Pi}$. *Moreover, the function* $\mu_{\chi,t}$ *is a Schwartz function of* ξ.

Finally, the sets

$$\left\{ \frac{\partial^{a+b}}{\partial\xi^a \partial\overline{\xi}^b} \mu^{(\mathsf{S};N)}_{\mathbf{a},t}[\mathbf{b}] \,:\, N \in \mathbb{N} \right\} \tag{11.4.28.a}$$

and

$$\left\{ \frac{\partial^{a+b}}{\partial\xi^a \partial\overline{\xi}^b} \mu^{(\mathsf{S};N)}_{\chi,t}[\mathbf{b}] \,:\, N \in \mathbb{N} \right\}, \tag{11.4.28.b}$$

are uniformly bounded for any $\mathbf{a} \in \widetilde{\Pi}$, $\chi \in \mathcal{D}(\mathbb{C})$, $t \geqslant 0$ *and* $a, b \geqslant 0$, *and moreover*

$$\lim_{N\to\infty} \frac{\partial^{a+b}}{\partial\xi^a \partial\overline{\xi}^b} \mu^{(\mathsf{S};N)}_{\mathbf{a},t}[\mathbf{b}] = \frac{\partial^{a+b}}{\partial\xi^a \partial\overline{\xi}^b} \mu_{\mathbf{a},t}[\mathbf{b}], \tag{11.4.29.a}$$

$$\lim_{N\to\infty} \frac{\partial^{a+b}}{\partial\xi^a \partial\overline{\xi}^b} \mu^{(\mathsf{S};N)}_{\chi,t}[\mathbf{b}] = \frac{\partial^{a+b}}{\partial\xi^a \partial\overline{\xi}^b} \mu_{\chi,t}[\mathbf{b}], \tag{11.4.29.b}$$

where this convergence is also uniform as $\mathbf{b}$ *varies over compacta in* $\widetilde{\Pi}$.

---

[23]This space has a natural topology, with respect to which it is complete. When we later consider distributions on $\mathcal{D}(\mathbb{C})$, these distributions are understood to be continuous with respect to this topology. The interested reader is referred to [224], amongst other authors, for a discussion of this topology.

We also note that

$$\frac{1}{2\pi}[\![\, \mathcal{F}T\,,\, \mu^{(\mathsf{S};N)}_{\mathbf{a},t}\,]\!] \;=\; \Psi^{(\mathsf{S};N)}_{\mathbf{a}}\Big(\mathfrak{U}^{(\mathsf{S};N)}_t\big(e^{i\mathbf{b}\cdot\mathfrak{s}^{(\mathsf{S};N)}}\mathfrak{D}^{(\mathsf{S};N)}[T]\big)\Big)\,, \qquad (11.4.30.a)$$

$$\frac{1}{2\pi}[\![\, \mathcal{F}T\,,\, \mu^{(\mathsf{S};N)}_{\chi,t}\,]\!] \;=\; \int_{\mathbb{C}} \chi(\vartheta)\Psi^{(\mathsf{S};N)}_{\mathbf{a}}\Big(\mathfrak{U}^{(\mathsf{S};N)}_t\big(e^{i\mathbf{b}\cdot\mathfrak{s}^{(\mathsf{S};N)}}\mathfrak{D}^{(\mathsf{S};N)}[T]\big)\Big)\, dA(\vartheta), \qquad (11.4.30.b)$$

for any $\mathbf{a} \in \widetilde{\Pi}$, $\chi \in \mathcal{D}(\mathbb{C})$ and $t \geqslant 0$, and that any distribution $T \in \mathcal{S}'(\Pi)$ gives rise to a distribution $T_{\mathbf{s},\mathbf{b},t} \in \mathcal{D}'(\mathbb{C})$ given by the formula

$$[\![\, T_{\mathbf{s},\mathbf{b},t}\,,\, \chi\,]\!] \;=\; \frac{1}{2\pi}[\![\, \mathcal{F}T\,,\, \mu_{\chi,t}\,]\!]\,, \qquad \chi \in \mathcal{D}(\mathbb{C})\,, \qquad (11.4.31)$$

observing that, if $T \in \mathcal{S}'(\Pi)$ is a function, $T_{\mathbf{s},\mathbf{b},t}$ is the function

$$T_{\mathbf{s},\mathbf{b},t}(\vartheta) \;=\; e^{i\mathbf{s}(t)\cdot\mathbf{b}}\, T\big(\vartheta(t)\big)\,. \qquad (11.4.32)$$

From the above results, we deduce the following:

Proposition 11.16 *If $T \in \mathcal{S}'(\Pi)$ is such that $\mathcal{F}\big((1+|\,\xi\,|^2)^{-M}T\big)$ belongs to $L^1(\mathbb{R}^2)$ for some $M \in \mathbb{N}$, then*

$$\lim_{N\to\infty} \Psi^{(\mathsf{S};N)}_{\mathbf{a}}\Big(\mathfrak{U}^{(\mathsf{S};N)}_t\big(e^{i\mathbf{b}\cdot\mathfrak{s}^{(\mathsf{S};N)}}\mathfrak{D}^{(\mathsf{S};N)}[T]\big)\Big) \;=\; T_{\mathbf{s},\mathbf{b},t}(\vartheta)\,, \qquad (11.4.33)$$

both pointwise and weakly as distributions in $\mathcal{D}'(\mathbb{C})$.

Proof: If T satisfies the given condition, then T must be a continuous function on Π, and hence

$$G(\xi) \;=\; (1+|\,\xi\,|^2)^{-M}T(\xi)$$

is also a continuous function of ξ. Then, since

$$(\mathcal{F}T)(\xi) \;=\; \Big(1 - \frac{\partial^2}{\partial\xi\partial\overline{\xi}}\Big)^M (\mathcal{F}G)(\xi)\,,$$

it follows that

$$\begin{aligned}\lim_{N\to\infty}[\![\, \mathcal{F}T\,,\, \mu^{(\mathsf{S};N)}_{\mathbf{a},t}\,]\!] \;&=\; \lim_{N\to\infty}[\![\, \mathcal{F}G\,,\, \Big(1 - \frac{\partial^2}{\partial\xi\partial\overline{\xi}}\Big)^M \mu^{(\mathsf{S};N)}_{\mathbf{a},t}\,]\!] \\ &=\; [\![\, \mathcal{F}G\,,\, \Big(1 - \frac{\partial^2}{\partial\xi\partial\overline{\xi}}\Big)^M \mu_{\mathbf{a},t}\,]\!] \\ &=\; \int_{\mathbb{C}} (\mathcal{F}G)(\xi)(1+|\,\vartheta(t)\,|^2)^M e^{i\tau_t(\mathbf{a})\cdot\mathbf{b}}\, dA(\xi)\end{aligned}$$

$$\begin{aligned} &= 2\pi(1+|\vartheta(t)|^2)^M e^{i\mathbf{s}(t)\cdot\mathbf{b}} G(\vartheta(t)) \\ &= 2\pi T_{\mathbf{s},\mathbf{b},t}(\vartheta)\,, \end{aligned}$$

as required. Similar considerations show us that

$$\begin{aligned} \lim_{N\to\infty} [\![\mathcal{F}T\,,\,\mu^{(\mathsf{S};N)}_{\chi,t}]\!] &= [\![\mathcal{F}G\,,\,\Big(1-\frac{\partial^2}{\partial\xi\partial\overline{\xi}}\Big)^M \mu_{\chi,t}]\!] \\ &= [\![\mathcal{F}G\,,\,\mu_{\tilde{\chi}_t,t}]\!] \\ &= 2\pi e^{i\mathbf{s}(t)\cdot\mathbf{b}} \int_{\mathbb{C}} G(\vartheta(t))\tilde{\chi}_t(\vartheta)\,dA(\vartheta) \\ &= 2\pi \int_{\mathbb{C}} \chi(\vartheta) T_{\mathbf{s},\mathbf{b},t}(\vartheta)\,dA(\vartheta)\,, \end{aligned}$$

again as required, where $\tilde{\chi}_t \in \mathcal{D}(\mathbb{C})$ is the function

$$\tilde{\chi}_t(\vartheta) = (1+|\vartheta(t)|^2)^M \chi(\vartheta)\,.$$

■

It is elementary to show that this result is valid for all distributions T in $\mathcal{S}'(\Pi)$ which are functions in $C^4(\Pi)$ whose partial derivatives up to the fourth order are polynomially bounded. Certainly, therefore, the above result holds for all $T \in \mathcal{S}(\Pi)$ (this result is to be found in [5]), but evidently holds for a much larger space of distributions, including all polynomial functions on Π. Since $\mathcal{S}(\Pi)$ is weakly dense in $\mathcal{S}'(\Pi)$, the above Proposition holds for a collection of distributions which is weakly dense in $\mathcal{S}'(\Pi)$, but since it is not clear that the procedure of taking a weak approximation in $\mathcal{S}'(\Pi)$ commutes with the operation of taking the thermodynamic limit, it is not yet possible to deduce that the result of the above Proposition holds for all distributions $T \in \mathcal{S}'(\Pi)$ — indeed, only the weak version of that Proposition can be true in general.

However, this partial result enables us to extend the quasi-classical interpretation of the thermodynamic limit to all positive times. For if $F \in \mathcal{S}(\Pi)$, then

$$X_t^{(N)}(F) = W^{(\mathsf{R})}[-i\sqrt{N}\vartheta]\cdot(\varpi^{(\mathsf{M};N)}\otimes 1)\Big(\mathfrak{U}_t^{(\mathsf{S};N)}\big(\mathfrak{D}^{(\mathsf{S};N)}[F]\big)\Big)\cdot W^{(\mathsf{R})}[i\sqrt{N}\vartheta]$$

is a radiation observable, and it can be shown to be of the form

$$\mathbf{\Delta}^{(\mathsf{R})}\big[\mathcal{E}_N^{\mathrm{tr}} F[\mathbf{a},N,t]\big]\,,$$

where $F[\mathbf{a}, N, t] \in \mathcal{S}(\Pi)$. Moreover, since

$$\begin{aligned}\lim_{N\to\infty} \omega^{(\mathsf{R})}\big[\mathbf{\Delta}^{(\mathsf{R})}\big[\mathcal{E}_N^{\mathrm{tr}} F[\mathbf{a}, N, t]\big]\big] &= \lim_{N\to\infty} \Psi_{\mathbf{a}}^{(\mathsf{S};N)}\Big(\mathfrak{U}_t^{(\mathsf{S};N)}\big(\mathfrak{D}^{(\mathsf{S};N)}[F]\big)\Big) \\ &= F_{\mathbf{s},\mathbf{0},t}(\vartheta) \;=\; F(\vartheta(t))\end{aligned}$$

for any state $\omega^{(\mathsf{R})} \in \mathfrak{A}_*^{(\mathsf{R})}$, it follows that

$$\varPi[F;t]_{\mathbf{a}} \;=\; F\big(\vartheta(t)\big)\,,$$

and also that

$$\lim_{N\to\infty} F[\mathbf{a}, N, t] \;=\; F(\vartheta(t))\,\mathfrak{i}$$

weakly in $\mathcal{S}'(\Pi)$. Moreover, the clustering and homogeneity properties of ϖ imply that

$$\begin{aligned}&\lim_{N\to\infty} \big(\varpi^{(\mathsf{M};N)} \otimes \omega^{(\mathsf{R})}\big)\Big(\mathcal{T}_\vartheta^{(\mathsf{S};N)}\big(\mathfrak{U}_{t_1}^{(\mathsf{S};N)}\big(\mathfrak{D}^{(\mathsf{S};N)}[F]\big)\mathfrak{U}_{t_2}^{(\mathsf{S};N)}\big(\mathfrak{D}^{(\mathsf{S};N)}[G]\big)\big)\Big) \\ &\qquad= \lim_{N\to\infty} \omega^{(\mathsf{R})}\Big(X_{t_1}^{(N)}(F)X_{t_2}^{(N)}(G)\Big) \\ &\qquad= \lim_{N\to\infty} \Psi_{\mathbf{a}}^{(\mathsf{S};N)}\Big(\mathfrak{D}^{(\mathsf{S};N)}\big[F[\mathbf{a}, N, t_1]\big]\mathfrak{D}^{(\mathsf{S};N)}\big[G[\mathbf{a}, N, t_2]\big]\Big) \\ &\qquad= \lim_{N\to\infty} \Psi_{\mathbf{a}}^{(\mathsf{S};N)}\Big(\mathfrak{D}^{(\mathsf{S};N)}\big(F[\mathbf{a}, N, t_1] \star_{(1/N)} G[\mathbf{a}, N, t_2]\big)\Big)\,,\end{aligned}$$

and this limit can be shown to exist weakly, and to be equal to

$$\varPi[F, t_1]_{\mathbf{a}}\, \varPi[F, t_2]_{\mathbf{a}} \;=\; F(\vartheta(t_1))\, G(\vartheta(t_2))\,.$$

We interpret these results by saying that the classical description for the microscopic radiation observable $\mathbf{\Delta}^{(\mathsf{R})}[F]$ at time t, with respect to the sequence of states $\big(\Psi_{\mathbf{a}}^{(\mathsf{S};N)}\big)_{N\geqslant 1}$, is the quantity $F\big(\vartheta(t)\big)$. If we view $\vartheta(t)$, for $t \geqslant 0$, as the orbit of the emergent classical dynamics which passes through the phase space point $\vartheta \in \Pi$ at time zero, as determined by the sequence of states, then the function $t \mapsto F\big(\vartheta(t)\big)$ gives the correct dynamical behaviour for a classical observable. Together with the factorization result, this shows that this interpretation of the variables in the thermodynamic limit as classical observables is consistent with the dynamics. As we have shown above, this interpretation of a microscopic radiation observable $\mathbf{\Delta}^{(\mathsf{R})}[T]$ is valid when $T \in \mathcal{S}(\Pi)$, but it is also valid when $T \in \mathcal{S}'(\Pi)$ satisfies the conditions of Proposition 11.16 and so, for all such T, we can write

$$\varPi[T;t]_{\mathbf{a}} \;=\; T\big(\vartheta(t)\big)\,. \tag{11.4.34}$$

This partial state of affairs is sufficient to derive physically interesting results for a large class of observables. Since they form the main burden of this book, let us consider radiation distributions of the angle in particular.

Given any function $\mathfrak{f} \in \mathcal{C}^4(\mathbb{T})$, the associated phase space distribution $\mathfrak{f}_{\text{ang}}$ does not satisfy the conditions of Proposition 11.16, since $\mathfrak{f}_{\text{ang}}$ is not even continuous at the origin. However, if we consider the function

$$\mathfrak{f}_\delta(r\cos\beta, r\sin\beta) \;=\; P_\delta(r)\,\mathfrak{f}(e^{i\beta})\,, \tag{11.4.35}$$

where P_δ is the smoothing function

$$P_\delta(r) \;=\; \begin{cases} 0\,, & r = 0\,, \\ \left(1 + e^{\frac{\delta - 2r}{r(\delta - r)}}\right)^{-1}, & 0 < r < \delta\,, \\ 1\,, & r \geqslant \delta\,, \end{cases} \tag{11.4.36}$$

then $\mathfrak{f}_\delta \in \mathcal{C}^4(\Pi)$ does satisfy the conditions of Proposition 11.16. We note that the function $\mathfrak{f}_\delta$ only differs from $\mathfrak{f}$ inside the disc $r < \delta$ in phase space, and takes the value 0 at the origin. Moreover, it is clear that

$$\lim_{\delta\to 0} \mathfrak{f}_\delta(p,q) \;=\; \mathfrak{f}_{\text{ang}}(p,q)\,, \qquad (p,q) \neq (0,0)\,, \tag{11.4.37.a}$$

and

$$\mathfrak{f}_\delta(p,q) \;=\; \mathfrak{f}_{\text{ang}}(p,q)\,, \qquad p^2 + q^2 \geqslant \delta^2\,. \tag{11.4.37.b}$$

In previous Chapters, we have sometimes described angular distributions through functions on $\mathbb{T}$, and sometimes through functions on $[-\pi, \pi]$. It is therefore necessary for us to be able to handle both types of description. Thus, given any function $\mathfrak{f} \in \mathcal{C}^4[-\pi, \pi]$, for any $0 < \delta < \pi$ we can (for example, by polynomial interpolation) find a function $\mathfrak{f}[\delta] \in \mathcal{C}^4(\mathbb{T})$ such that

$$\mathfrak{f}[\delta](e^{i\beta}) \;=\; \mathfrak{f}(\beta)\,, \qquad -\pi + \delta \leqslant \beta \leqslant \pi - \delta\,.$$

We can then consider the function $\mathfrak{f}_\delta \;=\; \mathfrak{f}[\delta]_\delta \in \mathcal{C}^4(\Pi)$, obtained from $\mathfrak{f}[\delta]$ as above. This function $\mathfrak{f}_\delta$ then satisfies the conditions of Proposition 11.16, and is such that

$$\lim_{\delta\to 0} \mathfrak{f}_\delta(p,q) \;=\; \mathfrak{f}_{\text{ang}}(p,q)\,, \tag{11.4.38.a}$$

for all (p,q) away from the negative p-axis, and

$$\mathfrak{f}_\delta(r\cos\beta, r\sin\beta) \;=\; \mathfrak{f}_{\text{ang}}(r\cos\beta, r\sin\beta) \;=\; \mathfrak{f}(\beta) \tag{11.4.38.b}$$

whenever $r \geqslant \delta$ and $|\beta| \leqslant \pi - \delta$.

Thus, whether $\mathfrak{f}$ belongs to $\mathcal{C}^4(\mathbb{T})$ or to $\mathcal{C}^4[-\pi, \pi]$, it is possible to approximate $\mathfrak{f}_{\text{ang}}$ by functions $\mathfrak{f}_\delta$ which satisfy the conditions of Proposition 11.16, and yet which agree with $\mathfrak{f}_{\text{ang}}$ everywhere, except upon a δ-dependent neighbourhood[24] of the region of discontinuity of $\mathfrak{f}_{\text{ang}}$.

As has been remarked previously, it is not possible to measure any (classical or quantum) observable exactly, if that observable has continuous spectrum. In particular, therefore, it is not possible to determine whether measurements are being made of the observable $\mathfrak{f}_{\text{ang}}$ or of the observable $\mathfrak{f}_\delta$, provided that δ is sufficiently small that the difference between $\mathfrak{f}_{\text{ang}}$ and $\mathfrak{f}_\delta$ is within the margin of experimental accuracy currently being employed. Thus it is entirely reasonable to suppose that a measurement of the classical observable $\mathfrak{f}_{\text{ang}}$ is, in effect, being obtained by a measurement of the observable $\mathfrak{f}_\delta$ for small enough $\delta > 0$. Since

$$\Pi[\mathfrak{f}_\delta; t]_{\mathbf{a}} = \lim_{N \to \infty} \Psi_{\mathbf{a}}^{(\mathsf{S};N)}\left(\mathfrak{U}_t^{(\mathsf{S};N)}\left(\mathfrak{D}^{(\mathsf{S};N)}[\mathfrak{f}_\delta]\right)\right) = \mathfrak{f}_\delta\left(\vartheta(t)\right), \tag{11.4.39}$$

we see that this limit is equal to $\mathfrak{f}_{\text{ang}}(\vartheta(t))$ so long as $|\vartheta(t)| \geqslant \delta$ and (if necessary) $|\operatorname{Arg}\vartheta(t)| \leqslant \pi - \delta$. Thus the thermodynamic limit of such an approximating observable for $\mathfrak{f}_{\text{ang}}$ yields a result which (under certain conditions) agrees with the expected result for $\mathfrak{f}_{\text{ang}}$ itself. In this approximating sense, therefore, we state that the classical description, in thermodynamic limit, of the microscopic radiation observable $\mathbf{\Delta}^{(\mathsf{R})}[\mathfrak{f}_{\text{ang}}]$ is the function

$$\Pi[\mathfrak{f}_{\text{ang}}; t]_{\mathbf{a}} \approx \mathfrak{f}_{\text{ang}}\left(\vartheta(t)\right). \tag{11.4.40}$$

We have used the symbol $\approx$, rather than equality $=$, to reflect the above proviso concerning this identification.

11.4.3 *Solutions, Phase Transitions And Lasing*

The true value of this model lies in the fact that varying one of its parameters results in a phase transition from a situation in which no coherent radiation is emitted to one in which such radiation is emitted.

To see this, note that equations (11.4.23.a), (11.4.23.b) and (11.4.23.c)

---

[24] It should be emphasized that the value of δ can be as small as we please, so that this approximation can be made arbitrarily good.

have the particular fixed point solution

$$\gamma(t) = 0\,, \qquad p(t) = \eta\,, \qquad \vartheta(t) = 0\,, \tag{11.4.41}$$

and analysis in [5] shows that this solution is stable if $0 < \eta < \eta_1$, where

$$\eta_1 = \frac{\kappa u}{\lambda^2}\Big[1 + \big(\frac{\varepsilon - \omega}{\kappa + u}\big)^2\Big]\,. \tag{11.4.42}$$

Since $\vartheta(t) = 0$ for all $t \geqslant 0$, the intensity $|\,\vartheta(t)\,|^2$ of the radiation is zero for all $t \geqslant 0$, and hence it follows that no coherent radiation is produced to be emitted from the cavity.

Although no coherent radiation is produced when η is less than this critical value η_1, the situation is very different when η exceeds η_1, since there is then a Hopf bifurcation corresponding to a periodic orbit. To be specific, when $\eta > \eta_1$, the equations (11.4.23.a), (11.4.23.b) and (11.4.23.c) have the solution

$$\gamma(t) = Ge^{-i\nu t}\,, \qquad p(t) = \eta_1\,, \qquad \vartheta(t) = He^{i\nu t}\,, \tag{11.4.43}$$

where the frequency ν of this solution is

$$\nu = \frac{\kappa\varepsilon + u\omega}{\kappa + u}\,, \tag{11.4.44}$$

and the coefficients G and H are determined by the formulæ

$$G = \frac{\kappa}{\lambda(\kappa + u)}\big(\kappa + u + i(\omega - \varepsilon)\big)\overline{H}\,, \tag{11.4.45.a}$$

$$H = -\frac{1}{2}i\sqrt{\frac{v(\eta - \eta_1)}{\kappa}}\,e^{i\beta}\,, \tag{11.4.45.b}$$

where β is some real constant. Moreover, it can be shown that this solution is stable while η lies in some interval (η_1, η_2), where $\eta_2 > \eta_1$. However, the determination of the exact value of η_2 is rather complicated. As an example, however, it can be shown that

$$\eta_2 = \frac{\kappa^2(\kappa + 3u + v)}{\lambda^2(\kappa - u - v)}$$

in the case that $\varepsilon = \omega$ and $\kappa > u + v$. If η exceeds this second critical value η_2, there is a second bifurcation, which yields a state of chaos of a Lorentz strange attractor type[25].

[25]This possibility is known from the theory of dynamical systems, cf [67].

From the above it is clearly appropriate to interpret η as the pumping parameter of the model. Thus, when the pumping parameter η exceeds η_1, and yet is not too large, an order-disorder transition from a stationary pure phase to a periodic orbit of pure states occurs. In this region, coherent radiation of time-independent intensity

$$|\vartheta(t)|^2 = |H|^2 = \frac{v(\eta - \eta_1)}{4\kappa} \tag{11.4.46}$$

is produced. Thus we observe that the intensity of the coherent radiation emitted by this model displays the same behaviour either side of the critical value of the pumping parameter as did that in the Lamb model[26], see equation (11.1.15). This change from zero to strictly positive intensity of coherent radiation implies that this phase transition is associated with a spontaneous breakdown of gauge symmetry, and one which takes place far from thermal equilibrium.

From preceding calculations we see that[27], if $\eta_1 < \eta < \eta_2$,

$$\varPi[\mathfrak{f}_\delta; t]\mathbf{a} = P_\delta\Big(\sqrt{\frac{v(\eta - \eta_1)}{2\kappa}}\Big)\, \mathfrak{f}\big(e^{i(\beta + \nu t)}\big) \tag{11.4.47.a}$$

for any $\mathfrak{f} \in \mathcal{C}^4(\mathbb{T})$ and $\delta > 0$, so that

$$\varPi[\mathfrak{f}_\delta; t]\mathbf{a} = \mathfrak{f}\big(e^{i(\beta + \nu t)}\big) \tag{11.4.47.b}$$

for any $\mathfrak{f} \in \mathcal{C}^4(\mathbb{T})$, provided that $0 < \delta < \sqrt{\frac{v(\eta - \eta_1)}{2\kappa}}$. Similarly, if $\mathfrak{f}$ is in $\mathcal{C}^4[-\pi, \pi]$, then

$$\varPi[\mathfrak{f}_\delta; t]\mathbf{a} = \mathfrak{f}(\beta + \nu t)\,, \tag{11.4.47.c}$$

provided that $|\,\beta + \nu t\,| \leqslant \pi - \delta$ and $0 < \delta < \sqrt{\frac{v(\eta - \eta_1)}{2\kappa}}$.

In particular, these results can be applied to the phase space observables $e^{\pm i\varphi}$ and φ respectively. We deduce that, when η exceeds the critical value η_1 (but is not so large that chaos ensues), then approximations to those phase space observables can be chosen which are sufficiently exact that the

[26]Since the thermodynamic limit yields a classical description of the QL-model, and since the Lamb model is semi-classical, this result is perhaps to be expected.

[27]These results are general, and not confined to the special case leading to the particular value of η_2 given above.

thermodynamic limits of their expectations yield the classical descriptions

$$e^{\pm i(\beta+\nu t)} \tag{11.4.48.a}$$

in the case of $e^{\pm i\varphi}$, and

$$\beta + \nu t \qquad\qquad \text{modulo } (-\pi, \pi]\,, \tag{11.4.48.b}$$

in the case of φ. More precisely, in this latter case, what is obtained is a smoothed version of this function which mediates the jump discontinuity in the function φ. Then the classical description $\Pi[\varphi; t]_{\mathbf{a}}$ of the phase operator $\boldsymbol{\Delta}^{(\mathsf{R})}[\varphi]$ in the thermodynamic limit has the time development

$$\Pi[\varphi; t]_{\mathbf{a}} \approx \beta + \nu t\,, \qquad\qquad \text{modulo } (-\pi, \pi]\,. \tag{11.4.49}$$

This result justifies our previous assertion that radiation is described in the thermodynamic limit as a classical oscillator, and that the phase observable $\boldsymbol{\Delta}^{(\mathsf{R})}[\varphi]$ determines the phase of that oscillator. This last equation also identifies the (as yet unspecified) coefficient β which appears in the definition of the parameter H as the initial phase of the coherent radiation oscillator.

Chapter 12

Weyl Dequantization

The attempt to extract from a purely arbitrary idea the existence of an object corresponding to it is a quite unnatural procedure and a mere innovation of scholastic subtlety.

– Immanuel Kant, *Critique of Pure Reason.*

12.1 Introduction

In this Chapter we discuss the theory of dequantization, which will tell us, for the Weyl scheme, what function or distribution in phase space to assign to a given observable as its symbol.

Because the Wigner transform is bijective, an abstract proof that dequantization is possible is not very difficult. Unfortunately, the proof does not provide a practicable method for evaluating T from knowledge of $\mathbf{\Delta}[T]$, and devising such methods is the core of the problem. Another sort of problem that requires attention is to determine the class of T given the class of $\mathbf{\Delta}[T]$, which is the obverse of the problem considered in connection with quantization. Before proceeding, we recall that Weyl quantization was based upon the smooth model, and hence Weyl dequantization must also be so based — we are therefore interested in "operators" and classes of "operators" in $\mathcal{L}(\mathcal{S}(\mathbb{R}), \mathcal{S}'(\mathbb{R}))$, and their dequantizations in $\mathcal{S}'(\Pi)$.

Note that Proposition 8.16 shows that Weyl quantization is a linear bijection to the space $\mathfrak{T}_0(L^2(\mathbb{R}))$ of finite rank operators from the subspace of $L^2(\Pi)$ consisting of the linear span of the functions $\{\Xi_{\phi,\psi} : \phi, \psi \in L^2(\mathbb{R})\}$, and hence Weyl dequantization is a linear bijection from $\mathfrak{T}_0(L^2(\mathbb{R}))$ to that space of functions. Similarly, Theorem 8.23 of Pool in Chapter 8 states that Weyl quantization is a linear bijection from $L^2(\Pi)$ to the space $\mathfrak{T}_2(L^2(\mathbb{R}))$ of Hilbert-Schmidt operators on $L^2(\mathbb{R})$, and consequently it is clear that Weyl dequantization is a linear bijection from $\mathfrak{T}_2(L^2(\mathbb{R}))$ to $L^2(\Pi)$. Moreover, an explicit formula for Weyl dequantization in this context is provided by equation (8.4.17.a). Thus, for these classes of observables, the problem of Weyl dequantization is completely solved, but these classes of observables are too restrictive for our purposes.

Most treatments of this topic begin with what was referred to in Chapter 8 as the *familiar formula*, namely the ansatz that the Weyl dequantization T of the observable $\boldsymbol{\Delta}[T]$ is to be given by the expression

$$T(p,q) \;=\; \operatorname{Tr}\left(\boldsymbol{\Delta}[p,q]\,\boldsymbol{\Delta}[T]\right). \tag{8.4.11}$$

Results concerning this formula were formalized in Theorem 8.21, where it was shown that this formula was valid for all trace-class operators $\boldsymbol{\Delta}[T]$, but only for such operators. Moreover, while the familiar formula provides an expression for the Weyl dequantization of an observable in $\mathcal{T}_1\big(L^2(\mathbb{R})\big)$, it tells us little about the properties of that dequantization as a function on Π, and moreover it is clear that the familiar formula will not handle all the observables permitted by Pool's Theorem. We noted in Chapter 8 that the familiar formula was not sufficiently general to provide a complete description of Weyl dequantization, and we suggested that it might be possible to extend it beyond trace-class operators by developing a summability method. However, aside from theoretical interest, such an extension is not likely to lead to explicit expressions for the symbols of many operators, for by the nature of things it is not easy to evaluate the trace formula except in some very simple cases. *Inter alia*, this is because the operators of interest in quantum theory are often known only in terms of comparatively singular integral kernels, or else in terms of their matrix coefficients with respect to some orthonormal basis for $L^2(\mathbb{R})$. To evaluate the trace formula in such cases thus requires the extremely difficult determination of multiple sums or integrals.

What we need, therefore, are good approximation methods. One approach might be to replace the trace sum in equation (8.4.11) by a sum over a finite number of basis vectors — this is a very natural first step. Provided that the errors created by this approximation are small, this approach could be used in numerical calculations to obtain estimates for the Weyl dequantization of observables. However, numerical analysis indicates that the trace sum in the above formula is likely to converge slowly, so that a large number of terms are required to give good approximations. Moreover, the method is likely to be unstable, so that small numerical errors will accumulate and magnify. Thus the answers obtained by this method are likely to be unreliable. For these reasons, and despite its rough utility as a formal expression, we shall say nothing more in this Chapter concerning the familiar trace formula.

Evidently, something new is necessary. A rigorous and widely applicable method of dequantization suggests itself through an important property of the laser model which was discussed in Chapter 11, namely that the thermodynamic limit of the scaled expectation of the scaled radiation variable $\mathfrak{D}^{(\mathrm{S};N)}[T]$ is $T(p,q)$, see equation (11.4.18.a).

More explicitly, by forgetting the physics of the previous Chapter, the state and observable scalings and taking the limit as $N \to \infty$ can be interpreted as a collection of transformations to be performed upon some observable $B \in \mathcal{L}(\mathcal{S}(\mathbb{R}), \mathcal{S}'(\mathbb{R}))$, yielding a distribution $T \in \mathcal{S}'(\Pi)$ for which $\boldsymbol{\Delta}[T] = B$. Of course, it may not always be possible to perform all of these transformations explicitly, giving a closed form for the observable B, but the various scalings for each finite value of N can always be performed. Consequently any $B \in \mathcal{L}(\mathcal{S}(\mathbb{R}), \mathcal{S}'(\mathbb{R}))$ gives rise to a sequence $(T_N)_{N \geqslant 1}$ of precisely known distributions in $\mathcal{S}'(\Pi)$ which converges (in some sense) to a distribution $T \in \mathcal{S}'(\Pi)$ for which $\boldsymbol{\Delta}[T] = B$. Thus we obtain, in a natural manner, an approximative scheme for Weyl dequantization which does not suffer from the stability problems of other, perturbative, techniques.

A property of the thermodynamical limit in the laser model is that it obliterates the details of the initial radiation state. This can be turned to our advantage here in that this initial state can be chosen on the grounds of mathematical convenience rather than physical necessity. Moreover, it does not even have to be a state — any normalized (smooth) density matrix will do, positive or not. Since this method arose from the theory of the laser, and because that origin is partially concealed from direct view, we have called this *the method of motes*, and the freely chosen density matrix (more precisely, its integral kernel) will be termed a *mote*.

In addition to the method of motes, we also show an exact form of Weyl dequantization can be derived, which expresses the symbol of an observable as a sum in terms of a particularly interesting Schauder basis for $\mathcal{S}'(\Pi)$. Partial sums of this series thus provide an alternative approximation scheme for dequantization. This second approach will be convenient for studying the symbols of Toeplitz operators, which will enable us to compare the symbols of operators such as X, E and E^* with those of $\boldsymbol{\Delta}[\varphi]$, $\boldsymbol{\Delta}\left[e^{-i\varphi}\right]$ and $\boldsymbol{\Delta}\left[e^{i\varphi}\right]$.

While this second method does not solve the problem of the dequantization of Toeplitz operators completely, it does enable us to prove that for a wide class of Toeplitz operators their symbols are of the form $R+S$, where R is a phase space distribution that can be evaluated explicitly — R is the phase space angular distribution naturally associated with the Toeplitz operator — and S is square integrable, but otherwise very little can be said about it. While this result is of limited practicality from a calculational point of view, it offers a reasonable starting-point for theorems about phase space symbols of Toeplitz operators.

12.2 Inverse Quantization

There is no particular difficulty in determining the theoretical inverse of quantization, since combining the formalism of Definition 8.7 with the results of Proposition 8.8 yields the answer in a straightforward manner. The result is not so much a method for calculation as a rigorous expression which is to form the basis for other, calculational, techniques.

Proposition 12.1 ***(Weyl Dequantization)*** *The Weyl quantization map $\boldsymbol{\Delta} : \mathcal{S}'(\Pi) \to \mathcal{L}(\mathcal{S}(\mathbb{R}), \mathcal{S}'(\mathbb{R}))$ is a linear bijection, with its inverse map $\boldsymbol{\Delta}^{-1} : \mathcal{L}(\mathcal{S}(\mathbb{R}), \mathcal{S}'(\mathbb{R})) \to \mathcal{S}'(\Pi)$ given by the formula*

$$\boldsymbol{\Delta}^{-1}[B] = \left(\mathcal{G}^{\mathrm{tr}}\right)^{-1} K_B, \qquad B \in \mathcal{L}(\mathcal{S}(\mathbb{R}), \mathcal{S}'(\mathbb{R})), \qquad (12.2.1)$$

where $K_B \in \mathcal{S}'\left(\mathbb{R}^2\right)$ is, as usual, the integral kernel of B.

Proof: We noted in Proposition 8.9 that the integral kernel $K_{\boldsymbol{\Delta}[T]}$ of $\boldsymbol{\Delta}[T]$ was related to its defining distribution $T \in \mathcal{S}'(\Pi)$ by the formula

$$K_{\boldsymbol{\Delta}[T]} = \mathcal{G}^{\mathrm{tr}}\, T. \qquad (8.3.13)$$

We have already noted that $\mathcal{G} : \mathcal{S}\left(\mathbb{R}^2\right) \to \mathcal{S}(\Pi)$ is a continuous linear bijection, and hence its transpose $\mathcal{G}^{\mathrm{tr}} : \mathcal{S}'(\Pi) \to \mathcal{S}'\left(\mathbb{R}^2\right)$ is also bijective, with inverse $\left(\mathcal{G}^{\mathrm{tr}}\right)^{-1} = \left(\mathcal{G}^{-1}\right)^{\mathrm{tr}}$, so we can rewrite the above formula as

$$T = \left(\mathcal{G}^{\mathrm{tr}}\right)^{-1} K_{\boldsymbol{\Delta}[T]}.$$

From this it is clear that the map $\boldsymbol{\Delta}^{-1} : \mathcal{L}(\mathcal{S}(\mathbb{R}), \mathcal{S}'(\mathbb{R})) \to \mathcal{S}'(\Pi)$ defined by equation (12.2.1) is such that $\boldsymbol{\Delta}^{-1}[\boldsymbol{\Delta}[T]] = T$ for any $T \in \mathcal{S}'(\Pi)$. On the other hand, given $B \in \mathcal{L}(\mathcal{S}(\mathbb{R}), \mathcal{S}'(\mathbb{R}))$, since

$$\begin{aligned} [\![\boldsymbol{\Delta}^{-1}[B] , \mathcal{G}(g \otimes f)]\!] &= [\![(\mathcal{G}^{\mathrm{tr}})^{-1} K_B , \mathcal{G}(g \otimes f)]\!] \\ &= [\![K_B , g \otimes f]\!] = [\![Bf , g]\!] \end{aligned}$$

for any f, $g \in \mathcal{S}(\mathbb{R})$, we see that $\boldsymbol{\Delta}\left[\boldsymbol{\Delta}^{-1}[B]\right]$ is equal to B for any such B. Hence we deduce that the maps $\boldsymbol{\Delta}$ and $\boldsymbol{\Delta}^{-1}$ are mutually inverse, as required. ■

We shall now present two calculational techniques for dequantization. Neither of the two methods is useful for all operators, but they are designed with applications in mind in that the first technique can be used for mappings $B \in \mathcal{L}(\mathcal{S}(\mathbb{R}), \mathcal{S}'(\mathbb{R}))$ whose integral kernel is known, while the second is useful when its matrix elements with respect to the Hermite-Gauss basis are known.

12.3 The Method Of Motes

As mentioned, the method of motes is essentially a transcription of the scaling procedure used in the QL-model. One consequence of this method will be (as promised previously) the limit formula (11.4.18.a) for $\mathfrak{D}^{(\mathrm{S};N)}[T]$ given in Proposition 11.12.

First, choose a mote. A mote is any function[1] $\mathcal{M} \in \mathcal{S}(\Pi)$ such that

$$\iint_{\Pi} \mathcal{M}(p, q)\, dpdq = 1\,. \tag{12.3.1}$$

To construct a sequence from the mote which will reproduce the effect of the scaled expectation values needs the scaling maps $\mathcal{E}_N$ and τ_ϑ considered in Chapter 11 (which are continuous endomorphisms of $\mathcal{S}(\Pi)$ or $\mathcal{S}(\mathbb{R}^2)$) and their transposes. We repeat their formulæ for convenience,

$$[\mathcal{E}_N F](p, q) = N\, F(p\sqrt{N}, q\sqrt{N})\,, \tag{11.2.21.a}$$

[1]While, except for the integral condition, the choice of mote is unrestricted, some motes are clearly better suited to a given problem than others.

$$\big[\tau_\vartheta F\big](w) \;=\; F(w+\vartheta)\,, \tag{11.2.32.b}$$

for any $F \in \mathcal{S}(\Pi)$.

Definition 12.2 *For any mote $\mathcal{M} \in \mathcal{S}(\Pi)$ and any $\vartheta \in \mathbb{C}$, define the function $\mathcal{M}_{\vartheta:N} \in \mathcal{S}\big(\mathbb{R}^2\big)$ by the formula*

$$\mathcal{M}_{\vartheta;N} \;=\; \mathcal{G}^{-1}\,\tau_{-\vartheta}\,\mathcal{E}_N\,\mathcal{M}. \tag{12.3.2}$$

The sequence $\{\,\mathcal{M}_{\vartheta;N} \,:\, N \in \mathbb{N}\,\}$ will be known as the mote sequence at the point $\vartheta \in \mathbb{C}$ obtained from $\mathcal{M}$.

For any mote $\mathcal{M}$ and any $B \in \mathcal{L}(\mathcal{S}(\mathbb{R}), \mathcal{S}'(\mathbb{R}))$, the function $\mathbf{\Delta}^{-1}_{\mathcal{M};N}[B]$ defined by

$$\big[\mathbf{\Delta}^{-1}_{\mathcal{M};N}[B]\big](\vartheta) \;=\; [\![\, K_B\,,\, \mathcal{M}_{\vartheta;N}\,]\!], \qquad \vartheta \in \mathbb{C}\,, \tag{12.3.3}$$

will be known as the Nth approximation to the symbol (Weyl dequantization) of B relative to $\mathcal{M}$.

It is relatively easy to show that the function $\mathbf{\Delta}^{-1}_{\mathcal{M};N}[B]$ is infinitely differentiable and polynomially bounded, and so is a well-defined element of $\mathcal{S}'(\Pi)$.

The justification for the above notation is to be found in the fact that $\big\{\mathbf{\Delta}^{-1}_{\mathcal{M};N}[B] \,:\, N \in \mathbb{N}\big\}$ is an approximating sequence for the dequantization symbol $\mathbf{\Delta}^{-1}[B]$ of B.

Proposition 12.3 ***(Mote Dequantization)*** *If $B \in \mathcal{L}(\mathcal{S}(\mathbb{R}), \mathcal{S}'(\mathbb{R}))$ and if $\mathcal{M}$ is any mote, then the sequence of functions $\big\{\mathbf{\Delta}^{-1}_{\mathcal{M};N}[B] \;:\; N \in \mathbb{N}\big\}$ approximates $\mathbf{\Delta}^{-1}\,[\,B\,]$, in that this sequence converges to $\mathbf{\Delta}^{-1}\,[\,B\,]$ weakly in $\mathcal{S}'(\Pi)$, so that*

$$\lim_{N\to\infty}\, [\![\,\mathbf{\Delta}^{-1}_{\mathcal{M};N}\,[B]\,,\, G\,]\!] \;=\; [\![\,\mathbf{\Delta}^{-1}\,[\,B\,]\,,\, G\,]\!], \qquad G \in \mathcal{S}(\Pi)\,. \tag{12.3.4}$$

Proof: Direct calculation shows that

$$\big(\mathbf{\Delta}^{-1}_{\mathcal{M};N}\,[B]\,\big)(\vartheta) \;=\; [\![\,\mathbf{\Delta}^{-1}\,[\,B\,]\,,\, \tau_{-\vartheta}\,\mathcal{E}_N\,\mathcal{M}\,]\!],$$

for any $N \in \mathbb{N}$ and $\vartheta \in \mathbb{C}$. It follows that

$$\begin{aligned} [\![\,\mathbf{\Delta}^{-1}_{\mathcal{M};N}\,[B]\,,\, G\,]\!] \;&=\; \int_{\mathbb{C}} [\![\,\mathbf{\Delta}^{-1}\,[\,B\,]\,,\, \tau_{-\vartheta}\,\mathcal{E}_N\,\mathcal{M}\,]\!]\, G(\vartheta)\, dA(\vartheta) \\ &=\; [\![\,\mathbf{\Delta}^{-1}\,[\,B\,]\,,\, \mathcal{E}_N\mathcal{M} \,*\, G\,]\!] \end{aligned}$$

for any $G \in \mathcal{S}(\Pi)$. It is simple to establish that the integration can be brought inside the pairing symbol in this manner. Standard analysis[2] shows that

$$\lim_{N\to\infty} \mathcal{E}_N \mathcal{M} * G = G$$

for any $G \in \mathcal{S}(\Pi)$, the convergence being with respect to the Fréchet topology on $\mathcal{S}(\Pi)$. This establishes the desired result. ■

To see that this result indeed establishes Proposition 11.12, combining the results of the preceding Chapter with the above discussion, and using Proposition 8.28, it follows that

$$\Psi_{\mathbf{a}}^{(\mathsf{S};N)}\big(\mathfrak{D}^{(\mathsf{S};N)}[T]\big) = \big(\mathbf{\Delta}_{\mathcal{M};N}^{-1}[\mathbf{\Delta}\,[T]]\big)(\vartheta)\,, \tag{12.3.5.a}$$

where $\mathcal{M} \in \mathcal{S}(\Pi)$ is the mote obtained from the radiation state $\omega^{(\mathsf{R})}$ by the formula

$$\mathcal{M} = \mathcal{G}\big(\mathcal{R}K_{\omega^{(\mathsf{R})}}\big)\,. \tag{12.3.5.b}$$

Here $\mathcal{R}$ is the coordinate reversal operator defined in equation (8.4.17.b). That $\mathcal{M}$ is indeed a mote is elementary, and Proposition 11.12 is now an elementary consequence of the above Proposition.

It is worth reiterating that a different approximating sequence is obtained for each mote, but the result is independent of this choice, providing the flexibility to simplify calculations by an appropriate choice of mote.

12.3.1 *Examples*

Using the method of motes requires a knowledge of the integral kernel K_B of $B \in \mathcal{L}(\mathcal{S}(\mathbb{R}), \mathcal{S}'(\mathbb{R}))$, but even if K_B is known explicitly, it cannot be expected that the method will always give a closed form for $\mathbf{\Delta}^{-1}[B]$. Nor will any other method, since most operators are not going to have symbols which are simply expressed in closed form. But there are examples for which a closed form can be found. Sometimes these are wholly contrived (as in the second example), sometimes the result is useful (as in the third example), and sometimes the result may even be important (as in our last example).

[2] Note how the mote disappears in the limit.

Example 12.4 If $B \in \mathcal{T}_2(L^2(\mathbb{R}))$ then, for any mote $\mathcal{M}$, the unitarity of the map $\sqrt{2\pi}\mathcal{G}$ implies that

$$\begin{aligned}
[\boldsymbol{\Delta}^{-1}_{\mathcal{M};N}[B]](\vartheta) &= [\![K_B \,,\, \mathcal{M}_{\vartheta;N}]\!] \\
&= 2\pi[\![\mathcal{G}\,\overline{\overline{K_B}} \,,\, \tau_{-\vartheta}\mathcal{E}_N\mathcal{M}]\!] \\
&= 2\pi[\![\mathcal{G}\mathcal{R}K_B \,,\, \tau_{-\vartheta}\mathcal{E}_N\mathcal{M}]\!] \\
&= 2\pi\big(\mathcal{G}\mathcal{R}K_B \,*\, \mathcal{E}_N\mathcal{P}\mathcal{M}\big)(\vartheta)\,,
\end{aligned}$$

where $\mathcal{P}$ is the parity operator on $L^2(\mathbb{R}^2)$, so $(\mathcal{P}\mathcal{M})(x,y) = \mathcal{M}(-x,-y)$. It is clear that all of these functions belong to $\mathcal{S}(\Pi)$, and moreover, letting $N \to \infty$,

$$\boldsymbol{\Delta}^{-1}[B] = 2\pi\mathcal{G}(\mathcal{R}K_B) = 2\pi\mathcal{G}\,\overline{\overline{K_B}}\,, \tag{12.3.6}$$

which result provides further confirmation of the veracity of (8.4.17.a) concerning the Weyl dequantization of Hilbert-Schmidt operators.

The manner in which this Example is formulated enables the proof of an important result. The expression *Wigner function* is found in the physics literature as a synonym for the symbol of a state (a positive normalized density matrix). An explicit connection can now be made between the present formalism and Wigner functions.

Proposition 12.5 ***(Wigner functions)*** *The distribution T is a member of $\mathcal{S}(\Pi)$ if and only if $\boldsymbol{\Delta}[T] \in \mathfrak{A}_*$. Thus any Wigner function of a smooth state is a test function, and any test function (satisfying the appropriate positivity and normalization conditions) is the Wigner function of a state.*

Proof: By equation (12.3.6), a distribution T belongs to $\mathcal{S}(\Pi)$ if and only if its integral kernel belongs to $\mathcal{S}(\mathbb{R}^2)$. By Proposition 8.27, this is equivalent to $\boldsymbol{\Delta}[T]$ being a smooth density matrix. ■

Consider again equation (8.5.1), in which the expectation of $\boldsymbol{\Delta}[T]$ in the state determined by the density matrix ρ can be rewritten as the pairing of T and $\mathcal{G}(\mathcal{R}K_\rho)$. Since ρ is a smooth observable, $\boldsymbol{\Delta}^{-1}[\rho]$ exists and belongs to $\mathcal{S}(\Pi)$. Thus we can extend equation (8.5.1) to obtain the intriguing formula

$$\mathrm{Tr}\,(\rho\,\boldsymbol{\Delta}[T]) = \frac{1}{2\pi}[\![T \,,\, \boldsymbol{\Delta}^{-1}[\rho]]\!]\,, \tag{12.3.7}$$

which, formally at least, has been known since Moyal's 1949 paper [169]. This formula is interesting for two reasons. The first is that it shows that expectation values can be calculated as integrals in phase space. Or, speaking metaphorically, they can be calculated classically — this was an original motive of Wigner for considering the possibility of Weyl dequantization. The second is a consequence of the fact that ρ is an observable, and so must be measurable (in any state). Considerations such as these have led Raymer and his co-workers to embark on a significant programme to measure the state of the electromagnetic field of coherent laser light [185].

Example 12.6 Our next example is a rather singular operator of no particular physical interest, but which can be dequantized in closed form with almost no effort. We discuss it, since it illustrates the meanings of the mappings, and it also demonstrates the ability of this technique to deal with general distributions.

Consider the mapping $B = \delta(Q) \in \mathcal{L}(\mathcal{S}(\mathbb{R}), \mathcal{S}'(\mathbb{R}))$, so that

$$Bf = f(0)\,\delta, \qquad\qquad f \in \mathcal{S}(\mathbb{R}),$$

which has the integral kernel $K_B = \delta \otimes \delta$. Direct calculation shows that for any mote $\mathcal{M}$,

$$\left(\mathbf{\Delta}_{\mathcal{M};N}^{-1}[B]\right)(\vartheta) = \sqrt{N}\mathcal{M}_1(\sqrt{N}\,q),$$

where $\sqrt{2}\vartheta = q - ip$ and $\mathcal{M}_1 \in \mathcal{S}(\mathbb{R})$ is the function

$$\mathcal{M}_1(v) = \int_{\mathbb{R}} \mathcal{M}(u, -v)\,du.$$

Standard analysis then shows us that

$$\mathbf{\Delta}^{-1}[B] = \mathrm{i} \otimes \delta,$$

so that $\mathbf{\Delta}^{-1}[B](p,q) = \delta(q)$, a result which can be readily checked. This Example extends to a particular distribution the result of Proposition 8.31 concerning the quantization of Q-marginals. Note that this calculation, like the previous one, does not require the choice of a specific mote.

Example 12.7 A slightly less artificial problem is to determine the dequantization of the Weyl group itself, since doing so begins to develop the techniques necessary to use this method. Of course, this is a result whose answer is already known, being fundamental to the development of Weyl quantization — see Proposition 8.32.

The integral kernel of the Weyl group element $W(a,b)$ for fixed a and b can be determined from the action of $W(a,b)$ on functions, and is given by

$$[\![\, K_{W(a,b)}\,,\, F\,]\!] \;=\; e^{\frac{1}{2}iab}\int_{\mathbb{R}} e^{ibx} F(x, x+a)\,dx\,, \qquad F \in \mathcal{S}(\mathbb{R}^2)\,, \quad (12.3.8.a)$$

or, symbolically,

$$K_{W(a,b)}(x,y) = e^{\frac{1}{2}iab} e^{ibx} \delta(y-x-a)\,. \qquad (12.3.8.b)$$

This result is evidently of independent interest. For a mote $\mathcal{M}$, it implies that

$$\begin{aligned}\left[\mathbf{\Delta}^{-1}_{\mathcal{M};N}\,[W(a,b)]\right](\vartheta) &= [\![\, K_{W(a,b)}\,,\, \mathcal{M}_{\vartheta;N}\,]\!] \\ &= e^{\frac{1}{2}iab}\int_{\mathbb{R}} e^{ibx}\,[\mathcal{M}_{\vartheta;N}](x,x+a)\,dx \\ &= 2\pi\, e^{i(ap+bq)}\,\big(\mathcal{E}^{\mathrm{tr}}_N \mathcal{F}^{-1}\mathcal{M}\big)(a,b)\,,\end{aligned}$$

where, as usual, $\sqrt{2}\vartheta = q - ip$. Then

$$\lim_{N\to\infty}\left[\mathbf{\Delta}^{-1}_{\mathcal{M};N}[W(a,b)]\right](\vartheta) \;=\; 2\pi\, e^{i(ap+bq)}\,\left[\mathcal{F}^{-1}\mathcal{M}\right](0,0)\,,$$

and so, using the normalization of the mote $\mathcal{M}$,

$$\left[\mathbf{\Delta}^{-1}\,[\,W(a,b)\,]\right](p,q) \;=\; e^{i(ap+bq)} \;=\; E_{a,b}(p,q)\,.$$

Example 12.8 In equation (10.3.63), an expression for the kernel of $\mathbf{\Delta}\,[\,\varphi\,]$ was given in a form that can be written

$$\mathbf{\Delta}\,[\,\varphi\,] \;=\; \frac{1}{2}\pi\,\mathrm{sgn}(Q) \;-\; \frac{1}{2}i\,S\,, \qquad (12.3.9)$$

defining $S \in \mathcal{L}(\mathcal{S}(\mathbb{R}), \mathcal{S}'(\mathbb{R}))$. From (10.3.63), the integral kernel of S is given by

$$[\![\, K_S\,,\, F\,]\!] \;=\; \lim_{L\to\infty}\iint_{\mathbb{R}^2} \mathrm{sgn}(y)\, e^{-|xy|}\, g_{I(L)}(x)\, F\big(y + \tfrac{1}{2}x, y - \tfrac{1}{2}x\big)\,dx\,dy \qquad (12.3.10)$$

for $F \in \mathcal{S}(\Pi)$, where $g_{I(L)}$ is the cut-off function introduced in equation (9.4.22).

To use the method of motes to dequantize equation (12.3.9), it is most convenient to use a particular mote, namely the Gaussian function

$$G(p,q) \;=\; \frac{1}{\pi}e^{-p^2-q^2}\,. \tag{12.3.11}$$

The mote sequence associated with G at the point $\vartheta \in \mathbb{C}$ can be shown to be

$$G_{\vartheta;N}(x,y) \;=\; \sqrt{\frac{N}{\pi}}\exp\Big[-\tfrac{1}{4}N(x+y-2q)^2 - \frac{1}{4N}(x-y)^2 - ip(x-y)\Big]\,, \tag{12.3.12}$$

from which it follows that $\big(\mathbf{\Delta}_{G;N}^{-1}\,[S]\big)(p,q)$ is equal to

$$\lim_{L\to\infty}\sqrt{\frac{N}{\pi}}\iint_{\mathbb{R}^2}\mathrm{sgn}(y)e^{-|xy|}g_{I(L)}(x)\exp\Big[-N(y-q)^2 - \frac{1}{4N}x^2 - ipx\Big]\,dxdy.$$

To complete the calculation from this point is not easy, and requires a fairly indirect approach. To begin with, the integral in this last expression can be rewritten as

$$-2i\int_{L^{-1}}^{L}\frac{\sin px}{x}\Big\{e^{-qx}\int_{-q+\frac{1}{2N}x}^{\infty}e^{-Ny^2}\,dy \;-\; e^{qx}\int_{q+\frac{1}{2N}x}^{\infty}e^{-Ny^2}\,dy\Big\}\,dx.$$

Applying the Dominated Convergence Theorem, this yields

$$-2i\sqrt{\frac{N}{\pi}}\int_0^{\infty}\frac{\sin px}{x}\Big\{e^{-qx}\int_{-q+\frac{1}{2N}x}^{\infty}e^{-Ny^2}\,dy - e^{qx}\int_{q+\frac{1}{2N}x}^{\infty}e^{-Ny^2}\,dy\Big\}dx,$$

in the limit as $L\to\infty$. Now the function $\mathbf{\Delta}_{G;N}^{-1}[S]$ is infinitely differentiable, and so it is legitimate to calculate its derivatives by differentiating the above formula inside the integral sign. In particular,

$$\Big[\frac{\partial}{\partial p}\mathbf{\Delta}_{G;N}^{-1}[S]\Big](p,q) \;=\;$$
$$-2i\sqrt{\frac{N}{\pi}}\int_0^{\infty}\cos px\Big\{e^{-qx}\int_{-q+\frac{1}{2N}x}^{\infty}e^{-Ny^2}\,dy - e^{qx}\int_{q+\frac{1}{2N}x}^{\infty}e^{-Ny^2}\,dy\Big\}\,dx$$

when $q \neq 0$, and after some lengthy (but elementary) manipulations, this can be rewritten as

$$\Big[\frac{\partial}{\partial p}\mathbf{\Delta}_{G;N}^{-1}[S]\Big](p,q) \;=\;$$
$$-2i\,\frac{q}{p^2+q^2} \;+\; 4i\sqrt{\frac{N}{\pi}}\frac{q}{p^2+q^2}e^{-Nq^2}\int_0^{\infty}e^{-Ny^2}\cos(2Npy)\,dy\,.$$

Since

$$\frac{\partial \varphi}{\partial p}(p,q) = -\frac{q}{p^2+q^2}, \qquad q \neq 0,$$

substituting this derivative of φ into the above equation yields the inequality

$$\begin{aligned}\left| \left[\frac{\partial}{\partial p} \left(\mathbf{\Delta}_{G;N}^{-1}[S] - 2i\varphi \right) \right](p,q) \right| &\leqslant 4\sqrt{\frac{N}{\pi}} \frac{|q|}{p^2+q^2}\, e^{-Nq^2} \int_0^\infty e^{-Ny^2}\, dy \\ &= \frac{2|q|}{p^2+q^2} e^{-Nq^2}.\end{aligned}$$

Since $\left(\mathbf{\Delta}_{G;N}^{-1}[S]\right)(0,q) = 0$ and $\varphi(0,q) = \frac{1}{2}\pi\,\mathrm{sgn}(q)$ it follows that

$$\begin{aligned}\left| \left(\mathbf{\Delta}_{G;N}^{-1}[S]\right)(p,q) + i\pi\,\mathrm{sgn}(q) - 2i\varphi(p,q) \right| &\leqslant 2|q| e^{-Nq^2} \int_0^{|p|} \frac{dt}{t^2+q^2} \\ &\leqslant \pi e^{-Nq^2}\end{aligned}$$

for all $p \in \mathbb{R}$ and $q \neq 0$. Letting N tend to infinity, we obtain

$$\lim_{N\to\infty} \left(\mathbf{\Delta}_{G;N}^{-1}[S]\right)(p,q) = 2i\varphi(p,q) - i\pi \mathrm{sgn}(q), \qquad p \in \mathbb{R},\ q \neq 0.$$

Now $\left(\mathbf{\Delta}_{G;N}^{-1}[S]\right)(p,0) = 0$ for all $p \in \mathbb{R}$ and $N \in \mathbb{N}$, and hence the above identity is valid for all points $(p,q) \in \Pi$ away from the cut along the negative p-axis. Hence, as distributions in $\mathcal{S}'(\Pi)$,

$$\mathbf{\Delta}^{-1}[S] = 2i\varphi - i\pi\, \mathrm{i} \otimes \mathrm{sgn},$$

which is the expected result.

12.4 Dequantization From Matrix Elements

It is clear from the above calculations that the method of motes is chiefly useful in circumstances where the precise form of the integral kernel of an element of $B \in \mathcal{L}(\mathcal{S}(\mathbb{R}), \mathcal{S}'(\mathbb{R}))$ is known. In some cases, however, the kernel of B may not be known, or may be too complicated to use in the mote method, while its matrix coefficients $[\![Bh_m, h_n]\!]$ with respect to the Hermite-Gauss functions are known for all $m, n \geqslant 0$. In such cases it is possible to obtain an expression for the dequantization of B in terms of a very remarkable family of functions, known as the special Hermite functions (although they are actually generalized Laguerre functions). These functions are often encountered in problems in theoretical quantum optics, phase theory, quantization or similar fields. Following the lead of the harmonic analysts, we are going to develop some of their important properties in a systematic fashion. As will be seen, considering these functions will uncover a remarkable relation between Weyl quantization and the topological structure of $\mathcal{S}'(\Pi)$.

12.4.1 *Special Hermite Functions*

The special Hermite functions are defined by the formula

$$\phi_{m,n} \;=\; 2\pi\, \mathcal{G}(h_m \otimes h_n)\,. \qquad m,\, n \geqslant 0, \tag{12.4.1}$$

Note that $\phi_{m,n} \in \mathcal{S}(\Pi)$ for all $m,\, n \geqslant 0$. This definition of the special Hermite functions differs from that of Folland [63] and Thangavelu [220] because our choices of scaling and normalization in the Wigner transform are different to theirs. Recall that the diagonal functions $\phi_{n,n}$ were first discussed in Proposition 9.8 in connection with radial functions.

Before considering the quantization properties of these functions, the following Proposition illustrates their connection to topological properties of $\mathcal{S}'(\Pi)$. For proofs of these results, the reader should refer to [220] and [109].

Proposition 12.9 *The set* $\{\phi_{m,n} \;:\; m,n \geqslant 0\}$ *is a Schauder basis for* $\mathcal{S}(\Pi)$*, and the sum*

$$\sum_{m,n\geqslant 0} \xi_{m,n}\, \phi_{m,n} \tag{12.4.2.a}$$

converges to an element of $\mathcal{S}(\Pi)$ *if and only if the sequence* $\xi = \big(\xi_{m,n}\big)_{m,n\geqslant 0}$ *is rapidly decreasing, in the sense that*

$$\sup_{m,n\geqslant 0}\; (m+1)^r (n+1)^s\; |\,\xi_{m,n}\,| \;<\; \infty\,, \qquad r, s \geqslant 0\,; \tag{12.4.2.b}$$

in other words, it is required that $\xi \in \mathfrak{s}^{(2)}$*, the space of rapidly decreasing complex one-sided sequences in two indices. If* $\mathfrak{s}^{(2)}$ *is equipped with its usual locally convex Fréchet topology, then this identification between elements of* $\mathcal{S}(\Pi)$ *and elements of* $\mathfrak{s}^{(2)}$ *is a topological isomorphism.*

The special Hermite functions satisfy the conjugation identity,

$$\overline{\phi_{m,n}} \;=\; \phi_{n,m}\,, \qquad m,n \geqslant 0\,, \tag{12.4.3.a}$$

and form an orthogonal collection in $\mathcal{S}(\Pi)$, *with*

$$\langle \phi_{j,k}\,,\, \phi_{m,n}\rangle \;=\; 2\pi\,\delta_{jm}\,\delta_{kn}\,, \qquad j,k,m,n \geqslant 0\,. \tag{12.4.3.b}$$

Regarding $\mathcal{S}(\Pi)$ *as a subspace of* $\mathcal{S}'(\Pi)$, *these identities imply*[3] *that*

$$\frac{1}{2\pi}[\![\, \phi_{k,j}\,,\, \phi_{m,n}]\!] \;=\; \delta_{jm}\,\delta_{kn}\,, \qquad j,k,m,n \geqslant 0\,, \tag{12.4.4}$$

and hence that $\{(2\pi)^{-1}\,\phi_{n,m} \;:\; m,n \geqslant 0\}$ *is a Schauder basis for* $\mathcal{S}'(\Pi)$, *dual to the basis* $\{\phi_{m,n} \;:\; m,n \geqslant 0\}$ *for* $\mathcal{S}(\Pi)$. *Moreover, the series*

$$\sum_{m,n\geqslant 0} \tau_{m,n}\,\phi_{n,m}$$

converges to an element of $\mathcal{S}'(\Pi)$ *if and only if* $(\tau_{m,n})_{m,n\geqslant 0}$ *belongs to the sequence space* $(\mathfrak{s}^{(2)})'$ *of polynomially bounded double sequences, which is dual to* $\mathfrak{s}^{(2)}$, *in that*

$$|\,\tau_{m,n}\,| \;\leqslant\; C(m+1)^r(n+1)^s\,, \qquad\qquad m,n \geqslant 0\,,$$

for some constant $C > 0$ *and some integers* $r,s \geqslant 0$.

Proof: Equations (12.4.3.a) and (12.4.3.b) are elementary consequences of the properties of the Wigner transform $\mathcal{G}$ and the orthonormality of the basis $\{h_n \;:\; n \geqslant 0\}$ for $\mathcal{S}(\mathbb{R})$. That $\{\phi_{m,n} \;:\; m,n \geqslant 0\}$ is a Schauder basis for $\mathcal{S}(\Pi)$ follows from the fact that the set $\{h_m \otimes h_n \;:\; m,n \geqslant 0\}$ is a Schauder basis for $\mathcal{S}(\mathbb{R}^2)$, and that the Wigner transform $\mathcal{G}$ is a bicontinuous linear bijection from $\mathcal{S}(\mathbb{R}^2)$ to $\mathcal{S}(\Pi)$. Moreover, since the implication of Proposition 4.24 is that the Schauder basis $\{h_m \otimes h_n \;:\; m,n \geqslant 0\}$ for $\mathcal{S}(\mathbb{R}^2)$ produces a topological isomorphism between $\mathcal{S}(\mathbb{R}^2)$ and $\mathfrak{s}^{(2)}$ of the above sort, the required topological identification between $\mathcal{S}(\Pi)$ and $\mathfrak{s}^{(2)}$ is now immediate, as is the identification between $\mathcal{S}'(\Pi)$ and the dual sequence space $(\mathfrak{s}^{(2)})'$. ∎

The rationale for introducing the special Hermite functions in this context is to be found in the following result. Although, as we have mentioned, our explicit formulation of the special Hermite functions is different to that found in Folland [63] and Thangavelu [220], the formula here derived is not.

[3] Note the reversal of index order.

Proposition 12.10 *For any $m, n \geqslant 0$, the mapping $\boldsymbol{\Delta}\left[\,\phi_{m,n}\,\right]$, which belongs to $\mathcal{L}(\mathcal{S}(\mathbb{R}), \mathcal{S}'(\mathbb{R}))$, is the bounded operator*

$$\boldsymbol{\Delta}\left[\,\phi_{m,n}\,\right] = |\, h_n\,\rangle\langle\, h_m\,|. \tag{12.4.5}$$

Proof: Since $\phi_{m,n} = \Xi_{h_n,h_m}$ for any $m, n \geqslant 0$, Proposition 8.16 establishes this result immediately. ■

12.4.2 *The Generating Function*

The generating function G_s of the Hermite-Gauss functions $\{\, h_n \,:\, n \geqslant 0\,\}$ has been seen to be of considerable utility. It will be similarly useful to have available to us a generating function for the special Hermite functions, and we shall define this function here. Since the special Hermite functions are indexed by a pair of integers, it is clear that the generating function for them will be indexed by a pair of variables.

Although not much was made of it at the time, this generating function of the special Hermite functions first made its appearance in equation (9.2.2) of Chapter 9, where it was used to calculate the values of the angular quantization matrix elements $g_{m,n}$.

In view of the connections between the special Hermite functions, the Hermite-Gauss functions and the Wigner transform, we are led to make the following definition. Consider the function $\mathcal{P}_{s,t}$ defined for any real numbers s, t by the formula

$$\mathcal{P}_{s,t}(p,q) \;=\; 2\pi\, \mathcal{G}(G_s \otimes G_t). \tag{12.4.6.a}$$

By direct calculation it can be established that

$$\mathcal{P}_{s,t}(p,q) \;=\; 2\,\exp\Big\{-p^2 - q^2 + is(p - iq) - it(p + iq) - \tfrac{1}{2}st\Big\}. \tag{12.4.6.b}$$

The next Proposition uses $\mathcal{P}_{s,t}$ to give an explicit expression for the special Hermite functions as Laguerre functions.

Proposition 12.11 *The function $\mathcal{P}_{s,t}$ is the generator of the special Hermite functions $\phi_{m,n}$ in accordance with the formula*

$$\mathcal{P}_{s,t}(p,q) \;=\; \sum_{m,n\geqslant 0} \frac{s^m t^n}{\sqrt{2^{m+n}\, m!\, n!}}\,\phi_{m,n}(p,q), \qquad s,\ t, \in \mathbb{R}. \tag{12.4.7}$$

By equating coefficients of $s^m t^n$ for all m, $n \geqslant 0$, the special Hermite functions can be identified as

$$\begin{aligned}\Phi_{m,n}(p,q) &= (-1)^{\min(m,n)} i^{m-n} 2^{1+\frac{1}{2}|m-n|} \sqrt{\frac{\min(m,n)!}{\max(m,n)!}} \\ &\quad \times e^{-r^2} r^{|m-n|} e^{i(n-m)\beta} L^{(|m-n|)}_{\min(m,n)}(2r^2) \qquad (12.4.8)\end{aligned}$$

for all m, $n \geqslant 0$, where $L^{(\alpha)}_m(x)$ denotes the usual generalized Laguerre polynomial, and $p + iq = re^{i\beta}$.

Referring back to the determination of the $g_{m,n}$ in Chapter 9, it can now be seen that they are the radial averages of the special Hermite functions, a result which goes some of the way to explaining their appearance in angular quantization theory. Equation (9.4.5.b) in particular may be given the following interpretation.

Proposition 12.12 *Recall the continuous linear map $\mathcal{A} : \mathcal{S}(\Pi) \to \mathcal{C}^\infty(\mathbb{T})$ given by*

$$[\mathcal{A}F](e^{i\beta}) = \int_0^\infty F(r\cos\beta, r\sin\beta)\, r\, dr, \qquad F \in \mathcal{S}(\Pi), \quad (9.4.5.a)$$

giving the radial average of a test function. Up to a phase factor, its action on the special Hermite functions yields the angular quantization coefficients $g_{m,n}$ for all m, $n \geqslant 0$:

$$[\mathcal{A}\Phi_{m,n}](e^{i\beta}) = i^{m-n}\, e^{i(n-m)\beta}\, g_{m,n}. \qquad (12.4.9)$$

Recall that in Proposition 9.8 we stated that the set $\{\Phi_{m,m} : m \geqslant 0\}$ was a Schauder basis for the space $\mathcal{S}_{\text{rad}}(\Pi)$ of radial test functions. We may now prove this result. In equation (9.3.1.a) we used the action of the rotation group SO(2) on $\mathcal{S}(\Pi)$ to define a continuous projection Σ from $\mathcal{S}(\Pi)$ to $\mathcal{S}_{\text{rad}}(\Pi)$. Since

$$\Sigma\, \Phi_{m,n} = \delta_{mn}\, \Phi_{m,n}, \qquad m, n \geqslant 0, \qquad (12.4.10)$$

it is clear that $\mathcal{S}_{\text{rad}}(\Pi)$ is spanned by the diagonal special Hermite functions. The results of Proposition 9.8 are now immediate.

Additionally, in Chapter 9 we claimed that the map $S \mapsto S_{\text{ang}}$ from $\mathcal{D}(\mathbb{T})$ to $\mathcal{S}'_{\text{ang}}(\Pi)$ was bijective. To see this (recalling the notation of that

Chapter), define the continuous linear map $\mathcal{J} : \mathcal{C}^\infty(\mathbb{T}) \to \mathcal{S}(\Pi)$

$$\mathcal{J}\chi_n = \begin{cases} i^n\, g_{0,n}^{-1}\, \phi_{0,n}\,, & n \geqslant 0\,, \\ i^n\, g_{-n,0}^{-1}\, \phi_{-n,0}\,, & n \leqslant 0\,. \end{cases} \tag{12.4.11}$$

Then it is clear that $\mathcal{A}\mathcal{J}\omega = \omega$ for all $\omega \in \mathcal{C}^\infty(\mathbb{T})$, and hence $\mathcal{J}^{\mathrm{tr}}$ from $\mathcal{S}'(\Pi)$ to $\mathcal{D}(\mathbb{T})$ is a linear map such that

$$[\![\mathcal{J}^{\mathrm{tr}} S_{\mathrm{ang}}\,, \omega]\!] = [\![S_{\mathrm{ang}}\,, \mathcal{J}\omega]\!] = [\![S\,, \mathcal{A}\mathcal{J}\omega]\!] = [\![S\,, \omega]\!] \tag{12.4.12}$$

for all $S \in \mathcal{D}(\mathbb{T})$ and $\omega \in \mathcal{C}^\infty(\mathbb{T})$. Hence we see that $\mathcal{J}^{\mathrm{tr}} S_{\mathrm{ang}} = S$ for all $S \in \mathcal{D}(\mathbb{T})$, and so the map $S \mapsto S_{\mathrm{ang}}$ from $\mathcal{D}(\mathbb{T})$ to $\mathcal{S}'_{\mathrm{ang}}(\Pi)$ is injective. On the other hand, if $T \in \mathcal{S}'_{\mathrm{ang}}(\Pi)$ is an angular distribution, then

$$\begin{aligned} g_{0,k}\, [\![T\,, \phi_{m,m+k}]\!] &= g_{m,m+k}\, [\![T\,, \phi_{0,k}]\!]\,, \\ g_{k,0}\, [\![T\,, \phi_{m+k,m}]\!] &= g_{m+k,m}\, [\![T\,, \phi_{k,0}]\!]\,, \end{aligned} \tag{12.4.13}$$

for all $m, k \geqslant 0$. This implies, if $S = \mathcal{J}^{\mathrm{tr}} T \in \mathcal{D}(\mathbb{T})$, that

$$[\![T\,, \phi_{m,n}]\!] = [\![S_{\mathrm{ang}}\,, \phi_{m,n}]\!]\,, \qquad m, n \geqslant 0\,, \tag{12.4.14}$$

and thus that $T = S_{\mathrm{ang}}$. Hence the map $S \mapsto S_{\mathrm{ang}}$ is bijective, as required.

12.4.3 *Differential Relations*

Special functions in mathematical physics are typically the solutions of second order ordinary differential equations. Also typically, many of these operators can be factorized into products of first order operators which act as generalized raising and lowering operators. Moreover, this structure is connected to infinite dimensional representations of certain Lie algebras and groups. A variant of this is true for the special Hermite functions, which satisfy two second order partial differential equations in two independent variables, and have two raising and two lowering operators which are independent of each other.

Definition 12.13 *The lowering and raising operators for the first index of the special Hermite functions are given by the formulæ*

$$\begin{aligned} L^{(-)} &= \tfrac{1}{2}\Big(\frac{\partial}{\partial p} + i\frac{\partial}{\partial q}\Big) + (p + iq), \\ L^{(+)} &= \tfrac{1}{2}\Big(\frac{\partial}{\partial p} - i\frac{\partial}{\partial q}\Big) - (p - iq), \end{aligned} \tag{12.4.15}$$

while the lowering and raising operators for the second index are given by the formulæ

$$R^{(-)} = \frac{1}{2}\Big(\frac{\partial}{\partial p} - i\frac{\partial}{\partial q}\Big) + (p - iq),$$
$$R^{(+)} = \frac{1}{2}\Big(\frac{\partial}{\partial p} + i\frac{\partial}{\partial q}\Big) - (p + iq). \tag{12.4.16}$$

The names of these operators have been chosen for the following reason:

Proposition 12.14 *The actions of the lowering and raising operators for the first index on the functions* $\phi_{m,n}$ *are*

$$L^{(-)}\phi_{m,n} = i\sqrt{2m}\,\phi_{m-1,n},\qquad m,\, n \geqslant 0,$$
$$L^{(+)}\phi_{m,n} = i\sqrt{2m+2}\,\phi_{m+1,n}, \tag{12.4.17}$$

while the actions of the lowering and raising operators for the second index on the functions $\phi_{m,n}$ *are*

$$R^{(-)}\phi_{m,n} = -i\sqrt{2n}\,\phi_{m,n-1},\qquad m,\, n \geqslant 0.$$
$$R^{(+)}\phi_{m,n} = -i\sqrt{2n+2}\,\phi_{m,n+1}. \tag{12.4.18}$$

The four differential operators $L^{(+)}$, $L^{(-)}$, $R^{(+)}$ and $R^{(-)}$ are endomorphisms of $\mathcal{S}(\Pi)$. Defining the linear combinations

$$Q_1 = \frac{1}{2}\big(L^{(+)} + L^{(-)}\big), \qquad Q_2 = \frac{1}{2}\big(R^{(+)} + R^{(-)}\big),$$
$$P_1 = \frac{1}{2}i\big(L^{(+)} - L^{(-)}\big), \qquad P_2 = \frac{1}{2}i\big(R^{(+)} - R^{(-)}\big), \tag{12.4.19.a}$$

these latter endomorphisms of $\mathcal{S}(\Pi)$ satisfy the canonical commutation relations

$$\big[\,Q_j\,,\,Q_k\,\big] = 0, \qquad \big[\,P_j\,,\,P_k\,\big] = 0, \qquad \big[\,Q_j\,,\,P_k\,\big] = i\,\delta_{jk}\,I \tag{12.4.19.b}$$

for $1 \leqslant j,\, k \leqslant 2$, where I is, of course, the identity operator on $\mathcal{S}(\Pi)$.

In Section 8.7, the Heisenberg group $\mathfrak{H}$ and its Lie algebra $\mathfrak{h}$ were discussed, and it was observed that certain representations of $\mathfrak{H}$ were related to representations of the CCR. Being then primarily interested in systems with one degree of freedom, only the Heisenberg group and algebra for one degree of freedom were considered there. But there is an obvious generalization to systems with n degrees of freedom for any $n \in \mathbb{N}$, in which the corresponding group and algebra are to be denoted $\mathfrak{H}_{n+1}$ and $\mathfrak{h}_{n+1}$ respectively. The group $\mathfrak{H}$ and the algebra $\mathfrak{h}$ of Section 8.7 are then $\mathfrak{H}_2$ and $\mathfrak{h}_2$ respectively. Equations (12.4.19.b) can be interpreted as showing that the differential operators for the special Hermite functions, together with the identity operator I, provide a representation of the Heisenberg Lie algebra $\mathfrak{h}_3$ for two degrees of freedom.

The lowering and raising operators for the two indices can be used to define associated second order elliptic differential operators as follows,

$$\begin{aligned} H_L &= -\frac{1}{2}\{ L^{(+)} , L^{(-)} \}_+ = -\frac{1}{2}(L^{(+)}L^{(-)} + L^{(-)}L^{(+)}), \\ H_R &= -\frac{1}{2}\{ R^{(+)} , R^{(-)} \}_+ = -\frac{1}{2}(R^{(+)}R^{(-)} + R^{(-)}R^{(+)}). \end{aligned} \tag{12.4.20}$$

Writing these operators out in polar coordinates yields

$$\begin{aligned} H_L &= -\frac{1}{4}\nabla^2 + r^2 + i\frac{\partial}{\partial\beta}, \\ H_R &= -\frac{1}{4}\nabla^2 + r^2 - i\frac{\partial}{\partial\beta}, \end{aligned} \tag{12.4.21}$$

where ∇^2 is the Laplacian on phase space and (r, β) are polar coordinates, so that $p + iq = r\, e^{i\beta}$.

Proposition 12.15 *The special Hermite functions satisfy the two sets of partial differential equations*

$$\begin{aligned} H_L\phi_{m,n} &= (2m+1)\phi_{m,n}, \\ H_R\phi_{m,n} &= (2n+1)\phi_{m,n}, \end{aligned} \qquad m,\, n \geqslant 0, \tag{12.4.22}$$

which show them to be eigenfunctions of H_L and H_R.

12.4.4 *The Dequantization Formula*

While the special Hermite functions are fascinating in themselves, it is their connection with dequantization which is of interest here. Suppose we are given a mapping $B \in \mathcal{L}(\mathcal{S}(\mathbb{R}), \mathcal{S}'(\mathbb{R}))$. Since the special Hermite functions form a Schauder basis for $\mathcal{S}'(\Pi)$ and $\mathbf{\Delta}^{-1}[B]$ belongs to that space, it must be possible to write $\mathbf{\Delta}^{-1}[B]$ as a series expansion with respect to this basis. The next result shows how the coefficients of this expansion may be determined.

Theorem 12.16 *For any $B \in \mathcal{L}(\mathcal{S}(\mathbb{R}), \mathcal{S}'(\mathbb{R}))$, its dequantization symbol $\boldsymbol{\Delta}^{-1}[B] \in \mathcal{S}'(\Pi)$ is given by the series expansion*

$$\boldsymbol{\Delta}^{-1}[B] = \sum_{m,n \geqslant 0} [\![Bh_m , h_n]\!] \, \phi_{m,n}, \tag{12.4.23}$$

which converges in the topology on $\mathcal{S}'(\Pi)$.

Proof: We know that

$$\boldsymbol{\Delta}^{-1}[B] = \sum_{m,n \geqslant 0} t_{m,n} \, \phi_{m,n} ,$$

where the double sequence $(t_{m,n})$ belongs to $(\mathfrak{s}^{(2)})'$. Moreover, the orthogonality of the special Hermite functions implies that

$$\begin{aligned} t_{m,n} &= \frac{1}{2\pi} [\![\boldsymbol{\Delta}^{-1}[B] , \phi_{n,m}]\!] \\ &= [\![\boldsymbol{\Delta}^{-1}[B] , \mathcal{G}(h_n \otimes h_m)]\!] = [\![Bh_m , h_n]\!] \end{aligned}$$

for any $m, n \geqslant 0$, as required. ■

There is nothing special from a mathematical point of view in the use of the Hermite-Gauss functions here. Taking any orthonormal Schauder basis for $\mathcal{S}(\mathbb{R})$ and using it to define functions in $\mathcal{S}(\Pi)$ by a procedure analogous to that found in equation (12.4.1), an orthogonal Schauder basis for $\mathcal{S}(\Pi)$ would be obtained. By duality, a basis for $\mathcal{S}'(\Pi)$ would result, with respect to which a result analogous to that found in Theorem 12.16 could be derived. However, it is unlikely to be as useful a characterization of elements of $\mathcal{S}(\Pi)$ and $\mathcal{S}'(\Pi)$ as the special Hermite functions give, since the Hermite-Gauss functions are naturally generated by the Schrödinger representation of the CCR. In addition, it is relatively easy to perform calculations with the Hermite-Gauss functions, so that we stand a fighting chance of being able to calculate the matrix coefficients for B explicitly, after which equation (12.4.23) is an infinite series we might be able to do business with.

Example 12.17 For illustrative purposes, suppose $B \in \mathcal{L}(\mathcal{S}(\mathbb{R}), \mathcal{S}'(\mathbb{R}))$ is a weighted shift operator

$$Bh_m = b_m h_{m+1}, \qquad m \geqslant 0, \tag{12.4.24}$$

where $(b_m)_{m\geqslant 0}$ is some sequence of constants. From the above Theorem,

$$\mathbf{\Delta}^{-1}[B] = \sum_{m\geqslant 0} b_m \Phi_{m,m+1},$$

and, in terms of the special Hermite functions, this reads

$$\mathbf{\Delta}^{-1}[B](p,q) = -i\,2^{\frac{3}{2}} r e^{-r^2} e^{i\beta} \sum_{m\geqslant 0} (-1)^m \frac{b_m}{\sqrt{m+1}} L_m^{(1)}(2r^2). \quad (12.4.25)$$

This series will have a closed form for certain "nice" sequences $(b_m)_{m\geqslant 0}$ but, in general, all we can say is that this series converges to a tempered distribution on Π. The interesting thing about this result is that it shows that the dequantization of a weighted shift is necessarily of the form $e^{i\beta}$ multiplied by a (generally nontrivial) radial distribution.

As is to be expected, equation (12.4.25) has a closed form when B is the raising operator A^+, in which case $b_m = \sqrt{m+1}$ for all $m \geqslant 0$. For then

$$\begin{aligned}\left[\mathbf{\Delta}^{-1}\left[A^+\right]\right](p,q) &= -i\,2^{\frac{3}{2}} r e^{-r^2} e^{i\beta} \sum_{m\geqslant 0} (-1)^m L_m^{(1)}(2r^2) \\ &= -\frac{i}{\sqrt{2}} r e^{i\beta} = \frac{1}{\sqrt{2}}(q - ip),\end{aligned}$$

which is the correct expression.

A more complicated example where equation (12.4.25) yields a closed form is given by

$$b_m = \frac{(-t)^m}{m!\sqrt{m+1}}, \qquad m \geqslant 0, \quad (12.4.26)$$

for some positive constant t. This example has been chosen to take advantage of another known identity concerning Laguerre functions, since

$$\begin{aligned}\left[\mathbf{\Delta}^{-1}[B]\right](p,q) &= -i\,2^{\frac{3}{2}}\, r e^{-r^2} e^{i\beta} \sum_{m\geqslant 0} \frac{t^m}{(m+1)!} L_m^{(1)}(2r^2) \\ &= -\frac{2i}{\sqrt{t}} e^t\, e^{-r^2} e^{i\beta} J_1\left(r\sqrt{8t}\right). \quad (12.4.27)\end{aligned}$$

However, whether this last example has any physical significance is moot.

12.5 Dequantization Of Toeplitz Operators

A characteristic feature of the quantization of angular distributions on phase space is the occurrence of the coefficients $g_{m,n}$ in the matrix elements with respect to the Hermite-Gauss functions. In a theoretical sense the deepest understanding of why they appear probably comes through Proposition 12.12. But it is their practical properties that are often the concern, since they are rather complicated, are difficult to handle analytically, and do not seem to have an obvious physical interpretation.

What happens if we try to drop these constants from our analysis? In other words, given some function $\omega \in L^\infty(\mathbb{T})$, rather than considering the observable $\mathbf{\Delta}[\omega_{\text{ang}}] \in \mathcal{L}(\mathcal{S}(\mathbb{R}), \mathcal{S}'(\mathbb{R}))$, where

$$[\![\mathbf{\Delta}[\omega_{\text{ang}}] h_n , h_m]\!] = i^{m-n} g_{m,n} \hat{\omega}_{m-n} , \qquad m, n \geqslant 0 , \qquad (9.4.9)$$

we choose to consider the Toeplitz operator $\overline{\mathcal{M}}(\omega) \in \mathbb{B}\big(L^2(\mathbb{R})\big)$ discussed in Chapter 10, where

$$\big\langle h_m , \overline{\mathcal{M}}(\omega) h_n \big\rangle = i^{m-n} \hat{\omega}_{m-n} , \qquad m, n \geqslant 0 . \qquad (10.3.10.c)$$

Since

$$\lim_{n\to\infty} g_{n+k,n} = 1, \qquad k \geqslant 0 ,$$

as was observed in the proof of Proposition 9.15, it is evident that (in some sense) $\overline{\mathcal{M}}(\omega)$ is an approximation to $\mathbf{\Delta}[\omega_{\text{ang}}]$, and hence results concerning one will help with the analysis of the other. Moreover, it should be remembered that the phase-related operators X, E, E^*, S and C derived from the London distribution are all Toeplitz operators of this form, and so the study of the Weyl dequantizations of such operators will provide us with insight into the amount by which these operators differ from being the Weyl quantizations of angular distributions. Moreover it is a hope (so far unfulfilled) that, since the spectral theory of Toeplitz operators is so completely known, an understanding of the relationship between $\overline{\mathcal{M}}(\omega)$ and $\mathbf{\Delta}[\omega_{\text{ang}}]$ will shed some light on the spectral properties of $\mathbf{\Delta}[\omega_{\text{ang}}]$.

One of the problems preventing progress in this direction is the surprising difficulty encountered when determining the dequantizations of Toeplitz operators. Indeed, the answers are not known in closed form even in simple cases. However, certain technical properties of these dequantizations can be isolated, and the remainder of this Chapter will be devoted to enumerating these.

We denote the dequantization of $\overline{\mathcal{M}}(\omega)$ by

$$\mathfrak{D}(\omega) = \mathbf{\Delta}^{-1}\left[\overline{\mathcal{M}}(\omega)\right]. \tag{12.5.1}$$

Applying the results of the preceding Section, we can obtain series expansions for ω_{ang} and for $\mathfrak{D}(\omega)$ in terms of the special Hermite functions.

Proposition 12.18 *For any $\omega \in L^\infty(\mathbb{T})$, the identities*

$$\omega_{\text{ang}} = \sum_{m,n\geqslant 0} i^{n-m}\, g_{m,n}\, \hat{\omega}_{n-m}\, \phi_{m,n} \tag{12.5.2.a}$$

and

$$\mathfrak{D}(\omega) = \sum_{m,n\geqslant 0} i^{n-m}\, \hat{\omega}_{n-m}\, \phi_{m,n}, \tag{12.5.2.b}$$

hold, with both series converging in $\mathcal{S}'(\Pi)$.

The above series expansion for $\mathfrak{D}(\omega)$ can be reordered to obtain an alternative expression which, at least formally, involves only a singly-infinite sum.

Proposition 12.19 *For any $\omega \in L^\infty(\mathbb{T})$, the distributional identity*

$$\left[\mathfrak{D}(\omega)\right](r\cos\beta, r\sin\beta) = \sum_{k\in\mathbb{Z}} \hat{\omega}_k\, e^{ik\beta}\, \mathfrak{F}_{|k|}(r), \tag{12.5.3}$$

holds where, for any $k \geqslant 0$, $\mathfrak{F}_k$ is the function defined by the formula

$$\begin{aligned} \mathfrak{F}_k(r) &= i^k e^{-ik\beta} \sum_{m\geqslant 0} \phi_{m,m+k}(r\cos\beta, r\sin\beta) \\ &= 2^{1+\frac{1}{2}k}\, r^k\, e^{-r^2} \sum_{m\geqslant 0} (-1)^m \sqrt{\frac{m!}{(m+k)!}}\, L_m^{(k)}(2r^2). \end{aligned} \tag{12.5.4}$$

Comparing this result with the distributional identity

$$\omega_{\text{ang}}(r\cos\beta, r\sin\beta) = \sum_{k\in\mathbb{Z}} \hat{\omega}_k\, e^{ik\beta},$$

it is clear that the radial dependence of the distribution $\mathfrak{D}(\omega)$ has been localized in the functions $\mathfrak{F}_k$.

Thus, in order to understand the distribution $\mathfrak{D}(\omega)$, we must study the functions $\mathfrak{F}_k(r)$. Aside from the elementary observation that $\mathfrak{F}_0(r) = 1$, very little is known about these functions. However,

Proposition 12.20 *Each $\mathfrak{F}_k$ is a smooth bounded function on $[0,\infty)$, and*

$$\lim_{r\to\infty} \mathfrak{F}_k(r) = 1\,, \tag{12.5.5}$$

for all $k \geqslant 0$.

A proof of this result, and of the subsequent results of this Section, can be found in the authors' paper [113].

It might seem at first sight that an integral representation for the Laguerre functions, or something similar, might be pushed hard enough to determine rather more about the functions $\mathfrak{F}_k$. After a certain amount of work, it can be shown that $\mathfrak{F}_1$ has the integral representation

$$\mathfrak{F}_1(r) = \frac{1}{\sqrt{2\pi}}\, r \int_0^\infty \exp\left[\, - r^2 \tanh\left(\tfrac{1}{2}\xi\right)\right] \operatorname{sech}^2\left(\tfrac{1}{2}\xi\right) \frac{d\xi}{\sqrt{\xi}}\,,$$

which is not going to yield any information easily. Expressions for higher-indexed functions $\mathfrak{F}_k$ can be obtained, but they are even more complicated.

Although simple concrete formulæ for the functions $\mathfrak{F}_k$ are not available, enough is known about these functions to be able to derive a number of results concerning the properties of the distributions $\mathfrak{D}(\omega)$. To begin with, we have the following result.

Proposition 12.21 *For any $\omega \in L^\infty(\mathbb{T})$, the distribution $\mathfrak{D}(\omega) \in \mathcal{S}'(\Pi)$ is a smooth function on Π.*

Moreover, the following statements concerning more detailed properties of the function $\mathfrak{D}(\omega)$ hold. These results are rather technical, and are conditional upon certain growth properties of the sequence of Fourier coefficients of the function ω, and parallel similar results for $\boldsymbol{\Delta}\,[\,\varphi\,]$.

Proposition 12.22 *If $\omega \in L^\infty(\mathbb{T})$, define the sequence $\omega(\alpha)$ to be equal to $\big(\,|\,k\,|^\alpha\,\hat{\omega}_k\big)_{k\in\mathbb{Z}}$. Then*

1. *if $\omega\big(9/16\big) \in \ell^1(\mathbb{Z})$, then $\mathfrak{D}(\omega)$ is a bounded function on Π,*
2. *if $\omega\big(5/8\big) \in \ell^2(\mathbb{Z})$, then $\omega_{\mathrm{ang}} - \mathfrak{D}(\omega)$ belongs to $L^2(\Pi)$, and so $\boldsymbol{\Delta}\,[\,\omega_{\mathrm{ang}}\,] - \overline{\mathcal{M}}(\omega)$ is a Hilbert-Schmidt operator on $L^2(\mathbb{R})$,*

3. *if* $\omega\big(11/16\big) \in \ell^1(\mathbb{Z})$*, then*

$$\lim_{r\to\infty} \big[\mathfrak{D}(\omega)\big](r\cos\beta, r\sin\beta) \;=\; \omega(e^{i\beta}) \tag{12.5.6}$$

uniformly in β*.*

Note that all of these conditions are strong enough to ensure that $\omega(0)$ belongs to $\ell^1(\mathbb{Z})$, and so that ω is a continuous function on $\mathbb{T}$. Hence none of these results are good enough to be applied to that particular function $\mathfrak{p}$ for which $\mathfrak{p}_{\text{ang}} = \varphi$. Thus none of these results gives any information concerning the difference between $\mathbf{\Delta}\left[\varphi\right]$ and X, although they do provide us with information concerning the difference between $\mathbf{\Delta}\left[e^{i\varphi}\right]$ and E^*, for example. However, these results are indicative, since they show that (for well-behaved functions ω), the angle function ω is the "limit at infinity" of the distribution $\mathfrak{D}(\omega)$, in that $\mathfrak{D}(\omega)$ converges to ω (in some sense) as the radius r tends to infinity.

Indeed, these heuristic considerations can be made precise by defining the continuous function $\mathfrak{D}^{(R)}(\omega) \in \mathcal{C}(\mathbb{T})$ for any $R > 0$ by setting

$$\big[\mathfrak{D}^{(R)}(\omega)\big](e^{i\beta}) \;=\; [\mathfrak{D}(\omega)](R\cos\beta, R\sin\beta). \tag{12.5.7}$$

As $\mathfrak{D}^{(R)}$ is a function of an angular variable, it can be used to construct the phase space angular distribution $\mathfrak{D}^{(R)}(\omega)_{\text{ang}}$. Admittedly this distribution is rather a long way from $\overline{\mathcal{M}}(\omega)$, but it has the virtue of satisfying the following limiting result:

Proposition 12.23 *For any* $\omega \in L^\infty(\mathbb{T})$*, the limit*

$$\lim_{R\to\infty} \mathfrak{D}^{(R)}(\omega)_{\text{ang}} \;=\; \omega_{\text{ang}}\,, \tag{12.5.8.a}$$

holds with respect to the weak topology on $\mathcal{S}'(\Pi)$*. Moreover, if the sequence* $\omega\big(11/16\big)$ *belongs to* $\ell^2(\mathbb{Z})$*, then*

$$\lim_{R\to\infty} \mathfrak{D}^{(R)}(\omega) \;=\; \omega\,, \tag{12.5.8.b}$$

where this convergence is with respect to the norm topology on $L^2(\mathbb{T})$*.*

Chapter 13

The Moyal Product

The truth is rarely pure, and never simple.
– Oscar Wilde, *The Importance Of Being Earnest.*

13.1 Introduction

It is surprising how many people believe that if two phase space functions F and G have Poisson bracket equal to unity,

$$\{F, G\} = \frac{\partial F}{\partial p}\frac{\partial G}{\partial q} - \frac{\partial F}{\partial q}\frac{\partial G}{\partial p} = 1, \qquad (13.1.1.a)$$

then their Weyl quantizations, $\mathbf{\Delta}[F]$ and $\mathbf{\Delta}[G]$ satisfy the commutator identity

$$[\mathbf{\Delta}[F], \mathbf{\Delta}[G]] = \mathbf{\Delta}[F]\mathbf{\Delta}[G] - \mathbf{\Delta}[G]\mathbf{\Delta}[F] = -iI. \qquad (13.1.1.b)$$

In the discussion leading up to Proposition 10.10 we addressed this issue, observing that while the above connection was valid for position and momentum observables, it is not the case for the number operator N and the phase operator $\mathbf{\Delta}[\varphi]$. The simple fact is that the Poisson bracket is not the phase space bracket whose quantization is the operator commutator; that honour goes to the Moyal bracket which we shall discuss below.

But why should this matter? Apart from being an intellectual curiosity, the correct bracket must be an interesting object on phase space. This is because it is a representation of the Lie algebra structure of quantum theory transported to phase space, and so it can be compared with the Lie algebra structure of classical mechanics.

There is a related structure to consider in this context. The quantum algebra of observables involves the nonabelian operator product, and the classical algebra of observables involves the abelian pointwise product of functions. By pulling the operator product back to phase space, a second, nonabelian product will be defined on phase space functions, termed the Moyal product[1]. These two products reflect two different geometries (in the group theoretic sense of Klein), classical and quantal. Now that we have established the formalism of quantization and dequantization, these ideas may be put on a rigorous footing, and that is the business of this Chapter.

The original motivation of Moyal [169] was to extend Wigner's semi-classical expansions of statistical quantities on phase space [241]. But to suppose that quantum mechanics might be classical mechanics plus some exotic probability distributions would be incorrect. If there were any doubt about it, Moyal's product shows that viewpoint to be geometrically untenable.

We need, therefore, a product $\star$ on some collection of phase space distributions so that

$$\boldsymbol{\Delta}[S \star T] = \boldsymbol{\Delta}[S]\,\boldsymbol{\Delta}[T] \tag{13.1.2.a}$$

for any pair of distributions S and T in the collection. This product could then be used to define the $\star$-bracket $\{\,,\,\}_\star$ through the formula

$$\{S, T\}_\star = i(S \star T - T \star S), \tag{13.1.2.b}$$

in which case the operator identity

$$\boldsymbol{\Delta}[\{S, T\}_\star] = i\Big(\boldsymbol{\Delta}[S]\,\boldsymbol{\Delta}[T] - \boldsymbol{\Delta}[T]\,\boldsymbol{\Delta}[S]\Big) \tag{13.1.2.c}$$

holds for any such pair S, T. The $\star$-bracket evidently gives the collection of distributions a structure of Lie algebra. In recognition of the defining work of Moyal in this area, the $\star$-product is usually referred to as the *Moyal product*, and the associated Lie bracket $\{\,,\,\}_\star$ is called the *Moyal bracket.*

It is worth noting that since the Moyal bracket is derived from the Moyal product by a commutation rule, and not by anything more complicated, there are useful identities which interrelate the Moyal product and bracket. For example, we can show that

$$\{R \star S, T\}_\star = R \star \{S, T\}_\star + \{R, T\}_\star \star S, \tag{13.1.3}$$

for all suitable distributions R, S, T.

If Planck's constant $\hbar$ were to be included in the quantization formalism explicitly, the Moyal product and bracket would be seen to depend on it

[1] The Moyal bracket of two functions is then i times the difference of the Moyal products of those functions, taken in opposite orders.

nontrivially, and would converge in an appropriate sense to the pointwise product and Poisson bracket, respectively, in the limit as $\hbar$ tends to zero[2], provided that the phase space quantities in question do not themselves depend on $\hbar$.

We shall formulate the theory of the Moyal product in a manner which most closely tallies with our development of the smooth model, and therefore our results will be extensions of those of Moyal [169]. Once again the problem of what class of distributions and operators to work with will reappear. It is evident from the above discussion that it will not be possible to define the Moyal product of two phase space observables unless it is possible to compose their quantizations in some sense. Thus it will not be possible to define the Moyal product of two general distributions in $\mathcal{S}'(\Pi)$. On the other hand, there is no problem defining the Moyal product of two test functions in $\mathcal{S}(\Pi)$. Our aim is to strike a balance between these two extremes, and find a subclass of $\mathcal{S}'(\Pi)$ (which contains interesting phase space observables) on which the Moyal product can be defined meaningfully. There are a number of such spaces, but our approach will enable us to see all of these spaces within a single, more general framework.

Before proceeding, we sound one note of notational warning. It is traditional to denote the Moyal product of observables by the symbol $\star$, but the similar symbol $*$ has been used in this book to denote the classical convolution of functions. Throughout this Chapter, however, no use will be made of the convolution symbol, and the Moyal product has only been referred to sparingly outside this Chapter, so no misunderstanding should occur.

13.2 The Moyal Product - The Analytic Approach

The foundation of the theory will be the definition of the Moyal product on the space $\mathcal{S}(\Pi)$ of test functions, after which it can be extended by continuity and/or transposition to larger spaces.

---

[2]This is another realization of the classical limit. As usual in such circumstances, the limit as $\hbar \to 0$ is purely formal.

13.2.1 Test Functions

In Chapter 3 we defined the twisted convolution[3] $\circ$ given by the formula

$$(F \circ G)(\xi) = \iint_{\mathbb{R}^2} F(\xi - \eta) G(\eta)\, e^{\frac{1}{2} i \Omega(\xi,\eta)}\, dA(\eta), \qquad F, G \in L^1(\mathbb{R}^2), \tag{3.3.10.a}$$

as well as the twisted involution * given by

$$F^*(\xi) = \overline{F(-\xi)}, \qquad F \in L^1(\mathbb{R}^2), \tag{3.3.11.a}$$

on the Banach space $L^1(\mathbb{R}^2)$. Then $L^1(\mathbb{R}^2)$ is a Banach *-algebra with respect to the product $\circ$ and the involution *. Unlike the classical convolution on $L^1(\mathbb{R}^2)$, the twisted convolution is not commutative. It was observed in Chapter 3 (although not in so many words) that the map $F \mapsto W[\![F]\!]$ was a norm-decreasing Banach *-algebra homomorphism from $L^1(\mathbb{R}^2)$ to $\mathbb{B}(L^2(\mathbb{R}))$.

It is elementary to show that the subspace $\mathcal{S}(\mathbb{R}^2)$ of $L^1(\mathbb{R}^2)$ is closed under the twisted convolution $\circ$ and the twisted involution *, and moreover that the map $\circ : \mathcal{S}(\mathbb{R}^2) \times \mathcal{S}(\mathbb{R}^2) \to \mathcal{S}(\mathbb{R}^2)$ is jointly continuous and bilinear, while the map $^* : \mathcal{S}(\mathbb{R}^2) \to \mathcal{S}(\mathbb{R}^2)$ is continuous and antilinear. In other words, $\mathcal{S}(\mathbb{R}^2)$ is a jointly continuous locally convex *-algebra. Since Schwartz functions behave well under the Fourier transform, we may make the following Definition.

Definition 13.1 *The Moyal product $\star$ is defined on the space $\mathcal{S}(\Pi)$ by the formula*

$$F \star G = \frac{1}{2\pi} \mathcal{F}^{-1}\big(\mathcal{F}F \circ \mathcal{F}G\big), \qquad F, G \in \mathcal{S}(\Pi). \tag{13.2.1}$$

There is bound to be the occasional uncertainty about which maps are meant to be acting on a given space so, when it is helpful, a notation like $[\mathcal{S}(\mathbb{R}^2), \circ, ^*]$ will indicate that $\mathcal{S}(\mathbb{R}^2)$ is meant to be equipped with the product $\circ$ and the involution *, and so on.

Proposition 13.2 *The Moyal product is a jointly continuous associative product on $\mathcal{S}(\Pi)$. Equipped with its usual involution of complex conjugation, $F \mapsto \overline{F}$, $[\mathcal{S}(\Pi), \star, ^-]$ is a jointly continuous locally convex *-algebra, and*

[3] For notational convenience, the original equations have here been rewritten in vector notation, so that dA denotes the Lebesgue measure on $\mathbb{R}^2$, and $\Omega(\xi, \eta) = \xi_1\eta_2 - \xi_2\eta_1$ is the symplectic form on $\mathbb{R}^2$ originally introduced in equation (2.6.9).

the mapping $(2\pi)^{-1}\mathcal{F} : \mathcal{S}(\Pi) \to \mathcal{S}(\mathbb{R}^2)$ *describes a continuous* **-algebra isomorphism between* $[\mathcal{S}(\Pi), \star, ^-]$ *and* $[\mathcal{S}(\mathbb{R}^2), \circ, ^*]$.

Proof: Since the Fourier transform $\mathcal{F} : \mathcal{S}(\Pi) \to \mathcal{S}(\mathbb{R}^2)$ is bicontinuous, it is clear that $\star$ is a jointly continuous associative bilinear map on $\mathcal{S}(\Pi)$ such that $2\pi\,\mathcal{F}(F \star G) = \mathcal{F}F \circ \mathcal{F}G$ for all $F, G \in \mathcal{S}(\Pi)$. Since it is elementary to show that $\mathcal{F}\overline{F} = (\mathcal{F}F)^*$ for any F in $\mathcal{S}(\Pi)$, it follows that $[\mathcal{S}(\Pi), \star, ^-]$ forms a jointly continuous locally convex *-algebra, and that $(2\pi)^{-1}\mathcal{F}$ is a *-algebra homomorphism between $[\mathcal{S}(\Pi), \star, ^-]$ and $[\mathcal{S}(\mathbb{R}^2), \circ, ^*]$. ∎

This is the correct definition of the Moyal product on $\mathcal{S}(\Pi)$, since

$$\frac{1}{2\pi} W[\![\mathcal{F}(F \star G)]\!] = \frac{1}{4\pi^2} W[\![\mathcal{F}F \circ \mathcal{F}G]\!] = \frac{1}{4\pi^2} W[\![\mathcal{F}F]\!]\, W[\![\mathcal{F}G]\!] \qquad (13.2.2.a)$$

for any $F, G \in \mathcal{S}(\Pi)$, and hence

$$\boldsymbol{\Delta}[\,F \star G\,] = \boldsymbol{\Delta}[\,F\,]\,\boldsymbol{\Delta}[\,G\,] \qquad (13.2.2.b)$$

for any F, $G \in \mathcal{S}(\Pi)$.

The next task is to obtain an explicit integral expression for the Moyal product on $\mathcal{S}(\Pi)$.

Proposition 13.3 *For any* F, $G \in \mathcal{S}(\Pi)$, *the Moyal product* $F \star G$ *is given by the formula*

$$(F \star G)(\xi) = \frac{1}{\pi^2} \iiiint_{\Pi \times \Pi} F(\eta) G(\zeta)\, e^{-2i\Psi(\xi,\eta,\zeta)}\, dA(\eta)\, dA(\zeta), \qquad (13.2.3.a)$$

where $\Psi : \mathbb{R}^2 \times \mathbb{R}^2 \times \mathbb{R}^2 \to \mathbb{R}$ *is the totally antisymmetric multilinear form*[4]

$$\Psi(\xi,\eta,\zeta) = \Omega(\xi,\eta) + \Omega(\eta,\zeta) + \Omega(\zeta,\xi), \qquad \xi,\, \eta,\, \zeta \in \mathbb{R}^2. \qquad (13.2.3.b)$$

Proof: For any $\vartheta \in \mathbb{C}$, recall the endomorphism τ_ϑ of $\mathcal{S}(\mathbb{R}^2)$ given by

$$(\tau_\vartheta F)(w) = F(w + \vartheta), \qquad F \in \mathcal{S}(\mathbb{R}^2) \qquad (11.2.32.a)$$

and also consider the endomorphism E_ϑ of $\mathcal{S}(\mathbb{R}^2)$ defined by

$$(E_\vartheta F)(w) = e^{i(\vartheta\overline{w} + \overline{\vartheta}w)}\, F(w), \qquad F \in \mathcal{S}(\mathbb{R}^2). \qquad (13.2.4)$$

[4] Ψ is sometimes written in the more compact form $\Psi(\xi,\eta,\zeta) = \Omega(\xi - \eta, \xi - \zeta)$.

With the usual complex parametrization $\sqrt{2}\vartheta = b - ia$, note that

$$\begin{aligned}(\tau_\vartheta F)(x,y) &= F(x+a, y+b)\,, \\ (E_\vartheta F)(x,y) &= e^{i(ax+by)}\,F(x,y) \;=\; E_{a,b}(x,y)F(x,y)\,,\end{aligned} \tag{13.2.5}$$

for any $F \in \mathcal{S}(\mathbb{R}^2)$, where the function $E_{a,b}$ was initially defined in Proposition 8.32. Since it can be shown that

$$(F \circ G)(a,b) \;=\; \langle \tau_{-\vartheta}F^*\,,\, E_{\frac{1}{2}i\vartheta}G\rangle \;=\; \langle E_\vartheta \overline{\mathcal{F}^{-1}F}\,,\, \tau_{\frac{1}{2}i\vartheta}\mathcal{F}^{-1}G\rangle$$

for any $F, G \in \mathcal{S}(\mathbb{R}^2)$, it follows that

$$\begin{aligned}\big[\mathcal{F}(F \star G)\big](a,b) \;&=\; \frac{1}{2\pi}\big(\mathcal{F}F \circ \mathcal{F}G\big)(a,b) \;=\; \frac{1}{2\pi}\langle E_\vartheta \overline{F}\,,\, \tau_{\frac{1}{2}i\vartheta}G\rangle \\ &=\; \frac{1}{2\pi}\iint_\Pi F(x,y)G(x-\tfrac{1}{2}b, y+\tfrac{1}{2}a)e^{-i(ax+by)}\,dxdy\end{aligned}$$

for any $F, G \in \mathcal{S}(\Pi)$. Taking the Fourier transform of this identity completes the proof. ■

Since the "familiar" identities

$$\begin{aligned}\boldsymbol{\Delta}[F] \;&=\; \frac{1}{2\pi}\iint_\Pi F(\xi)\boldsymbol{\Delta}[\xi]\,dA(\xi)\,, \\ F(\xi) \;&=\; \mathrm{Tr}\,(\boldsymbol{\Delta}[\xi]\,\boldsymbol{\Delta}[F])\,,\end{aligned}$$

are valid quantization and dequantization formulæ for any $F \in \mathcal{S}(\Pi)$, the above Proposition tempts us to write down the "trace formula"

$$\mathrm{Tr}\,(\boldsymbol{\Delta}[\xi]\,\boldsymbol{\Delta}[\eta]\,\boldsymbol{\Delta}[\zeta]) \;=\; 4\,e^{-2i\Psi(\xi,\eta,\zeta)}\,, \qquad \xi,\eta,\zeta \in \mathbb{R}^2\,. \tag{13.2.6}$$

However, this formula is suspect, since none of the operators involved is trace-class. Hence, while suggestive, this statement should be treated with the same caution that we have accorded to similar statements which we have discussed previously. Like other such statements, however, judicious use of this formula usually gives correct answers.

13.2.2 *Square Integrable Functions*

Pool's Theorem 8.23 gives a complete correspondence between the Hilbert space $L^2(\Pi)$ of square integrable functions on Π and the class $\mathcal{T}_2\big(L^2(\mathbb{R})\big)$ of Hilbert-Schmidt operators on $L^2(\mathbb{R})$. Since this latter space is an algebra under the operator product, we expect $L^2(\Pi)$ to be an algebra under the Moyal product, and this is indeed the case.

Proposition 13.4 *The Moyal product on $\mathcal{S}(\Pi)$ extends to a unique jointly continuous associative bilinear product on $L^2(\Pi)$ such that*

$$\| \Phi \star \Psi \| \leqslant \frac{1}{\sqrt{2\pi}} \| \Phi \| \, \| \Psi \| , \qquad \Phi, \Psi \in L^2(\Pi), \tag{13.2.7}$$

so that $\left[L^2(\Pi), \star, ^- \right]$ is a normed *-algebra.

Proof: If $\Phi, \Psi \in L^2(\Pi)$ then $\boldsymbol{\Delta}[\Phi], \boldsymbol{\Delta}[\Psi] \in \mathcal{T}_2\big(L^2(\mathbb{R})\big)$. Thus the product $\boldsymbol{\Delta}[\Phi]\boldsymbol{\Delta}[\Psi]$ is also Hilbert-Schmidt, so there exists a unique function $\Phi \star \Psi \in L^2(\Pi)$ such that $\boldsymbol{\Delta}[\Phi \star \Psi] = \boldsymbol{\Delta}[\Phi]\boldsymbol{\Delta}[\Psi]$. In view of equation (13.2.2.b), this defines a bilinear associative product $\star$ on $L^2(\Pi)$ which extends the Moyal product on $\mathcal{S}(\Pi)$. Since, for any $\Phi, \Psi \in L^2(\Pi)$,

$$\begin{aligned} \| \boldsymbol{\Delta}[\Phi \star \Psi] \|_2 &= \| \boldsymbol{\Delta}[\Phi]\boldsymbol{\Delta}[\Psi] \|_2 \\ &\leqslant \| \boldsymbol{\Delta}[\Phi] \|_2 \| \boldsymbol{\Delta}[\Psi] \|_2 \\ &= \frac{1}{2\pi} \| \Phi \| \, \| \Psi \| , \end{aligned}$$

it follows that $\sqrt{2\pi} \| \Phi \star \Psi \| \leqslant \| \Phi \| \, \| \Psi \|$ for $\Phi, \Psi \in L^2(\Pi)$, as required. ■

By applying the inverse Fourier transform, this result has the following consequence. Although the integral formula (3.3.10.a) defining the twisted convolution $\circ$ is given for functions in $L^1(\mathbb{R}^2)$, it is clear that a function $F \circ G$ can be defined for any two functions $F, G \in L^2(\mathbb{R}^2)$.

Corollary 13.5 *The function $F \circ G$ belongs to $L^2(\mathbb{R}^2)$ for any F, G in $L^2(\mathbb{R}^2)$, and $\left[L^2(\mathbb{R}^2), \circ, ^* \right]$ is a normed *-algebra. Moreover the bounded operator $W[\![F]\!] = 2\pi \boldsymbol{\Delta}\left[\mathcal{F}^{-1} F \right]$ exists for any $F \in L^2(\mathbb{R}^2)$, and the map which sends F to $W[\![F]\!]$ defines a continuous *-algebra homomorphism from $\left[L^2(\mathbb{R}^2), \circ, ^* \right]$ to $\mathcal{T}_2\big(L^2(\mathbb{R})\big)$.*

A key result in what follows is the relationship that exists between the Moyal product on $L^2(\Pi)$ and that space's inner product. The following result can be established by direct calculation, and we omit the proof[5].

Proposition 13.6 *The Moyal product has the adjoint property*

$$\langle \Phi \star \Psi , \Upsilon \rangle = \langle \Psi , \overline{\Phi} \star \Upsilon \rangle = \langle \Phi , \Upsilon \star \overline{\Psi} \rangle , \qquad \Phi, \Psi, \Upsilon \in L^2(\Pi). \quad (13.2.8)$$

As discussed in Section 8.4.1, the functions $\Xi_{\phi,\psi}$, where $\phi, \psi \in L^2(\mathbb{R})$, span the subspace $\mathcal{O}\big[\mathcal{T}_0\big(L^2(\mathbb{R})\big)\big]$ of $L^2(\Pi)$, the preimage under Weyl quantization of the finite-rank operators on $L^2(\mathbb{R})$. Moyal products involving these functions, and in particular those involving the special Hermite functions, are particularly easy to calculate, and form an important base for understanding the general structure of the Moyal product.

Proposition 13.7 *If $\Phi \in L^2(\Pi)$ and $\phi, \psi \in L^2(\mathbb{R})$ then*

$$\Phi \star \Xi_{\phi,\psi} = \Xi_{\mathbf{\Delta}[\Phi]\phi,\psi} . \quad (13.2.9.a)$$

In particular, this implies that

$$\Xi_{\alpha,\beta} \star \Xi_{\phi,\psi} = \langle \beta , \phi \rangle \, \Xi_{\alpha,\psi} , \qquad \alpha, \beta, \phi, \psi \in L^2(\mathbb{R}) , \quad (13.2.9.b)$$

and so

$$\Phi_{j,k} \star \Phi_{m,n} = \delta_{jn} \, \Phi_{m,k} , \qquad j, k, m, n \geqslant 0 . \quad (13.2.9.c)$$

Proof: The first identity holds, since

$$\begin{aligned} \mathbf{\Delta}[\Phi \star \Xi_{\phi,\psi}] &= \mathbf{\Delta}[\Phi]\,\mathbf{\Delta}[\Xi_{\phi,\psi}] = \mathbf{\Delta}[\Phi]\,|\phi\rangle\langle\psi| \\ &= |\mathbf{\Delta}[\Phi]\phi\rangle\langle\psi| = \mathbf{\Delta}\big[\Xi_{\mathbf{\Delta}[\Phi]\phi,\psi}\big] \end{aligned}$$

for any $\Phi \in L^2(\Pi)$ and $\phi, \psi \in L^2(\mathbb{R})$. The second identity is an immediate consequence of the first, and the third is true because $\Phi_{m,n} = \Xi_{h_n,h_m}$ for any $m, n \geqslant 0$. ∎

Since the special Hermite functions $\{\Phi_{m,n} : m, n \geqslant 0\}$ form a Schauder basis for $\mathcal{S}(\Pi)$ as well as an orthogonal basis for $L^2(\Pi)$, the results in this

[5]Without wishing to pursue the matter, we note that this Proposition shows that $[L^2(\Pi), \star, ^{-}]$ is a *generalized Hilbert algebra*, which is a notion connected to noncommutative integration theory and the condition for thermal equilibrium for continuous quantum systems. See [28], [35] and further references there.

Proposition tell us, in principle, all that there is to be known about the Moyal product on the spaces $\mathcal{S}(\Pi)$ and $L^2(\Pi)$.

The function $\phi_{0,0}(p,q) = 2e^{-(p^2+q^2)}$ has a special rôle to play in any study of the Moyal product, as can be seen from the next result, which is reminiscent of the projection property (3.3.12.b) of $W[\![H_0]\!]$, where H_0 is the two dimensional Gaussian.

Corollary 13.8 *The identity*

$$\phi_{0,0} \star \Phi \star \phi_{0,0} \;=\; \frac{1}{2\pi} \left\langle \phi_{0,0} \,,\, \Phi \right\rangle \phi_{0,0} \tag{13.2.10}$$

holds for any $\Phi \in L^2(\Pi)$.

Proof: Since the special Hermite functions form an orthogonal basis for $L^2(\Pi)$ and the Moyal product is jointly continuous on $L^2(\Pi)$, the fact that $\phi_{0,0} \star \phi_{m,n} \star \phi_{0,0} \;=\; \delta_{m0}\,\delta_{n0}\,\phi_{0,0}$ for all $m,n \geqslant 0$ implies the truth of the desired identity. ■

13.2.3 *Quantization In Phase Space*

Since the Moyal product is inextricably linked to Weyl quantization, it is reasonable to ask to what extent is it possible to represent the formalism of quantum mechanics entirely within phase space Π, without explicitly introducing Weyl quantization. Formulating just such a classical statistical mechanics was the stated motivation behind Moyal's work in this area. Pool's Theorem can be used to achieve this for observables which are Hilbert-Schmidt operators, and the work for these observables provides us with the formalism to extend the theory to cover more interesting observables. However, even the Hilbert-Schmidt case is not trivial, and requires a bit of *-algebra representation theory.

Define $\mathcal{H} = \left\{ \Phi \star \phi_{0,0} \,:\, \Phi \in L^2(\Pi) \right\}$ to be the left ideal of the *-algebra $\left[L^2(\Pi), \star, ^- \right]$ generated by the element $\phi_{0,0}$. Since $\phi_{0,0} \star \phi_{0,0} \;=\; \phi_{0,0}$ is idempotent, $\mathcal{H}$ is the image of a continuous projection (that of right multiplication by $\phi_{0,0}$) and is thus a closed linear subspace of $L^2(\Pi)$.

Remark There is a standard theory for obtaining representations of a normed *-algebra $\mathcal{A}$ through the study of its left ideals. If $\mathcal{I}$ is a left ideal of $\mathcal{A}$ and ω is a continuous positive linear functional on $\mathcal{A}$, then the map

$$(x, y) \mapsto \omega(x^* y), \qquad x, y \in \mathcal{I}$$

becomes a pre-inner product on $\mathcal{I}$. Taking the quotient of the space $\mathcal{I}$ by the kernel of this pre-inner product, and then calculating the Hilbert space completion of the resulting inner product space, leads to a Hilbert space on which a continuous *-algebra representation of $\mathcal{A}$ can be defined. The construction of this Hilbert space is essentially the same as the one mentioned in Theorem 6.1 in respect of the GNS representation. ■

The structure of the ideal $\mathcal{H}$ is sufficiently simple that the full generality of normed *-algebra theory is not needed, as will be seen from the following result.

Lemma 13.9 *If* $\Phi,\ \Psi \in \mathcal{H}$ *then*

$$\overline{\Phi} \star \Psi = \frac{1}{2\pi} \langle \Phi, \Psi \rangle\, \phi_{0,0}. \tag{13.2.11}$$

Proof: It is clear from Corollary 13.8 that $\overline{\Phi} \star \Psi = \frac{1}{2\pi} \langle \phi_{0,0}, \overline{\Phi} \star \Psi \rangle\, \phi_{0,0}$ for any $\Phi,\ \Psi \in \mathcal{H}$. But $\langle \phi_{0,0}, \overline{\Phi} \star \Psi \rangle = \langle \Phi \star \phi_{0,0}, \Psi \rangle = \langle \Phi, \Psi \rangle$, so we are done. ■

Choosing the continuous positive[6] linear functional on $L^2(\Pi)$ to be

$$\omega(\Phi) = \langle \phi_{0,0}, \Phi \rangle, \qquad \Phi \in L^2(\Pi),$$

the construction outlined in the above Remark has the effect of equipping the ideal $\mathcal{H}$ with the same inner product that it naturally inherits as a subspace of $L^2(\Pi)$. Thus there is no need to take any quotients, or to go to any completions, since $\mathcal{H}$ is already complete. In this context, the results of the general theory indicated in the above Remark can thus be summarized as follows.

Theorem 13.10 *The formula*

$$\mathcal{R}(\Phi)\Psi = \Phi \star \Psi, \qquad \Phi \in L^2(\Pi),\ \Psi \in \mathcal{H}, \tag{13.2.12}$$

[6] Corollary 13.8 assures us that this functional is indeed positive.

*defines a norm-decreasing *-algebra representation $\mathcal{R}$ of the Moyal *-algebra $\left[L^2(\Pi), \star, ^-\right]$ on $\mathbb{B}(\mathcal{H})$.*

The real utility of this construction will only become apparent after the representation $\mathcal{R}$ has been identified explicitly. This we now proceed to do. It is clear from the way in which the special Hermite functions multiply with respect to the Moyal product that the set

$$\left\{ \frac{1}{\sqrt{2\pi}} \phi_{0,n} \;:\; n \geqslant 0 \right\}$$

is an orthonormal basis for $\mathcal{H}$. Consequently there is a unitary isomorphism $\mathcal{V} : L^2(\mathbb{R}) \to \mathcal{H}$ such that

$$\mathcal{V}\, h_n = \frac{1}{\sqrt{2\pi}}\, \phi_{0,n}, \qquad n \geqslant 0. \tag{13.2.13.a}$$

It is also clear that $\mathcal{V}$ is defined by the integral formula

$$\mathcal{V}\phi = \sqrt{2\pi}\, \mathcal{G}(h_0 \otimes \phi) = \frac{1}{\sqrt{2\pi}}\, \Xi_{\phi,h_0}, \qquad \phi \in L^2(\mathbb{R}). \tag{13.2.13.b}$$

With this in hand, we can show that the unitary map $\mathcal{V}$ intertwines the representation $\mathcal{R}$ and Weyl quantization.

Theorem 13.11 *For any $\Phi \in L^2(\Pi)$, the map $\mathcal{R}(\Phi)$ is a Hilbert-Schmidt operator on $\mathcal{H}$, and moreover*

$$\mathcal{V}^{-1}\mathcal{R}(\Phi)\, \mathcal{V} = \mathbf{\Delta}\,[\,\Phi\,]. \tag{13.2.14}$$

Proof: Since $\Phi \star \Xi_{\phi,\psi} = \Xi_{\mathbf{\Delta}[\,\Phi\,]\phi,\psi}$ for all $\phi,\ \psi \in L^2(\mathbb{R})$, it follows that

$$\mathcal{R}(\Phi)\mathcal{V}\phi = \frac{1}{\sqrt{2\pi}} \Phi \star \Xi_{\phi,h_0} = \frac{1}{\sqrt{2\pi}} \Xi_{\mathbf{\Delta}[\,\Phi\,]\phi,h_0} = \mathcal{V}\mathbf{\Delta}\,[\,\Phi\,]\,\phi$$

for any $\phi \in L^2(\mathbb{R})$, as required. ∎

Thus the operator $\mathcal{R}(\Phi)$ of left multiplication by Φ on $\mathcal{H}$ is unitarily equivalent to the Hilbert-Schmidt operator $\mathbf{\Delta}\,[\,\Phi\,]$ on $L^2(\mathbb{R})$, and so we have succeeded in describing Hilbert-Schmidt observables entirely in terms of the Moyal product on phase space.

Since the Hilbert-Schmidt operators constitute only a fraction of the observables of interest in quantum theory, our next concern is to extend the definition of the Moyal product from $L^2(\Pi)$ to a wider class of distributions.

13.2.4 *Extending The Moyal Product To Distributions*

Extending the Moyal product to distributions is a delicate business, as we have indicated. In essence, however, we have already introduced the general approach when in Chapter 10 we defined the generalized commutator $\tau_{A,B}$ of two operators $A, B \in \mathcal{L}^+\big(\mathcal{S}(\mathbb{R}), L^2(\mathbb{R})\big)$, as well as its representing map $X_{A,B} \in \mathcal{L}(\mathcal{S}(\mathbb{R}), \mathcal{S}'(\mathbb{R}))$ given by equation (10.3.70). Our extension of the Moyal product will result in our being able to construct a bilinear map $\bar{\star} : \mathcal{O}\big[\mathcal{L}^+\big(\mathcal{S}(\mathbb{R}), L^2(\mathbb{R})\big)\big] \times \mathcal{O}\big[\mathcal{L}^+\big(\mathcal{S}(\mathbb{R}), L^2(\mathbb{R})\big)\big] \to \mathcal{S}'(\Pi)$ such that

$$\mathbf{\Delta}^{-1}[X_{A,B}] = \mathbf{\Delta}^{-1}[A] \,\bar{\star}\, \mathbf{\Delta}^{-1}[B] - \mathbf{\Delta}^{-1}[B] \,\bar{\star}\, \mathbf{\Delta}^{-1}[A], \quad (13.2.15)$$

for $A, B \in \mathcal{L}^+\big(\mathcal{S}(\mathbb{R}), L^2(\mathbb{R})\big)$, implementing the generalized commutator as an algebraic commutator with respect to the Moyal product.

Finding subspaces of $\mathcal{O}\big[\mathcal{L}^+\big(\mathcal{S}(\mathbb{R}), L^2(\mathbb{R})\big)\big]$ which are closed and associative under the Moyal product $\bar{\star}$ (and the involution $^-$) is then an important matter, since such spaces are equipped with a full *-algebraic structure with respect to the Moyal product. Two such spaces are $\mathcal{O}\big[\mathbb{B}\big(L^2(\mathbb{R})\big)\big]$ and $\mathcal{O}\big[\mathcal{L}^+\big(\mathcal{S}(\mathbb{R})\big)\big]$, which spaces are (of course) of particular relevance to the bounded and smooth models, respectively. It will also prove possible to provide a complete characterization of these two spaces with respect to the Moyal product. This is an important result, in view of the problems discussed in Chapter 8 concerning the characterizations of these spaces.

However, it should be remembered that the results in Chapter 8 differ from those in this Chapter in that the results there are intended to be of practical utility, providing specific conditions that a distribution must satisfy in order that its quantization have particular properties. The results of this Chapter, on the other hand, demonstrate that the problems of quantization can be recast solely in the terminology of the Moyal product. However, it is not to be supposed that the Moyal product terminology will provide a complete, and practical, solution more readily than the previous terminology — the two approaches are, of necessity, two faces of the same coin.

The process of extension of the Moyal product needs to be done carefully, and this requires us to take detailed account of the order in which operators are considered. Consequently, the first stage is to introduce two different extensions of the Moyal product. Proofs of the various stages in the argument will be omitted, and the reader is referred to [100] and [108] for the necessary details.

Definition 13.12 *The Moyal product can be extended to define the continuous bilinear maps* $\star_1 : \mathcal{S}'(\Pi) \times \mathcal{S}(\Pi) \to \mathcal{S}'(\Pi)$, $\star_2 : \mathcal{S}(\Pi) \times \mathcal{S}'(\Pi) \to \mathcal{S}'(\Pi)$ *through the formulæ:*

$$\begin{aligned} [\![T \star_1 F \,,\, G]\!] &= [\![T \,,\, F \star G]\!], & (13.2.16.a)\\ [\![F \star_2 T \,,\, G]\!] &= [\![T \,,\, G \star F]\!], & (13.2.16.b) \end{aligned}$$

for $T \in \mathcal{S}'(\Pi)$ and $F, G \in \mathcal{S}(\Pi)$. These are extensions of the product on $\mathcal{S}(\Pi)$ in the sense that $\Phi \star_1 F = \Phi \star F$ and $F \star_2 \Phi = F \star \Phi$ whenever $\Phi \in L^2(\Pi)$ and $F \in \mathcal{S}(\Pi)$. These two products are not independent, since they are connected through the identity[7]

$$\overline{F \star_2 T} = \overline{T} \star_1 \overline{F}\,, \qquad T \in \mathcal{S}'(\Pi),\ F \in \mathcal{S}(\Pi)\,, \qquad (13.2.17)$$

and they are as associative as might be expected, in the sense that

$$\begin{aligned} &(F \star_2 T) \star_1 G = F \star_2 (T \star_1 G), \\ &(T \star_1 F) \star_1 G = T \star_1 (F \star G), \qquad T \in \mathcal{S}'(\Pi),\ F, G \in \mathcal{S}(\Pi). \quad (13.2.18)\\ &F \star_2 (G \star_2 T) = (F \star G) \star_2 T, \end{aligned}$$

As was the case when we discussed extending the Poisson bracket, useful results arise through restricting the above two products to appropriate domains. Two pairs of choices will turn out to be of significance for quantum theory.

Definition 13.13 *The subspaces* N, $\overline{N}$, Q *and* $\overline{Q}$ *of* $\mathcal{S}'(\Pi)$ *are defined by the formulæ:*

$$\begin{aligned} N &= \{\, T \in \mathcal{S}'(\Pi) \,:\, T \star_1 F \in \mathcal{S}(\Pi)\ \ \forall F \in \mathcal{S}(\Pi) \,\}\,, & (13.2.19.a)\\ \overline{N} &= \{\, T \in \mathcal{S}'(\Pi) \,:\, F \star_2 T \in \mathcal{S}(\Pi)\ \ \forall F \in \mathcal{S}(\Pi) \,\}\,, & (13.2.19.b)\\ Q &= \{\, T \in \mathcal{S}'(\Pi) \,:\, T \star_1 F \in L^2(\Pi)\ \forall F \in \mathcal{S}(\Pi) \,\}\,, & (13.2.19.c)\\ \overline{Q} &= \{\, T \in \mathcal{S}'(\Pi) \,:\, F \star_2 T \in L^2(\Pi)\ \forall F \in \mathcal{S}(\Pi) \,\}\,. & (13.2.19.d) \end{aligned}$$

We note that $N \subseteq Q$, *that* $\mathcal{S}(\Pi) \subseteq N \cap \overline{N}$ *and that* $L^2(\Pi) \subseteq Q \cap \overline{Q}$. *Moreover, a distribution* T *belongs to* $\overline{N}$ *if and only if* $\overline{T} \in N$, *while* $T \in \overline{Q}$ *if and only if* $\overline{T} \in Q$.

---

[7]The overline indicates complex conjugation, defined weakly for distributions.

An application of the Closed Graph Theorem shows that the linear map $F \mapsto T\star_1 F$ is continuous from $\mathcal{S}(\Pi)$ to $L^2(\Pi)$ if $T \in Q$, and is a continuous endomorphism of $\mathcal{S}(\Pi)$ if $T \in N$.

The fundamental result of this analysis, from which all else follows, is the following.

Theorem 13.14 *If $S \in \overline{Q}$ and $T \in Q$ we can define a distribution $S \overline{\star} T$ in $\mathcal{S}'(\Pi)$ such that*

$$[\![S \overline{\star} T \,,\, F \star \overline{G}]\!] \;=\; \langle \overline{S} \star_1 G \,,\, T \star_1 F \rangle\,, \qquad F,\, G \in \mathcal{S}(\Pi). \qquad (13.2.20)$$

Proof: Integers $j,\, k \in \mathbb{N}$ and constants $A,\, B > 0$ can be found such that

$$\| T \star_1 \phi_{0,n} \| \;\leqslant\; A(n+1)^j, \qquad \| \overline{S} \star_1 \phi_{0,n} \| \;\leqslant\; B(n+1)^k,$$

for all $n \geqslant 0$. Thus it follows that

$$\langle \overline{S} \star_1 \phi_{0,m} \,,\, T \star_1 \phi_{0,n} \rangle \;\leqslant\; AB\,(m+1)^k (n+1)^j$$

for all $m,\, n \geqslant 0$, and hence a distribution $S \overline{\star} T \in \mathcal{S}'(\Pi)$ can be found such that

$$[\![S \overline{\star} T \,,\, \phi_{m,n}]\!] \;=\; \langle \overline{S} \star_1 \phi_{0,m} \,,\, T \star_1 \phi_{0,n} \rangle\,, \qquad m,\, n \geqslant 0.$$

It can now be shown that

$$\begin{aligned}
[\![S \overline{\star} T \,,\, \phi_{j,k} \star \phi_{n,m}]\!] &= \delta_{jm} [\![S \overline{\star} T \,,\, \phi_{n,k}]\!] \\
&= \delta_{jm} \langle \overline{S} \star_1 \phi_{0,n} \,,\, T \star_1 \phi_{0,k} \rangle \\
&= \langle \overline{S} \star_1 \phi_{0,n} \,,\, T \star_1 (\phi_{j,k} \star \phi_{0,m}) \rangle \\
&= \langle \overline{S} \star_1 \phi_{0,n} \,,\, (T \star_1 \phi_{j,k}) \star \phi_{0,m} \rangle \\
&= \langle (\overline{S} \star_1 \phi_{0,n}) \star \phi_{m,0} \,,\, T \star_1 \phi_{j,k} \rangle \\
&= \langle \overline{S} \star_1 \phi_{m,n} \,,\, T \star_1 \phi_{j,k} \rangle
\end{aligned}$$

for all $j,\, k,\, m,\, n \geqslant 0$, from which it follows that

$$[\![S \overline{\star} T \,,\, F \star \overline{G}]\!] \;=\; \langle \overline{S} \star_1 G \,,\, T \star_1 F \rangle$$

for all $F,\, G \in \mathcal{S}(\Pi)$, as required. ∎

This is an extension of previous products, in that

$$\Phi \overline{\star} \Psi \;=\; \Phi \star \Psi, \qquad S \overline{\star} F \;=\; S \star_1 F, \qquad F \overline{\star} T \;=\; F \star_2 T \qquad (13.2.21)$$

for all Φ, $\Psi \in L^2(\Pi)$, $F \in \mathcal{S}(\Pi)$, $S \in \overline{Q}$ and $T \in Q$. Moreover, the product $\overline{\star}$ behaves well with respect to the involution on $\mathcal{S}'(\Pi)$, because

$$\overline{S\overline{\star}T} = \overline{T}\overline{\star}\overline{S}, \qquad S \in \overline{Q}, T \in Q. \tag{13.2.22}$$

It is important to note at this stage that the map $\overline{\star} : \overline{Q} \times Q \to \mathcal{S}'(\Pi)$ involves three different subspaces of $\mathcal{S}'(\Pi)$, and so any questions as to the associativity or commutativity of $\overline{\star}$ are meaningless at present. It will be the business of the next Section to identify subspaces of $\overline{Q} \cap Q$ which are invariant under $\overline{\star}$ (and the involution $^-$), and with respect to which this product is associative.

Before proceeding, it is useful to give an alternative description of the space $Q \cap \overline{Q}$ using the unitary map $\mathcal{V} : L^2(\mathbb{R}) \to \mathcal{H}$ discussed above. To do this, let $\mathcal{S}$ denote the subspace of $\mathcal{H}$ consisting of all functions of the form $F \star \phi_{0,0}$, where $F \in \mathcal{S}(\Pi)$. It is clear that $\mathcal{S}$ is a dense linear subspace of $\mathcal{H}$, and that $\mathcal{S}$ is a closed linear subspace of $\mathcal{S}(\Pi)$. Moreover, the unitary map $\mathcal{V}$ maps $\mathcal{S}(\mathbb{R})$ bijectively and bicontinuously onto $\mathcal{S}$, whence the notation.

Proposition 13.15 *For any $T \in Q \cap \overline{Q}$, the function $T\overline{\star}G = T \star_1 G$ belongs to $\mathcal{H}$ whenever $G \in \mathcal{S}$, and the map*

$$\mathcal{R}(T)G = T\overline{\star}G, \qquad G \in \mathcal{S}, \tag{13.2.23}$$

belongs to $\mathcal{L}^+(\mathcal{S}, \mathcal{H})$. Moreover, the mapping $\mathcal{R} : Q \cap \overline{Q} \to \mathcal{L}^+(\mathcal{S}, \mathcal{H})$ is a linear bijection. The space $Q \cap \overline{Q}$ is then equal to the symbol space $\mathcal{O}\big[\mathcal{L}^+\big(\mathcal{S}(\mathbb{R}), L^2(\mathbb{R})\big)\big]$, with

$$\mathcal{R}(T)\mathcal{V}f = \mathcal{V}\boldsymbol{\Delta}[T]f, \qquad f \in \mathcal{S}(\mathbb{R}), \tag{13.2.24}$$

for any $T \in Q \cap \overline{Q}$. Thus, Weyl quantization provides a linear bijection between $Q \cap \overline{Q}$ and $\mathcal{L}^+\big(\mathcal{S}(\mathbb{R}), L^2(\mathbb{R})\big)$.

We can now confirm the validity of equation (13.2.15), showing that the above definition of the Moyal product indeed provides the correct interpretation of the generalized commutator of elements of $\mathcal{L}^+\big(\mathcal{S}(\mathbb{R}), L^2(\mathbb{R})\big)$.

Corollary 13.16 *If S, $T \in Q \cap \overline{Q}$, then the identity*

$$[\![\widehat{\boldsymbol{\Delta}[S]}f \, , \boldsymbol{\Delta}[T]g]\!] = [\![S\overline{\star}T \, , \mathcal{G}(f \otimes g)]\!] = [\![\boldsymbol{\Delta}[S\overline{\star}T]g \, , f]\!], \tag{13.2.25.a}$$

for $f, g \in \mathcal{S}(\mathbb{R})$, allows the generalized commutator $X_{\boldsymbol{\Delta}[S],\boldsymbol{\Delta}[T]}$ of $\boldsymbol{\Delta}[S]$ and $\boldsymbol{\Delta}[T]$ to be identified as follows:

$$X_{\boldsymbol{\Delta}[S],\boldsymbol{\Delta}[T]} = \boldsymbol{\Delta}[S\bar{\star}T - T\bar{\star}S]. \qquad (13.2.25.\mathrm{b})$$

Proof: For any f, $g \in \mathcal{S}(\mathbb{R})$ we have that

$$\begin{aligned} [\![\widehat{\boldsymbol{\Delta}[S]}f, \boldsymbol{\Delta}[T]g]\!] &= \langle \boldsymbol{\Delta}[\overline{S}]\overline{f}, \boldsymbol{\Delta}[T]g \rangle \\ &= \langle \mathcal{R}(\overline{S})\mathcal{V}\overline{f}, \mathcal{R}(T)\mathcal{V}g \rangle \\ &= [\![S\bar{\star}T, \mathcal{V}g \star \overline{\mathcal{V}\overline{f}}]\!]. \end{aligned}$$

Since direct calculation yields $\mathcal{V}g \star \overline{\mathcal{V}\overline{f}} = \mathcal{G}(f \otimes g)$, we have

$$[\![\widehat{\boldsymbol{\Delta}[S]}f, \boldsymbol{\Delta}[T]g]\!] = [\![S\bar{\star}T, \mathcal{G}(f \otimes g)]\!] = [\![\boldsymbol{\Delta}[S\bar{\star}T]g, f]\!],$$

for any f, $g \in \mathcal{S}(\mathbb{R})$. Thus it follows that

$$\tau_{\boldsymbol{\Delta}[S],\boldsymbol{\Delta}[T]}(f,g) = [\![\boldsymbol{\Delta}[S\bar{\star}T - T\bar{\star}S]g, f]\!]$$

for all f, $g \in \mathcal{S}(\mathbb{R})$, and hence that

$$X_{\boldsymbol{\Delta}[S],\boldsymbol{\Delta}[T]} = \boldsymbol{\Delta}[S\bar{\star}T - T\bar{\star}S],$$

as required. ■

It is conventional to drop all the suffices and superscripts used to distinguish between the various forms of the Moyal product, and simply to denote all of them by the one symbol $\star$, and we shall do so. But a word of caution is necessary: anyone working in this field must think carefully about which Moyal product of any two distributions is being used.

13.3 Moyal Algebras

After this preliminary work of definition, in this Section we shall consider how to find *-algebras[8] of phase space observables with respect to the Moyal product $\star$. The problem is not just one of finding a subspace of $Q \cap \overline{Q}$ invariant under $\star$ and $^-$, but of also ensuring that the product is associative there. These conditions are summarized in the next Definition.

[8] It is worth noting that an alternative approach to dealing with the problems caused by the structure of the Moyal product on $Q \cap \overline{Q}$ is to cast the whole problem in the framework of the theory of partial *-algebras, for which see [6] and references therein.

Definition 13.17 *A Moyal algebra is a* $^-$*-invariant subspace of* $Q \cap \overline{Q}$ *which is closed and associative with respect to the Moyal product. A Moyal algebra is therefore a* $*$*-algebra.*

Three Moyal algebras have been identified by the preceding analysis: the class $\mathcal{O}[\mathfrak{J}_0\left(L^2(\mathbb{R})\right)]$ corresponding to the finite rank operators on $L^2(\mathbb{R})$, the space of test functions $\mathcal{S}(\Pi)$, and the space of square-integrable functions $L^2(\Pi) = \mathcal{O}[\mathfrak{J}_2\left(L^2(\mathbb{R})\right)]$. In this section, other examples will be identified, extending the theory in useful directions.

13.3.1 *Moyal-Bounded Distributions*

Evidently, a necessary part of the modelling problem that the Moyal product has presented is to find classes of distributions closely adapted to it. In particular, we have claimed to be able to determine the class of distributions whose quantizations were bounded operators. As we shall see below, this class is distinguished by the following property:

Definition 13.18 *A distribution* $T \in \mathcal{S}'(\Pi)$ *is called* Moyal-bounded *if there exists a constant* $K > 0$ *such that*

$$\left|[\![T\,,\,F\star G]\!]\right| \leqslant K\,\|\,F\,\|\,\|\,G\,\|\,, \qquad\qquad F, G \in \mathcal{S}(\Pi), \tag{13.3.1}$$

in which case the mapping $(F,G) \mapsto [\![T\,,\,F\star G]\!]$ *extends to a jointly continuous bilinear functional on* $L^2(\Pi)$*. The norm* $\|\,T\,\|$ *of a Moyal-bounded distribution is defined to be the least positive number* K *for which the above inequality holds. The collection of all Moyal-bounded distributions is denoted by* $\mathfrak{B}$*.*

It is clear that $L^2(\Pi) \subseteq \mathfrak{B} \subseteq Q \cap \overline{Q}$, and that $\mathfrak{B}$ is closed under the involution of $\mathcal{S}'(\Pi)$. The following result can be established.

Proposition 13.19 *The product* $S \star T \in \mathfrak{B}$*, with* $\|\,S\star T\,\| \leqslant \|\,S\,\|\,\|\,T\,\|$*, for any* $S,T \in \mathfrak{B}$*. Moreover the product* $\star$ *is associative on* $\mathfrak{B}$*, and hence* $\mathfrak{B}$ *is a Moyal algebra. With respect to the norm* $\|\cdot\|$*, the linear space* $\mathfrak{B}$ *is a* C^**-algebra.*

Of course, it is of little use identifying the algebra $\mathfrak{B}$ unless something more can be said about which distributions it contains. In the previous Section,

$Q \cap \overline{Q}$ was identified with $\mathcal{O}[\mathcal{L}^+(\mathcal{S}(\mathbb{R}), L^2(\mathbb{R}))]$ using the unitary map $\mathcal{V}$. By restricting that analysis to $\mathfrak{B}$, it too can be identified.

Proposition 13.20 *For any $T \in \mathfrak{B}$, the map $\mathcal{R}(T) \in \mathcal{L}^+(\mathcal{S}, \mathcal{H})$ defined in equation (13.2.23) extends to a bounded linear operator (also denoted $\mathcal{R}(T)$) on $\mathcal{H}$, with norm $\| \mathcal{R}(T) \| = \| T \|$. The map $\mathcal{R} : \mathfrak{B} \to \mathbb{B}(\mathcal{H})$ is an isometric isomorphism between C^*-algebras which extends the *-representation $\mathcal{R}$ of $L^2(\Pi)$ which was defined in Theorem 13.10.*

Corollary 13.21 *The space $\mathfrak{B}$ is equal to $\mathcal{O}[\mathbb{B}(L^2(\mathbb{R}))]$, and*

$$\mathcal{V}^{-1} \mathcal{R}(T) \mathcal{V} = \mathbf{\Delta}[T] \tag{13.3.2}$$

for all $T \in \mathfrak{B}$. Moreover, Weyl quantization provides an isometric *-isomorphism between the C^*-algebras $\mathfrak{B}$ and $\mathbb{B}(L^2(\mathbb{R}))$.

Thus an operator on $L^2(\mathbb{R})$ is bounded[9] if and only if its symbol is in $\mathfrak{B}$.

13.3.2 *Smooth Observables*

The spaces N and $\overline{N}$ defined in the previous Section are even more important for smooth observables than Q and $\overline{Q}$ are for bounded observables.

Proposition 13.22 *The space $N \cap \overline{N}$ is closed under the Moyal product, so if $S, T \in N \cap \overline{N}$, then $S \star T \in N \cap \overline{N}$ as well. Moreover, $S \star T$ is then defined by the formula*

$$[\![S \star T, F]\!] = [\![S, T \star F]\!], \qquad F \in \mathcal{S}(\Pi). \tag{13.3.3}$$

Additionally, the product $\star$ is associative on $N \cap \overline{N}$, and hence $N \cap \overline{N}$ is a Moyal algebra.

Not only does this result show that $N \cap \overline{N}$ is a Moyal algebra, but also equation (13.3.3) provides us with a useful calculational tool for determining the Moyal products in $N \cap \overline{N}$. Just as we have been able to identify the spaces $\mathfrak{B}$ and $Q \cap \overline{Q}$, we can now identify the space $N \cap \overline{N}$.

[9] This complete characterization of $\mathcal{O}[\mathbb{B}(L^2(\mathbb{R}))]$ should be compared with the partial results in Chapter 8, such as the one due to Calderón & Vaillancourt. As already mentioned, we have to make a choice between precision and utility — *You pays your money and takes your choice.*

Proposition 13.23 *If $T \in N \cap \overline{N}$, the map $\mathcal{R}(T)$ belongs to $\mathcal{L}^+(\mathcal{S})$, and the mapping $\mathcal{R}$ from $N \cap \overline{N}$ to $\mathcal{L}^+(\mathcal{S})$ is a *-algebra isomorphism.*

Corollary 13.24 *The space $N \cap \overline{N}$ is equal to the phase space symbol class $\mathcal{O}[\mathcal{L}^+(\mathcal{S}(\mathbb{R}))]$ of smooth observables, and*

$$\mathcal{R}(T)\mathcal{V}f = \mathcal{V}\boldsymbol{\Delta}[T]f, \qquad f \in \mathcal{S}(\mathbb{R}), \tag{13.3.4}$$

*for any $T \in N \cap \overline{N}$. Thus Weyl quantization provides a *-isomorphism between the algebras $N \cap \overline{N}$ and $\mathcal{L}^+(\mathcal{S}(\mathbb{R}))$.*

Thus we have identified the phase space observable space $\mathcal{O}[\mathcal{L}^+(\mathcal{S}(\mathbb{R}))]$ wholly in terms of the Moyal product[10].

13.3.3 *The Moyal Product In Polar Coordinates*

Since we have been much interested in the quantization of distributions which are expressible solely in terms of either the phase space radius or angle, it is clearly of interest to us to investigate the properties of the Moyal product on such distributions.

13.3.3.1 *Radial Distributions*

In Chapter 9, it was shown that the Weyl quantization $\boldsymbol{\Delta}[T]$ of a radial distribution $T \in \mathcal{S}'_{\text{rad}}(\Pi)$ belongs to $\mathcal{L}^+(\mathcal{S}(\mathbb{R}))$, and is diagonal with respect to the Hermite-Gauss functions. As the product of two such diagonal operators is also a diagonal element of $\mathcal{L}^+(\mathcal{S}(\mathbb{R}))$, it must be that the radial distributions $\mathcal{S}'_{\text{rad}}(\Pi)$ form a commutative Moyal subalgebra of $N \cap \overline{N}$.

This does not of itself guarantee that any special expression for the Moyal product on $\mathcal{S}'_{\text{rad}}(\Pi)$ can be found, but there is one. In Thangavelu's book [220] what is effectively such a formula is given in terms of the twisted convolution. Applying the Fourier transform, this can be rewritten entirely in terms of the Moyal product.

Proposition 13.25 *If $\mathfrak{f}, \mathfrak{g} \in \mathcal{O}^\infty(\mathbb{R})$ are such that $\mathfrak{f}_{\text{rad}}, \mathfrak{g}_{\text{rad}} \in \mathcal{S}_{\text{rad}}(\Pi)$, then*

$$\mathfrak{f}_{\text{rad}} \star \mathfrak{g}_{\text{rad}} = (\mathfrak{f} \times \mathfrak{g})_{\text{rad}}, \tag{13.3.5.a}$$

[10] The same distinction between completeness and utility must be drawn between the above results and those of Chapter 8 concerning smooth observables.

where $\mathfrak{f} \times \mathfrak{g} \in \mathcal{O}^\infty(\mathbb{R})$ *is the function defined by the integral formula*

$$(\mathfrak{f} \times \mathfrak{g})(r) = \int_0^\infty \int_0^\infty \mathfrak{f}(s)\mathfrak{g}(t)\, K(r,s,t)\, st\, ds\, dt \tag{13.3.5.b}$$

for $r \geqslant 0$ *(the values that* $\mathfrak{f} \times \mathfrak{g}$ *might take for negative values of* r *are not relevant). Here* K *is the integral kernel*

$$K(r,s,t) = \frac{1}{2\pi^3} \int_{-\pi}^{\pi} \int_{-\pi}^{\pi} \int_{-\pi}^{\pi} \exp\left[2i\, F(r,s,t : \alpha,\beta,\gamma)\right] d\alpha\, d\beta\, d\gamma\,,$$

where

$$F(r,s,t : \alpha,\beta,\gamma) = rs\sin(\beta-\alpha) + st\sin(\gamma-\beta) + rt\sin(\alpha-\gamma)\,,$$

which expression is a bounded totally symmetric function in the three radial variables r*,* s *and* t*. Properties of the Racah coefficients [195] can be used to determine a power series expansion for the integral kernel* K*, namely*

$$K(r,s,t) = 4\sum_{n\geqslant 0} \frac{(-1)^n}{(n!)^2} K_n(r,s,t), \tag{13.3.6.a}$$

where $K_n(r,s,t)$ *is the symmetric polynomial*

$$K_n(r,s,t) = \sum_{\substack{a,b,c\geqslant 0\\ a+b+c=2n}} \binom{n}{a}\binom{n}{b}\binom{n}{c} r^{2a}\, s^{2b}\, t^{2c}. \tag{13.3.6.b}$$

The connection with the work of Thangavelu can be seen from the identity

$$K(r,s,t) = \frac{4}{\pi}\int_{|s-t|}^{s+t} J_0(2ru)\, \frac{\cos(2\rho(u,s,t))}{\rho(u,s,t)}\, u\, du, \tag{13.3.6.c}$$

where $\rho(x,y,z)$ *is the function*

$$\rho(x,y,z) = \frac{1}{2}\left[2(x^2y^2 + x^2z^2 + y^2z^2) - (x^4+y^4+z^4)\right]^{\frac{1}{2}}. \tag{13.3.6.d}$$

It remains to discover a more practicable expression for the integral kernel K with which the quantities $\mathfrak{f}_{\mathrm{rad}} \star \mathfrak{g}_{\mathrm{rad}}$ can be calculated.

The radial distributions $\mathfrak{f}^{(k)}_{\mathrm{rad}}(p,q) = (p^2+q^2)^{k/2} = r^k$ do not belong to $\mathcal{S}_{\mathrm{rad}}(\Pi)$ and so they are not covered by the previous Proposition. It is possible, however, to derive formulæ for their Moyal products, and these are particularly simple when the integers $k \geqslant 0$ are even.

Proposition 13.26 *For any j, $k \geqslant 0$ the following identity holds:*

$$\mathfrak{f}^{(2j)}_{\mathrm{rad}} \star \mathfrak{f}^{(2k)}_{\mathrm{rad}} = \sum_{t=0}^{\min(j,k)} (-1)^t \frac{j!\,k!}{(j-t)!\,(k-t)!} \binom{j+k-t}{t} \mathfrak{f}^{(2(j+k-2t))}_{\mathrm{rad}}. \tag{13.3.7}$$

Proof: Since $\mathfrak{f}^{(2)}_{\mathrm{rad}}(p,q) = p^2 + q^2$, it is possible to use the polynomial techniques of Subsection 13.3.4 to show that

$$\mathfrak{f}^{(2)}_{\mathrm{rad}} \star F = H_R\,F, \qquad F \in \mathcal{S}(\Pi),$$

where H_R is one of the elliptic differential operators introduced in Section 12.4.3, in relation to the special Hermite functions. From this it is elementary to show that

$$\mathfrak{f}^{(2)}_{\mathrm{rad}} \star \mathfrak{f}^{(2k)}_{\mathrm{rad}} = \mathfrak{f}^{(2(k+1))}_{\mathrm{rad}} - k^2\,\mathfrak{f}^{(2(k-1))}_{\mathrm{rad}}, \qquad k \geqslant 0,$$

and the general result follows by induction. ∎

Corollary 13.27 *The algebra of polynomials in the Hamiltonian function ν for the harmonic oscillator forms a Moyal subalgebra of $\mathcal{S}'_{\mathrm{rad}}(\Pi)$ (and also a subalgebra of the Moyal algebra of all polynomials in p and q to be considered below).*

This Corollary is not surprising, since it has already been observed that $\mathbf{\Delta}\big[\mathfrak{f}^{(2k)}_{\mathrm{rad}}\big]$ is a polynomial in the number operator N for any integer $k \geqslant 0$. However the Weyl quantization of a polynomial distribution $\mathfrak{g}(p^2+q^2)$ is not equal to the operator $\mathfrak{g}(2N+I)$ for any but linear polynomials — this is reflected in the fact that $\mathfrak{f}^{(2j)}_{\mathrm{rad}} \star \mathfrak{f}^{(2k)}_{\mathrm{rad}} \neq \mathfrak{f}^{(2(j+k))}_{\mathrm{rad}}$.

While it is possible to display formulæ for $\mathfrak{f}^{(j)}_{\mathrm{rad}} \star \mathfrak{f}^{(k)}_{\mathrm{rad}}$ which include odd integers, these will be omitted on the grounds of complexity. It is easy to show, for example, that $\mathfrak{f}^{(1)}_{\mathrm{rad}} \star \mathfrak{f}^{(1)}_{\mathrm{rad}}$ is a radial distribution which is not a simple function of r — it is truly a distribution. However, restricting attention to even powers of r keeps us within the province of functions and much simpler calculations. But it might be the case that more general calculations lead to identities amongst hypergeometric functions which are of interest elsewhere.

13.3.3.2 *Angular Distributions*

Unlike the above case of radial distributions, the class of angular distributions $\mathcal{S}'_{\text{ang}}(\Pi)$ is not a good source for Moyal algebras. This is despite the interesting fact that the quantizations $U_k = \mathbf{\Delta}\big[(\chi_k)_{\text{ang}}\big]$, where $\chi_k(e^{i\beta}) = e^{ik\beta}$, satisfy the identity,

$$U_{2j}\, U_{2k} = U_{2(j+k)}, \qquad j,\, k \geqslant 0, \tag{13.3.8.a}$$

and hence the angular distributions $(\chi_k)_{\text{ang}}$ are such that

$$(\chi_{2j})_{\text{ang}} \star (\chi_{2k})_{\text{ang}} = (\chi_{2(j+k)})_{\text{ang}}, \qquad j,\, k \geqslant 0. \tag{13.3.8.b}$$

Such an identity does not hold, however, for odd indices — U_1^2 is not the quantization of an angular distribution, and is certainly not equal to U_2, for example. Thus the finite linear span $\mathbb{S}_+$ of the set of functions $\{\, (\chi_{2k})_{\text{ang}} \,:\, k \geqslant 0 \,\}$ is a subspace of $\mathcal{S}'_{\text{ang}}(\Pi)$ which is invariant and associative with respect to the Moyal product. However, this space is not invariant under the involution of $\mathcal{S}'(\Pi)$, and hence is not a Moyal algebra. Similarly, the finite linear span $\mathbb{S}_-$ of the set of functions $\{\, (\chi_{-2k})_{\text{ang}} \,:\, k \geqslant 0 \,\}$ is a subspace of $\mathcal{S}'_{\text{ang}}(\Pi)$ which is invariant and associative under the Moyal product, but is not a Moyal algebra. This last observation follows since $\mathbb{S}_- = \overline{\mathbb{S}_+}$. Nor is the space $\mathbb{S}_+ \cup \mathbb{S}_-$ a Moyal algebra, even though it is closed under the involution of $\mathcal{S}'(\Pi)$. This is because the operator $U_2^*\, U_2$ is not the Weyl quantization of an angular distribution (and in particular is not equal to $I = U_0$), and so $(\chi_{-2})_{\text{ang}} \star (\chi_2)_{\text{ang}}$ is not an angular distribution — since its quantization $U_2^*\, U_2$ is diagonal with respect to the Hermite-Gauss functions, it is in fact a radial distribution!

For these reasons, except for the trivial case of the algebra of constant functions, we do not expect that Moyal algebras of angular distributions can be found — which is itself interesting.

13.3.4 *Polynomials*

When considering the Moyal products of the $\mathfrak{f}_{\text{rad}}^{(2k)}$, it was observed that these distributions are all polynomials in the phase space coordinate functions p and q, and that the properties of the $\mathfrak{f}_{\text{rad}}^{(k)}$ were obtained from those of polynomial distributions in general. Many treatments of the Moyal product begin with this topic, and often do not go much further. We have chosen to end with it so as to give the greater significance to the analytic and algebraic structural themes. From a historical perspective, the Moyal product was in fact derived from a consideration of the identities to be found below.

The generating function $E_{a,b}(p,q) = e^{i(ap+bq)}$ of the coordinate functions was originally introduced in Proposition 8.32, and it was noted in that Proposition that $\mathbf{\Delta}[E_{a,b}] = W(a,b)$ for all a, $b \in \mathbb{R}$. This implies the Moyal product result

$$E_{a,b} \star E_{c,d} = e^{\frac{1}{2}i(ad-bc)} E_{a+c,b+d}, \qquad a,\, b,\, c,\, d \in \mathbb{R}, \tag{13.3.9}$$

so that the linear span of the set $\{ E_{a,b} : a,\, b \in \mathbb{R} \}$ forms a rather basic Moyal algebra contained in both $\mathfrak{B}$ and $N \cap \overline{N}$. More importantly for our purposes, this identity can be integrated, yielding

$$\big[E_{a,b} \star F\big](p,q) = e^{i(ap+bq)} F\big(p - \tfrac{1}{2}b, q + \tfrac{1}{2}a\big), \qquad F \in \mathcal{S}(\Pi),\ a,\, b \in \mathbb{R}. \tag{13.3.10}$$

Differentiation leads to the expression

$$\begin{aligned} \big(p^m q^n \star F\big)(p,q) &= (-i)^{m+n} \frac{\partial^{m+n}}{\partial a^m \partial b^n} \big(E_{a,b} \star F\big)\Big|_{a=b=0} \\ &= \sum_{j=0}^{m} \sum_{k=0}^{n} \binom{m}{j} \binom{n}{k} \frac{i^{k-j}}{2^{j+k}} p^{m-j} q^{n-k} \frac{\partial^{j+k} F}{\partial p^k \partial q^j}(p,q), \end{aligned} \tag{13.3.11}$$

for any m, $n \geqslant 0$. This implies in particular that the space $\mathcal{P}$ of all polynomial functions in the coordinate functions p and q forms a Moyal algebra contained in $N \cap \overline{N}$. It is now trivial to derive the fact that $\mathfrak{f}^{(2)}_{\mathrm{rad}} \star F = H_R F$ for any $F \in \mathcal{S}(\Pi)$, as was stated in the earlier Subsection on radial distributions.

It is our intention now to investigate the algebra structure of the Moyal algebra $\mathcal{P}$. To do so, the Moyal product of p and q is necessary. It follows from (13.3.11) that

$$p \star q = pq - \tfrac{1}{2} i\, \mathfrak{i}, \qquad q \star p = pq + \tfrac{1}{2} i\, \mathfrak{i}, \tag{13.3.12}$$

from which we deduce that the Moyal bracket of p and q is given by the formula

$$\{p\,,\,q\}_\star = i\big[p \star q - q \star p\big] = \mathfrak{i}. \tag{13.3.13}$$

If we recall the Heisenberg Lie algebra $\mathfrak{h}$ discussed in Chapter 8, namely the three-dimensional Lie algebra with basis $X(1)$, $X(2)$ and $X(3)$ satisfying the identities

$$[X(1), X(2)] = X(3), \quad [X(1), X(3)] = [X(2), X(3)] = 0, \qquad (8.7.3.a)$$

then the following can be shown.

Proposition 13.28 *The map defined by the identifications*

$$X(1) \mapsto p, \qquad X(2) \mapsto q, \qquad X(3) \mapsto \mathrm{i}, \qquad (13.3.14)$$

defines a bijective algebra isomorphism between the universal enveloping algebra of $\mathfrak{h}$ *and the Moyal algebra* $\mathcal{P}$ *of polynomial functions in* p *and* q.

So now we have a direct derivation of the equality of Moyal and Poisson brackets for p and q, a fundamental tenet of quantization. Going further, inspection of the results of the following Section will show that

$$\{\mathfrak{g}, F\}_\star = \{\mathfrak{g}, F\}, \qquad F \in \mathcal{S}(\Pi) \qquad (13.3.15)$$

for any polynomial $\mathfrak{g}$ which is no more than quadratic in p and q. In particular, therefore, this result holds for the harmonic oscillator Hamiltonian $\nu = \frac{1}{2}\mathfrak{f}^{(2)}_{\mathrm{rad}}$. In this context, it will be recalled that it has been shown that the Moyal and Poisson brackets of ν and the phase space angle function φ coincide. This is in accord with these results, but nonetheless one which needed proving, given the singular nature of the distribution φ.

Remark That equation (13.3.15) is only valid for polynomials $\mathfrak{g}$ of low degree in p and q is no accident, and is confirmed by a number of more general results. For example, a theorem of Groenewald [90] states that there is no linear map $\Upsilon : \mathcal{P}_4 \to \mathcal{L}^+(\mathcal{S}(\mathbb{R}))$ such that $\Upsilon(p) = P$, $\Upsilon(q) = Q$, and $\Upsilon(\{F, G\}) = i[\Upsilon(F), \Upsilon(G)]$ for all $F, G \in \mathcal{P}_3$, where (for any $n \in \mathbb{N}$) $\mathcal{P}_n$ is the space of polynomials in p and q of degree at most n.

A cognate result is due to van Hove, [121], [34], [86], [63]. Let $\mathcal{X}$ denote the set of real infinitely differentiable functions on phase space which generate global (not just local) one-parameter flows[11]. Then there exists no dense subspace $\mathcal{D}$ of $L^2(\mathbb{R})$ and map $\Upsilon : \mathcal{X} \to \mathcal{L}^+(\mathcal{D})$ such that

1. $\Upsilon(F) = \Upsilon(F)^+$ is symmetric for all $F \in \mathcal{X}$,

[11] It is a result of classical analysis that $\mathcal{X}$ is neither a linear space nor a Lie algebra.

2. the operator $e^{it\Upsilon(F)}$ exists and preserves $\mathcal{D}$ for any $t \in \mathbb{R}$ and $F \in \mathcal{X}$,
3. if $F, G \in \mathcal{X}$ and $a, b \in \mathbb{R}$ are such that $aF + bG \in \mathcal{X}$, then $\Upsilon(aF + bG) = a\Upsilon(F) + b\Upsilon(G)$,
4. $\Upsilon(p) = P$ and $\Upsilon(q) = Q$ (and hence $\mathcal{S}(\mathbb{R}) \subseteq \mathcal{D}$),
5. if $F, G \in \mathcal{X}$ are such that $\{F, G\} \in \mathcal{X}$, then

$$\Upsilon(\{F, G\}) = i\big[\Upsilon(F), \Upsilon(G)\big],$$

6. if $F, G \in \mathcal{X}$ are such that there exists $H \in \mathcal{X}$ such that

$$\Phi(F; s)\Phi(G; t)\Phi(F; -s) = \Phi(H; t)$$

for any $s, t \in \mathbb{R}$, where $\Phi(F; t)$ denotes the flow at time t generated by $F \in \mathcal{X}$, then

$$e^{is\Upsilon(F)}\, e^{it\Upsilon(G)}\, e^{-is\Upsilon(F)} = e^{it\Upsilon(H)}$$

for any $s, t \in \mathbb{R}$.

These results confirm the view that the Poisson bracket is of limited validity in quantum mechanics, and should be replaced by the Moyal bracket. ■

13.4 The Moyal Product As A Deformation

Deformation is the name mathematicians give to what might be called perturbation theory for algebraic and geometric structures. The basic idea is to establish a formalism for analyzing structures which are similar, but not identical, and which approximate some simpler structure. If some control can be placed on the manner and extent to which a deformed algebraic structure differs from the simpler one, it is possible that properties of the simpler structure can be used to deduce corresponding properties of the more complex family. A good general discussion of these ideas can be found in [74].

As an example, consider the Euclidean space $\mathbb{R}^2$. Topologically, $\mathbb{R}^2$ can (almost) be identified as the limit of spheres of radius R as $R \to \infty$. After appropriate scalings, this means that geometric transformations of $\mathbb{R}^2$ can be approximated, for any $R > 0$, by the (translations to $\mathbb{R}^2$ of the) geometric transformations of the sphere of radius R. Moreover, these approximations are ever more accurate as R increases. Hence algebraic structures on $\mathbb{R}^2$ can be approximated by algebraic structures based upon the group SO(2) and dependent upon a parameter R, with the original structure being recovered by taking the limit as $R \to \infty$.

This position is typical of deformation theory; the deformation is parametrized by some real parameter t (in the above case, $t = R^{-1}$). Typically, it is necessary to demand that the deformed algebra structure varies smoothly with this parameter t.

In concrete examples, this approach is full of problems relating to the convergence of series and the continuity or smoothness of relevant functions. On the other hand, the advantage of this approach is that topological problems can be separated from algebraic ones[12].

Our aim is to describe quantum mechanics as a deformation of classical mechanics by means of a scaling of the Moyal product. This is a special case of the theory of algebra deformations, which we outline very briefly. Let us suppose, therefore, that $\mathcal{A}$ is a complex unital subalgebra of $\mathcal{C}^\infty(\Pi)$, which we now regard as a subalgebra of the algebra $\mathbb{C}[\![\mathcal{A}]\!]$ of all formal power series in the indeterminate t with coefficients in the algebra $\mathcal{A}$.

Definition 13.29 *A formal deformation of $\mathcal{A}$ is determined by a family $\{\mu_n : n \in \mathbb{N}\}$ of bilinear maps from $\mathcal{A} \times \mathcal{A}$ to $\mathcal{A}$ such that the formula*

$$f \star_{(t)} g \;=\; fg + \mu_1(f,g)t + \mu_2(f,g)t^2 + \dots\,, \qquad f,\, g \in \mathcal{A}, \quad (13.4.1)$$

defines a structure of associative algebra on $\mathbb{C}[\![\mathcal{A}]\!]$.

---

[12] It should be remembered that, in physical applications, the deformation parameter t is often of crucial physical significance, so that changing the value of this parameter results in different physics. For example, we shall be interested in deformation systems in which classical mechanics is described when t takes the value 0, while quantum mechanics results when t is $\hbar$. Thus deformation theory should be seen as a mathematical tool for interpolating between classical and quantum mechanics — caution should be exercised before assigning any physical interpretation to the intermediate systems that it describes.

Equation (13.4.1) defines a bilinear map $\star_{(t)} : \mathcal{A} \times \mathcal{A} \to \mathcal{A}$ via the formula

$$\left(\sum_{n \geqslant 0} f_n \, t^n\right) \star_{(t)} \left(\sum_{n \geqslant 0} g_n \, t^n\right) = \sum_{N \geqslant 0} \left(\sum_{m+n+k=N} \mu_k(f_m, g_n)\right) t^N, \tag{13.4.2}$$

where, for convenience, we define the map $\mu_0(f, g) = fg$ for any $f,\, g \in \mathcal{A}$. For obvious reasons, the map μ_1 is called the *derivative* of the product $\star_{(t)}$.

Complications arise from the requirement that the resulting map $\star_{(t)}$ on $\mathbb{C}[\![\mathcal{A}]\!]$ be associative. For this to be the case, the maps μ_n must satisfy a countable collection of increasingly complex interrelationships. For example, the first two of these requirements demand that

$$\begin{aligned} \mu_1(f, g)h + \mu_1(fg, h) &= f\mu_1(g, h) + \mu_1(f, gh), \\ \mu_2(f, g)h + \mu_1\big(\mu_1(f, g), h\big) + \mu_2(fg, h) &= f\mu_2(g, h) + \mu_1\big(f, \mu_1(g, h)\big) + \mu_2(f, gh), \end{aligned}$$

for all $f,\, g,\, h \in \mathcal{A}$. The first of these is well-known to mathematicians. Technically, it states that μ_1 must be a 2-cocycle in the Hochschild cohomology of the algebra $\mathcal{A}$ when acting upon itself. Interpretation of the second, or any subsequent, requirement is even more complicated.

One of the problems of interest to mathematicians is to ask, given a particular algebra $\mathcal{A}$, how to classify all the formal deformations of $\mathcal{A}$ which have particular properties. For example, it would be interesting to classify all the formal deformations of the algebra $\mathcal{A}$ for which the first function μ_1 is given in advance[13].

For purposes of application to physics, the above algebraic considerations must be combined with topology. In other words, we must seek an algebra $\mathcal{A}$ equipped with a formal derivation $\star_{(t)}$ such that the series in (13.4.1) converges to some element of $\mathcal{A}$ for any $f, g \in \mathcal{A}$ and $t > 0$, thereby providing $\mathcal{A}$ with a different algebraic structure for each $t > 0$. Moreover, these algebraic structures should vary smoothly with the parameter t. To be able to do this, all the formal manipulations beloved of deformation theorists must be subjected to detailed consideration concerning their convergence and validity, and such considerations will be different for each choice of algebra $\mathcal{A}$. Fortunately, for our purposes, all of these questions

[13] This latter question is of particular relevance for quantum mechanics where the initial function μ_1 is equal to $-\frac{1}{2}i$ times the Poisson bracket.

can be answered positively, at least when $\mathcal{A}$ is either the space $\mathcal{S}(\Pi)$ of test functions or else the space $\mathcal{P}$ of polynomials in p and q. Thus, at least, a complete deformation theory can be established for these spaces.

An interesting aspect of this formalism is that the resulting algebra structures $\left[\mathcal{A}, \star_{(t)}\right]$ are topologically isomorphic for all $t > 0$, and hence in particular are all isomorphic to the algebra of quantum observables, which is traditionally described when $t = \hbar$. However, the algebraic structure collapses to one of a quite different sort when $t = 0$, namely to the commutative classical algebra structure of pointwise multiplication. Such drastic structure changes at a limiting point occur elsewhere in mathematics and physics.

While this formulation of quantum mechanics is instructive and has its interest, it is not clear to us that it gives insights into the deep unsolved problems of quantum theory which the analytic approach cannot. For this reason, we have chosen in this book to concentrate on the analytic approach, and therefore we shall content ourselves with showing how the Moyal product can be expressed as a deformation of the pointwise product on phase space.

In order to be certain that all the series we consider are convergent, we choose $\mathcal{A}$ to be equal to $\mathcal{S}(\Pi)$. If Planck's constant were included explicitly in equation (13.2.3.a) for the Moyal product, we would naturally be led to consider the products

$$(F \star_{(t)} G)(\xi) = \frac{1}{(\pi t)^2} \iint_\Pi \iint_\Pi F(\eta)G(\zeta)\, e^{-\frac{2i}{t}\Psi(\xi,\eta,\zeta)}\, dA(\eta)\, dA(\zeta), \tag{13.4.4.a}$$

for any F, $G \in \mathcal{S}(\Pi)$ and $t > 0$, with the actual Moyal product being obtained when $t = \hbar$.

This product formula is most readily analyzed through its Fourier transform,

$$\left[\mathcal{F}(F \star_{(t)} G)\right](\xi) = \frac{1}{2\pi} \iint_{\mathbb{R}^2} (\mathcal{F}F)(\xi - \eta)\, (\mathcal{F}G)(\eta)\, e^{\frac{1}{2}it\Omega(\xi,\eta)}\, dA(\eta), \tag{13.4.4.b}$$

for F, $G \in \mathcal{S}(\Pi)$ and $t > 0$. This equation is clearly a reparametrized version of equation (3.3.10.a) describing the twisted convolution, and shows that the apparent singularity in equation (13.4.4.a) at $t = 0$ is illusory.

Expanding this integral as a power series[14] in t yields the identity

$$\big[\mathcal{F}(F \star_{(t)} G)\big](\xi) \;=\; \sum_{n\geqslant 0} \frac{i^n\, t^n}{2^n\, n!}\big[I_n(F,G)\big](\xi) \tag{13.4.5.a}$$

for all F, $G \in \mathcal{S}(\Pi)$ and $t > 0$, where the functions $I_n(F,G)$ are defined as follows:

$$\big[I_n(F,G)\big](\xi) \;=\; \frac{1}{2\pi}\iint_{\mathbb{R}^2} (\mathcal{F}F)(\xi-\eta)\,(\mathcal{F}G)(\eta)\,\Omega(\xi,\eta)^n\, dA(\eta), \qquad n \geqslant 0. \tag{13.4.5.b}$$

In order to identify the functions $I_n(F,G)$, it is convenient to introduce the differential operator $\mathfrak{P}$ on $\mathcal{S}(\mathbb{R}^4)$ defined by

$$\mathfrak{P}H \;=\; \frac{\partial^2 H}{\partial\xi_1\partial\xi_4} - \frac{\partial^2 H}{\partial\xi_2\partial\xi_3}, \qquad H \in \mathcal{S}(\mathbb{R}^4)\,. \tag{13.4.6}$$

For any F, $G \in \mathcal{S}(\Pi)$ and $n \geqslant 0$,

$$\big[I_n(F,G)\big](\xi) \;=\; \frac{1}{2\pi}(-1)^n \iint_{\mathbb{R}^2} \Big[(\mathcal{F}\otimes\mathcal{F})\mathfrak{P}^n(F\otimes G)\Big](\xi-\eta,\eta)\, dA(\eta), \tag{13.4.7}$$

in an obvious notation, from which it is immediately clear that

$$\big[\mathcal{F}^{-1}I_n(F,G)\big](\xi) \;=\; (-1)^n\,\big[\mathfrak{P}^n(F\otimes G)\big](\xi,\xi) \tag{13.4.8}$$

for all $n \geqslant 0$. Thus we have derived the following (convergent) power series expansion for the Moyal product.

Proposition 13.30 *The parametrized Moyal product $F \star_{(t)} G$ of two test functions F, $G \in \mathcal{S}(\Pi)$ is given by the convergent power series*

$$(F \star_{(t)} G)(\xi) \;=\; \sum_{n\geqslant 0} \frac{(-i)^n\, t^n}{2^n\, n!}\,\big[\mathfrak{P}^n(F\otimes G)\big](\xi,\xi)\,, \tag{13.4.9.a}$$

which formula can be abbreviated to

$$(F \star_{(t)} G)(\xi) \;=\; \Big[\exp\big[-\tfrac{1}{2}it\,\mathfrak{P}\big]\,(F\otimes G)\Big](\xi,\xi)\,. \tag{13.4.9.b}$$

The fact that

$$\big[\mathfrak{P}(F\otimes G)\big](\xi,\xi) \;=\; \{\,F\,,\,G\,\}(\xi),$$

[14]Since F, $G \in \mathcal{S}(\Pi)$, we can be sure that all the functions in this series belong to $\mathcal{S}(\Pi)$, and moreover that this series converges in the topology of $\mathcal{S}(\Pi)$.

where $\{\,,\,\}$ denotes the Poisson bracket as usual, leads many writers to express this power series expansion in the form

$$F \star_{(t)} G \;=\; \exp\left[-\tfrac{1}{2}it\{\,,\,\}\right](F \otimes G). \tag{13.4.10}$$

Like many other similar formulæ, this expression is convenient, but needs to be handled with caution.

What is remarkable is the robustness of this power series expansion for the Moyal product. The functions

$$\mu_n(F,G) \;=\; \frac{i^n}{2^n\, n!}\mathcal{F}^{-1} I_n(F,G), \qquad n \geqslant 1,$$

satisfy the requirements for defining a formal deformation of the algebra $\mathcal{S}(\Pi)$, but equation (13.4.9.a) is only certain to converge when $F, G \in \mathcal{S}(\Pi)$. However, the observed fact is that, in nearly all cases, when equation (13.4.9.a) converges, the resulting function turns out to be the Moyal product of F and G.

For example, by taking transposes, equation (13.4.9.a) can be used to define the Moyal product of an element of $\mathcal{S}(\Pi)$ and an element of $\mathcal{S}'(\Pi)$ (in either order), with the series in equation (13.4.9.a) now being convergent in the topology of $\mathcal{S}'(\Pi)$. This change of topology, however, makes it difficult to take this result much further — for example, we cannot use it to define a parametrized Moyal product on the space $N \cap \overline{N}$, even though an unparametrized Moyal product exists on that space[15]. However, we can define the parametrized Moyal product of an element of $\mathcal{S}(\Pi)$ and an element of the polynomial algebra $\mathcal{P}$, or even the parametrized Moyal product of two elements of $\mathcal{P}$, since, in either of these cases, the series in equation (13.4.9.a) is a finite series.

Dropping the subscript notation, but including $\hbar$ explicitly in our formalism, we have shown that the (true) Moyal product $\star$ on $\mathcal{S}(\Pi)$ has the asymptotic form[16]

$$F \star G \;\sim\; FG \,-\, \tfrac{1}{2}i\hbar\{\,F\,,G\,\} \,+\, O(\hbar^2), \qquad \hbar \to 0, \tag{13.4.11}$$

for any $F,\, G \,\in\, \mathcal{S}(\Pi)$, giving the rate of convergence to the algebra of classical mechanics in terms of $\hbar$, at least for test functions.

[15]This is, perhaps, not surprising, since $N \cap \overline{N}$ is not an algebra with respect to pointwise multiplication.

[16]Assuming F and G are independent of $\hbar$.

When $\hbar$ is included explicitly in the formalism, equation (13.1.2.b) for the Moyal bracket must be modified to read

$$\{F, G\}_\star = \frac{i}{\hbar}(F\star G - G\star F). \tag{13.4.12}$$

Since $I_n(F,G)$ is symmetric in F and G when n is even, but antisymmetric when n is odd, we obtain the interesting (and well-known) expression

$$\{F, G\}_\star(\xi) = \frac{2}{\hbar}\Big[\sin\big[\tfrac{1}{2}\,\hbar\,\mathfrak{P}\big]\,(F\otimes G)\Big](\xi,\xi), \tag{13.4.13.a}$$

for the Moyal bracket, again at least when F, G belong to either $\mathcal{S}(\Pi)$ or $\mathcal{P}$. This gives the following asymptotic expression for the Moyal bracket,

$$\{F, G\}_\star \sim \{F, G\} + O(\hbar^2), \qquad \hbar\to 0, \tag{13.4.13.b}$$

when $F, G \in \mathcal{S}(\Pi)$. Both the sine formula for the Moyal bracket and its asymptotic extension enjoy the same calculational robustness that the comparable formulæ for the Moyal product do, and so can be extended, for example, to $\mathcal{P}$.

Chapter 14

Ordered Quantization

There are nine and sixty ways of constructing tribal lays
And—every—single—one—of—them—is—right!
– Rudyard Kipling, *In the Neolithic Age*

14.1 Prologue

When developing the theory of Weyl quantization in Chapter 8, we observed that there were many possible associations between classical and quantum mechanics. In that Chapter, we chose to work with the association of Weyl, not least on the grounds that that approach treated position and momentum observables on an equal footing — for example, the Weyl quantization of the phase space observable pq is the operator $\frac{1}{2}(PQ+QP)$. However, because quantum mechanics is properly more general than classical mechanics, there are, in fact, too many possible connections between the two for comfort, though the number can be cut down by the application of certain general principles. We shall discuss how these principles are to be applied later.

As was emphasized in Chapter 8, each connection between classical and quantum mechanics corresponds to a choice of spectral theory for noncommuting operators. However, we shall not formulate our discussion in this way. Rather, we shall adopt the usual approach, which is a carry-over from quantum field theory.

In order to eliminate certain spurious divergences which result from the prescription adopted in second quantization, the notion of *operator ordering* was invented. The most familiar example of this procedure arises concerning the zero point energy. When quantizing the free electromagnetic field Hamiltonian, the field is decomposed into modes, as discussed in Chapter 11. A naïve approach would result in the presence of a term $\frac{1}{2}\hbar\omega$ for each mode. But this would yield an infinite contribution from the collection of all modes. This infinity is eliminated by writing all polynomials in the annihilation and creation operators with the creation operators to the left of the annihilation operators. This convention for the order in which operators are to be considered is termed *normal ordering*. The utility of this algorithm rests on the fact that the Fock vacuum is annihilated by the lowering operator of each mode. For interacting states, there is no such possibility. A partial substitute based on subtracting vacuum expectation values, as in a linked cluster expansion, is sometimes useful — but will not be considered here. The legitimacy of using such an ordering is that, since second quantization is a construct whose definition is at our disposal to a certain extent, it might as well be defined in this way.

Once normal ordering has been considered, various other orderings come to mind. This is true even of systems with one degree of freedom, which will be the ones discussed in this Chapter. For reasons of space, and because they seem to represent the cases of particular interest to quantum optics, only two families of ordered quantizations will be considered.

14.1.1 *Ordered Weyl Group Quantization*

Any scheme of quantization can, ultimately, be characterized in terms of the choice that it makes for the operator to be associated with the classical generating function $E_{a,b}(p,q) = e^{i(ap+bq)}$ introduced in Proposition 8.32 — Weyl quantization, in particular, associates this function with the Weyl group. While, for reasons that we have already discussed, we feel that Weyl quantization provides the most natural choice for this association, there are others that might be made. We shall concentrate on what are (to some extent) the four most natural alternative choices for this association (together with the choice for Weyl quantization, for the purposes of completeness). These families are based on the respective formal associations:

$$E_{a,b} \mapsto e^{ia\overline{P}}\, e^{ib\overline{Q}}, \tag{14.1.1.a}$$

$$E_{a,b} \mapsto e^{ib\overline{Q}}\, e^{ia\overline{P}}, \tag{14.1.1.b}$$

$$E_{a,b} \mapsto W(a,b) = e^{i(\overline{aP+bQ})}, \tag{14.1.1.c}$$

$$E_{a,b} \mapsto e^{i\overline{z}A^+}\, e^{izA}, \tag{14.1.1.d}$$

$$E_{a,b} \mapsto e^{izA}\, e^{i\overline{z}A^+}, \tag{14.1.1.e}$$

where, as usual, $\sqrt{2}z = b - ia$. These choices will lead to five ordering schemes for quantization which we shall term Q-ordering, P-ordering, Weyl ordering, normal (or Wick) ordering, and antinormal (or anti-Wick) ordering, respectively. It is not necessary to study these five quantization schemes separately, since they can each be regarded as special cases of a two-parameter family of orderings, as will be shown below.

However, before embarking upon any unified study of the various quantization schemes, it is important to take a step back and to see how any choice of an association between the classical generating function $E_{a,b}$ with some ordered variant $W_\natural(a,b)$ of the Weyl group yields a full quantization scheme[1].

Given any such choice, the expression

$$\boldsymbol{\Delta}_\natural\,[T] \;=\; \frac{1}{2\pi}\iint_{\mathbb{R}^2}[\mathcal{F}T](a,b)\,W_\natural(a,b)\,da\,db \qquad (14.1.2)$$

determines (at least formally) what will be called $\natural$-*ordered quantization*, or simply $\natural$-*quantization*. As with Weyl quantization, it is necessary to make this heuristic formula rigorous and capable of application to a wide range of functions or distributions T by constructing the ordered analogue of the Wigner transform, the $\natural$-Wigner transform $\mathcal{G}_\natural$.

This leads back to the fundamental problem faced when constructing the smooth model. The strength of the smooth model lies in the fact that the Wigner transform $\mathcal{G}(g \otimes f)$ can usefully be defined for all functions $f, g \in \mathcal{S}(\mathbb{R})$. While it may be possible to define the transform $\mathcal{G}_\natural(g \otimes f)$ for all such functions in some other quantization schemes, this is not the case in all of them. Consequently, a rigorous implementation of some quantization schemes may require the development of some new analogue of the smooth model, and there is no guarantee that the result will satisfy the *desiderata* that we have suggested are necessary in any such model.

To formalize the possibilities, all choices of $W_\natural$ that we shall consider will be such that $W_\natural(a,b) \in \mathcal{L}^+(\mathcal{S}(\mathbb{R}))$ for all $a,\, b \in \mathbb{R}$. Now define

$$\big[\mathcal{W}_\natural(g\otimes f)\big](a,b) \;=\; \langle \overline{g}\,,\, W_\natural(a,b)f\rangle\,, \qquad a,\, b \in \mathbb{R}. \qquad (14.1.3)$$

for any $f, g \in \mathcal{S}(\mathbb{R})$. When $\natural$-quantization works, it will be possible to

[1]There is no reason, however, to suppose that the operators $W_\natural(a,b)$ will obey a (symplectic) group relation in general — that is a property enjoyed by Weyl ordering in particular.

identify a function class $\mathcal{S}_\sharp$, a dense linear subspace of $L^2(\mathbb{R})$ contained in $\mathcal{S}(\mathbb{R})$, for which $\mathcal{W}_\sharp(g\otimes f)$ belongs to $\mathcal{S}(\mathbb{R}^2)$ whenever $f,\, g \in \mathcal{S}_\sharp$. The space $\mathcal{S}_\sharp$ is the common dense domain for the smooth $\sharp$-model.

The $\sharp$-Wigner transform $\mathcal{G}_\sharp$ is then the bilinear map from $\mathcal{S}_\sharp \times \mathcal{S}_\sharp$ to $\mathcal{S}(\Pi)$ given by

$$\mathcal{G}_\sharp(g\otimes f) \;=\; \frac{1}{2\pi}\,\mathcal{F}\big(\mathcal{W}_\sharp(g\otimes f)\big), \qquad f,\, g\in\mathcal{S}_\sharp. \tag{14.1.4}$$

For any distribution $T \in \mathcal{S}'(\Pi)$, then, the map

$$(g,f) \;\mapsto\; [\![\, T\,,\, \mathcal{G}_\sharp(g\otimes f)\,]\!]$$

is a bilinear functional on $\mathcal{S}_\sharp$, and so can be identified with a linear map $\mathbf{\Delta}_\sharp\,(T)$ from $\mathcal{S}_\sharp$ to its algebraic dual $\mathcal{S}_\sharp^*$ by the formula

$$[\![\, \mathbf{\Delta}_\sharp\,[T]\, f\,,\, g\,]\!] = [\![\, T\,,\, \mathcal{G}_\sharp(g\otimes f)\,]\!], \tag{14.1.5}$$

for any $f, g \in \mathcal{S}_\sharp$. Ideally, we would like to be able to topologize $\mathcal{S}_\sharp$ so that $\mathbf{\Delta}_\sharp\,(T)$ becomes a continuous linear map from $\mathcal{S}_\sharp$ to its (strong) topological dual $\mathcal{S}'_\sharp$.

For many, but not all, of the orderings to be considered here, it is possible to choose $\mathcal{S}_\sharp$ to be equal to $\mathcal{S}(\mathbb{R})$. The $\sharp$-Wigner transform $\mathcal{G}_\sharp$ can then be extended to an integral transform from $\mathcal{S}(\mathbb{R}^2)$ to $\mathcal{S}(\Pi)$, as was the case for Weyl quantization. Such ordered quantizations can be considered fully within the smooth model as we have defined it. It is notable, however, that this cannot be done for normal ordering[2].

Returning in particular to the five quantization schemes indicated above, we introduce the two-parameter family[3] of operators $W_{(\lambda,\mu)}(a,b)$, where

$$W_{(\lambda,\mu)}(a,b) \;=\; e^{\frac{1}{4}\lambda(a^2+b^2)}e^{\frac{1}{2}i(\mu+1)ab}V(b)U(a), \qquad a,b\in\mathbb{R}. \tag{14.1.6}$$

The constants λ, μ are both taken to be real and to lie in the interval $[-1,1]$. Comparing equation (14.1.6) with equations (14.1.1.a), (14.1.1.b),

[2] The above comments have been based on the premise that we wish to maximize the space of phase space observables that can be quantized, to the end that we regard a "good" ordered quantization scheme as one that admits the quantization of all elements of $\mathcal{S}'(\Pi)$. It may well be necessary to vary the class of desired phase space observables at the same time as varying the common smooth domain $\mathcal{S}_\sharp$, but we have not done this here. In any event it does not seem as though doing so would change our conclusions about normal ordering.

[3] Because $W_{(\lambda,\mu)}(a,b)^*W_{(\lambda,\mu)}(a,b) \;=\; W_{(\lambda,\mu)}(a,b)W_{(\lambda,\mu)}(a,b)^* \;=\; e^{\frac{1}{2}\lambda(a^2+b^2)}I$, these operators $W_{(\lambda,\mu)}(a,b)$ are unitary only when $\lambda = 0$.

(14.1.1.c), (14.1.1.d) and (14.1.1.e), it is clear that $W_{(0,1)}(a,b) = e^{ia\overline{P}}e^{ib\overline{Q}}$, $W_{(0,-1)}(a,b) = e^{ib\overline{Q}}e^{ia\overline{P}}$, $W_{(0,0)}(a,b) = W(a,b)$, $W_{(1,0)}(a,b) = e^{i\overline{z}A^+}e^{izA}$ and $W_{(-1,0)}(a,b) = e^{izA}e^{i\overline{z}A^+}$ for any $a,b \in \mathbb{R}$, so that P-ordered, Q-ordered, Weyl-ordered, normal ordered and anti-normal ordered quantization can be subsumed in this single formalism. The operator $W_{(\lambda,\mu)}$ is closely related to the Weyl group, in that

$$W_{(\lambda,\mu)}(a,b) = \mathcal{E}_{\lambda,\mu}(a,b)\, W(a,b)\,, \tag{14.1.7}$$

where $\mathcal{E}_{\lambda,\mu}$ is the function

$$\mathcal{E}_{\lambda,\mu}(a,b) = \exp\left\{\frac{1}{4}\lambda(a^2+b^2) + \frac{1}{2}i\mu ab\right\}. \tag{14.1.8}$$

At least formally, then, the (λ,μ)-Wigner transform can be obtained from the standard Wigner transform $\mathcal{G}$ by the formula

$$\mathcal{G}_{(\lambda,\mu)}F = \frac{1}{2\pi}\mathcal{F}(\mathcal{E}_{\lambda,\mu}) * \mathcal{G}F\,, \qquad F \in \mathcal{S}(\mathbb{R}^2)\,, \tag{14.1.9}$$

where the symbol $*$ denotes convolution. This formula makes sense within the space $\mathcal{S}(\mathbb{R}^2)$ when $\lambda < 0$, and can be understood within the space $\mathcal{S}'(\mathbb{R}^2)$ when $\lambda = 0$. However, this formula is meaningless when $\lambda > 0$ since, in that case, $\mathcal{E}_{\lambda,\mu}$ does not belong to $\mathcal{S}'(\mathbb{R}^2)$. More refined techniques are needed to handle ordered quantizations for positive values of λ, and even these techniques fail when $\lambda = 1$. We see this situation as indicating that normal ordering is, in some sense, incompatible with quantization of the smooth model. This incompatibility might be overcome by some other (as yet unenunciated) axiomatization of quantum mechanics, possibly in terms of analytic functions, which could describe normal ordered quantization satisfactorily[4].

In practice we shall not consider the quantization family $W_{(\lambda,\mu)}$ in full generality when both λ and μ are nonzero; to do so simply obscures the features of some of the arguments. A sufficient flavour of the theory will be provided by considering the two subfamilies $W_{(\lambda,0)}$ and $W_{(0,\mu)}$. The first of these families will be called the *Wick/anti-Wick family* (WAW family for

[4]We do not mean by this simply that the smooth model should be expressed in terms of analytic functions — as in the Bargmann-Segal representation — but rather that quantum mechanics might need to be based upon a different space of test functions to $\mathcal{S}(\mathbb{R})$, providing a larger dual space of distributions including functions which are analytic in their arguments. Replacing distributions by hyperfunctions, say, might provide a basis for normal ordering, but we have not investigated this.

short), since it interpolates between the Wick and the anti-Wick orderings. The second will be called the *PQ-family* for an analogous reason. Note that Weyl ordering belongs both of these families.

14.1.2 *Linear Quantization*

Before considering particular ordered quantizations, it is instructive to consider the connection that these orderings bear to a broad classification of orderings introduced by Berezin & Shubin, which will now be briefly described. For elaboration and proofs, see their monograph [19].

Berezin & Shubin study that class of quantizations, which they call *linear*, which satisfy a version of the correspondence principle. While some authors consider certain nonlinear quantizations as physically justified in appropriate circumstances, in view of the fact that quantization in general can denote an almost totally arbitrary phase space function/operator association, it is reasonable first to study those quantizations which enjoy more regular properties.

In the theory of linear quantization, it is required that the algebra of polynomials in Q and P (or A and A^+) should be dense in the set of all observables (in some technical sense that we shall not bother to specify here). This is enough to imply that the quantization $\mathbf{\Delta}_\sharp$ scheme be completely specified by the four endomorphisms S_1, S_2, S_3 and S_4 of $\mathcal{P}$ (the algebra of polynomial functions on Π) given by the formulæ:

$$S_1(T) = \mathbf{\Delta}_\sharp^{-1}[P\mathbf{\Delta}_\sharp[T]], \qquad (14.1.10.a)$$

$$S_2(T) = \mathbf{\Delta}_\sharp^{-1}[Q\mathbf{\Delta}_\sharp[T]], \qquad (14.1.10.b)$$

$$S_3(T) = \mathbf{\Delta}_\sharp^{-1}[\mathbf{\Delta}_\sharp[T]P], \qquad (14.1.10.c)$$

$$S_4(T) = \mathbf{\Delta}_\sharp^{-1}[\mathbf{\Delta}_\sharp[T]Q], \qquad (14.1.10.d)$$

for any $T \in \mathcal{P}$. Berezin & Shubin impose their version of the correspondence principle by requiring that the operators S_j are of the form

$$S_j(T) = \mathcal{D}_j T, \qquad T \in \mathcal{P}, \qquad 1 \leqslant j \leqslant 4, \qquad (14.1.11)$$

where the $\mathcal{D}_j$ are first order linear differential operators[5]. In a certain sense, this requirement ensures that the Moyal bracket be not too wildly deformed from the Poisson bracket.

[5]Hence the term *linear* quantization.

It is also required that the quantization of the constant function be the identity operator, $\mathbf{\Delta}_\sharp\,[\mathrm{i}] = I$. Finally, it is assumed that the quantization procedure provides a representation of the CCR, in that

$$[\,\mathbf{\Delta}_\sharp\,[p]\,,\,\mathbf{\Delta}_\sharp\,[q]\,] = -iI$$

with $\mathbf{\Delta}_\sharp\,[p]$ and $\mathbf{\Delta}_\sharp\,[q]$ being essentially self-adjoint operators.

That these various requirements should be consistent leads to the following relationship between the differential operators,

$$\begin{pmatrix} \mathcal{D}_1 \\ \mathcal{D}_2 \\ \mathcal{D}_3 \\ \mathcal{D}_4 \end{pmatrix} = \begin{pmatrix} A & AB \\ & \\ A & AB^{\mathrm{T}} \end{pmatrix} \begin{pmatrix} p \\ q \\ \partial_p \\ \partial_q \end{pmatrix}, \tag{14.1.12}$$

where $A \in \mathrm{Sp}\,(2;\mathbb{R})$, and where $B \in M_2(\mathbb{C})$ must satisfy the identity[6] $\mathrm{Tr}\,(\,BJ\,) = -i$. Otherwise, the matrices A and B may be freely chosen[7]. When we wish to emphasize that the quantization $\mathbf{\Delta}_\sharp$ is linear (in the sense of Berezin & Shubin) and determined by A and B, we will use the notation $\mathbf{\Delta}_{A,B}$ for $\mathbf{\Delta}_\sharp$.

All the quantizations determined by the family of operators $W_{(\lambda,\mu)}$ are linear in this sense, and it may be verified that the matrices A and B for $W_{(\lambda,\mu)}$ are given by

$$A = \begin{pmatrix} 1 & 0 \\ 0 & 1 \end{pmatrix}, \qquad B = \begin{pmatrix} \frac{1}{2}\lambda & -\frac{1}{2}i(1-\mu) \\ \frac{1}{2}i(1+\mu) & \frac{1}{2}\lambda \end{pmatrix}. \tag{14.1.13}$$

There is a double action of the symplectic group $\mathrm{Sp}\,(2;\mathbb{R})$ on the collection of linear quantizations in that if $A, B \in M_2(\mathbb{R})$ are matrices which determine a linear quantization $\mathbf{\Delta}_{A,B}$ and if $X, Y \in \mathrm{Sp}\,(2;\mathbb{R})$, then the matrices

$$XAY^{-1} \qquad \text{and} \qquad YBY^{\mathrm{T}} \tag{14.1.14.a}$$

also determine a linear quantization. In other words, the map

$$\big((X,Y)\,,\,\mathbf{\Delta}_{A,B}\big) \mapsto \mathbf{\Delta}_{A,B}^{(X,Y)} = \mathbf{\Delta}_{XAY^{-1},YBY^{\mathrm{T}}} \tag{14.1.14.b}$$

defines a action of $\mathrm{Sp}\,(2;\mathbb{R}) \times \mathrm{Sp}\,(2;\mathbb{R})$ on linear quantizations. Moreover, this action is implemented via the metaplectic representation π of $\mathrm{Sp}\,(2;\mathbb{R})$

[6] Here J is the symplectic matrix introduced in equation (2.5.3).

[7] This formalism can be extended satisfactorily to cover cases with n degrees of freedom.

discussed in Chapter 8, since it can be shown that

$$\mathbf{\Delta}_{A,B}^{(X,Y)}\left[T\right] \;=\; \pi(X)\cdot\mathbf{\Delta}_{A,B}\left[Y^{-1}\circ T\right]\cdot\pi(X)^{-1}\,, \quad T\in\mathcal{P}\,, \qquad (14.1.14.c)$$

for any $X, Y \in \mathrm{Sp}\,(2;\mathbb{R})$ and any linear quantization $\mathbf{\Delta}_{A,B}$. Here $\circ$ denotes the standard action of $\mathrm{Sp}\,(2;\mathbb{R})$ on $\mathcal{S}'(\Pi)$ defined in equation (8.7.8).

Two linear quantizations will be regarded as equivalent if one can be obtained from the other by a double action of the symplectic group of the form indicated above. The problem of determining the equivalence classes of linear quantizations is then essentially that of determining the congruence classes of 2×2 complex matrices B such that $\mathrm{Tr}\,(\,BJ\,) \;=\; -i$. Without going into details, most (but not all) of the equivalence classes of linear quantizations are represented by linear quantizations where A and B are of the form given in equation (14.1.13), provided that λ and μ are allowed to be complex. Thus analysis of $W_{(\lambda,\mu)}$ quantizations provides the basic framework for the study of all linear quantizations.

Notice that Weyl quantization is the linear quantization described by the matrices $A \;=\; I$ and $B \;=\; \frac{1}{2}iJ$, and hence the linear quantizations equivalent to Weyl quantization are of the form

$$\mathbf{\Delta}_{X,\frac{1}{2}iJ}\,, \qquad\qquad X\in\mathrm{Sp}\,(2;\mathbb{R})\;,$$

since $YJY^{\mathrm{T}} \;=\; J$ for any $Y \in \mathrm{Sp}\,(2;\mathbb{R})$. We note in passing that Weyl quantization is invariant under the action of any element of the group $\mathrm{Sp}\,(2;\mathbb{R})\times\mathrm{Sp}\,(2;\mathbb{R})$ of the form (X, X), where $X \in \mathrm{Sp}\,(2;\mathbb{R})$. This result is not at all surprising, since comparison of equations (8.7.13) and (14.1.14.c) shows that this is simply the defining property of the metaplectic representation itself.

14.2 The P- And Q- Orderings

Putting the general theory of linear quantization aside, we commence our study of one of the two key subfamilies of the $W_{(\lambda,\mu)}$, namely the PQ-family, which is described by the operators

$$W_{(0,\mu)}(a,b) \;=\; e^{\frac{1}{2}i(\mu+1)ab}V(b)U(a), \qquad a,\, b\in\mathbb{R},\; |\,\mu\,|\leqslant 1\,. \qquad (14.2.1)$$

14.2.1 *Existence Of PQ-Ordered Quantization*

To say that a proposed ordered quantization exists means that a choice of common domain is possible (in the sense stated above) so that the ordered Wigner and Weyl maps can be constructed rigorously, as was done previously for the case of Weyl quantization. The existence of PQ-ordered quantization could be established through a study of the Fourier transform of the function

$$\mathcal{E}_{0,\mu}(a,b) \;=\; e^{\frac{1}{2}i\mu ab}\,, \tag{14.2.2}$$

but, because of the simple form of $W_{(0,\mu)}$, it is easier to follow the treatment for the Weyl case closely.

For this family, the definition given in equation (14.1.4) for the $(0,\mu)$-Wigner transform is easily made rigorous, and results in a simple generalization of the Wigner transform $\mathcal{G}$.

Definition 14.1 ***(PQ-Wigner Transform)*** *By the $(0,\mu)$-Wigner transform is meant the map $\mathcal{G}_{(0,\mu)} : \mathcal{S}(\mathbb{R}^2) \to \mathcal{S}(\Pi)$ given by*

$$\left[\mathcal{G}_{(0,\mu)}F\right](p,q) \;=\; \frac{1}{2\pi}\int_{\mathbb{R}} F\left(q+\tfrac{1}{2}(1+\mu)u, q-\tfrac{1}{2}(1-\mu)u\right)\, e^{ipu}\, du\,, \tag{14.2.3}$$

for $F \in \mathcal{S}(\mathbb{R}^2)$.

Most of the properties of $\mathcal{G}$ can be generalized to apply to the PQ-family of $(0,\mu)$-transforms $\mathcal{G}_{(0,\mu)}$. Key to this observation is the following bicontinuity result.

Lemma 14.2 *The $(0,\mu)$-Wigner transform $\mathcal{G}_{(0,\mu)}$ is a bicontinuous linear bijection from $\mathcal{S}(\mathbb{R}^2)$ to $\mathcal{S}(\Pi)$.*

Proof: Extending the result of Proposition 8.5, we can write

$$\mathcal{G}_{(0,\mu)} \;=\; \frac{1}{\sqrt{2\pi}}(M\otimes I)\,\mathcal{F}_1^{-1}\,\tau_\mu,$$

where $\tau_\mu : \mathcal{S}(\mathbb{R}^2) \to \mathcal{S}(\mathbb{R}^2)$ is the bicontinuous linear bijection

$$[\tau_\mu F](x,y) \;=\; F\big(y+(1+\mu)x, y-(1-\mu)x\big), \qquad F \in \mathcal{S}(\mathbb{R}^2)\;.$$

The result of this Lemma is now elementary. Note that in Proposition 8.5, the endomorphism τ_0 was referred to as τ in equation (8.3.8). ∎

Moreover, it can be shown that, for any $f,\, g \in \mathcal{S}(\mathbb{R})$,

$$\begin{aligned} \left[\mathcal{F}^{-1}\mathcal{G}_{(0,\mu)}(\overline{g} \otimes f)\right](a,b) &= \frac{1}{2\pi}\left\langle g\,,\, W_{(0,\mu)}(a,b)f\right\rangle, & (14.2.4.a)\\ \left[\mathcal{G}_{(0,\mu)}(\overline{g} \otimes f)\right](a,b) &= \left[\mathcal{G}_{(0,-\mu)}(\overline{\mathcal{F}g} \otimes \mathcal{F}f)\right](-b,a), & (14.2.4.b) \end{aligned}$$

and that

$$\overline{\mathcal{G}_{(0,\mu)}(\overline{g} \otimes f)} = \mathcal{G}_{(0,-\mu)}(\overline{f} \otimes g). \qquad (14.2.5)$$

These results imply in particular that

$$\begin{aligned} \int_{\mathbb{R}} \left[\mathcal{G}_{(0,\mu)}(\overline{g} \otimes f)\right](p,q)\,dp &= \overline{g(q)}f(q), & (14.2.6.a)\\ \int_{\mathbb{R}} \left[\mathcal{G}_{(0,\mu)}(\overline{g} \otimes f)\right](p,q)\,dq &= \overline{(\mathcal{F}g)(p)}(\mathcal{F}f)(p), & (14.2.6.b) \end{aligned}$$

for any $f,\, g \in \mathcal{S}(\mathbb{R})$ and $-1 \leqslant \mu \leqslant 1$. These observations yield the basic results for PQ-quantization.

Proposition 14.3 ***(PQ-Quantization)*** *For a distribution $T \in \mathcal{S}'(\Pi)$, the $(0,\mu)$-quantization $\boldsymbol{\Delta}_{(0,\mu)}[T]$ of T is that element of $\mathcal{L}(\mathcal{S}(\mathbb{R}),\mathcal{S}'(\mathbb{R}))$ defined by the formula*

$$[\![\, \boldsymbol{\Delta}_{(0,\mu)}[T]f\,,\, g\,]\!] = [\![\, T\,,\, \mathcal{G}_{(0,\mu)}(g \otimes f)\,]\!], \qquad f,\, g \in \mathcal{S}(\mathbb{R}). \qquad (14.2.7)$$

The integral kernel of $\boldsymbol{\Delta}_{(0,\mu)}[T]$ is given by the formula

$$K_{\boldsymbol{\Delta}_{(0,\mu)}[T]} = \mathcal{G}_{(0,\mu)}^{\mathrm{tr}}\, T, \qquad (14.2.8)$$

and so $\boldsymbol{\Delta}_{(0,\mu)} : \mathcal{S}'(\Pi) \to \mathcal{L}(\mathcal{S}(\mathbb{R}),\mathcal{S}'(\mathbb{R}))$ is a linear bijection. Moreover, the same marginal distributions hold for $(0,\mu)$-quantization as for Weyl quantization, in that

$$\begin{aligned} \boldsymbol{\Delta}_{(0,\mu)}[\mathfrak{i} \otimes \mathfrak{f}] &= \mathfrak{f}(\overline{Q}), & (14.2.9.a)\\ \boldsymbol{\Delta}_{(0,\mu)}[\mathfrak{f} \otimes \mathfrak{i}] &= \mathfrak{f}(\overline{P}), & (14.2.9.b) \end{aligned}$$

for any function $\mathfrak{f} \in \mathcal{O}^{\infty}(\mathbb{R})$. As usual, $\mathfrak{i}$ is the constant function $\mathfrak{i}(x) = 1$.

However, PQ-quantization does not enjoy all the properties that Weyl quantization does, the main defect being poor behaviour with respect to the adjoint operation. This is a consequence of equation (14.2.5), which implies that

$$\llbracket \mathbf{\Delta}_{(0,\mu)}[\overline{T}]f \, , \, \overline{g} \rrbracket \;=\; \overline{\llbracket \mathbf{\Delta}_{(0,-\mu)}[T]g \, , \, \overline{f} \rrbracket}, \tag{14.2.10.a}$$

for any $f, g \in \mathcal{S}(\mathbb{R})$ and $T \in \mathcal{S}'(\Pi)$.

Proposition 14.4 ***(PQ-Adjoints)*** *If $T \in \mathcal{S}'(\Pi)$ is such that $\mathbf{\Delta}_{(0,\mu)}[T]$ belongs to $\mathcal{L}^+\big(\mathcal{S}(\mathbb{R}), L^2(\mathbb{R})\big)$, then so does $\mathbf{\Delta}_{(0,-\mu)}[\overline{T}]$ and*

$$\mathbf{\Delta}_{(0,\mu)}[T]^+ \;=\; \mathbf{\Delta}_{(0,-\mu)}[\overline{T}] \, . \tag{14.2.10.b}$$

Thus taking the adjoint of $\mathbf{\Delta}_{(0,\mu)}[T]$ requires not only conjugating the distribution T but also changing the sign of μ.

14.2.2 *Wigner Functions Revisited*

We know from equation (12.3.7) how expectations can be calculated entirely in phase space, a result that can be generalized to cover the whole family of PQ-quantization. To be specific, if $\rho \in \mathfrak{A}_*$ is a (smooth) density matrix, then

$$\begin{aligned} \mathbf{Exp}_\rho \left[\mathbf{\Delta}_{(0,\mu)}[T]\right] &= \operatorname{Tr}\left(\rho\, \mathbf{\Delta}_{(0,\mu)}[T]\right) && (14.2.11.a)\\ &= \frac{1}{2\pi}\llbracket T \, , \, \mathbf{\Delta}^{-1}_{(0,-\mu)}[\rho] \rrbracket, && (14.2.11.b) \end{aligned}$$

holds for all $T \in \mathcal{S}'(\Pi)$, where

$$\mathbf{\Delta}^{-1}_{(0,-\mu)}[\rho] \;=\; 2\pi \mathcal{G}_{(0,\mu)}(\mathcal{R}K_\rho) \, . \tag{14.2.12}$$

The usual understanding holds here, in that the trace expression (14.2.11.a) only makes sense when $\mathbf{\Delta}_{(0,\mu)}[T] \in \mathcal{L}^+(\mathcal{S}(\mathbb{R}))$, while the second expression (14.2.11.b) holds in all cases. In other words, the quantity

$$\Psi^\mu_\rho \;=\; \frac{1}{2\pi}\mathbf{\Delta}^{-1}_{(0,-\mu)}[\rho] \tag{14.2.13}$$

is that element of $\mathcal{S}(\Pi)$ whose pairing with any distribution $T \in \mathcal{S}'(\Pi)$ represents the expectation of the observable $\mathbf{\Delta}_{(0,\mu)}[T]$ in the state ρ. It will be recognized that Ψ^μ_ρ is the Wigner function for the state ρ in the $(0,\mu)$-quantization: the phase space symbol of the state.

Suppose now that ρ is a pure state defined by some unit vector $f \in \mathcal{S}(\mathbb{R})$, so that $\rho = |\, f \,\rangle\langle\, f \,|$, $K_\rho = f \otimes \overline{f}$, and $\Psi^\mu_\rho = \mathcal{G}_{(0,\mu)}(\overline{f} \otimes f)$. The equations

(14.2.6.a) and (14.2.6.b) for the marginals then imply that

$$\int_{\mathbb{R}} \Psi^{\mu}_{\rho}(p,q)\,dp = \left|f(q)\right|^2, \tag{14.2.14.a}$$

$$\int_{\mathbb{R}} \Psi^{\mu}_{\rho}(p,q)\,dq = \left|(\mathcal{F}f)(p)\right|^2. \tag{14.2.14.b}$$

These equations provide the correct probability density functions for the distributions of position and momentum, respectively, for the pure state ρ, a fact which has prompted many people to consider to what extent the quantity Ψ^{μ}_{ρ} is a joint probability density[8] for some classical random variables corresponding in some sense to P and Q.

For all orderings, the noncommutativity of P and Q renders finding such a joint probability distribution impossible[9]. Amongst other reasons, the test function Ψ^{μ}_{ρ} is not, in general, a non-negative function. A theorem of Hudson [123] shows that (in the case of Weyl quantization) if ρ is a pure state corresponding to a normalized vector in $L^2(\mathbb{R})$, then Ψ^0_{ρ} is non-negative, and is thus open to being interpreted as a joint probability distribution, only when that vector is a Gaussian function. Notwithstanding these problems, the idea of a classical probability density function is so attractive to some people that the phase space test functions Ψ^{μ}_{ρ} are referred to as quasi-probability distributions (or something similar). As a piece of terminology, that is acceptable, and we shall adopt it — but there are dangers in carrying the interpretation too far, as we shall see.

Consider a general state $\rho \in \mathfrak{A}_*$ and its associated quasi-probability distribution Ψ^{μ}_{ρ}. There is no problem in considering its Q- and P-marginals,

$$\Psi^{\mu}_{\rho,Q}(q) = \int_{\mathbb{R}} \Psi^{\mu}_{\rho}(p,q)\,dp, \tag{14.2.15.a}$$

$$\Psi^{\mu}_{\rho,P}(p) = \int_{\mathbb{R}} \Psi^{\mu}_{\rho}(p,q)\,dq, \tag{14.2.15.b}$$

[8] As noted in our discussion of the Moyal product, the motivation of Wigner in developing the much used Wigner transform Ψ^0_{ρ} was to effect a semi-classical expansion of fully quantum statistical quantities in what he regarded as the simplest way.

[9] If we allow ourselves to work with approximations to the operators P and Q, rather than these operators themselves, a greater degree of flexibility is possible, and joint probability distributions can then be defined. This approach to the problem is discussed in the book of Davies [41]; also see Folland (*ibid*).

for it can be shown that

$$\int_{\mathbb{R}} \mathfrak{f}(q)\, \Psi^{\mu}_{\rho,Q}(q)\, dq = \mathbf{Exp}_{\rho}\left[\mathfrak{f}(\overline{Q})\right], \qquad (14.2.16.a)$$

$$\int_{\mathbb{R}} \mathfrak{f}(p)\, \Psi^{\mu}_{\rho,P}(p)\, dp = \mathbf{Exp}_{\rho}\left[\mathfrak{f}(\overline{P})\right], \qquad (14.2.16.b)$$

for any function $\mathfrak{f} \in \mathcal{O}^{\infty}(\mathbb{R})$. This is because equations (14.2.9.a) and (14.2.9.b) show that the PQ-family of quantizations provides the correct marginal distributions for position and momentum. However, the situation is less happy for polar quantization. Consider what happens if we define "marginal distributions" for functions of the radius or of the angle,

$$\Psi^{\mu}_{\rho,R}(r) = r \int_{-\pi}^{\pi} \Psi^{\mu}_{\rho}(r\cos\vartheta, r\sin\vartheta)\, d\vartheta, \qquad (14.2.17.a)$$

and

$$\Psi^{\mu}_{\rho,\Theta}(\vartheta) = \int_{0}^{\infty} \Psi^{\mu}_{\rho}(r\cos\vartheta, r\sin\vartheta)\, r\, dr\,. \qquad (14.2.17.b)$$

These functions give the correct expectations,

$$\int_{0}^{\infty} \mathfrak{f}(r)\, \Psi^{\mu}_{\rho,R}(r)\, dr = \mathbf{Exp}_{\rho}\left[\boldsymbol{\Delta}_{(0,\mu)}[\mathfrak{f}_{\mathrm{rad}}]\right], \qquad (14.2.18.a)$$

for $\mathfrak{f} \in \mathcal{O}^{\infty}(\mathbb{R})$, and

$$\int_{-\pi}^{\pi} \mathfrak{g}(\vartheta)\, \Psi^{\mu}_{\rho,\Theta}(\vartheta)\, d\vartheta = \mathbf{Exp}_{\rho}\left[\boldsymbol{\Delta}_{(0,\mu)}[\mathfrak{g}_{\mathrm{ang}}]\right], \qquad (14.2.18.b)$$

for $\mathfrak{g} \in L^{\infty}[-\pi,\pi]$, for the PQ-ordered quantizations of radial and angular distributions, respectively. However, $\Psi^{\mu}_{\rho,R}$ and $\Psi^{\mu}_{\rho,\Theta}$ cannot be regarded as classical probability distributions for the quantum observables $\boldsymbol{\Delta}_{(0,\mu)}[r]$ or $\boldsymbol{\Delta}_{(0,\mu)}[\varphi]$ (or any other posited phase operator) corresponding to radius and angle, since

$$\boldsymbol{\Delta}_{(0,\mu)}[(\mathfrak{f}^n)_{\mathrm{rad}}] \neq \left(\boldsymbol{\Delta}_{(0,\mu)}[\mathfrak{f}_{\mathrm{rad}}]\right)^n, \qquad (14.2.19.a)$$

for general functions $\mathfrak{f} \in \mathcal{O}^{\infty}(\mathbb{R})$ and integers $n \geqslant 2$. Similarly,

$$\boldsymbol{\Delta}_{(0,\mu)}[(\mathfrak{g}^n)_{\mathrm{ang}}] \neq \left(\boldsymbol{\Delta}_{(0,\mu)}[\mathfrak{g}_{\mathrm{ang}}]\right)^n, \qquad (14.2.19.b)$$

for general functions $\mathfrak{g} \in L^\infty[-\pi, \pi]$ and integers $n \geqslant 2$. Hence, for example, while the expression

$$\int_0^\infty \mathfrak{f}(r)^2 \Psi^\mu_{\rho,R}(r)\, dr$$

is certainly the expectation of the observable $\boldsymbol{\Delta}_{(0,\mu)}[(\mathfrak{f}^2)_{\mathrm{rad}}]$ in the state ρ, it is not the expectation of $\boldsymbol{\Delta}_{(0,\mu)}[\mathfrak{f}_{\mathrm{rad}}]^2$ in that state, which would need to be the case were $\Psi^\mu_{\rho,R}$ to be interpreted as a classical probability distribution. Similar observations are true of $\Psi^\mu_{\rho,\Theta}$.

These results eliminate $\Psi^\mu_{\rho,R}$ and $\Psi^\mu_{\rho,\Theta}$ as potential probability distributions. These infelicities apply for all values of μ, and in particular to the Weyl scheme (for which $\mu = 0$). But for nonzero μ, there are extra difficulties with radial and angular marginals since, for example, radial observables are then no longer diagonal with respect to the Hermite-Gauss functions.

As a general rule, any attempt to interpret these quasi-probabilities as probabilities effectively defines the Moyal product to be equal to the pointwise product. The results of any such interpretation are therefore, at best, semi-classical. This underlines the conclusion that while probability distributions, and joint probability distributions, are a natural component of quantum mechanics, they are not a natural part of any classical interpretation of quantum mechanics once two or more noncommuting observables are involved. Theories based on Wigner functions, then, cannot be used to determine processes in which purely quantum correlations are important. Since there are proposals of this sort in the literature concerning phase operators, a certain care must be exercised when interpreting their results. Provided no moment calculus has been assumed, the conclusions should be correct in the semi-classical region, and should coincide with the results obtained from the phase operators detailed in Chapter 10 in the classical limit. Some examples of this will be seen in the discussion of asymptotics in Chapter 15.

14.2.3 *P-Quantization*

One consequence of equation (14.2.10.a) is that it is no longer necessary to consider the entirety of the PQ-family of quantizations — it is sufficient to study that half of the family for which $-1 \leqslant \mu \leqslant 0$, obtaining the properties of the other half of the family by taking adjoints. In particular, in this Subsection, we shall concentrate on P-quantization, $\mathbf{\Delta}_{(0,-1)}$, which is particularly simple to work with. This is because the P-Wigner transform $\mathcal{G}_{(0,-1)}$ can be written

$$\big[\mathcal{G}_{(0,-1)}F\big](p,q) \;=\; \frac{1}{\sqrt{2\pi}}\, e^{ipq}\,\big(\mathcal{F}_2 F\big)(q,p), \qquad F \in \mathcal{S}\big(\mathbb{R}^2\big)\,, \qquad (14.2.20)$$

which has the following immediate consequence.

Proposition 14.5 *P-quantization has the integral kernel formulation*

$$\big[\mathbf{\Delta}_{(0,-1)}[T]f\big](x) \;=\; \frac{1}{\sqrt{2\pi}} \int_{\mathbb{R}} T(p,x)\,(\mathcal{F}f)(p)\, e^{ipx}\,dx, \qquad (14.2.21)$$

for $T \in \mathcal{S}'(\Pi)$ and $f \in \mathcal{S}(\mathbb{R})$, where this integral should be interpreted weakly where necessary.

We already know that P-quantization has the correct P- and Q-marginals — this result is also clear from the kernel identity. As has already been indicated, most of what can be said about the properties of P-quantization with respect to radial and angular distributions is negative. For example, direct calculation gives

$$\big[\mathbf{\Delta}_{(0,-1)}[(h_0)_{\mathrm{rad}}]\,h_0\big](x) \;=\; \frac{1}{\sqrt{2\pi}}\, e^{-\frac{3}{4}x^2}\,,$$

so that $\mathbf{\Delta}_{(0,-1)}[(h_0)_{\mathrm{rad}}]\,h_0$ is not a scalar multiple of h_0, and hence it is clear that $\mathbf{\Delta}_{(0,-1)}[(h_0)_{\mathrm{rad}}]$ is not diagonal with respect to the Hermite-Gauss functions — this result is generally true of P-ordered quantizations of radial distributions. On the other hand, we have already remarked that the adjoint of the operator $\mathbf{\Delta}_{(0,-1)}[\varphi]$ is $\mathbf{\Delta}_{(0,1)}[\varphi]$, and hence it is to be expected that $\mathbf{\Delta}_{(0,-1)}[\varphi]$ is not self-adjoint. This can indeed be verified by direct calculation, since

$$\big\langle h_0\,,\,\mathbf{\Delta}_{(0,-1)}[\varphi]\,h_0\big\rangle = -\frac{i}{2}\int_0^\infty \Big[1 - e^{x^2}\,\mathrm{erfc}(x)\Big]\; e^{-\frac{1}{2}x^2}\,\frac{dx}{x}$$

is nonzero and purely imaginary.

Not all is gloom, however, since it is possible to extend the Method of Wedges of Chapter 9 to show that the operators $\mathbf{\Delta}_{(0,-1)}\big[\mathcal{D}^{(\alpha)}\big]$ are uniformly bounded for $0 \leqslant \alpha \leqslant \pi$. This implies, in particular, that the operator $\mathbf{\Delta}_{(0,-1)}[\varphi]$ is uniformly bounded. This is even true for the whole family of quantizations:

Proposition 14.6 *The family of operators* $\{\, \mathbf{\Delta}_{(0,\mu)}[\varphi] \,:\, -1 \leqslant \mu \leqslant 1 \,\}$ *is uniformly bounded in* $\mathbb{B}\big(L^2(\mathbb{R})\big)$.

The proof of this result is complex, and the reader is referred to [110] for details.

14.3 Anti-Wick Quantization

We now proceed to consider the WAW-family of orderings, which interpolate between Wick and anti-Wick ordering.

As observed in equation (14.1.9), the (λ, μ)-Wigner transform $\mathcal{G}_{(\lambda,\mu)}$ is obtained from the Weyl Wigner transform by convolution with the function

$$\begin{aligned} Q_{\lambda,\mu}(x,y) &= \frac{1}{2\pi}\big(\mathcal{F}\mathcal{E}_{\lambda,\mu}\big)(x,y) \\ &= \frac{1}{\pi\sqrt{\lambda^2+\mu^2}}\exp\left[\frac{\lambda}{\lambda^2+\mu^2}(x^2+y^2) - \frac{2i\mu}{\lambda^2+\mu^2}xy\right] \end{aligned} \tag{14.3.1}$$

whenever $-1 \leqslant \lambda < 0$ since, in that case, $\mathcal{E}_{\lambda,\mu} \in \mathcal{S}\big(\mathbb{R}^2\big)$ and hence its Fourier transform belongs to $\mathcal{S}(\Pi)$. The function $Q_{\lambda,\mu}$ has the property that

$$\int_{\mathbb{R}} Q_{\lambda,\mu}(x,y)\,dy = q_\lambda(x)\,, \tag{14.3.2.a}$$

for any $-1 \leqslant \lambda < 0$ and $|\,\mu\,| \leqslant 1$, where q_λ is the Gaussian

$$q_\lambda(x) = \frac{1}{\sqrt{\pi\,|\,\lambda\,|}}\,e^{x^2/\lambda}\,; \tag{14.3.2.b}$$

moreover we note that $Q_{\lambda,0} = q_\lambda \otimes q_\lambda$.

The smoothing effect of convolution by $Q_{\lambda,\mu}$ makes the half of the family $W_{(\lambda,\mu)}$ for which $\lambda < 0$ particularly simple to analyze. We shall therefore begin by studying this "good" half, and in particular will concentrate on the corresponding "good" half of the WAW-family of orderings, namely the orderings provided by the operators $W_{(\lambda,0)}$ where $-1 \leqslant \lambda < 0$. We shall refer to this family as the *anti-Wick family* (or *AW-family*) of orderings. For the remainder of this Section, therefore, we shall assume without further comment that $-1 \leqslant \lambda < 0$. The value of μ does not affect the problem of existence, so need not be restricted.

14.3.1 *Existence Of Anti-Wick Quantization*

The existence problem for anti-Wick, and related, quantizations can be summarized in the following Proposition, which is an elementary generalization of some of the results of Chapter 8.

Proposition 14.7 ***(AW-Quantization)*** *$\mathcal{G}_{(\lambda,\mu)}$, the (λ,μ)-Wigner transform, is derived from the Weyl-Wigner transform $\mathcal{G}$ by the formula*

$$\mathcal{G}_{(\lambda,\mu)}F \;=\; Q_{\lambda,\mu} * \mathcal{G}F, \qquad F \in \mathcal{S}(\mathbb{R}^2)\,, \tag{14.3.3}$$

and is a continuous linear map from $\mathcal{S}(\mathbb{R}^2)$ to $\mathcal{S}(\Pi)$ such that

$$\left[\mathcal{F}^{-1}\mathcal{G}_{(\lambda,\mu)}(\overline{g}\otimes f)\right](a,b) \;=\; \frac{1}{2\pi}\left\langle g, W_{(\lambda,\mu)}(a,b)f\right\rangle, \quad f,g \in \mathcal{S}(\mathbb{R}). \tag{14.3.4}$$

Thus (λ,μ)-ordered quantization $\mathbf{\Delta}_{(\lambda,\mu)}$ is a linear map from $\mathcal{S}'(\Pi)$ to $\mathcal{L}(\mathcal{S}(\mathbb{R}),\mathcal{S}'(\mathbb{R}))$ given by the formula

$$[\![\, \mathbf{\Delta}_{(\lambda,\mu)}[T]f\,,\, g\,]\!] \;=\; [\![\, T\,,\, \mathcal{G}_{(\lambda,\mu)}(g\otimes f)\,]\!], \qquad f,\, g \in \mathcal{S}(\mathbb{R}). \tag{14.3.5}$$

Consequently the integral kernel of the operator $\mathbf{\Delta}_{(\lambda,\mu)}[T]$ is given by the formula

$$K_{\mathbf{\Delta}_{(\lambda,\mu)}[T]} \;=\; \mathcal{G}^{\mathrm{tr}}_{(\lambda,\mu)}T \;=\; \mathcal{G}^{\mathrm{tr}}(Q_{\lambda,\mu} * T), \qquad T \in \mathcal{S}'(\Pi), \tag{14.3.6}$$

and (λ,μ)-quantization is related to Weyl quantization by the formula

$$\mathbf{\Delta}_{(\lambda,\mu)}[T] \;=\; \mathbf{\Delta}\,[\,Q_{\lambda,\mu} * T\,]\,, \qquad T \in \mathcal{S}'(\Pi). \tag{14.3.7}$$

In these last two equations the standard convolution $*$ of two elements of $\mathcal{S}(\Pi)$ has been extended to define the convolution of $Q_{\lambda,\mu} \in \mathcal{S}(\Pi)$ with an element T of $\mathcal{S}'(\Pi)$ by the formula

$$[\![\, Q_{\lambda,\mu} * T\,,\, F\,]\!] \;=\; [\![\, T\,,\, Q_{\lambda,\mu} * F\,]\!], \qquad F \in \mathcal{S}(\Pi). \tag{14.3.8}$$

Since taking the convolution with $Q_{\lambda,\mu}$ is an injective linear map on $\mathcal{S}'(\Pi)$, it is clear that (λ,μ)-quantization is an injective linear map from $\mathcal{S}'(\Pi)$ to $\mathcal{L}(\mathcal{S}(\mathbb{R}),\mathcal{S}'(\mathbb{R}))$. However, it is not surjective, since not every distribution in $\mathcal{S}'(\Pi)$ is the convolution of $Q_{\lambda,\mu}$ with another tempered distribution.

To determine the basic properties of these quantizations, we calculate that

$$\overline{\mathcal{G}_{(\lambda,\mu)}(\overline{g}\otimes f)} \;=\; \mathcal{G}_{(\lambda,-\mu)}(\overline{f}\otimes g), \qquad f,\, g \in \mathcal{S}(\mathbb{R}), \tag{14.3.9}$$

and also the partial integral formulæ:

$$\int_{\mathbb{R}} \big[\mathcal{G}_{(\lambda,\mu)}(\overline{f}\otimes f)\big](p,q)\,dp \;=\; \big(g_\lambda * |\,f\,|^2\big)(q), \qquad (14.3.10.\text{a})$$

$$\int_{\mathbb{R}} \big[\mathcal{G}_{(\lambda,\mu)}(\overline{f}\otimes f)\big](p,q)\,dq \;=\; \big(g_\lambda * |\,\mathcal{F}f\,|^2\big)(p), \qquad (14.3.10.\text{b})$$

for all $f,\ g \in \mathcal{S}(\mathbb{R})$. From these technical calculations, the following adjoint and marginal properties can be shown.

Proposition 14.8 ***(AW Adjoints)*** *For any $T \in \mathcal{S}'(\Pi)$ and $f,\ g \in \mathcal{S}(\mathbb{R})$,*

$$[\![\,\mathbf{\Delta}_{(\lambda,\mu)}\,\big[\overline{T}\big]\,f\,,\,\overline{g}\,]\!] = \overline{[\![\,\mathbf{\Delta}_{(\lambda,-\mu)}\,[T]\,g\,,\,\overline{f}\,]\!]}. \qquad (14.3.11)$$

Thus, if $T \in \mathcal{S}'(\Pi)$ is such that $\mathbf{\Delta}_{(\lambda,\mu)}[T] \in \mathcal{L}^+\big(\mathcal{S}(\mathbb{R}), L^2(\mathbb{R})\big)$, then the map $\mathbf{\Delta}_{(\lambda,-\mu)}[\overline{T}] \in \mathcal{L}^+\big(\mathcal{S}(\mathbb{R}), L^2(\mathbb{R})\big)$ as well, and

$$\Big(\mathbf{\Delta}_{(\lambda,\mu)}\,[T]\Big)^* = \mathbf{\Delta}_{(\lambda,-\mu)}\,\big[\overline{T}\big]\,. \qquad (14.3.12)$$

Thus, while (λ,μ)-quantization in general suffers from the same defect with respect to adjoints as does PQ-quantization, AW-quantization does not, since the $(\lambda,0)$-quantization of a "real" distribution $T=\overline{T}\in\mathcal{S}'(\Pi)$ is symmetric. On the other hand, all (λ,μ)-quantizations have the same (slightly imperfect) P- and Q- marginals.

Proposition 14.9 ***(AW Marginals)*** *For any $\mathfrak{f} \in \mathcal{O}^\infty(\mathbb{R})$ the identities*

$$\mathbf{\Delta}_{(\lambda,\mu)}\,[\mathfrak{i}\otimes\mathfrak{f}] \;=\; \big(g_\lambda * \mathfrak{f}\big)(\overline{Q}), \qquad (14.3.13.\text{a})$$

$$\mathbf{\Delta}_{(\lambda,\mu)}\,[\mathfrak{f}\otimes\mathfrak{i}] \;=\; \big(g_\lambda * \mathfrak{f}\big)(\overline{P}), \qquad (14.3.13.\text{b})$$

hold. Thus $(\lambda,0)$-quantization does not lead to the correct P- and Q-marginals, but it does map functions of p (respectively q) to functions of $\overline{P}$ (respectively $\overline{Q}$).

On the other hand, AW-quantization will be shown to have the correct marginals for a different pair of phase space independent variables.

We could now proceed with a further study of the family of (λ,μ)-quantizations just defined, but we choose not to do so, and shall concentrate our attention on the AW-family. We do this firstly in the interests of clarity, but secondly because this family of orderings enjoys particularly interesting properties.

In view of the definition of the WAW-family of orderings, we might expect that the lowering and raising operators A and A^+ are more fundamental to this family than are the position and momentum operators Q and P, and this is indeed the case, for if $T \in \mathcal{S}'(\Pi)$ is a polynomial function of the complex number $z = (q - ip)/\sqrt{2}$, then it can be shown that

$$T \star Q_{\lambda,0} = T,$$

and hence that

$$\mathbf{\Delta}_{(\lambda,0)}[T] = \mathbf{\Delta}[T] = T(A^+), \tag{14.3.14.a}$$

while

$$\mathbf{\Delta}_{(\lambda,0)}[S] = \mathbf{\Delta}[S] = S(A), \tag{14.3.14.b}$$

for any $S \in \mathcal{S}'(\Pi)$ which is a polynomial in the complex number $\overline{z}$. In other words, the AW-family of quantizations provides the correct A- and A^+-marginals. Unfortunately, it is not possible to consider such marginals for functions which are much more complicated than polynomials in z and $\overline{z}$, since we are working in the smooth model, and any element of $\mathcal{S}'(\Pi)$ which is either an entire analytic or an entire anti-analytic function is also polynomially bounded, and hence is in fact a polynomial.

14.3.2 *The Bargmann-Segal Representation Revisited*

When we discussed the phase observable $\Xi(\varphi)$ suggested by the Bargmann-Segal representation of the CCR in Chapter 10, we indicated that it was not self-evidently connected to any particular classical phase space angular distribution. Certainly $\Xi(\varphi)$ is not the Weyl quantization of an angular distribution. However, viewed in the light of ordered quantization — to be specific, anti-Wick quantization — the operator $\Xi(\varphi)$ can be connected in another way with the angle function in phase space. This result has been discovered by a number of authors, if not enunciated in the formal framework that we shall use (for example, see [19], §5.2).

Simple calculations show that

$$\big(q_{-1} * \mathcal{G}F\big)(p,q) = \frac{1}{2\pi}\int_{\mathbb{R}} e^{-\frac{1}{4}u^2+ipu}\, F\big(q+\tfrac{1}{2}u, q-\tfrac{1}{2}u\big)\,du\,,$$

for any $F \in \mathcal{S}(\mathbb{R}^2)$, and hence

$$\begin{aligned}(\mathcal{G}_{(-1,0)}F)(p,q) &= (Q_{-1,0} * \mathcal{G}F)(p,q) \\ &= \frac{1}{2\pi^{\frac{3}{2}}} \iint_{\mathbb{R}^2} \exp\left[-\tfrac{1}{2}(a^2+b^2) + a(q+ip) + b(q-ip) - q^2\right] F(a,b)\, da\, db \\ &= \frac{1}{2\pi} e^{-|w|^2} \iint_{\mathbb{R}^2} \overline{\mathcal{U}(w,a)}\, \mathcal{U}(w,b)\, F(a,b)\, da\, db\,, \qquad (14.3.15)\end{aligned}$$

for any $F \in \mathcal{S}(\mathbb{R}^2)$, where $w = (q-ip)/\sqrt{2}$ and $\mathcal{U}(w,a)$ is the integral kernel for the unitary map $\mathfrak{U}_{BS} : L^2(\mathbb{R}) \to \mathcal{B}$, defined in equation (5.6.4.a), which intertwines the Schrödinger and the Bargmann-Segal representations of the CCR discussed in Subsection 5.6. It follows that

$$\left[\mathcal{G}_{(-1,0)}(\overline{g} \otimes f)\right](p,q) = \frac{1}{2\pi} e^{-|w|^2}\, \overline{\left[\mathfrak{U}_{BS}g\right](w)}\, \left[\mathfrak{U}_{BS}f\right](w)\,, \qquad (14.3.16)$$

for any $f, g \in \mathcal{S}(\mathbb{R})$, which in turn implies that

$$\begin{aligned}[\![\, \mathbf{\Delta}_{(-1,0)}[F]f\,,\, \overline{g}\,]\!] &= \frac{1}{\pi}\int_{\mathbb{C}} e^{-|w|^2} F(w)\, \overline{\left[\mathfrak{U}_{BS}g\right](w)}\left[\mathfrak{U}_{BS}f\right](w)\, dA(w) \\ &= \langle g\,,\, \Xi(F)f \rangle\,, \qquad (14.3.17)\end{aligned}$$

for any $f, g \in \mathcal{S}(\mathbb{R})$ and $F \in L^\infty(\Pi)$, where $\Xi(F)$ is the operator defined in equation (10.3.32). We summarize these observations in the following Proposition.

Proposition 14.10 *If F belongs to $L^\infty(\Pi)$, then its anti-Wick quantization $\mathbf{\Delta}_{(-1,0)}[F]$ is a bounded operator, with $\|\, \mathbf{\Delta}_{(-1,0)}[F] \,\| \leqslant \|\, F\,\|_\infty$, being equal to the Bargmann-Segal operator $\Xi(F)$.*

It is worth noting the identity

$$\omega_w\big(\Xi(F)\big) = \langle \Phi_w\,,\, \Xi(F)\Phi_w \rangle = \frac{1}{\pi}\int_{\mathbb{C}} F(z) e^{-|z-w|^2}\, dA(z)\,, \qquad (14.3.18)$$

which describes the expectation of the operator $\Xi(F)$ in the pure state ω_w (which is determined by the coherent vector Φ_w, as introduced in Subsection 10.3.3) directly in terms of the function F. This identity should be compared with the formula for the Weyl symbol D_F of the operator $\Xi(F)$,

$$D_F(p,q) = \frac{2}{\pi}\int_{\mathbb{C}} F(z)\, e^{-2|z-w|^2}\, dA(z), \quad w = \frac{1}{\sqrt{2}}(q-ip), \qquad (10.3.35)$$

since together these equation imply that

$$\omega_w(\Xi(F)) = \frac{2}{\pi}\int_{\mathbb{C}} D_F(z)e^{-2|z-w|^2}\,dA(z)\,, \tag{14.3.19}$$

so that $D_F(z)$ (the Weyl-ordering symbol of $\Xi(F)$) is obtained from $F(z)$ (the anti-Wick symbol of $\Xi(F)$), and $\omega_z(\Xi(F))$ is obtained from $D_F(z)$, by successive convolution by the same Gaussian function on $\mathbb{C}$. This relationship between these three functions has been noted in [63]. Moreover Folland (*ibid*) and Berezin & Shubin [19] both regard the function $\omega_z(\Xi(F))$ as the Wick-ordered symbol $\mathbf{\Delta}_{(1,0)}^{-1}[\Xi(F)]$ of the operator $\Xi(F)$, displaying a particularly symmetric relationship between the normal-ordered, Weyl-ordered and antinormal-ordered symbols of the operator $\Xi(F)$. However, as we shall see, the smooth model does not seem to support a detailed analysis of normal ordering, so we content ourselves by simply noting this point.

It is also interesting that these relationships provide another very simple relationship between the Weyl symbol $D_F(z)$ and the posited normal-ordered symbol $\omega_z(\Xi(F))$ of the Bargmann-Segal operator $\Xi(F)$ for angular distributions $F \in L^\infty(\Pi)$, since in that case

$$D_F(z) = \omega_{\sqrt{2}z}(\Xi(F))\,, \tag{14.3.20}$$

and so although F, the antinormal-ordered symbol of $\Xi(F)$, is an angular distribution, neither the Weyl symbol nor the posited normal-ordered symbol of $\Xi(F)$ is an angular distribution.

14.3.3 *Polar AW-Quantization*

As was the case for Weyl quantization, the simplest device for studying the quantization of polar observables is to consider the effect of the Wigner transform on the generating function for the Hermite-Gauss functions. In the case of AW-quantization, therefore, we need to calculate the function

$$\mathcal{G}_{(\lambda,0)}\left(\overline{G_s}\otimes G_t\right) = Q_\lambda * \mathcal{G}\left(\overline{G_s}\otimes G_t\right). \tag{14.3.21}$$

Since all the functions involved are Gaussians, the necessary calculations are elementary, if lengthy, and the result is as follows:

$$\begin{aligned}&\left[\mathcal{G}_{(\lambda,0)}(\overline{G_s}\otimes G_t)\right](p,q) = \\ &\frac{1}{\pi(1-\lambda)}\exp\left\{-\frac{1}{2}\frac{1+\lambda}{1-\lambda}st + \frac{(q+ip)s}{1-\lambda} + \frac{(q-ip)t}{1-\lambda} - \frac{p^2+q^2}{1-\lambda}\right\}.\end{aligned} \tag{14.3.22}$$

Note how the factor $1-\lambda$ appears in all the denominators here. Given this expression, the next step is to calculate the partial polar integrals of this function, yielding a simplified expression which can then be used to analyze polar quantizations.

For radial quantization, the crucial formula is

$$\int_{-\pi}^{\pi} \left[\mathcal{G}_{(\lambda,0)}(\overline{G_s} \otimes G_t)\right](r\cos\beta, r\sin\beta)\, d\beta \tag{14.3.23}$$
$$= \frac{2}{1-\lambda} \exp\left\{-\frac{r^2}{1-\lambda}\right\} \sum_{N=0}^{\infty} \frac{1}{N!}\left(-\frac{1}{2}\frac{1+\lambda}{1-\lambda}st\right)^N L_N\left(\frac{2r^2}{1-\lambda^2}\right)$$

for any $-1<\lambda<0$. While it may seem as if the presence of the factor $1-\lambda^2$ in the denominator of the Laguerre polynomials renders this formula unusable in the case of anti-Wick ordering $\lambda=-1$, the additional factors of powers of $(1+\lambda)$ are precisely sufficient to cancel out these singularities, so the above expression can be made valid for the whole range of AW-ordering.

It is clear that, for any $\mathfrak{f} \in \mathcal{O}^{\infty}(\mathbb{R})$, the integrals

$$\rho_{\lambda,n} = (-1)^n \left(\frac{1+\lambda}{1-\lambda}\right)^n \int_0^{\infty} e^{-u} L_n\left(\frac{2u}{1+\lambda}\right) \mathfrak{f}\left(\sqrt{(1-\lambda)u}\right) du \tag{14.3.24}$$

exist for all $-1<\lambda<0$ and $n \geqslant 0$. These integrals are then the eigenvalues of the operator $\mathbf{\Delta}_{(\lambda,0)}[\mathfrak{f}_{\mathrm{rad}}]$, and just as for the Weyl case, the spectrum of $\mathbf{\Delta}_{(\lambda,0)}[\mathfrak{f}_{\mathrm{rad}}]$ consists only of these eigenvalues, each of which is (in general) nondegenerate. The corresponding eigenvalues are the Hermite-Gauss functions,

$$\mathbf{\Delta}_{(\lambda,0)}[\mathfrak{f}_{\mathrm{rad}}]\, h_n = \rho_{\lambda,n}\, h_n, \qquad n \geqslant 0,\ -1<\lambda<0. \tag{14.3.25}$$

and so the $(\lambda,0)$-quantization of a radial function is, as was the case for Weyl quantization, diagonal with respect to the Hermite-Gauss functions. Taking due care of the opposing factors of $1+\lambda$, it is possible to take the limit as $\lambda \to -1$ in the above formalism, obtaining thereby the appropriate expression for anti-Wick ordering,

$$\mathbf{\Delta}_{(-1,0)}[\mathfrak{f}_{\mathrm{rad}}]\, h_n = \rho_{-1,n}\, h_n, \qquad n \geqslant 0, \tag{14.3.26.a}$$

where

$$\rho_{-1,n} = \frac{1}{n!}\int_0^{\infty} e^{-u}\, \mathfrak{f}(\sqrt{2u})\, u^n\, du, \qquad n \geqslant 0, \tag{14.3.26.b}$$

which is perfectly well-defined.

How does angular AW-quantization compare with angular Weyl quantization? By going through the involved algebraic rearrangements for the coefficients, it is found that the characteristic coefficients $g_{m,n}$ for the Weyl matrix elements must be replaced by new coefficients $g^{(\lambda)}_{m,n}$, which are fairly complicated functions of λ,

$$\begin{aligned} g^{(\lambda)}_{m,n} &= \frac{1}{(1-\lambda)^{\frac{1}{2}(m+n)}} \left[\frac{\max(m,n)!\,2^{\min(m,n)}}{\min(m,n)!\,2^{\max(m,n)}}\right]^{\frac{1}{2}} \\ &\quad \times \left[\sum_{j=0}^{\min(m,n)} \binom{\min(m,n)}{j} (-\lambda)^{\min(m,n)-j} \frac{\Gamma(\frac{1}{2}j+s_j)}{\Gamma(\frac{1}{2}|m-n|+\frac{1}{2}j+s_j)}\right], \end{aligned} \tag{14.3.27}$$

where $s_j = \frac{1}{2}$ if j is even, while $s_j = 1$ if j is odd[10].

The angular matrix elements in the AW-ordering for an angular distribution $\mathfrak{f}_{\text{ang}}$, where $\mathfrak{f} \in L^\infty(\mathbb{T})$, can now be written down:

$$[\![\, \mathbf{\Delta}_{(\lambda,0)}\,[\mathfrak{f}_{\text{ang}}]\, h_n\,,\, h_m \,]\!] = g^{(\lambda)}_{m,n}\, i^{m-n}\, \hat{\mathfrak{f}}_{m-n}, \qquad m,\, n \geqslant 0. \tag{14.3.28}$$

The meaning of the coefficients $g^{(\lambda)}_{m,n}$ is similar to their meaning for Weyl quantization: they are essentially the matrix elements of the anti-Wick quantization of the distribution $\mathfrak{d}_{\text{ang}}$, where $\mathfrak{d} \in \mathcal{D}(\mathbb{T})$ is the delta distribution concentrated at 1, so that

$$[\![\, \mathfrak{d}\,,\, \mathfrak{g} \,]\!] = \mathfrak{g}(1), \qquad \mathfrak{g} \in C^\infty(\mathbb{T}),$$

in that

$$[\![\, \mathbf{\Delta}_{(\lambda,0)}\,[\mathfrak{d}_{\text{ang}}]\, h_n\,,\, h_m \,]\!] = i^{m-n}\, g^{(\lambda)}_{m,n}, \qquad m,\, n \geqslant 0. \tag{14.3.29}$$

14.3.4 *The AW-Phase Operator*

Before beginning to discuss the AW-phase operator, let us consider its occasional rival, the Toeplitz phase operator X. As shown previously, there is no distribution $\mathfrak{f}$ on the circle such that $\mathbf{\Delta}\,[\,\mathfrak{f}_{\text{ang}}\,] = X$, but perhaps there is an AW-ordering and a distribution $\mathfrak{f}$ for which $\mathbf{\Delta}_{(\lambda,0)}\,[\mathfrak{f}_{\text{ang}}] = X$? The argument we used for the Weyl case was to note that its Hermite-Gauss

[10] Reassuringly, we note that $g^{(0)}_{m,n} = g_{m,n}$, as expected.

matrix elements

$$\langle h_m, X h_n \rangle = i^{m-n}\hat{\mathfrak{p}}_{m-n} = \begin{cases} \dfrac{i^{n-m+1}}{m-n}, & m \neq n, \\ 0, & m = n, \end{cases} \tag{10.3.15}$$

cannot be written as $g_{m,n}$ multiplied by the Fourier coefficients of some function $\mathfrak{f}$. This same argument still holds, unaffected by the substitution of $g_{m,n}^{(\lambda)}$ for $g_{m,n}$. Thus

Proposition 14.11 *The Toeplitz operator X has no AW-symbol which is an angular distribution.*

Thus, unlike the case of the Bargmann-Segal phase operator $\Xi(\varphi)$, we do not know of any way in which X can be associated with an angular distribution through any of the ordered quantization procedures considered in this book.

Consider the AW-phase operator, $\boldsymbol{\Delta}_{(\lambda,0)}[\varphi]$. We know that its matrix coefficients are given by the formula

$$[\![\, \boldsymbol{\Delta}_{(\lambda,0)}[\varphi]\, h_n\,,\, h_m \,]\!] = \begin{cases} g_{m,n}^{(\lambda)} \dfrac{i^{n-m+1}}{m-n}, & m \neq n\,, \\ 0\,, & m = n\,. \end{cases} \tag{14.3.30}$$

Since the operator properties of $\boldsymbol{\Delta}_{(\lambda,0)}[\varphi]$ are reflected in its matrix elements, at least in principle, and since the matrix elements have changed, perhaps the operator properties of $\boldsymbol{\Delta}_{(\lambda,0)}[\varphi]$ are different to those of $\boldsymbol{\Delta}[\varphi]$. Thus we need to ask whether $\boldsymbol{\Delta}_{(\lambda,0)}[\varphi]$ is bounded for all (or any) λ in $[-1,0]$, what its integral kernel is, whether it belongs to $\mathcal{L}^+(\mathcal{S}(\mathbb{R}))$ for any λ, and how its spectrum has changed. These and other questions are more difficult to answer than for $\boldsymbol{\Delta}[\varphi]$, but some results are known. Leaving the proofs to the references, [53], [110], the principal results are these. Firstly, we have an explicit expression for the integral kernel of $\boldsymbol{\Delta}_{(\lambda,0)}[\varphi]$.

Proposition 14.12 *For $-1 \leqslant \lambda < 0$, the operator $\boldsymbol{\Delta}_{(\lambda,0)}[\varphi]$ has integral kernel representation*

$$\begin{aligned} [\![\, \boldsymbol{\Delta}_{(\lambda,0)}[\varphi] f\,,\, g \,]\!] &= \tfrac{1}{2}\pi \int_{\mathbb{R}} \operatorname{erf}\Big(\frac{y}{\sqrt{|\lambda|}}\Big)\, g(y)\, f(y)\, dy \\ &- \tfrac{1}{2} i \lim_{L\to\infty} \iint_{\mathbb{R}^2} \operatorname{sgn}(y) \Big\{ e^{-|xy|} - \frac{2}{\sqrt{\pi}} \int_{|y/\lambda|}^{\infty} \exp\Big(-\xi^2 - \frac{x^2y^2}{4\xi^2}\Big)\, d\xi \Big\} \\ &\qquad \times g_{I(L)}(x)\, g(y+\tfrac{1}{2}x)\, f(y-\tfrac{1}{2}x)\, dx dy, \end{aligned} \tag{14.3.31}$$

for any f, $g \in \mathcal{S}(\mathbb{R})$, where $g_{I(L)}$ is the convergence factor introduced in equation (9.4.22), and all integrals are meant in the sense of Lebesgue.

An extension of the Method of Wedges introduced in Subsection 9.4.5 enables us to rewrite this result, expressing $\mathbf{\Delta}_{(\lambda,0)}[\varphi]$ in terms of $\mathbf{\Delta}[\varphi]$.

Proposition 14.13 *The matrix elements of $\mathbf{\Delta}_{(\lambda,0)}[\varphi]$ with respect to arbitrary f, $g \in \mathcal{S}(\mathbb{R})$ can be written as*

$$\Big\langle g, \mathbf{\Delta}_{(\lambda,0)}[\varphi]\, f\Big\rangle = \Big\langle g, \mathbf{\Delta}[\varphi]\, f\Big\rangle \tag{14.3.32.a}$$
$$-\pi\Big\langle g, \chi_{(0,\infty)}(P) \circ \big[\operatorname{sgn}(Q) - \operatorname{erf}\big(\tfrac{Q}{\sqrt{|\lambda|}}\big)\big] f\Big\rangle + i\Big\langle \mathcal{F}g, \mathcal{K}_{h(\lambda)}\mathcal{F}f\Big\rangle$$
$$+\frac{i}{8}\int_0^{|\lambda|}\Big(\iint_{\mathbb{R}^2}\Big[x e^{-\frac{1}{4}\mu x^2}\big[\operatorname{sgn}(y) - \operatorname{erf}\big(\tfrac{y}{\sqrt{\mu}}\big)\big] V_{f,g}(x,y)\Big]\,dx\,dy\Big)\,d\mu\,,$$

where

$$V_{f,g}(x,y) = \overline{(\mathcal{F}g)(y+\tfrac{1}{2}x)}\,(\mathcal{F}f)(y-\tfrac{1}{2}x)\,. \tag{14.3.32.b}$$

The operator $\mathcal{K}_h$ for any function $h \in L^\infty[0,\infty)$ has been defined in equation (9.4.17.d). Here we take $h(\lambda)$ to be the function

$$[h(\lambda)](x) = 1 - e^{\frac{1}{4}\lambda x^2}, \qquad x > 0. \tag{14.3.33}$$

From this we can deduce that all of the operators $\mathbf{\Delta}_{(\lambda,0)}[\varphi]$ are bounded.

Corollary 14.14 *The operator $\mathbf{\Delta}_{(\lambda,0)}[\varphi]$ is bounded, with*

$$\big\|\mathbf{\Delta}_{(\lambda,0)}[\varphi]\big\| \leqslant 3\pi + \sqrt{|\lambda|}\,2^{-\frac{5}{4}}, \tag{14.3.34}$$

and the results of the previous Proposition can be extended to give the matrix elements for $\mathbf{\Delta}_{(\lambda,0)}[\varphi]$ with respect to any two elements of $L^2(\mathbb{R})$.

The technicalities involved in proving these results are such that it is likely that they could be improved. The upper bound on the norm might be lowered, perhaps even to π, independent of λ. This would be in line with the idea that all the $\mathbf{\Delta}_{(\lambda,0)}[\varphi]$ have the (continuous) spectrum $[-\pi,\pi]$, consonant with their connections with quantum phase. Differences between these operators would presumably still be apparent in the generalized eigendistributions. However, since the above results were not easy to obtain, extending them is probably not simply a matter of sharpening some of the inequalities used, but rather of finding new approaches to the problem.

Since convolution with Q_λ has a smoothing effect on distributions, some aspects of AW-quantization will have better properties than Weyl quantization. For example, more quantum mechanical observables will belong to $\mathcal{L}^+(\mathcal{S}(\mathbb{R}))$. In particular, the matrix coefficients for the anti-Wick-ordered quantization $\mathbf{\Delta}_{(-1,0)}[\varphi] = \Xi(\varphi)$ of φ are given in equation (10.3.39), and study of this equation leads to the following result for the Bargmann-Segal phase operator.

Proposition 14.15 *The map* $\mathbf{\Delta}_{(-1,0)}[\varphi] = \Xi(\varphi)$ *is a symmetric operator in* $\mathcal{L}^+(\mathcal{S}(\mathbb{R}))$.

The next step is to compare the matrix coefficients of $\mathbf{\Delta}_{(\lambda,0)}[\varphi]$ with those of $\mathbf{\Delta}_{(-1,0)}[\varphi] = \Xi(\varphi)$ in the same way that we previously compared them with $\mathbf{\Delta}[\varphi]$.

Lemma 14.16 *The integral representation*

$$\begin{aligned}\langle g, \mathbf{\Delta}_{(\lambda,0)}[\varphi] f\rangle &= \langle g, \mathbf{\Delta}_{(-1,0)}[\varphi] f\rangle + \frac{\pi}{2}\langle g, \left[\operatorname{erf}\left(\frac{Q}{\sqrt{|\lambda|}}\right) - \operatorname{erf}(Q)\right] f\rangle \\ &\quad + \frac{i}{8}\int_{|\lambda|}^{1} \langle \mathcal{F}g, Y_\mu \mathcal{F}f\rangle \, d\mu, \end{aligned} \tag{14.3.35}$$

holds for matrix elements for all $f,\ g \in \mathcal{S}(\mathbb{R})$, *where the operator* Y_μ, *defined through its integral kernel,*

$$[Y_\mu g](p) = \int_{\mathbb{R}} (p-q)e^{-\frac{1}{4}\mu(p-q)^2} \operatorname{erf}\left(\frac{p+q}{2\sqrt{\mu}}\right) g(q)\, dq \tag{14.3.36}$$

for any $\mu > 0$ *and* $g \in \mathcal{S}(\mathbb{R})$, *maps* $\mathcal{S}(\mathbb{R})$ *continuously into itself.*

Putting these results together, we deduce the following.

Proposition 14.17 *The operator* $\mathbf{\Delta}_{(\lambda,0)}[\varphi]$ *belongs to* $\mathcal{L}^+(\mathcal{S}(\mathbb{R}))$ *for all* $-1 \leqslant \lambda < 0$, *but not when* $\lambda = 0$.

14.4 Wick Quantization

Having considered the PQ and AW-families, we now turn to the so-called *Wick family* (or *W-family*), namely the other half of the WAW-family for which $0 < \lambda \leqslant 1$. The difference here is that since the function $\mathcal{E}_{\lambda,0}$ no longer belongs to $\mathcal{S}'(\Pi)$, we can no longer calculate its Fourier transform within the smooth model, and the upshot of this is that the $(\lambda, 0)$-Wigner transform can no longer have the whole of $\mathcal{S}(\mathbb{R}^2)$ as its domain. As indicated previously, we can work around this difficulty by considering a suitable subspace of $\mathcal{S}(\mathbb{R}^2)$, but losing the use of the whole of $\mathcal{S}(\mathbb{R}^2)$ is a significant blow, since this space is fundamental to the structure of the smooth model. As we indicated above in our discussion of marginals for the AW-family of orderings, we suspect that a variant of the smooth model is required, working with a new subspace of $\mathcal{S}(\mathbb{R}^2)$, which would enable us not only to consider A- and A^+-marginals for the AW-quantizations for more than just polynomials, but would also enable us to extend the formalism into the W-family of quantizations.

Because of these technical problems the theory of W-quantization, as we shall term it, is much less well studied from the point of view of rigour, and all we shall do is remark on the existence question. Proofs can be found in [111]. We begin by defining a substitute for the space $\mathcal{S}(\mathbb{R})$.

Proposition 14.18 *By the space $\Sigma(\mathbb{R})$ we shall mean the following subset of $\mathcal{S}(\mathbb{R})$,*

$$\Sigma(\mathbb{R}) = \left\{ f \in \mathcal{S}(\mathbb{R}) \,:\, e_\beta \mathcal{F} e_\alpha f \in \mathcal{S}(\mathbb{R}) \,, \quad \begin{matrix} 0 \leqslant \alpha < 1,\, \beta \geqslant 0, \\ (1-\alpha)\beta < 1, \end{matrix} \right\} \tag{14.4.1}$$

where for any $\alpha \geqslant 0$ the linear map $e_\alpha : \mathcal{S}(\mathbb{R}) \to \mathcal{C}^\infty(\mathbb{R})$ is defined by the formula

$$[e_\alpha f](x) \;=\; e^{\frac{1}{2}\alpha x^2} f(x), \qquad\qquad f \in \mathcal{S}(\mathbb{R}). \tag{14.4.2}$$

The space $\Sigma(\mathbb{R})$ is a dense linear subspace of $\mathcal{S}(\mathbb{R})$ which contains all the Hermite-Gauss functions and their translates. In particular, $\Sigma(\mathbb{R})$ contains the generating function G_t of the Hermite-Gauss functions.

This space $\Sigma(\mathbb{R})$ is now small enough to permit the quantization procedure that we need, and yet is large enough (just) to contain useful functions. The properties of $\Sigma(\mathbb{R})$ as a topological vector space have not been established in detail. In particular, it is not known whether it has a natural topology of its own with respect to which it is complete and (if that topology exists) whether it is metrizable, or nuclear.

Given the absence of a topology on $\Sigma(\mathbb{R})$, our approach to quantization involving this space has to be more circumspect. For all orderings so far considered, $\mathbf{\Delta}_\sharp\,[T]$ has been defined as a mapping in $\mathcal{L}(\mathcal{S}(\mathbb{R}),\mathcal{S}'(\mathbb{R}))$. Since we now have to restrict our "wave functions" to $\Sigma(\mathbb{R})$, analogy leads us to want to define $\mathbf{\Delta}_{(\lambda,0)}\,[T]$ as a continuous linear mapping from $\Sigma(\mathbb{R})$ to $\Sigma(\mathbb{R})'$. But without a topology on $\Sigma(\mathbb{R})$, there is no topological dual $\Sigma(\mathbb{R})'$, and so we have to proceed in a more algebraic fashion, making use of the notion of sesquilinear forms. If and when a topology is imposed on $\Sigma(\mathbb{R})$, it can be checked whether or not such a form is continuous, but until then the following will have to suffice.

Proposition 14.19 *The function $\mathcal{W}_{(\lambda,0)}(\overline{g}\otimes f)$ belongs to $\mathcal{S}(\mathbb{R}^2)$ for all $f,\, g \in \Sigma(\mathbb{R})$ and $0<\lambda<1$. Hence, for any $f,\, g\in\Sigma(\mathbb{R})$ and $0<\lambda<1$ the W-Wigner function $\mathcal{G}_{(\lambda,0)}(\overline{g}\otimes f)$ belongs to $\mathcal{S}(\Pi)$. Thus any $T\in\mathcal{S}'(\Pi)$ defines a sesquilinear form on $\Sigma(\mathbb{R})$, denoted $\mathbf{\Delta}_{(\lambda,0)}\,[T]$, by the formula*

$$\big[\,\mathbf{\Delta}_{(\lambda,0)}\,[T]\,\big](g,f) \;=\; [\![T\,,\,\mathcal{G}_{(\lambda,0)}(\overline{g}\otimes f)\,]\!], \qquad f,g\in\Sigma(\mathbb{R})\,, \tag{14.4.3}$$

for any $0<\lambda<1$.

The fact that, in this case, $\mathbf{\Delta}_{(\lambda,0)}\,[T]$ is a form, and neither an operator nor a mapping on $\Sigma(\mathbb{R})$, must be borne in mind in any application of this theory. We can apply this Proposition to angular quantization by proceeding in the usual way, letting f and g be generating functions of the Hermite-Gauss functions, introducing polar coordinates and then integrating out the radial variable. Doing so, we obtain the following close analogue of equation (14.3.28).

Proposition 14.20 *If $\mathfrak{f}\in L^\infty(\mathbb{T})$, then*

$$\big[\mathbf{\Delta}_{(\lambda,0)}\,[\mathfrak{f}_{\mathrm{ang}}]\big](h_m,h_n) \;=\; g^{(\lambda)}_{m,n}\,i^{m-n}\,\hat{\mathfrak{f}}_{m-n}, \qquad \lambda\in(0,1). \tag{14.4.4}$$

Since the form of the matrix elements of $\mathbf{\Delta}_{(\lambda,0)}\,[\varphi]$ remains the same as for AW-quantization, we might hope that $\mathbf{\Delta}_{(\lambda,0)}\,[\varphi]$ is also a bounded operator. But this is not to be.

Proposition 14.21 *The sesquilinear form $\mathbf{\Delta}_{(\lambda,0)}\,[\varphi]$ is not a linear operator from $\Sigma(\mathbb{R})$ to $L^2(\mathbb{R})$ for any $\lambda\in(0,1)$.*

Proof: Suppose that we could find a linear map $Z:\Sigma(\mathbb{R})\to L^2(\mathbb{R})$ such that

$$\big[\mathbf{\Delta}_{(\lambda,0)}\,[\varphi]\big](g,f) \;=\; \langle g\,,\,Zf\rangle\,,$$

for $f,g \in \Sigma(\mathbb{R})$. Then we would deduce that

$$\begin{aligned}\langle h_n\,,\, Zh_0\rangle &= i^n\, g_{n,0}^{(\lambda)}\,\hat{\mathfrak{p}}_n = \frac{i^{1-n}}{n}\, g_{n,0}^{(\lambda)} \\ &= \frac{i^{1-n}}{n}\,(1-\lambda)^{-\frac{1}{2}n}\sqrt{\frac{\pi\, n!}{2^n}}\,\Gamma\big(\tfrac{1}{2}n+\tfrac{1}{2}\big)^{-1},\end{aligned}$$

for any nonzero integer n, and hence that

$$\langle h_n\,,\, \mathbf{\Delta}_{(\lambda,0)}[\varphi]h_0\rangle \sim \big(\tfrac{\pi}{2}\big)^{\frac{1}{4}}\, i^{1-n}\,(1-\lambda)^{-\frac{1}{2}n}\, n^{-\frac{3}{4}}\,, \qquad n \to \infty\,.$$

This asymptotic formula implies that the sequence $\big(\langle h_n, Zh_0\rangle\big)_{n\geqslant 0}$ does not belong to ℓ^2, and so we deduce that the desired operator Z does not exist. ■

Thus, for $\lambda \in (0,1)$, things are much less favourable than for any of the other orderings. For Wick ordering itself, $\lambda = 1$, we do not even have a Schwartz type space to define quantization within the smooth model[11].

Now that we have defined these families of ordered quantization we can ask, as did the London cabman of Bertrand Russell, *"What does it all mean, then, Guv?"*. Famously, Bertrand Russell could offer the cabman no advice, and we can only do a little better here. What is clear is that we have many different ways at our disposal for assigning a quantum operator to a classical observable. Some, but not all, of these ways have the property of ensuring that real classical observables yield symmetric quantum operators, and hence physically observable quantum mechanical observables. Equally well some, but not all, of these ways yield the correct marginals for position and momentum (and other quantities).

More subtle, and very much still an open question, is the question of the nature of and relationship between the various physical qualities so represented for a given phase space distribution. At a minimum, to be able to answer this question requires precise knowledge of the nature of the various phase operator candidates from an experimental standpoint, and this knowledge is not available.

Conversely, any given quantum mechanical observable yields a two-parameter family of phase space distributions, these being its dequantization symbols with respect to the various orderings. Since the symbol of

---

[11] It must be possible, however, to define some rigorous form of Wick ordered quantization, since the polynomial functions on Π behave well under formal Wick quantization.

an operator represents some sort of classical limit, or representation, of the operator, the different orderings available, and hence the variety of possible classical analogues to quantum mechanical observables, indicates the complexity of the relationship between quantum and classical mechanics. We would like to be able to shed some light on this problem, but we are not even clear in what sense an answer can be given, and whether progress will come from mathematics, theoretical physics considerations, or from experiments.

In summary we see that, to a greater or lesser extent, there is a respectable quantization theory for the PQ, AW- and W-families ($\lambda = 1$ excepted). Each has good features and bad, *but the one choice which behaves well in all respects is that of Weyl ordering.* We take the moral of this Chapter on orderings to be an affirmation of Weyl quantization, unless there is a particular physical reason to choose otherwise, because it has the richest structure.

CHAPTER 15

ASYMPTOTICS

The two extremes, of too much stiffness in refusing, and of too much easiness in admitting any variation.

– The Book of Common Prayer

15.1 Introduction

As part of our discussion of the theory of phase operators, we have introduced three key families of pure states, the family defined by the Hermite-Gauss functions $\{h_n \,:\, n \geqslant 0\}$, the coherent states given by the functions $\{\Phi_\zeta \,:\, \zeta \in \mathbb{C}\}$, and finally the transformed LHW states of the Barnett & Pegg theory obtained from the vectors $\{\eta_s[\theta] \,:\, \theta \in \mathbb{R}\,,\, s \geqslant 1\}$, which were defined in equation (10.3.42.a). A function in any of these families is, at least in part, described by a parameter, namely n for a Hermite-Gauss function h_n, $|\,\zeta\,|$ for a coherent state Φ_ζ and s for a transformed LHW state $\eta_s[\theta]$, and the behaviour of quantum mechanical observables in states with large values of that parameter is supposed to describe some aspect of the classical limit for those observables. Thus calculating the expectation and variance of a quantum mechanical observable in each of these states, and determining the asymptotic behaviour of these quantities as the relevant parameter tends to infinity, will presumably provide us with information concerning the classical qualities of that observable.

In this Chapter we shall outline these calculations for the various quantum phase observables that we have considered in previous Chapters. We have already made it clear that our preferred quantum phase observable is the operator $\mathbf{\Delta}\,[\varphi]$, but other competitors are the Toeplitz phase operator X, the Bargmann-Segal phase operator $\Xi(\varphi)$, and the construction of Barnett & Pegg. In the hope that it will eventually become possible to conduct experiments to distinguish between these observables, it is important to indicate how they differ from each other, and in this Chapter we shall discuss their asymptotic behaviour.

In the interests of brevity, we shall condense our presentation. Since there are three collections of states and four types of quantum phase observable (with three key phase-related operators of each type), determining the asymptotic behaviour of both the expectation and variance in all cases would require us to present a total of 72 different results. We will not do this, but shall be content with a sample of the results which are (in our opinion) representative of the most interesting problems. However, what will become clear from the mathematics is that all four classes of observables exhibit asymptotic behaviours which are consistent with their being interpreted as quantum phase observables, although the exact nature of the asymptotic requirement for this to be the case is disputed amongst authors — where they differ is in the detailed nature of the asymptotic behaviour required by the physics[1]. Consequently, any experiments made to distinguish between these phase observables will of necessity have to be very subtle.

Sometimes it is comparatively simple to determine these asymptotic limits, but in other cases the calculations are extremely delicate. This is due to the fairly complicated nature of any of the quantum phase observables, resulting in our having to determine the asymptotic behaviour of rather involved integrals and sums. Asymptotic analysis abounds with problems of instability, and a family of functions which varies smoothly with some parameter may have asymptotic behaviour which does not vary smoothly, or even continuously, with that parameter[2]. Thus any attempt to integrate (or sum) such a family of asymptotic expansions with respect to this parameter, in the hope of obtaining the asymptotic expansion of the corresponding integral (or sum) of the family of functions, is often fraught with problems, unless some control can be placed upon those asymptotic expansions which is uniform with respect to the parameter.

Unfortunately, the classical methods of asymptotic analysis, as exemplified by those found in Whittaker & Watson [239] and in Copson [36] may not be well-known nowadays, but we shall assume them to be known to the reader without detailed comment. Moreover, most of the proofs of our results will be omitted, since otherwise this Chapter would assume the

---

[1] They also differ in the ways that have been discussed in previous Chapters.

[2] A classic example is the fact that the asymptotic behaviour of the Bessel coefficient $J_\nu(\nu a)$, as $\nu \to \infty$, is discontinuous at $a = 1$.

length of a book on its own. The reader will be referred to the literature for the proofs. However some proofs will be included, since we feel it important to give a few examples of detailed and rigorous calculation. Calculations in the physics literature are frequently rough-and-ready, and dependent loosely upon physical intuition. That is not to say that the results obtained by these calculations may not be correct, but the arguments given to justify them are often heuristic rather than rigorous (often based upon making approximations to functions which are not valid uniformly over the domain of their application) and should really be replaced by detailed and exact mathematical analysis.

In view of the fact that the states that we shall consider are themselves equipped with indices, it is no longer suitable to describe the expectation of the quantum mechanical observable X in the state ρ by the symbol $\mathsf{Exp}_\rho[X]$, or by the symbol $\mathsf{Exp}_f[X]$ if the state ρ is pure, determined by the unit vector f. In the interests of clarity, in this Chapter we shall denote this expectation by the expression $\mathsf{Exp}\{X;\, \rho\}$ instead, and we shall make an analogous adjustment to our previous notation for the variance and uncertainty of observables.

15.2 Asymptotics For Hermite-Gauss States

The first class of states for which we shall consider the asymptotic behaviour of phase observables is that determined by the Hermite-Gauss functions $\{h_n : n \geqslant 0\}$. Since the excitation of the system is greater the larger the value of n, it is to be expected that a "good" quantum phase observable displays the characteristics of classical behaviour in the limit as $n \to \infty$. Since heuristic classical considerations would lead us to expect the phase to be uniformly distributed over its range $[-\pi, \pi]$, a "good" quantum phase observable should have expectations and variances in the Hermite-Gauss states which approach 0 and $\frac{1}{3}\pi^2$ asymptotically[3] in the limit as $n \to \infty$.

15.2.1 *Barnett & Pegg Operators*

The asymptotic behaviour with respect to the Hermite-Gauss states of the operators X_s of Barnett & Pegg theory is particularly easy to determine,

[3] Whether a certain rate of approach to these values is required by the physics and, if so, what that rate is, is one of the causes of disagreement to be found in the literature.

since a consequence of equation (10.3.46.b) is that all of the expectations and variances concerned depend very simply on n, and the limiting "expectations" and "variances" required by that theory are independent of n (indeed, that this should be so is one of the major motivations leading to the development of that theory). To be specific, it can be shown that

$$\mathsf{Exp}\,\{X_s;\, h_n\} \;=\; \begin{cases} -\dfrac{\pi}{s+1}\,, & n \leqslant s\,, \\ 0\,, & n > s\,, \end{cases} \tag{15.2.1.a}$$

$$\mathsf{Var}\,\{X_s;\, h_n\} \;=\; \begin{cases} \dfrac{1}{3}\pi^2\dfrac{s(s+2)}{(s+1)^2}\,, & n \leqslant s\,, \\ 0\,, & n > s\,, \end{cases} \tag{15.2.1.b}$$

which equations yield the limiting identities:

$$BP_1(h_n) \;=\; \lim_{s\to\infty} \mathsf{Exp}\,\{X_s;\, h_n\} \;=\; 0\,, \tag{15.2.2.a}$$

and

$$V_{BP}(h_n) \;=\; \lim_{s\to\infty} \mathsf{Var}\,\{X_s;\, h_n\} \;=\; \tfrac{1}{3}\pi^2\,. \tag{15.2.2.b}$$

These results are certainly in line with the theory of Barnett & Pegg representing some aspects of phase, but are of no interest as regards asymptotic analysis, and do *not* represent the expectation and variance of a single phase operator.

15.2.2 *Toeplitz Operators*

The matrix coefficients for the Toeplitz phase operator X with respect to the Hermite-Gauss states have been determined in equation (10.3.15), and the particularly simple form of this equation makes determining the relevant asymptotic properties simple. Indeed, it was shown in Subsection 10.3.4.3 that the expectation of X in the state h_n is zero and that

$$\mathsf{Var}\,\{X;\, h_n\} \;=\; \tfrac{1}{3}\pi^2 - \sum_{k=n+1}^{\infty} k^{-2}\,. \tag{10.3.55}$$

Thus, while there is no interesting asymptotic behaviour for the expectations of X, the variances of X are such that

$$\mathsf{Var}\,\{X;\, h_n\} \;\sim\; \tfrac{1}{3}\pi^2 - \frac{1}{n}\,, \qquad n \to \infty\,. \tag{10.3.56}$$

Thus $\mathsf{Exp}\{X;\, h_n\}$ certainly converges to $\frac{1}{3}\pi^2$ as $n \to \infty$, and we have a degree of control upon the (comparatively slow) rate of convergence.

Similar calculations can be performed for the cosine and sine operators, yielding

$$\mathsf{Exp}\{C;\, h_n\} \;=\; \mathsf{Exp}\{S;\, h_n\} \;=\; 0 \tag{15.2.3}$$

for the expectations, and

$$\mathsf{Var}\{C;\, h_n\} \;=\; \mathsf{Var}\{S;\, h_n\} \;=\; \begin{cases} \frac{1}{4}\,, & n=0\,, \\ \frac{1}{2}\,, & n \geqslant 1\,. \end{cases} \tag{15.2.4}$$

From these identities it follows that

$$\lim_{n\to\infty} \mathsf{Var}\{C;\, h_n\} \;=\; \lim_{n\to\infty} \mathsf{Var}\{S;\, h_n\} \;=\; \frac{1}{2}\,, \tag{15.2.5}$$

which "classical" limiting behaviour is, once again, to be expected for the variances of quantum mechanical phase observables.

15.2.3 *The Bargmann-Segal Phase Observables*

While equation (10.3.39) for the matrix coefficients of the Bargmann-Segal phase observable $\Xi(\varphi)$ with respect to the Hermite-Gauss functions makes the calculation of the expectations

$$\mathsf{Exp}\{\Xi(\varphi);\, h_n\} \;=\; 0\,, \qquad\qquad n \geqslant 0\,, \tag{15.2.6}$$

particularly simple, the same cannot be said of the variances

$$\mathsf{Var}\{\Xi(\varphi);\, h_n\} \;=\; \sum_{m\neq n} \frac{1}{(m-n)^2}\, \frac{\Gamma\big(\frac{1}{2}m+\frac{1}{2}n+1\big)^2}{m!\, n!}\,, \tag{15.2.7}$$

since the additional Gamma function term renders this expression very difficult to analyze. However, some information is available concerning these quantities, since elementary properties of the Gamma function show us that

$$\frac{\Gamma\big(\frac{1}{2}m+\frac{1}{2}n+1\big)^2}{m!\, n!} \;\leqslant\; 1$$

for all $m, n \geqslant 0$, which in turn implies that

$$\mathsf{Var}\{\Xi(\varphi);\, h_n\} \;\leqslant\; \mathsf{Var}\{X;\, h_n\} \;\leqslant\; \frac{1}{3}\pi^2\,, \tag{15.2.8}$$

for all $n \geqslant 0$. We are not currently in possession of techniques which will permit us to control these variances from below, ensuring that they converge to $\frac{1}{3}\pi^2$ in the limit as $n \to \infty$, but we have no reason to suspect that they will not. Besides the evidence of numerical calculations, the reasons for our confidence are two-fold. Firstly, the analogue of equation (15.2.7) for the Weyl phase observable $\mathbf{\Delta}[\varphi]$ is yet more complicated, while yielding the desired result, as we shall see in the next Subsection[4]. Additionally, the Bargmann-Segal analogues of C and S, namely $\Xi(\cos\varphi)$ and $\Xi(\sin\varphi)$, exhibit the correct asymptotic behaviour, since it can be shown that

$$\mathsf{Exp}\left\{\Xi(e^{\pm i\varphi});\, h_n\right\} = 0\,, \qquad n \geqslant 0\,, \tag{15.2.9}$$

while

$$\left\| \Xi(e^{i\varphi})h_n \right\|^2 = \frac{\Gamma\left(n+\frac{3}{2}\right)^2}{n!\,(n+1)!}\,, \qquad n \geqslant 0\,, \tag{15.2.10.a}$$

$$\left\| \Xi(e^{-i\varphi})h_n \right\|^2 = \begin{cases} \dfrac{\Gamma\left(n+\frac{1}{2}\right)^2}{n!\,(n-1)!}\,, & n \geqslant 1\,, \\ 0\,, & n = 0\,, \end{cases} \tag{15.2.10.b}$$

so it follows that

$$\left\| \Xi(e^{\pm i\varphi})h_n \right\|^2 = 1 - \frac{1}{4n} + O\left(\frac{1}{n^2}\right), \qquad n \to \infty\,, \tag{15.2.11}$$

and these results imply that

$$\begin{aligned} \mathsf{Var}\left\{\Xi(\cos\varphi);\, h_n\right\} &= \mathsf{Var}\left\{\Xi(\sin\varphi);\, h_n\right\} \\ &= \frac{1}{2} - \frac{1}{8n} + O\left(\frac{1}{n^2}\right), \qquad n \to \infty\,, \end{aligned} \tag{15.2.12}$$

which results certainly accord with $\Xi(\cos\varphi)$ and $\Xi(\sin\varphi)$ being interpreted as phase observables in some sense.

15.2.4 *The Weyl Phase Observable* $\mathbf{\Delta}[\varphi]$

When we begin to consider the problem of the behaviour of the Weyl phase observable $\mathbf{\Delta}[\varphi]$ and its relatives with respect to the Hermite-Gauss states,

---

[4]However the proofs for $\mathbf{\Delta}[\varphi]$ involve a detailed study of the integral kernel formulation of this operator, and the integral kernel for $\Xi(\varphi) = \mathbf{\Delta}_{(-1,0)}[\varphi]$ is much more complex (if smoother), so it is not clear that the calculations for $\mathbf{\Delta}[\varphi]$ can be adapted to deal with $\Xi(\varphi)$.

matters become extremely complicated. While it is elementary to show from expression in equation (10.3.61) that

$$\mathsf{Exp}\{\boldsymbol{\Delta}[\varphi];\, h_n\} \;=\; 0\,, \qquad\qquad n \geqslant 0\,, \tag{15.2.13}$$

the corresponding expression for the variance,

$$\mathsf{Var}\{\boldsymbol{\Delta}[\varphi];\, h_n\} \;=\; \|\,\boldsymbol{\Delta}[\varphi]\,h_n\,\|^2 \;=\; \sum_{m\neq n} \frac{1}{(m-n)^2}\, g_{m,n}^2$$

is much less tractable. We choose to take another approach to the problem. Detailed analysis[5] of the integral kernel expression for $\boldsymbol{\Delta}[\varphi]$ yields the identity

$$\langle h_m\,,\, \boldsymbol{\Delta}[\varphi]\,h_n\rangle \;=\; \langle h_m\,,\, \mathcal{A}h_n\rangle \;+\; i\mathrm{sgn}(n-m)\langle h_m\,,\, \mathcal{B}h_n\rangle\,, \tag{15.2.14}$$

for $m, n \geqslant 0$, where $\mathcal{A}, \mathcal{B} \in \mathcal{L}(\mathcal{S}(\mathbb{R}), \mathcal{S}'(\mathbb{R}))$ are the unbounded maps

$$\mathcal{A} \;=\; \frac{1}{2}\pi\mathrm{sgn}(\overline{Q})\,, \tag{15.2.15.a}$$

$$\mathcal{B} \;=\; \frac{1}{2}\gamma\, I \;+\; \log(2\,|\,\overline{Q}\,|)\,, \tag{15.2.15.b}$$

and $\gamma \;=\; 0.5772\cdots$ is the Euler-Mascheroni constant. We note in passing that this identity also confirms that the expectation of $\boldsymbol{\Delta}[\varphi]$ in the Hermite-Gauss states is zero. More important to us is the observation that the variance of $\boldsymbol{\Delta}[\varphi]$ in the Hermite-Gauss states can be expressed solely in terms of the operator $\mathcal{B}$, since

$$\mathsf{Var}\{\boldsymbol{\Delta}[\varphi];\, h_n\} \;=\; \frac{1}{4}\pi^2 \;+\; \mathsf{Var}\{\mathcal{B};\, h_n\}\,, \tag{15.2.16}$$

for any $n \geqslant 0$. Evaluating the variance of $\mathcal{B}$ in the Hermite-Gauss states requires knowledge of the integrals

$$I_n^{(k)} \;=\; \int_0^\infty [\log(2s)]^k\, h_n(s)^2\, ds\,, \qquad k = 1, 2\,,\ n \geqslant 0\,, \tag{15.2.17}$$

since

$$\langle h_m\,,\, \mathcal{B}h_n\rangle \;=\; \frac{1}{2}\gamma \;+\; 2I_n^{(1)}\,, \tag{15.2.18.a}$$

$$\|\,\mathcal{B}h_n\,\|^2 \;=\; \frac{1}{4}\gamma^2 \;+\; 2\gamma I_n^{(1)} \;+\; 2I_n^{(2)}. \tag{15.2.18.b}$$

[5] This analysis can be found in [113].

Aside from the difficulties inherent in having a log term in the integrand, the occurrence of the square of the Hermite-Gauss functions is our principal problem, since there are only a few integrals involving this factor known in closed form. However, there is an intimate connection between Hermite and Laguerre polynomials (which is at heart geometrical, since it arises from representations of the Heisenberg group [63]). Thus if we consider the imaginary part of the integral

$$\int_0^\infty 2s\frac{\log(\varepsilon+2is)}{\varepsilon+2is}\,h_n(s)^2\,ds\,,$$

where $\varepsilon > 0$, and use the identity

$$\int_0^\infty 2se^{-s^2}\,H_n(s)^2\sin(2st)\,ds \;=\; \sqrt{\pi}2^n n!\,te^{-t^2}\left[L_n(2t^2)-2L_n'(2t^2)\right]$$

which interrelates the Hermite and Laguerre polynomials, as well as the relationship

$$\frac{\log(\varepsilon+2is)}{\varepsilon+2is} \;=\; -\int_0^\infty (\gamma+\log t)\,e^{-(\varepsilon+2is)t}\,dt\,,$$

then by letting $\varepsilon \to 0$ we can prove that

$$I_n^{(1)} \;=\; \frac{1}{2}\int_0^\infty \left(\gamma+\tfrac{1}{2}\log u\right)\frac{d}{du}\left[L_n(2u)e^{-u}\right]du\,, \qquad (15.2.19.a)$$

while similar considerations show that

$$I_n^{(2)} \;=\; \frac{1}{24}\pi^2 - \frac{1}{2}\int_0^\infty \left(\gamma+\tfrac{1}{2}\log u\right)^2\frac{d}{du}\left[L_n(2u)e^{-u}\right]du\,. \qquad (15.2.19.b)$$

These new formulations are a distinct improvement on the old ones, since the index n is now associated with a simple Laguerre polynomial, rather than with the square of a Hermite-Gauss function. Consequently it is possible to obtain generating functions for the first and second moments of the operator $\mathcal{B}$ with respect to the Hermite-Gauss states, since

$$\sum_{n\geqslant 0}\langle h_n\,,\,\mathcal{B}h_n\rangle\xi^n \;=\; -\tfrac{1}{2}(1-\xi)^{-1}\log\left(\frac{1-\xi}{1+\xi}\right), \qquad (15.2.20.a)$$

$$\sum_{n\geqslant 0}\|\,\mathcal{B}h_n\,\|^2\,\xi^n \;=\; (1-\xi)^{-1}\left\{\tfrac{1}{8}\pi^2+\tfrac{1}{4}\left[\log\left(\frac{1-\xi}{1+\xi}\right)\right]^2\right\} \qquad (15.2.20.b)$$

are absolutely convergent series whenever $|\,\xi\,| < 1$.

Equating powers of ξ yields explicit formulæ for the first and second moments of $\mathcal{B}$,

$$\begin{aligned}\langle h_n, \mathcal{B}h_n\rangle &= \sum_{m=0}^{\lfloor(n-1)/2\rfloor} \frac{1}{2m+1}, \\ \| \mathcal{B}h_n \|^2 &= \frac{\pi^2}{8} + \sum_{\substack{l,m\geqslant 0 \\ l+m\leqslant\lfloor(n-2)/2\rfloor}} \frac{1}{(2l+1)(2m+1)},\end{aligned}$$

where, as usual, $\lfloor x \rfloor$ denotes the integer part of the real number x. These series enable us to give (relatively) simple expressions for the variances of $\mathbf{\Delta}[\varphi]$ in the Hermite-Gauss states, and it is not difficult to establish their asymptotic behaviour.

Proposition 15.1 *The variance of* $\mathbf{\Delta}[\varphi]$ *in a Hermite-Gauss state is given by*

$$\mathsf{Var}\{\mathbf{\Delta}[\varphi]; h_{2n}\} = \frac{3}{8}\pi^2 - \sum_{\substack{0\leqslant l,m\leqslant n-1 \\ l+m\geqslant n}} \frac{1}{(2l+1)(2m+1)} \tag{15.2.21.a}$$

and

$$\mathsf{Var}\{\mathbf{\Delta}[\varphi]; h_{2n+1}\} = \frac{3}{8}\pi^2 - \sum_{\substack{0\leqslant l,m\leqslant n \\ l+m\geqslant n}} \frac{1}{(2l+1)(2m+1)}. \tag{15.2.21.b}$$

for all $n \geqslant 0$.

The sequence of variances for even index n *is monotonically decreasing, while that for odd index* n *is monotonically increasing. The asymptotic order is*

$$\mathsf{Var}\{\mathbf{\Delta}[\varphi]; h_n\} = \frac{1}{3}\pi^2 + \mathrm{O}\left(\frac{\log n}{n}\right). \tag{15.2.22}$$

The asymptotic limit $\frac{1}{3}\pi^2$ can be found by transforming the above sums into integrals. For example,

$$\lim_{n\to\infty} \mathsf{Var}\{\mathbf{\Delta}[\varphi]; h_{2n}\} = \frac{3}{8}\pi^2 + \frac{1}{4}\int_0^1 \frac{\log(1-x)}{x}\,dx = \frac{1}{3}\pi^2. \tag{15.2.23}$$

The Weyl quantized exponentials $U_{\pm 1} = \mathbf{\Delta}\left[e^{\pm i\varphi}\right]$ were first introduced in Chapter 9, and in Chapter 10 their matrix coefficients with respect to the Hermite-Gauss functions were given in equations (10.3.57.a) and

(10.3.57.b). These imply that the expectations of $\mathbf{\Delta}\left[e^{\pm i\varphi}\right]$ in the Hermite-Gauss states are all zero, and moreover that

$$\left\| \mathbf{\Delta}\left[e^{i\varphi}\right] \right\|^2 = g_{n,n+1}^2, \qquad n \geqslant 0, \tag{15.2.24.a}$$

$$\left\| \mathbf{\Delta}\left[e^{-i\varphi}\right] \right\|^2 = \begin{cases} g_{n,n-1}^2, & n \geqslant 1, \\ 0, & n = 0. \end{cases} \tag{15.2.24.b}$$

The asymptotic behaviour of these expressions can be determined readily, and for this purpose the following Lemma is useful.

Lemma 15.2 *The coefficients $g_{n,n+1}$ have the asymptotic expansion*

$$g_{n,n+1} = 1 + \frac{(-1)^n}{4n} + \mathrm{O}\left(\tfrac{1}{n^2}\right), \qquad n \to \infty. \tag{15.2.25}$$

Proof: An elementary consequence of Stirling's formula is that

$$\frac{\Gamma(n+a)}{\Gamma(n+b)} = n^{a-b}\left\{1 - \frac{(b-a)(b+a-1)}{2n} + \mathrm{O}\left(\tfrac{1}{n^2}\right)\right\}, \; n \to \infty,$$

for any $a, b > 0$, from which the result is immediate. ■

We can then summarize the asymptotic properties of the Weyl quantized exponential operators as follows.

Proposition 15.3 *Expectations* $\mathsf{Exp}\left\{\mathbf{\Delta}\left[e^{\pm i\varphi}\right]; h_n\right\}$ *of the Weyl quantized exponential operators in the Hermite-Gauss states vanish for all $n \geqslant 0$, and*

$$\left\| \mathbf{\Delta}\left[e^{\pm i\varphi}\right] h_n \right\|^2 = 1 \pm \frac{(-1)^n}{2n} + \mathrm{O}\left(\tfrac{1}{n^2}\right), \qquad n \to \infty. \tag{15.2.26}$$

Casting these results in terms of the associated (self-adjoint) Weyl quantized cosine and sine operators $\mathbf{\Delta}[\cos\varphi]$ and $\mathbf{\Delta}[\sin\varphi]$, we have

$$\mathsf{Exp}\left\{\mathbf{\Delta}[\cos\varphi]; h_n\right\} = \mathsf{Exp}\left\{\mathbf{\Delta}[\sin\varphi]; h_n\right\} = 0, \tag{15.2.27.a}$$

for $n \geqslant 0$, and

$$\begin{aligned} \mathsf{Var}\left\{\mathbf{\Delta}[\cos\varphi]; h_n\right\} &= \mathsf{Var}\left\{\mathbf{\Delta}[\sin\varphi]; h_n\right\} \\ &= \tfrac{1}{2} + \mathrm{O}\left(\tfrac{1}{n^2}\right), \end{aligned} \tag{15.2.27.b}$$

as $n \to \infty$.

Again, these are results are consistent with the standard expectation for the asymptotic behaviour of quantum phase observables.

15.3 Asymptotics For Coherent States

The conventional interpretation of coherent states is that, when radiation is described by the coherent state Φ_ζ, the parameter $|\zeta|$ is related to the intensity of that radiation, while the argument $\mathrm{Arg}\zeta$ describes some aspect of the phase of that radiation. Thus it is to be expected that a quantum phase observable should exhibit asymptotic behaviour in coherent states which relates the quantum phase observable directly to the argument $\mathrm{Arg}\zeta$ of the state parameter ζ in the limit as $|\zeta| \to \infty$. All candidate quantum phase observables have this property, as we shall see. It has been noted above that there is a substantial debate in the physics literature as to what the exact nature of the asymptotic behaviour ought to be, and it is interesting that the various quantum phase observables each have different behaviour. That the theory provides us with different behaviours in the asymptotic limit of the intensity of the coherent state tending to infinity is particularly interesting, since it offers two areas for experimentation. Primarily, it is to be hoped that an experimental apparatus might be designed which would help to determine which of the various quantum phase observables was the "right" one. Secondarily, and perhaps more pessimistically, it might help us to determine what the various experiments concerning quantum phase are actually measuring, since it is not always clear that this is known.

15.3.1 *Barnett & Pegg Operators*

There are a number of easy-going calculations concerning the asymptotic behaviour of the Barnett & Pegg operators in the coherent states defined in equation (10.3.26.a). In particular, the behaviour most frequently discussed concerns (with our notational conventions) the coherent state $\Phi_{-iR/\sqrt{2}}$ as $R \to \infty$. However, to obtain a rigorous derivation of this asymptotic behaviour requires a great deal of care, and we shall do this here.

Most standard calculations involving the Barnett & Pegg operators evaluating expectations for the operators X_s, and then taking the limit as $s \to \infty$. However, doing this explicitly introduces a further difficulty into asymptotic calculations, since any study of the asymptotic behaviour for X_s has to be considered in the limit as $s \to \infty$ and, in general, asymptotic properties do not go through limiting procedures well.

Fortunately, we can avoid this problem. In Chapter 10 it was observed that, for any $k \in \mathbb{N}$, the sequence of operators $\left(X_s^k\right)_{s\geqslant 1}$ converges weakly to the operator $\overline{\mathcal{M}}(\mathfrak{p}_k)$, so that

$$\begin{aligned} BP_k(f) &= \langle f\,,\,\overline{\mathcal{M}}(\mathfrak{p}_k)f\rangle \;=\; \langle \mathfrak{U}_T\mathcal{F}f\,,\,M(\mathfrak{p}_k)\mathfrak{U}_T\mathcal{F}f\rangle \\ &= \frac{1}{2\pi}\int_{-\pi}^{\pi}\beta^k\big|\,[\mathfrak{U}_T\mathcal{F}f](e^{i\beta})\,\big|^2\,d\beta\,, \end{aligned} \tag{15.3.1}$$

for all test functions $f \in \mathcal{S}(\mathbb{R})$, where the meanings of the operators referred to here are given in Section 10.3.2. What we need is the asymptotic form of $BP_k(\Phi_{-iR/\sqrt{2}})$ for $k = 1,\,2$ and real R as $R \to \infty$, from which we can obtain the asymptotic form of the Barnett & Pegg variance $V_{BP}(\Phi_{-iR/\sqrt{2}})$ as $R \to \infty$. Now the function $\mathfrak{U}_T\mathcal{F}\Phi_{-iR/\sqrt{2}} \in H^2(\mathbb{T})$ does not have a simple closed form, since its power series expansion

$$\big[\mathfrak{U}_T\mathcal{F}\Phi_{-iR/\sqrt{2}}\big](e^{i\beta}) \;=\; e^{-\frac{1}{4}R^2}\sum_{n=0}^{\infty}\frac{R^n}{\sqrt{2^n n!}}\,e^{in\beta}\,, \qquad R > 0\,,$$

cannot be readily summed. Our first task, then, is to obtain a somewhat more tractable integral representation of this function.

Lemma 15.4 *We have the identity*

$$\begin{aligned} \sum_{n\geqslant 0}\frac{R^n}{\sqrt{n!}}\,e^{in\beta} \;&=\; \int_{-\frac{1}{2}}^{\infty}\frac{R^t e^{it\beta}}{\Gamma(t+1)^{\frac{1}{2}}}\,dt \\ &+\frac{1}{2i\sqrt{R}e^{\frac{1}{2}i\beta}}\int_0^{\infty}\frac{e^{-\pi s}}{\cosh(\pi s)}\left[\frac{R^{is}e^{-s\beta}}{\Gamma(\frac{1}{2}+is)^{\frac{1}{2}}}-\frac{R^{-is}e^{s\beta}}{\Gamma(\frac{1}{2}-is)^{\frac{1}{2}}}\right]ds. \end{aligned} \tag{15.3.2}$$

Applying Stirling's formula to this representation, it follows that

$$[\mathfrak{U}_T\mathcal{F}\Phi_{-iR/\sqrt{2}}](e^{i\beta}) = e^{-\frac{1}{4}R^2}\int_0^{\infty}\Big(\frac{R}{\sqrt{2}}\Big)^t\,\frac{e^{it\beta}}{\Gamma(t+1)^{\frac{1}{2}}}\,dt + O\left(e^{-\frac{1}{4}R^2}\right), \tag{15.3.3}$$

as $R \to \infty$, uniformly for $\beta \in [-\pi,\pi]$.

Proof: If $C(n)$ is the positively-oriented rectangular contour in the complex plane with vertices $-\frac{1}{2} \pm i(n+\frac{1}{2})$, $n-\frac{1}{2} \pm i(n+\frac{1}{2})$, then it is clear that

$$\sum_{k=0}^{n-1}\frac{R^k}{\sqrt{k!}}e^{ik\beta} = \frac{1}{2i}\oint_{C(n)}\frac{R^z}{\Gamma(z+1)^{\frac{1}{2}}}e^{iz\beta}\cot(\pi z)\,dz.$$

Considering this integral along each of the four sides, and taking the limit as $n \to \infty$ results in the given integral representation. ■

Analyzing this integral requires a somewhat exotic change of variable. Full details of this analysis can be found in [112], but the argument can be summarized as follows. For any $R > 0$ there exists a unique value $T_R > 0$ such that $\psi(T_R + 1) = \log \frac{1}{2}R^2$, where ψ is the logarithmic derivative of the Gamma function. It can be shown that

$$T_R = \frac{1}{2}(R^2 - 1) + \mathrm{O}(R^{-2}), \qquad R \to \infty.$$

We then define the function

$$F_R(t) = \frac{1}{2}\big(t \log \tfrac{1}{2}R^2 - \log \Gamma(t+1)\big), \qquad t > 0, \tag{15.3.4}$$

and let $A_R^2 = F_R(T_R)$. Then the function $w : [0,\infty) \to [0,\infty)$ given by the formula[6]

$$w(t) = A_R + \operatorname{sgn}(t - T_R)\sqrt{A_R^2 - F_R(t)}, \qquad t \geqslant 0 \tag{15.3.5}$$

is a continuously differentiable strictly monotonic increasing linear bijection, and hence invertible, so we can regard $t = t(w)$ as a function of w. This change of variables leads to the expression

$$[\mathfrak{U}_T \mathcal{F} \Phi_{-iR/\sqrt{2}}](e^{i\beta}) =$$
$$\frac{\pi^{\frac{1}{2}} t'(A_R) e^{A_R^2 - \frac{1}{4}R^2}}{\left(1 - \frac{1}{2}it''(A_R)\beta\right)^{\frac{3}{2}}} \exp\left[iT_R\beta - \frac{t'(A_R)^2\beta^2}{4\left(1 - \frac{1}{2}it''(A_R)\beta\right)}\right] + \mathrm{O}\left(R^{-\frac{3}{2}}\right)$$

as $R \to \infty$, uniformly in $\beta \in [-\pi, \pi]$.

It follows from the series expansion for $\mathfrak{U}_T \mathcal{F} \Phi_{-iR/\sqrt{2}}$ that its modulus is an even function of β, and hence that $BP_1(\Phi_{-iR/\sqrt{2}}) = 0$. We can approximate $BP_2(\Phi_{-iR/\sqrt{2}})$ using the above uniform asymptotic expression, with the result that

$$\begin{aligned} BP_2(\Phi_{-iR/\sqrt{2}}) &= \sqrt{\frac{\pi}{2}} \frac{1}{t'(A_R)} e^{2A_R^2 - \frac{1}{2}R^2} + \mathrm{O}\left(\frac{1}{R^3}\right) \\ &= \frac{1}{2R^2} + \mathrm{O}\left(\frac{1}{R^3}\right), \qquad R \to \infty. \end{aligned} \tag{15.3.6}$$

[6]It is clear that the maximum value of F_R is A_R^2, achieved when $t = T_R$.

Summarizing these results,

Proposition 15.5 *The Barnett & Pegg expectation in coherent states is*

$$E_{BP}(\Phi_{-iR/\sqrt{2}}) = BP_1(\Phi_{-iR/\sqrt{2}}) = 0, \tag{15.3.7}$$

and Barnett & Pegg variance has the behaviour

$$V_{BP}(\Phi_{-iR/\sqrt{2}}) = \frac{1}{2R^2} + O\left(\frac{1}{R^3}\right), \qquad R \to \infty. \tag{15.3.8}$$

This justifies (and adds bounds to) the statements found in the literature concerning the asymptotic form of the Barnett & Pegg variance in the state $\Phi_{-iR/\sqrt{2}}$ for large R (see [14], equation (47)).

15.3.2 *The Toeplitz Phase Operator X*

Unfortunately, there are gaps in the known results concerning exact asymptotic expansions for the Toeplitz phase operator $X = \overline{\mathcal{M}}(\mathfrak{p}_1)$. It was clear that the analysis for the Barnett & Pegg operators in coherent states was complicated enough, and a study of the second moments of X will be even more complex, due to the additional presence of the Riesz-Szëgo projection in the formalism. However, it is still possible to obtain some information concerning X, for it is clear that $\mathsf{Exp}\,\{X;\,\phi\} = BP_1(\phi)$ for any $\phi \in L^2(\mathbb{R})$, and so equation (15.3.7) tells us that

$$\mathsf{Exp}\left\{X;\,\Phi_{-iR/\sqrt{2}}\right\} = 0\,, \qquad R > 0\,, \tag{15.3.9}$$

while inequality (10.3.53), in conjunction with equation (15.3.8), implies that

$$\mathsf{Var}\left\{X;\,\Phi_{-iR/\sqrt{2}}\right\} = \mathrm{O}\left(\tfrac{1}{R^2}\right), \qquad R \to \infty\,. \tag{15.3.10}$$

However, this result does not specify the exact nature of the way in which the variance of X in coherent states tends to zero as $R \to \infty$ — the rate of convergence might be much faster than $\mathrm{O}(R^{-2})$.

15.3.3 *The Bargmann-Segal Phase Operator $\Xi(\varphi)$*

Results concerning the Bargmann-Segal phase operator $\Xi(\varphi)$ are limited in the same way as are those for the Toeplitz phase operator, in that we do not have a precise description of the asymptotic behaviour of the variance of $\Xi(\varphi)$ for coherent states. We do not have such a description for a very similar reason to that given for the Toeplitz operator X — the additional projection in the formalism of the Bargmann-Segal operator makes analysis rather difficult. However, we can obtain an upper bound on the asymptotic behaviour of $\Xi(\varphi)$, and moreover in this case we can establish this upper bound for a much wider range of values of the parameter that defines the coherent state.

The key identity in this analysis is equation (14.3.18), which states that

$$\mathsf{Exp}\,\{\Xi(F);\,\Phi_w\} = \langle \Phi_w, \Xi(F)\Phi_w \rangle = \frac{1}{\pi}\int_{\mathbb{C}} F(z)e^{-|z-w|^2}\,dA(z), \quad (14.3.18)$$

for any $w \in \mathbb{C}$. With the parametrization $w = -iRe^{i\beta}/\sqrt{2}$, this formula reads

$$\mathsf{Exp}\,\{\Xi(\varphi);\,\Phi_w\} = \frac{1}{2\pi}\iint_{\mathbb{R}^2} \varphi(p\cos\beta - q\sin\beta, p\sin\beta + q\cos\beta)\, e^{-\frac{1}{2}(p-R)^2 - \frac{1}{2}q^2}\,dp\,dq$$

for any $-\pi \leqslant \beta \leqslant \pi$. This expression is an odd function of β, and so we shall restrict our attention to the case $0 \leqslant \beta < \pi$, for then

$$\begin{aligned}\mathsf{Exp}\,\{\Xi(\varphi);\,\Phi_w\} &= \frac{1}{2\pi}\iint_{\mathbb{R}^2} \big[\varphi(p,q) + \beta - 2\pi E_{\pi-\beta}(p,q)\big]\, e^{-\frac{1}{2}(p-R)^2 - \frac{1}{2}q^2}\,dp\,dq \\ &= \beta - \iint_{\mathbb{R}^2} E_{\pi-\beta}(p+R,q)\, e^{-\frac{1}{2}(p^2+q^2)}\,dp\,dq\,, \qquad (15.3.11)\end{aligned}$$

where $E_{\pi-\beta} \in \mathcal{S}'(\Pi)$ is the function

$$E_{\pi-\beta}(r\cos\gamma, r\sin\gamma) = \begin{cases} 1, & r > 0\,,\ \pi - \beta < \gamma < \pi\,, \\ 0, & \text{otherwise}\,. \end{cases} \qquad (15.3.12)$$

If we define the function $k : (-\pi, \pi) \to (0, 1]$ by the formula

$$k(\beta) = \begin{cases} 1, & |\beta| \leqslant \frac{1}{2}\pi\,, \\ |\sin\beta|\,, & \frac{1}{2}\pi < |\beta| < \pi\,, \end{cases} \qquad (15.3.13)$$

then it follows that $E_{\pi-\beta}(p+R,q)$ vanishes whenever $p^2 + q^2 < R^2k(\beta)^2$, and hence

$$\begin{aligned} 0 \leqslant \iint_{\mathbb{R}^2} E_{\pi-\beta}(p+R,q)e^{-\frac{1}{2}(p^2+q^2)}\,dp\,dq &\leqslant 2\pi\int_{Rk(\beta)}^{\infty} e^{-\frac{1}{2}r^2}\,r\,dr \\ &= 2\pi e^{-\frac{1}{2}R^2k(\beta)^2}\,, \end{aligned}$$

which implies the following result.

Proposition 15.6 *If* $w = -\frac{i}{\sqrt{2}}Re^{i\beta}$, *where* $|\beta| < \pi$, *then*

$$\mathbf{Exp}\,\{\Xi(\varphi);\,\Phi_w\} \;=\; \beta + \mathrm{O}\left(e^{-\frac{1}{2}R^2k(\beta)^2}\right), \qquad R\to\infty. \tag{15.3.14}$$

In Chapter 10 we introduced the formula

$$D_\varphi(p,q) \;=\; \frac{2}{\pi}\int_{\mathbb{C}}\varphi(z)\,e^{-|z-w|^2}\,dA(z), \qquad w = \frac{1}{\sqrt{2}}(q-ip), \tag{10.3.35}$$

for the Weyl dequantization symbol D_φ of the operator $\Xi(\varphi)$, and later showed in equation (14.3.20) that this was equal to $\mathbf{Exp}\left\{\Xi(\varphi);\,\sqrt{2}w\right\}$. Hence we can now deduce that

$$D_\varphi(p,q) \;=\; \varphi(p,q) + \mathrm{O}\left(e^{-r^2k(\beta)^2}\right), \qquad r\to\infty, \tag{15.3.15}$$

for any $|\beta| < \pi$, where $p+iq = re^{i\beta}$ — a result which was announced in Section 10.3.3.

It is possible to extend these calculations to obtain information concerning the variance of the operator $\Xi(\varphi)$ in coherent states, for we observe that

$$\begin{aligned} \|\,\Xi(\varphi)\,\Phi_w\,\|^2 &= \|\,\mathfrak{M}_\varphi\mathfrak{U}_{BS}\Phi_w\,\|^2 \;=\; \|\,\mathcal{P}_{BS}\mathcal{M}_\varphi\mathfrak{U}_{BS}\Phi_w\,\|^2 \\ &\leqslant \|\,\mathcal{M}_\varphi\mathfrak{U}_{BS}\Phi_w\,\|^2 \\ &= \frac{1}{\pi}\int_{\mathbb{C}}\varphi(z)^2\,e^{-|z-w|^2}\,dA(z), \end{aligned} \tag{15.3.16}$$

and so we deduce that

$$\begin{aligned} \|\,\Xi(\varphi)\Phi_w\,\|^2 &\leqslant \frac{1}{\pi}\iint_{\mathbb{R}^2}\left[\varphi(p,q)+\beta-2\pi E_{\pi-\beta}(p,q)\right]^2 e^{-\frac{1}{2}(p-R)^2-\frac{1}{2}q^2}\,dp\,dq \\ &\leqslant \beta^2 + \frac{1}{2\pi}\iint_{\mathbb{R}^2}\varphi(p,q)^2 e^{-\frac{1}{2}(p-R)^2-\frac{1}{2}q^2}\,dp\,dq \\ &\quad + 4\pi^2\,e^{-\frac{1}{2}R^2k(\beta)^2}, \end{aligned} \tag{15.3.17}$$

for any $0 \leqslant \beta < \pi$, where $w = -iRe^{i\beta}/\sqrt{2}$, and hence

$$\mathbf{Var}\,\{\Xi(\varphi);\,\Phi_w\} \;\leqslant\; \frac{1}{2\pi}\iint_{\mathbb{R}^2}\varphi(p,q)^2 e^{-\frac{1}{2}(p-R)^2-\frac{1}{2}q^2}\,dp\,dq + \mathrm{O}\left(e^{-\frac{1}{2}R^2k(\beta)^2}\right), \tag{15.3.18}$$

as $R \to \infty$, for any $|\beta| < \pi$. We therefore need to analyze the integral in this last expression.

Lemma 15.7 *We have the asymptotic formula*

$$\frac{1}{\pi}\iint_{\mathbb{R}^2} \varphi(p,q)^2 e^{-(p-R)^2-q^2}\,dp\,dq \;\sim\; \frac{1}{2R^2}\,, \qquad R \to \infty\,. \qquad (15.3.19)$$

Proof: Elementary considerations show that

$$\frac{1}{\pi}\int_{-\infty}^{0}\left(\int_{\mathbb{R}} \varphi(p,q)^2 e^{-(p-R)^2-q^2}\,dq\right)dp \;\leqslant\; \tfrac{1}{2}\pi^2 e^{-R^2}\,,$$

and hence this half of the integral is asymptotically small, and thus negligible. To deal with the other half of the integral, the substitution $q = pu$ yields the identity

$$\begin{aligned}
&\frac{1}{\pi}\int_{0}^{\infty}\left(\int_{\mathbb{R}} \varphi(p,q)^2 e^{-(p-R)^2-q^2}\,dq\right)dp \\
&= \frac{1}{\pi}\int_{0}^{\infty} p\left(\int_{\mathbb{R}} (\tan^{-1}u)^2 e^{-(p-R)^2-p^2u^2}\,du\right)dp \\
&= \frac{1}{24}\pi^2 e^{-R^2} \\
&\quad + \frac{R}{\pi}\int_{\mathbb{R}} \frac{(\tan^{-1}u)^2}{(1+u^2)^{\frac{3}{2}}}\exp\left[-\frac{R^2u^2}{1+u^2}\right]\left(\int_{-R/\sqrt{1+u^2}}^{\infty} e^{-p^2}\,dp\right)du\,,
\end{aligned}$$

from which it follows that the difference between the desired integral and

$$\frac{R}{\sqrt{\pi}}\int_{\mathbb{R}} \frac{(\tan^{-1}u)^2}{(1+u^2)^{\frac{3}{2}}}\exp\left[-\frac{R^2u^2}{1+u^2}\right]du$$

is exponentially small as $R \to \infty$. Thus, applying Laplace's method to this last integral, we obtain

$$\frac{1}{\pi}\iint_{\mathbb{R}^2} \varphi(p,q)^2 e^{-(p-R)^2-q^2}\,dp\,dq \;\sim\; \frac{1}{2R^2}\,, \qquad R \to \infty\,,$$

as required. ■

Since φ is an angular function, it is clear that the integral in equation (15.3.18) is simply a scaled version of that in the preceding Lemma. Thus, putting these results together, we obtain the following result.

Proposition 15.8 *If* $w = -\frac{i}{\sqrt{2}} Re^{i\beta}$ *and* $|\beta| < \pi$*, then*

$$\mathsf{Var}\,\{\Xi(\varphi);\,\Phi_w\} \;=\; \mathrm{O}\big(\tfrac{1}{R^2}\big)\,, \qquad\qquad R \to \infty\,. \tag{15.3.20}$$

Thus, although we do not have an exact asymptotic formula for the variance of $\Xi(\varphi)$ in coherent states, we do have sufficient control on this variance for our purposes. As already remarked, this situation is similar to that of our current level of knowledge concerning the Toeplitz phase operator X.

It should be noted that the result of Lemma 15.7 is often cited as justification for the notion that a quantum phase observable should have variance in the coherent state Φ_w which behaves asymptotically like $(2R^2)^{-1}$ as $R \to \infty$. However, such a justification is suspect for two reasons. Firstly the integral in equation (15.3.19) is not the second moment of any known phase observable and thus, secondly, the justification can only be valid if the quantity

$$\frac{1}{\pi} e^{-(p-R)^2 - q^2}$$

(being the Wigner function[7] for the state Φ_w) is treated as a classical probability distribution for the supposed phase observable in the state Φ_w. However, we have already noted the dangers inherent in any interpretation of Wigner functions as classical probability distributions for phase observables.

15.3.4 *Weyl Quantized Phase Space Operators*

Turning to the Weyl phase observable $\mathbf{\Delta}[\varphi]$, we note that much of our work has already been done, for the formula

$$\big[\mathcal{G}(\overline{\Phi_w} \otimes \Phi_w)\big](u,v) \;=\; \frac{1}{\pi}\, e^{-2|\,z-w\,|^2}\,, \qquad z \;=\; \frac{1}{\sqrt{2}}(v - iu)\,, \tag{10.3.34}$$

which identifies the Wigner function associated with the coherent state Φ_w, leads to the identity

$$\mathsf{Exp}\,\{\mathbf{\Delta}[\varphi]\,;\,\Phi_w\} \;=\; \frac{2}{\pi}\int_{\mathbb{C}} \varphi(z)\, e^{-2|\,z-w\,|^2}\, dA(z)\,, \tag{15.3.21}$$

giving the expectation of $\mathbf{\Delta}[\varphi]$ in the state Φ_w. We note the curiosity that $\mathsf{Exp}\,\{\mathbf{\Delta}[\varphi]\,;\,\Phi_w\}$ coincides with the Weyl dequantization $D_\varphi(p,q)$ of the

[7] This quantity is a positive normalized function because Φ_w is a Gaussian [123].

Bargmann-Segal operator $\Xi(\varphi)$ (writing $w = (q - ip)/\sqrt{2}$). The following is therefore a consequence of results in the previous Subsection.

Proposition 15.9 *The asymptotic identity*

$$\mathsf{Exp}\{\mathbf{\Delta}[\varphi];\Phi_w\} = \beta + \mathrm{O}\left(e^{-R^2k(\beta)^2}\right), \qquad R \to \infty, \tag{15.3.22}$$

holds for all $|\beta| < \pi$, *where* $w = -\frac{i}{\sqrt{2}}Re^{i\beta}$.

Dealing with the second moment of $\mathbf{\Delta}[\varphi]$ takes greater care. For any $f, g \in L^2(\mathbb{R})$ we note that

$$\left[\mathcal{G}(\overline{f} \otimes g)\right](p,q) = \frac{e^{2ipq}}{\pi}\langle V(2p)U(-2q)\mathcal{P}f\,,\,g\rangle\,, \tag{15.3.23}$$

where U and V are the usual subgroups of the Weyl group W, and $\mathcal{P}$ is the parity operator. Hence, if $T \in \mathcal{S}'(\Pi)$ is any function such that $\mathbf{\Delta}[T] \in \mathcal{L}^+\big(\mathcal{S}(\mathbb{R}), L^2(\mathbb{R})\big)$, we have

$$\begin{aligned}\left[\mathcal{G}(\overline{f} \otimes \mathbf{\Delta}[T]g)\right](p,q) &= \frac{e^{2ipq}}{\pi}\,[\![T\,,\,\mathcal{G}(\overline{V(2p)U(-2q)\mathcal{P}f} \otimes g)]\!]\\ &= \frac{1}{\pi}\iint_\Pi T(\xi,\eta)e^{2i(\xi q - \eta p)}\left[\mathcal{G}(\overline{\mathcal{P}f} \otimes g)\right](\xi - p, \eta - q)\,d\xi\,d\eta\,,\end{aligned} \tag{15.3.24}$$

and hence it follows that $\|\,\mathbf{\Delta}[T]f\,\|^2$ is equal to

$$\frac{1}{\pi}\iint_\Pi \overline{T(p,q)}\left(\iint_\Pi T(\xi,\eta)e^{2i(\xi q-\eta p)}\left[\mathcal{G}(\overline{\mathcal{P}f} \otimes g)\right](\xi - p, \eta - q)\,d\xi\,d\eta\right)dp\,dq. \tag{15.3.25}$$

We note that the integrand in this expression may well not be an element of $L^1(\Pi \times \Pi)$. Nonetheless, the indicated iterated integral converges — we must, however, take care if we wish to change the order of integration at any stage.

Specifying these calculations for coherent states, we observe that

$$\left[\mathcal{G}(\overline{\mathcal{P}\Phi_w} \otimes \Phi_w)\right](p,q) = \frac{1}{\pi}\,e^{-2|u|^2 - 2\overline{u}w + 2u\overline{w}}\,, \tag{15.3.26}$$

writing $u = (q - ip)/\sqrt{2}$, and this leads to the identity

$$\|\,\mathbf{\Delta}[T]\Phi_w\,\|^2 = \tag{15.3.27}$$
$$\frac{4}{\pi^2}\int_{\mathbb{C}}\overline{T(u+w)}\left(\int_{\mathbb{C}}T(v+w)e^{-2|u|^2-2|v|^2+4u\overline{v}}\,dA(v)\right)dA(u)$$

for any function $T \in \mathcal{S}'(\Pi)$ such that $\mathbf{\Delta}[T] \in \mathcal{L}^+\big(\mathcal{S}(\mathbb{R}), L^2(\mathbb{R})\big)$. Moreover, if $T \in L^\infty(\Pi)$, this equation implies that

$$\| \mathbf{\Delta}[T]\Phi_w \|^2 = \sum_{n \geqslant 0} \frac{4^n}{n!} \, | U_n(T,w) |^2 , \qquad (15.3.28.a)$$

where the coefficients $U_n(T,w)$ are given by the formula

$$U_n(T,w) = \frac{2}{\pi} \int_{\mathbb{C}} T(v+w) \, e^{-2|v|^2} \, \overline{v}^n \, dA(v) , \qquad n \geqslant 0 , \qquad (15.3.28.b)$$

so, in particular,

$$U_0(T,w) = \mathsf{Exp}\,\{\mathbf{\Delta}[T];\, \Phi_w\} . \qquad (15.3.28.c)$$

If we make the usual identification $w = -iRe^{i\beta}/\sqrt{2}$, and (for simplicity) restrict attention to the range $0 \leqslant \beta < \pi$, we see that equation (15.3.27) implies that

$$\begin{aligned}
\| \mathbf{\Delta}[\varphi]\Phi_w \|^2 &= \Big\| \big(\mathbf{\Delta}[\varphi] + \beta - 2\pi \mathbf{\Delta}[E_{\pi-\beta}]\big)\Phi_{-iR/\sqrt{2}} \Big\|^2 \\
&= \Big\| \mathbf{\Delta}[\varphi]\Phi_{-iR/\sqrt{2}} \Big\|^2 + \beta^2 + 4\pi^2 \Big\| \mathbf{\Delta}[E_{\pi-\beta}]\Phi_{-iR/\sqrt{2}} \Big\|^2 \\
&\quad - 4\pi\beta \mathsf{Exp}\Big\{\mathbf{\Delta}[E_{\pi-\beta}];\, \Phi_{-iR/\sqrt{2}}\Big\} \\
&\quad - 4\pi \mathrm{Re}\big\langle \mathbf{\Delta}[\varphi]\Phi_{-iR/\sqrt{2}} ,\, \mathbf{\Delta}[E_{\pi-\beta}]\Phi_{-iR/\sqrt{2}} \big\rangle \\
&\leqslant \beta^2 + 4\pi \Big| \mathsf{Exp}\Big\{\mathbf{\Delta}[E_{\pi-\beta}];\, \Phi_{-iR/\sqrt{2}}\Big\} \Big| \\
&\quad + \bigg(\Big\| \mathbf{\Delta}[\varphi]\Phi_{-iR/\sqrt{2}} \Big\|^2 + 2\pi \Big\| \mathbf{\Delta}[E_{\pi-\beta}]\Phi_{-iR/\sqrt{2}} \Big\|^2 \bigg)^2 ,
\end{aligned}$$

so we can obtain an upper bound on $\| \mathbf{\Delta}[\varphi]\Phi_w \|$ by finding upper bounds for $\big\|\mathbf{\Delta}[\varphi]\Phi_{-iR/\sqrt{2}}\big\|$ and for $\big\|\mathbf{\Delta}[E_{\pi-\beta}]\Phi_{-iR/\sqrt{2}}\big\|$.

Elementary observations show us that

$$\begin{aligned}
\Big| U_n\big(E_{\pi-\beta}, -\frac{i}{\sqrt{2}}R\big) \Big| &\leqslant \frac{2^{1-\frac{1}{2}n}}{\pi n} \int_{k(\beta)R}^{\infty} r^{n+1} e^{-r^2} \, dr \\
&= \frac{1}{\pi n 2^{\frac{1}{2}n}} k(\beta)^n R^n e^{-k(\beta)^2 R^2} + \frac{1}{\pi 2^{\frac{1}{2}n}} \int_{k(\beta)R}^{\infty} r^{n-1} e^{-r^2} \, dr ,
\end{aligned}$$

for any $n \geqslant 1$, and hence it follows that

$$\left\| \boldsymbol{\Delta}[E_{\pi-\beta}]\Phi_{-iR/\sqrt{2}} \right\|^2 \leqslant \frac{2}{\pi^2 k(\beta)^2 R^2} + \frac{2}{\pi^{\frac{3}{2}} k(\beta) R} + \left| \mathsf{Exp}\left\{ \boldsymbol{\Delta}[E_{\pi-\beta}];\, \Phi_{-iR/\sqrt{2}} \right\} \right|^2, \quad (15.3.29)$$

which in turn implies that

$$\left\| \boldsymbol{\Delta}[E_{\pi-\beta}]\Phi_{-iR/\sqrt{2}} \right\|^2 = \mathrm{O}\left(\tfrac{1}{R}\right), \qquad R \to \infty. \quad (15.3.30)$$

Similarly we can show that

$$U_n\left(\varphi, -\frac{i}{\sqrt{2}}R\right) = \begin{cases} 2^{-1-\frac{1}{2}n} i^{1-n} \left[\dfrac{\gamma(n, R^2)}{R^n} + \Gamma\left(\tfrac{1}{2}n, R^2\right) \right], & n \geqslant 1, \\ 0, & n = 0, \end{cases}$$

where $\Gamma(a, x)$ and $\gamma(a, x)$ (for $\mathrm{Re}\, a > 0$ and $x > 0$) are the incomplete Gamma functions,

$$\Gamma(a, x) = \int_x^\infty e^{-t} t^{a-1}\, dt, \qquad \gamma(a, x) = \int_0^x e^{-t} t^{a-1}\, dt. \quad (15.3.31)$$

Using standard techniques of asymptotic analysis,

$$\left\| \boldsymbol{\Delta}[\varphi]\, \Phi_{-iR/\sqrt{2}} \right\|^2 \sim \frac{1}{4} \sum_{n \geqslant 1} \frac{2^n}{n!} \Gamma\left(\tfrac{1}{2}n, R^2\right)^2 \sim \frac{\sqrt{\pi}}{R}, \quad (15.3.32)$$

as $R \to \infty$. Summarizing our results so far:

Proposition 15.10 *If* $w = -\frac{i}{\sqrt{2}} R e^{i\beta}$ *with* $|\beta| < \pi$, *then*

$$\| \boldsymbol{\Delta}[\varphi]\, \Phi_w \|^2 = \beta^2 + \mathrm{O}\left(\tfrac{1}{R}\right), \qquad R \to \infty, \quad (15.3.33.a)$$

and hence

$$\mathsf{Var}\{ \boldsymbol{\Delta}[\varphi];\, \Phi_w \} = \mathrm{O}\left(\tfrac{1}{R}\right), \qquad R \to \infty. \quad (15.3.33.b)$$

Moreover, these statements are sharp when $\beta = 0$, *since*

$$\mathsf{Var}\left\{ \boldsymbol{\Delta}[\varphi];\, \Phi_{-iR/\sqrt{2}} \right\} \sim \frac{\sqrt{\pi}}{R}, \qquad R \to \infty. \quad (15.3.33.c)$$

This result differs from previous ones — see, for example, equations (15.3.8), (15.3.10) and (15.3.19) — which show asymptotic behaviour of order R^{-2} or better as $R \to \infty$. Given this different result, we should consider the experimental results which have been interpreted as requiring asymptotic behaviour of order R^{-2} for the variance of the phase operator in coherent states [160]. It is our view that, heretofore, the experimental arrangements described in [160] do not unambiguously distinguish exactly which "function" of quantum phase is being measured. Even though angular data can always be presented in terms of some favoured function (say the cosine of the angle), doing so does not guarantee that (for example) $\cos\varphi$ is being measured rather than φ. Now quantum theory tells us that a maximal measurement (of spectral values and output data) is tantamount to defining an observable, while a partial measurement (involving spectral values only, for example) gives less information and so certainly does not define a unique operator. If it should happen that it was an operator corresponding (in some sense) to the cosine of the phase angle that was being measured (in a coherent state), then all the phase operator candidates, $\mathbf{\Delta}[\varphi]$ included, have asymptotic behaviour of order R^{-2} or better (see below). Moreover, since the phase space symbols of both the Toeplitz and the Bargmann-Segal phase operators converge (in the sense of distributions as $R \to \infty$) to the angle function, distinctions between them are valid in the quantum, but not the semi-classical, domain. For the present, then, the situation would seem to be that the experimental information currently available does not distinguish between the phase operators.

This asymptotic analysis for $\mathbf{\Delta}[\varphi]$ has an unexpected benefit, since it yields information concerning the spectrum of the operator $\mathbf{\Delta}[\varphi]$.

Corollary 15.11 *Since*

$$\| (\mathbf{\Delta}[\varphi] - \beta)\Phi_w \|^2 = \mathrm{O}\left(\tfrac{1}{R}\right), \qquad R \to \infty, \tag{15.3.34}$$

when $w = -iRe^{i\beta}/\sqrt{2}$ *with* $|\beta| < \pi$, *we deduce that the interval* $[-\pi, \pi]$ *is contained in the spectrum of* $\mathbf{\Delta}[\varphi]$.

Proof: The identity in equation (15.3.34) is an elementary consequence of what has gone before. What matters here is its interpretation. For a fixed value of $\beta \in (-\pi, \pi)$, we have shown that the functions

$$\{\Xi_R^\beta = \Phi_{-iRe^{i\beta}/\sqrt{2}} : R > 0\}$$

form a collection of unit vectors in $L^2(\mathbb{R})$ such that the norm $\|(\mathbf{\Delta}[\varphi]-\beta)\Xi_R^\beta\| \to 0$ as $R\to\infty$. Technically, this states that β is an approximate eigenvalue[8] for $\mathbf{\Delta}[\varphi]$, with $\{\Xi_R^\beta : R>0\}$ being a sequence of approximating eigenvectors for β. Since every approximate eigenvalue of a self-adjoint operator belongs to its spectrum, we deduce that the spectrum of $\mathbf{\Delta}[\varphi]$ contains the open interval $(-\pi,\pi)$, and so its closure $[-\pi,\pi]$, as required. ■

From elementary spectral theory, this implies that the norm of $\mathbf{\Delta}[\varphi]$ is greater than π. Thus we can now state that

$$\pi \leqslant \|\mathbf{\Delta}[\varphi]\| \leqslant \frac{3}{2}\pi. \tag{15.3.35}$$

Our belief is that the spectrum of $\mathbf{\Delta}[\varphi]$ is simply the interval $[-\pi,\pi]$, a conjecture that is supported by numerical studies. A proof that the norm of $\mathbf{\Delta}[\varphi]$ was equal to π would confirm this.

If we now consider instead the exponentiated Weyl phase observables $\mathbf{\Delta}\left[e^{i\varphi}\right]$ and $\mathbf{\Delta}\left[e^{-i\varphi}\right]$, it is relatively simple to determine their behaviour in coherent states. This first result concerning the expectation is due in the first instance to Freyberger & Schleich [64].

Proposition 15.12 *If* $w = -\frac{i}{\sqrt{2}}Re^{i\beta}$, *we have the identity*

$$\mathsf{Exp}\left\{\mathbf{\Delta}\left[e^{\pm i\varphi}\right];\,\Phi_w\right\} = \tfrac{1}{2}\sqrt{\pi}\,Re^{-\frac{1}{2}R^2}\left[I_0(\tfrac{1}{2}R^2)+I_1(\tfrac{1}{2}R^2)\right]e^{\pm i\beta}, \tag{15.3.36}$$

and the asymptotic formulæ:

$$\mathsf{Exp}\left\{\mathbf{\Delta}\left[e^{\pm i\varphi}\right];\,\Phi_w\right\} = e^{\pm i\beta}\left(1-\tfrac{1}{4R^2}\right)+O\left(\tfrac{1}{R^4}\right), \quad R\to\infty, \tag{15.3.37}$$

which are valid uniformly for $\beta\in[-\pi,\pi]$.

Proof: The first identity follows since

$$\begin{aligned}&\mathsf{Exp}\left\{\mathbf{\Delta}\left[e^{\pm i\varphi}\right];\,\Phi_w\right\}\\ &\quad= \frac{1}{\pi}\int_0^\infty\int_{-\pi}^{\pi} e^{\pm i\gamma}e^{-r^2+2rR\cos(\gamma-\beta)-R^2}\,r\,d\gamma\,dr\end{aligned}$$

[8] Approximate eigenvalues and sequences of approximating eigenvectors associated with them are considered in detail in Chapter 16.

$$
\begin{aligned}
&= \frac{2}{\pi} e^{\pm i\beta} \int_0^\infty \int_0^\pi \cos\gamma \, e^{-r^2+2rR\cos\gamma - R^2} \, r \, d\gamma \, dr \\
&= e^{\pm i\beta} e^{-R^2} \int_0^\infty e^{-r^2} I_1(2Rr) \, r \, dr \\
&= \frac{1}{2}\sqrt{\pi} R e^{-\frac{1}{2}R^2} \left[I_0(\tfrac{1}{2}R^2) + I_1(\tfrac{1}{2}R^2) \right] e^{\pm i\beta} \,,
\end{aligned}
$$

as required. Evaluating this quantity differently leads to the expression

$$
\begin{aligned}
&\mathsf{Exp}\left\{ \mathbf{\Delta}\left[e^{\pm i\varphi}\right]; \Phi_w \right\} \\
&= \frac{1}{\pi} e^{\pm i\beta} \int_0^\pi \cos\gamma \left(\int_{\mathbb{R}} e^{-r^2+2rR\cos\gamma - R^2} \, r \, dr \right) d\gamma \\
&= \frac{R}{\sqrt{\pi}} e^{\pm i\beta} \int_0^\pi \cos^2\gamma \, e^{-R^2 \sin^2\gamma} \, d\gamma \\
&= \frac{2R}{\sqrt{\pi}} e^{\pm i\beta} \int_0^1 \sqrt{1-u^2} \, e^{-R^2 u^2} \, du \,,
\end{aligned}
$$

and the desired asymptotic formulæ follow from this equation. ■

To study the second moments of the operators $\mathbf{\Delta}\left[e^{i\varphi}\right]$ and $\mathbf{\Delta}\left[e^{-i\varphi}\right]$ in coherent states, it is simplest to revert to considering power series rather than integral formulations, for

$$
\| \mathbf{\Delta}\left[e^{i\varphi}\right] \Phi_w \|^2 = e^{-|w|^2} \sum_{n \geqslant 0} \frac{|w|^{2n}}{n!} g_{n,n+1} \,, \qquad (15.3.38.a)
$$

$$
\| \mathbf{\Delta}\left[e^{-i\varphi}\right] \Phi_w \|^2 = e^{-|w|^2} \sum_{n \geqslant 1} \frac{|w|^{2n}}{n!} g_{n,n-1} \,, \qquad (15.3.38.b)
$$

for any $w \in \mathbb{C}$, and we can describe their asymptotic form using the expansion for the coefficients $g_{n,n+1}$ given in Lemma 15.2.

Lemma 15.13 *We have the following asymptotic behaviour as* $|w| \to \infty$*:*

$$
\| \mathbf{\Delta}\left[e^{i\varphi}\right] \Phi_w \|^2 = 1 + O\left(\frac{1}{|w|^4}\right), \qquad (15.3.39.a)
$$

$$
\| \mathbf{\Delta}\left[e^{-i\varphi}\right] \Phi_w \|^2 = 1 + O\left(\frac{1}{|w|^4}\right). \qquad (15.3.39.b)
$$

Proof: From the results of Lemma 15.2, we can find a constant $A > 0$ such that

$$\left| g_{n,n+1}^2 - 1 - \frac{(-1)^n}{2(n+1)} \right| \leqslant \frac{A}{(n+1)(n+2)}, \qquad n \geqslant 0,$$

which inequality implies that

$$\left| \| \mathbf{\Delta}\left[e^{i\varphi} \right] \Phi_w \|^2 - 1 - \frac{1}{2|w|^2} e^{-|w|^2}(1 - e^{-|w|^2}) \right| \leqslant \frac{A}{|w|^4},$$

for all nonzero w. This establishes the first of the two formulæ, and the second is derived similarly. ■

Putting these results together yields the following:

Proposition 15.14 *If* $w = -\frac{i}{\sqrt{2}} R e^{i\beta}$, *then*

$$\| (\mathbf{\Delta}\left[e^{\pm i\varphi}\right] - e^{\pm i\beta}) \Phi_w \|^2 = \frac{1}{2R^2} + \mathrm{O}\left(\tfrac{1}{R^4}\right), \qquad R \to \infty. \qquad (15.3.40)$$

Indeed, it is clear from the above calculations that $\|(\mathbf{\Delta}\left[e^{\pm i\varphi}\right] - e^{\pm i\beta})\Phi_w\|$ is independent of β. Thus it follows that $e^{\pm i\beta}$ is an approximate eigenvalue for the operator $\mathbf{\Delta}\left[e^{\pm i\varphi}\right]$, with the set $\left\{\Xi_R^\beta \,:\, R > 0\right\}$ forming a set of approximating eigenvectors, for any $\beta \in [-\pi, \pi]$. Thus every element of the boundary of the spectrum of $\mathbf{\Delta}\left[e^{\pm i\beta}\right]$ is an approximate eigenvalue for that operator — recall that such elements of the spectrum of these operators are not eigenvalues.

These results can, of course, be cast in terms of the Weyl quantized cosine and sine phase operators $\mathbf{\Delta}[\cos\varphi]$ and $\mathbf{\Delta}[\sin\varphi]$. Doing so,

$$\mathsf{Exp}\{\mathbf{\Delta}[\cos\varphi]; \Phi_w\} = \cos\beta\left(1 - \tfrac{1}{4R^2}\right) + \mathrm{O}\left(\tfrac{1}{R^4}\right), \qquad (15.3.41.\mathrm{a})$$

$$\mathsf{Exp}\{\mathbf{\Delta}[\sin\varphi]; \Phi_w\} = \sin\beta\left(1 - \tfrac{1}{4R^2}\right) + \mathrm{O}\left(\tfrac{1}{R^4}\right), \qquad (15.3.41.\mathrm{b})$$

as $R \to \infty$, where $w = -\frac{i}{\sqrt{2}} r e^{i\beta}$. Moreover, since

$$\mathbf{\Delta}[\cos\varphi]^2 + \mathbf{\Delta}[\sin\varphi]^2 = \tfrac{1}{2}\Big[\mathbf{\Delta}\left[e^{i\varphi}\right]\mathbf{\Delta}\left[e^{-i\varphi}\right] + \mathbf{\Delta}\left[e^{-i\varphi}\right]\mathbf{\Delta}\left[e^{i\varphi}\right]\Big],$$

we deduce that

$$\begin{aligned}
&\mathsf{Exp}\left\{\boldsymbol{\Delta}\left[\cos\varphi\right]^2;\,\Phi_w\right\} + \mathsf{Exp}\left\{\boldsymbol{\Delta}\left[\sin\varphi\right]^2;\,\Phi_w\right\} \\
&\quad = \frac{1}{2}\Big[\big\|\,\boldsymbol{\Delta}\left[e^{i\varphi}\right]\Phi_w\,\big\|^2 + \big\|\,\boldsymbol{\Delta}\left[e^{-i\varphi}\right]\Phi_w\,\big\|^2\Big] \\
&\quad = 1 + \mathrm{O}\big(\tfrac{1}{R^4}\big),
\end{aligned}$$

as $R \to \infty$. Moreover,

$$\begin{aligned}
\mathsf{Exp}\left\{\boldsymbol{\Delta}\left[\cos\varphi\right]^2;\,\Phi_w\right\} &\geqslant \mathsf{Exp}\left\{\boldsymbol{\Delta}\left[\cos\varphi\right];\,\Phi_w\right\}^2 \\
&= \cos^2\beta\big(1 - \tfrac{1}{2R^2}\big) + \mathrm{O}\big(\tfrac{1}{R^4}\big), \\
\mathsf{Exp}\left\{\boldsymbol{\Delta}\left[\sin\varphi\right]^2;\,\Phi_w\right\} &\geqslant \mathsf{Exp}\left\{\boldsymbol{\Delta}\left[\sin\varphi\right];\,\Phi_w\right\}^2 \\
&= \sin^2\beta\big(1 - \tfrac{1}{2R^2}\big) + \mathrm{O}\big(\tfrac{1}{R^4}\big),
\end{aligned}$$

as $R \to \infty$, which implies that

$$\mathsf{Exp}\left\{\boldsymbol{\Delta}\left[\cos\varphi\right]^2;\,\Phi_w\right\} = \cos^2\beta + \mathrm{O}\big(\tfrac{1}{R^2}\big), \qquad (15.3.42.a)$$

$$\mathsf{Exp}\left\{\boldsymbol{\Delta}\left[\sin\varphi\right]^2;\,\Phi_w\right\} = \sin^2\beta + \mathrm{O}\big(\tfrac{1}{R^2}\big), \qquad (15.3.42.b)$$

as $R \to \infty$, so we see that the expectations of $\boldsymbol{\Delta}\left[\cos\varphi\right]$ and $\boldsymbol{\Delta}\left[\sin\varphi\right]$ behave like (classical) trigonometric functions only asymptotically. Thus

Proposition 15.15 *If* $w = -\frac{i}{\sqrt{2}}Re^{i\beta}$, *then*

$$\mathsf{Var}\left\{\boldsymbol{\Delta}\left[\cos\varphi\right];\,\Phi_w\right\} = \mathrm{O}\big(\tfrac{1}{R^2}\big), \qquad (15.3.43.a)$$

$$\mathsf{Var}\left\{\boldsymbol{\Delta}\left[\sin\varphi\right];\,\Phi_w\right\} = \mathrm{O}\big(\tfrac{1}{R^2}\big), \qquad (15.3.43.b)$$

as $R \to \infty$.

Finally, we recall that the quantity

$$\mathsf{Var}\left\{\boldsymbol{\Delta}\left[\cos\varphi\right];\,\Phi_w\right\} + \mathsf{Var}\left\{\boldsymbol{\Delta}\left[\sin\varphi\right];\,\Phi_w\right\}$$

has been studied by Freyberger & Schleich (*ibid*), who consider it to be a useful measure of the dispersion of radiation. The above identities enable us to retrieve their formula

$$\mathbf{Var}\{\mathbf{\Delta}[\cos\varphi]\,;\,\Phi_w\} + \mathbf{Var}\{\mathbf{\Delta}[\sin\varphi]\,;\,\Phi_w\} \;=\; \frac{1}{4|\,w\,|^2} + \mathrm{O}\big(\frac{1}{|\,w\,|^4}\big)\,, \tag{15.3.44}$$

as $|\,w\,| \to \infty$.

15.4 Asymptotics For LHW States

The transformed LHW states $\eta_s[\theta]$, defined in equation (10.3.42.a), are a particular set of solutions to the angular shift equation (10.3.2.a). Further on in Chapter 10, in Section 10.3.4, the states $\eta_{s,j}$ central to the scheme of Barnett & Pegg were introduced. These two families of states are accounted by some to be pure phase states. Points in favour of this view are that they arise from the angular shift equation — the LHW states are sequences of approximating eigenvectors for various operators as we shall see — and the Barnett & Pegg states are an attempt to "distribute" angle evenly over the Hermite-Gauss functions. While we have reservations about just how fundamental they are, these families certainly deserve consideration with respect to asymptotic analysis.

When considering asymptotic behaviour with respect to LHW states, we are interested in taking the expectation and variance of observables in these states, and considering the asymptotic behaviour of these quantities in the limit as $s \to \infty$. This procedure presents no conceptual problems for the states $\eta_s[\theta]$, given a particular value of θ, but it is clear that considering the limit as $s \to \infty$ of moments of observables in the state $\eta_{s,j}$ (for fixed j) is not likely to provide us with much information of interest, since the value of $\theta_{s,j}$ clearly then converges to $-\pi$ as $s \to \infty$. Thus we shall be particularly interested in the behaviour of our various phase observables with respect to the general transformed LHW states $\eta_s[\theta]$, and not with respect to the Barnett & Pegg states $\eta_{s,j}$.

We have chosen not to investigate the properties of the Bargmann-Segal phase operator in the LHW states, the analysis of which we leave to interested readers.

15.4.1 *Barnett & Pegg Operators*

Recall that any function $\omega \in \mathcal{C}[-\pi,\pi]$ defines the function $\omega(X_s)$ via the formula

$$\omega(X_s) \;=\; \sum_{j=0}^{s} \omega(\theta_{s,j})\, P_{s,j} \;+\; \omega(0)\big(I - P^{(s)}\big)\,, \qquad (10.3.46.a)$$

and the interpretation of Barnett & Pegg theory is that the matrix coefficients of $\omega(X_s)$, in the limit as $s \to \infty$, represent the matrix coefficients of the operator which is understood to be the function ω acting on the Barnett & Pegg phase "observable". As was shown in equation (10.3.46.c), the weak limit of the sequence of observables $\big(\omega(X_s)\big)_{s\geqslant 1}$ is the Toeplitz operator $\overline{\mathcal{M}}(\omega)$. Thus, if we are to study the behaviour of "functions of the Barnett & Pegg operator" in the LHW states $\eta_s[\theta]$, we need to study the behaviour of the quantities

$$\big\langle \eta_s[\theta]\,,\, \overline{\mathcal{M}}(\omega)\eta_s[\theta]\big\rangle\,, \qquad (15.4.1)$$

for any $|\,\theta\,| < \pi$ and $\omega \in \mathcal{C}[-\pi,\pi]$ as $s \to \infty$. This is relatively simply to do, since

$$\big\langle \eta_s[\theta]\,,\, \overline{\mathcal{M}}(\omega)\eta_s[\theta]\big\rangle \;=\; \frac{1}{s+1}\sum_{k=-s}^{s}\big(s+1-|\,k\,|\big)\hat{\omega}_k\, e^{ik\theta} \;=\; \Sigma_s(\omega,\theta)$$

is the $s^{\rm th}$ Cesàro sum[9] of the function ω at the point θ. Consequently, it follows that

$$\lim_{s\to\infty}\big\langle \eta_s[\theta]\,,\, \omega(X_s)\eta_s[\theta]\big\rangle \;=\; \omega(\theta)\,, \qquad |\,\theta\,| < \pi\,. \qquad (15.4.2)$$

[9]Recall that the $s^{\rm th}$ Fourier sum of the function ω is given by the expression

$$S_s(\omega,\theta) \;=\; \sum_{k=-s}^{s} \hat{\omega}_k\, e^{ik\theta}\,,$$

and the $s^{\rm th}$ Cesàro sum of ω is then given by the expression

$$\Sigma_s(\omega,\theta) \;=\; \frac{1}{s+1}\sum_{k=0}^{s} S_k(\omega,\theta)\,.$$

There are many functions ω, for example continuous nondifferentiable ones, for which the sequence of Cesàro sums for ω converges to the function ω, but for which the sequence of Fourier sums does not.

In particular, standard Fourier analysis shows that

$$\begin{aligned} \left| BP_1\bigl(\eta_s[\theta]\bigr) - \theta \right| &= \frac{2}{s+1}\int_{\frac{1}{2}(\pi-\theta)}^{\frac{1}{2}\pi} \left(\frac{\sin(s+1)\phi}{\sin\phi}\right)^2 d\phi \\ &\leqslant \frac{\pi}{(s+1)\cos^2\frac{1}{2}\theta} \end{aligned}$$

for any $|\theta| < \pi$, so that

$$BP_1\bigl(\eta_s[\theta]\bigr) = \theta + \mathrm{O}\bigl(\tfrac{1}{s+1}\bigr), \qquad s \to \infty,\ |\theta| < \pi, \tag{15.4.3.a}$$

while similar calculations show that

$$BP_2\bigl(\eta_s[\theta]\bigr) = \theta^2 + \mathrm{O}\bigl(\tfrac{1}{s+1}\bigr), \qquad s \to \infty,\ |\theta| < \pi, \tag{15.4.3.b}$$

so that

$$V_{BP}\bigl(\eta_s[\theta]\bigr) = \mathrm{O}\bigl(\tfrac{1}{s+1}\bigr), \qquad s \to \infty,\ |\theta| < \pi. \tag{15.4.3.c}$$

Once again, results of this nature are what are expected of phase observables. It should be noted that all of these limiting results, or asymptotic formulæ, are uniformly true if θ is restricted to any compact subset of $(-\pi,\pi)$.

However other calculations might be considered. This is because expectations are calculated in this theory for finite s, and then the limit as $s \to \infty$ is taken. In most cases, only the observables $\omega(X_s)$ depend upon this parameter s, and so there is no cause for confusion. However, when considering the LHW states, the states $\eta_s[\theta]$ depend upon the very same parameter s which is used to determine the "phase observable" of this theory. Therefore it might be argued that it was appropriate to take all the limits as $s \to \infty$ at the same time. In other words, instead of calculating $\bigl\langle \eta_s[\theta]\,,\, \overline{\mathcal{M}}(\omega)\eta_s[\theta]\bigr\rangle$, we might consider the quantity

$$\rho_{\theta,s}(\omega) = \bigl\langle \eta_s[\theta]\,,\, \omega(X_s)\eta_s[\theta]\bigr\rangle, \tag{15.4.4}$$

for any $\omega \in \mathcal{C}[-\pi,\pi]$ and $|\theta| < \pi$, and investigate its properties in the limit as $s \to \infty$. Although $\omega(X_s)$ converges to $\overline{\mathcal{M}}(\omega)$ as $s \to \infty$, it only does so weakly, and so there is no *a priori* reason to suppose that the limit (if any) and the asymptotic behaviour in that limit of $\rho_{\theta,s}(\omega)$ is the same as that of $\bigl\langle \eta_s[\theta]\,,\, \overline{\mathcal{M}}(\omega)\eta_s[\theta]\bigr\rangle$. What is surprising, therefore, is that the two limits are the same, and the limiting behaviours similar.

Before proving these assertions, we need the following technical result.

Lemma 15.16 *If the function $f \in C^1[-\pi,\pi]$ is continuously differentiable, and if the function $g : [1,\infty) \to [0,\infty)$ is monotonic decreasing, then the set $\{ \cos\frac{1}{2}\theta \, X_{f,g,s}(\theta) \,:\, s \in \mathbb{N}, |\theta| \leqslant \pi \}$ is bounded, where*

$$X_{f,g,s}(\theta) \;=\; \sum_{k=1}^{s} f\Big(\frac{k}{s+1}\Big)\, g(s+1-k)\,(-1)^k\, e^{ik\theta}\,. \tag{15.4.5}$$

Proof: Since

$$X_{1,1,s}(\theta) \;=\; \sum_{k=1}^{s}(-1)^k e^{ik\theta} \;=\; \frac{1}{2}e^{\frac{1}{2}i\theta}\big(e^{is(\theta+\pi)}-1\big)\sec\tfrac{1}{2}\theta\,,$$

for any $|\theta| < \pi$, it is clear that

$$\big|\cos\tfrac{1}{2}\theta\, X_{1,1,s}(\theta)\big| \;\leqslant\; 1\,, \qquad s \in \mathbb{N}\,,\; |\theta| \leqslant \pi\,.$$

In general we see that

$$\begin{aligned} X_{f,g,s}(\theta) \;&=\; f\Big(\frac{s}{s+1}\Big)g(1)X_{1,1,s}(\theta) \\ &+\sum_{k=1}^{s-1}\Big[f\Big(\frac{k}{s+1}\Big)g(s+1-k) - f\Big(\frac{k+1}{s+1}\Big)g(s-k)\Big]\,X_{1,1,k}(\theta), \end{aligned}$$

and so, since

$$\begin{aligned} &\Big| f\Big(\frac{k}{s+1}\Big)g(s+1-k) - f\Big(\frac{k+1}{s+1}\Big)g(s-k)\Big| \\ &\quad\leqslant\; \Big|f\Big(\frac{k}{s+1}\Big)\Big|\,\big[g(s-k)-g(s+1-k)\big] \\ &\quad\quad+\Big|f\Big(\frac{k}{s+1}\Big)-f\Big(\frac{k+1}{s+1}\Big)\Big|\,g(s-k) \\ &\quad\leqslant\; \|f\|_\infty\,\big[g(s-k)-g(s+1-k)\big] \;+\; \frac{1}{s+1}\,\|f'\|_\infty\, g(1)\,, \end{aligned}$$

it follows that

$$\big|\cos\tfrac{1}{2}\theta\, X_{f,g,s}(\theta)\big| \;\leqslant\; \big(\|f\|_\infty + \|f'\|_\infty\big)g(1)$$

for $|\theta| \leqslant \pi$ and $s \in \mathbb{N}$, establishing the result. ■

For any $l \in \mathbb{Z}$ it is easy to show that

$$|\rho_{\theta,s}(\chi_l) \;-\; \chi_l(\theta)| \;\leqslant\; \frac{2}{s+1}\,, \qquad |\theta| < \pi\,,\; s \geqslant |l|\,, \tag{15.4.6}$$

where we are here interpreting the standard function χ_l as a function on $[-\pi,\pi]$. Consequently it is clear that $\rho_{\theta,s}(\omega) \to \omega(\theta)$ as $s \to \infty$ uniformly for $\theta \in [-\pi,\pi]$ for any trigonometric polynomial ω. Since the family of maps $\{\rho_{\theta,s} : |\theta| \leqslant \pi\,,\ s \in \mathbb{N}\}$ is a uniformly bounded set in the Banach space dual $\mathcal{C}[-\pi,\pi]^*$, it follows that $\rho_{\theta,s}(\omega) \to \omega(\theta)$ as $s \to \infty$ uniformly on $[-\pi,\pi]$ for any ω in the closure in $\mathcal{C}[-\pi,\pi]$ of the set of trigonometric polynomials, namely for any function $\omega \in \mathcal{C}[-\pi,\pi]$ for which $\omega(\pi) = \omega(-\pi)$.

To complete the analysis, we need to study the function $\mathfrak{p} \in \mathcal{C}[-\pi,\pi]$. We can show that

$$\begin{aligned}\rho_{\theta,s}(\mathfrak{p}) \;&=\; \Sigma_s(\mathfrak{p},\theta) - \frac{2}{s+1}\operatorname{Im} X_{f,1,s}(\theta)\\ &\quad - \frac{\pi}{2(s+1)^2\cos^2\frac{1}{2}\theta}\left[1 + (-1)^s\cos(s+1)\theta\right], \end{aligned}\tag{15.4.7.a}$$

where $f \in \mathcal{C}^1[-\pi,\pi]$ is the function given by

$$f(x) \;=\; \frac{1-x}{x}\left[\pi x\cot \pi x \;-\; 1\right], \qquad 0 < x < 1\,. \tag{15.4.7.b}$$

It is therefore clear that

$$\rho_{\theta,s}(\mathfrak{p}) \;=\; \theta + \mathrm{O}\left(\frac{1}{s+1}\right), \qquad s \to \infty\,,\ |\theta| < \pi\,, \tag{15.4.8}$$

where this asymptotic formula is valid uniformly if θ is restricted to any compact subset of $(-\pi,\pi)$.

Proposition 15.17 *For any $\omega \in \mathcal{C}[-\pi,\pi]$ the limit*

$$\lim_{s\to\infty} \rho_{\theta,s}(\omega) \;=\; \omega(\theta)\,, \qquad |\theta| < \pi\,, \tag{15.4.9}$$

holds, and is valid uniformly if θ is restricted to any compact subset of $(-\pi,\pi)$.

Proof: Any $\omega \in \mathcal{C}[-\pi,\pi]$ can be written as $\omega = \tilde{\omega} + \alpha\mathfrak{p}$ where $\alpha \in \mathbb{C}$ and $\tilde{\omega} \in \mathcal{C}[-\pi,\pi]$ is such that $\tilde{\omega}(\pi) = \tilde{\omega}(-\pi)$. The result is now immediate. ■

While equation (15.4.8) gives us, not only the limit as $s \to \infty$, but also some information as to the rate of convergence of $\rho_{\theta,s}(\mathfrak{p})$ to θ, the argument given above does not yield a convergence rate for general functions ω, but this information can often be obtained directly. For example, we can show that

$$\rho_{\theta,s}(\mathfrak{p}_2) \;=\; \Sigma_s(\mathfrak{p}_2,\theta) + \frac{2\pi^2}{3(s+1)^2} + \frac{4}{s+1}\operatorname{Re} X_{F,G,s}(\theta)\,, \tag{15.4.10.a}$$

for $|\theta| < \pi$, where $F \in C^1[-\pi,\pi]$ is given by the formula

$$F(x) \;=\; \Big(\frac{1-x}{x}\Big)^2 \big[\pi^2 x^2 \operatorname{cosec}^2 \pi x \;-\; 1\big]\,, \qquad 0 < x < 1\,, \qquad (15.4.10.b)$$

and G is the monotonic decreasing function

$$G(x) \;=\; \frac{1}{x}\,, \qquad x \geqslant 1\,.$$

From these observations it is clear that

$$\rho_{\theta,s}(\mathfrak{p}_2) \;=\; \theta^2 + \mathrm{O}\big(\tfrac{1}{s+1}\big)\,, \qquad s \to \infty\,,\; |\theta| < \pi\,, \qquad (15.4.11)$$

with this formula being valid uniformly if θ is restricted to any compact subset of $(-\pi,\pi)$.

So we see that this approach, while significantly more complicated, yields the same limits, and the same type of asymptotic behaviour, as the previous technique. For example, equation (15.4.8) implies that

$$\mathsf{Exp}\,\{X_s;\,\eta_s[\theta]\} \;=\; \theta + \mathrm{O}\big(\tfrac{1}{s+1}\big)\,, \qquad s \to \infty\,, \qquad (15.4.12.a)$$

while equations (15.4.8) and (15.4.11) together imply that

$$\mathsf{Var}\,\{X_s;\,\eta_s[\theta]\} \;=\; \mathrm{O}\big(\tfrac{1}{s+1}\big)\,, \qquad s \to \infty, \qquad (15.4.12.b)$$

where both of these equations are valid for any $|\theta| < \pi$, and uniformly so if θ is restricted to any compact subset of $(-\pi,\pi)$. We observe that these results correspond closely to those previously established for $BP_1\big(\eta_s[\theta]\big)$ and $V_{BP}\big(\eta_s[\theta]\big)$.

While there is no obvious reason for performing the calculations in this order (except that calculations in the model of Barnett & Pegg are often done in this manner), this second approach to the problem can be interpreted as providing information concerning the asymptotic behaviour of the Barnett & Pegg phase "observable" in the LHW states.

There is one further approach to these calculations which might be considered. Given $|\theta| < \pi$, for any $s \in \mathbb{N}$ we can choose $0 \leqslant j(\theta,s) \leqslant s$ such that

$$\big|\,\theta_{s,j(\theta,s)} \;-\; \theta\,\big| \;\leqslant\; \frac{2\pi}{s+1}\,.$$

We might then consider the quantities

$$\begin{aligned} &\big\langle \eta_s[\theta_{s,j(\theta,s)}]\,,\, X_s\, \eta_s[\theta_{s,j(\theta,s)}]\big\rangle\,, \\ &\big\langle \eta_s[\theta_{s,j(\theta,s)}]\,,\, (X_s)^2\, \eta_s[\theta_{s,j(\theta,s)}]\big\rangle\,, \end{aligned} \tag{15.4.13}$$

and their limits as $s \to \infty$. Since the vectors $\eta_s[\theta_{s,j}]$ are eigenvectors of the operator X_s, these quantities are readily calculated, yielding $\theta_{s,j(\theta,s)}$ and $\theta^2_{s,j(\theta,s)}$ respectively, and it is clear that they converge to θ and θ^2 respectively in the limit as $s \to \infty$. These calculations are evidently the simplest of all, but involve not only the whole sequence of operators X_s, but also a sequence of states $\big(\eta_s[\theta_{s,j(\theta,s)}]\big)_{s\geqslant 1}$ which successively approach the transformed LHW state $\eta_s[\theta]$.

15.4.2 *Toeplitz Operators*

Asymptotic properties for the exponentiated Toeplitz phase operators E and E^* with respect to the transformed LHW states are particularly easy to establish, since

$$E\,\eta_s[\theta] = \sqrt{\frac{s}{s+1}}\, e^{-i\theta}\, \eta_{s-1}[\theta]\,, \tag{15.4.14.a}$$

$$E^*\,\eta_s[\theta] = \sqrt{\frac{s+2}{s+1}}\, e^{i\theta}\, \eta_{s+1}[\theta] - \frac{e^{i\theta}}{\sqrt{s+1}}\, h_0\,, \tag{15.4.14.b}$$

for any $|\,\theta\,| \leqslant \pi$ and $s \in \mathbb{N}$. We state the relevant results without proof.

Proposition 15.18 *The identities*

$$\mathsf{Exp}\,\{E;\, \eta_s[\theta]\} = \frac{s}{s+1}\, e^{-i\theta}\,, \tag{15.4.15.a}$$

$$\|\, E\,\eta_s[\theta]\,\|^2 = \frac{s}{s+1}\,, \tag{15.4.15.b}$$

$$\|\,(E - e^{-i\theta})\eta_s[\theta]\,\|^2 = \frac{1}{s+1}\,, \tag{15.4.15.c}$$

$$\mathsf{Exp}\,\{E^*;\, \eta_s[\theta]\} = \frac{s}{s+1}\, e^{i\theta}\,, \tag{15.4.16.a}$$

$$\|\, E^*\,\eta_s[\theta]\,\|^2 = 1\,, \tag{15.4.16.b}$$

$$\|\,(E^* - e^{i\theta})\eta_s[\theta]\,\|^2 = \frac{2}{s+1}\,, \tag{15.4.16.c}$$

describe the behaviour of E and E^* with respect to the transformed LHW states $\eta_s[\theta]$. Consequently $\big(\eta_s[\theta]\big)_{s\geqslant 1}$ is evidently a sequence of approximat-

ing eigenvectors for the approximate eigenvalues $e^{-i\theta}$ and $e^{i\theta}$ of E and E^* respectively.

These identities yield information concerning the expectations and variances of the operators C and S in these states.

If we consider the Toeplitz phase operator X, part of our work has already been done, since $\mathsf{Exp}\,\{X;\,\eta_s[\theta]\}$ is equal to $BP_1\bigl(\eta_s[\theta]\bigr)$ for any $|\,\theta\,| \leqslant \pi$ and $s \in \mathbb{N}$. To analyze the second moment of X, we consider that

$$\begin{aligned}
\|\, X\eta_s[\theta] \,\|^2 &= \frac{1}{s+1}\sum_{n=0}^{s}\Bigl(\tfrac{1}{3}\pi^2 - \sum_{m=n+1}^{\infty}\frac{1}{m^2}\Bigr) \\
&\quad + \frac{2}{s+1}\sum_{k=1}^{s}\frac{(-1)^k\cos k\theta}{k}\sum_{j=0}^{s-k}\Bigl\{\frac{2}{k} - \sum_{m=j+1}^{j+k}\frac{1}{m}\Bigr\} \\
&= \Sigma_s(\mathfrak{p}_2,\theta) - \frac{1}{s+1}\sum_{m=1}^{\infty}\frac{1+\min(m-1,s)}{m^2} \\
&\quad - \frac{2}{s+1}\sum_{m=1}^{s}\sum_{j=0}^{m-1}\sum_{k=m-j}^{s-j}\frac{(-1)^k\cos k\theta}{mk},
\end{aligned}$$

and since

$$\sum_{m=1}^{\infty}\frac{1+\min(m-1,s)}{m^2} = \sum_{m=1}^{s}\frac{1}{m} + (s+1)\sum_{m=s+1}^{\infty}\frac{1}{m^2} \leqslant 1 + \log s + \frac{s+1}{s},$$

while

$$\begin{aligned}
\left|\sum_{m=1}^{s}\sum_{j=0}^{m-1}\sum_{k=m-j}^{s-j}\frac{(-1)^k\cos k\theta}{mk}\right| &\leqslant \frac{2}{\cos\frac{1}{2}\theta}\sum_{m=1}^{s}\sum_{j=0}^{m-1}\frac{1}{m(m-j)} \\
&\leqslant \frac{2(1+\log s)^2}{\cos\frac{1}{2}\theta},
\end{aligned}$$

for any $|\,\theta\,| < \pi$, we arrive at the following:

Proposition 15.19 *We have that*

$$\mathsf{Exp}\,\{X;\,\eta_s[\theta]\} = \theta + O\bigl(\tfrac{1}{s+1}\bigr), \tag{15.4.17.a}$$

$$\mathsf{Exp}\,\bigl\{X^2;\,\eta_s[\theta]\bigr\} = \theta^2 + O\bigl(\tfrac{(\log s)^2}{s+1}\bigr), \tag{15.4.17.b}$$

$$\text{Var}\{X;\,\eta_s[\theta]\} \;=\; \mathrm{O}\big(\frac{(\log s)^2}{s+1}\big), \qquad (15.4.17.c)$$

as $s \to \infty$ for any $|\,\theta\,| < \pi$, where these asymptotic identities are valid uniformly if θ is restricted to any compact subset of $(-\pi, \pi)$.

15.4.3 *Quantized Phase Space Operators*

The actions of the Weyl quantized exponential phase operators $\boldsymbol{\Delta}\left[e^{i\varphi}\right]$ and $\boldsymbol{\Delta}\left[e^{-i\varphi}\right]$ on the transformed LHW states can be determined readily. After that it is a simple matter to discover their asymptotic behaviour, using Lemma 15.2 to control the behaviour of the coefficients $g_{n,n+1}$. From the action of $\boldsymbol{\Delta}\left[e^{\pm i\varphi}\right]$ on h_n, it follows that

$$\boldsymbol{\Delta}\left[e^{i\varphi}\right]\eta_s[\theta] \;=\; \frac{e^{i\theta}}{\sqrt{s+1}}\sum_{n=1}^{s+1} i^n\, e^{-in\theta}\, g_{n,n-1}\, h_n, \qquad (15.4.18.a)$$

$$\boldsymbol{\Delta}\left[e^{-i\varphi}\right]\eta_s[\theta] \;=\; \frac{e^{-i\theta}}{\sqrt{s+1}}\sum_{n==0}^{s-1} i^n\, e^{-in\theta}\, g_{n,n+1}\, h_n, \qquad (15.4.18.b)$$

for any $|\,\theta\,| \leqslant \pi$.

Proposition 15.20 *We have that*

$$\text{Exp}\left\{\boldsymbol{\Delta}\left[e^{\pm i\varphi}\right];\,\eta_s[\theta]\right\} \;=\; e^{\pm i\theta} + \mathrm{O}\big(\frac{1}{s+1}\big), \qquad (15.4.19.a)$$

$$\left\|\,\boldsymbol{\Delta}\left[e^{\pm i\varphi}\right]\right\|^2 \;=\; 1 + \mathrm{O}\big(\frac{1}{s+1}\big), \qquad (15.4.19.b)$$

$$\left\|\,\left(\boldsymbol{\Delta}\left[e^{\pm i\varphi}\right] - e^{\pm i\theta}\right)\eta_s[\theta]\,\right\|^2 \;=\; \mathrm{O}\big(\frac{1}{s+1}\big), \qquad (15.4.19.c)$$

as $s \to \infty$, which identities are true uniformly for $|\,\theta\,| \leqslant \pi$. Consequently $\big(\eta_s[\theta]\big)_{s\geqslant 1}$ is a sequence of approximating eigenvectors for the approximate eigenvalues $e^{\pm i\theta}$ of $\boldsymbol{\Delta}\left[e^{\pm i\varphi}\right]$.

Proof: The above identities for the action of $\boldsymbol{\Delta}\left[e^{i\varphi}\right]$ and $\boldsymbol{\Delta}\left[e^{-i\varphi}\right]$ on $\eta_s[\theta]$ show that

$$\text{Exp}\left\{\boldsymbol{\Delta}\left[e^{\pm i\varphi}\right];\,\eta_s[\theta]\right\} \;=\; \frac{e^{\pm i\theta}}{s+1}\sum_{n=0}^{s-1} g_{n,n+1},$$

$$\left\| \mathbf{\Delta}\left[e^{i\varphi}\right] \eta_s[\theta] \right\|^2 - \frac{1}{s+1} g_{s,s+1}^2 = \left\| \mathbf{\Delta}\left[e^{-i\varphi}\right] \eta_s[\theta] \right\|^2 = \frac{1}{s+1} \sum_{n=0}^{s-1} g_{n,n+1}^2 ,$$

$$\begin{aligned} \left\| (\mathbf{\Delta}\left[e^{i\varphi}\right] - e^{i\theta}) \eta_s[\theta] \right\|^2 - \frac{1}{s+1} g_{s,s+1}^2 &= \left\| (\mathbf{\Delta}\left[e^{-i\varphi}\right] - e^{-i\theta}) \eta_s[\theta] \right\|^2 \\ &= \frac{1}{s+1} \Big\{ 1 + \sum_{n=0}^{s-1} (g_{n,n+1} - 1)^2 \Big\}. \end{aligned}$$

Since Lemma 15.2 shows that $g_{n,n+1} \sim 1 + \frac{(-1)^n}{4n} + \mathrm{O}\left(\frac{1}{n^2}\right)$ as $n \to \infty$, the results of this Proposition are immediate. ∎

Of course, these results could be translated to provide information concerning the expectations and variance of the observables $\mathbf{\Delta}[\cos\varphi]$ and $\mathbf{\Delta}[\sin\varphi]$ in the transformed LHW states.

We shall only touch briefly on the theory for the phase observable $\mathbf{\Delta}[\varphi]$ itself, since it is very detailed. The identities

$$\mathsf{Exp}\,\{\mathbf{\Delta}[\varphi]\,;\, \eta_s[\theta]\} = -\frac{2}{s+1} \sum_{N=1}^{s} \sum_{n=0}^{s-N} \frac{(-1)^N}{N} g_{n,n+N} \sin N\theta , \tag{15.4.20.a}$$

$$\left\| \mathbf{\Delta}[\varphi]\, \eta_s[\theta] \right\|^2 = \frac{1}{s+1} \sum_{m,n=0}^{s} (-1)^{m-n} e^{i(m-n)\theta} \sum_{\substack{k \geqslant 0 \\ k \neq m,n}} \frac{g_{k,m}\, g_{k,n}}{(k-m)(k-m)} , \tag{15.4.20.b}$$

are extremely difficult to analyze, in view of the complicated behaviour of the coefficients $g_{m,n}$. The method that we employ in [112] requires introducing the coefficients

$$Y_{m,n} = \left(\frac{\min(m,n)+1}{\max(m,n)+1} \right)^{s(m,n)-\frac{3}{4}} , \qquad m,n \geqslant 0 , \tag{15.4.21}$$

and defining the quantities

$$Z_1(\theta) = -\frac{2}{s+1} \sum_{N=1}^{s} \sum_{n=0}^{s-N} \frac{(-1)^N}{N} Y_{n,n+N} \sin N\theta , \tag{15.4.22.a}$$

$$Z_2(\theta) \;=\; \frac{1}{s+1} \sum_{m,n=0}^{s} (-1)^{m-n} e^{i(m-n)\theta} \sum_{\substack{k\geqslant 0\\ k\neq m,n}} \frac{Y_{k,m}\,Y_{k,n}}{(k-m)(k-n)}\,. \tag{15.4.22.b}$$

The reason for doing this is that we can write

$$g_{m,n} \;=\; Y_{m,n}\Big\{1 + \frac{C(m,n)}{\min(m,n)+1}\Big\}, \qquad m,n \geqslant 0\,, \tag{15.4.23}$$

where $\big\{C(m,n)\,:\, m,n \geqslant 0\big\}$ is a bounded set. Thus the coefficients $Y_{m,n}$, while still quite complicated, are simpler than the coefficients $g_{m,n}$, and the difference between $g_{m,n}$ and $Y_{m,n}$ can be controlled adequately. It then remains to compare $\mathsf{Exp}\,\{\mathbf{\Delta}\,[\varphi]\,;\,\eta_s[\theta]\}$ and $\|\,\mathbf{\Delta}\,[\varphi]\,\eta_s[\theta]\,\|^2$ with $Z_1(\theta)$ and $Z_2(\theta)$ respectively, and then in turn compare $Z_1(\theta)$ and $Z_2(\theta)$ with $\mathsf{Exp}\,\{X;\,\eta_s[\theta]\}$ and $\|\,X\eta_s[\theta]\,\|^2$ respectively — of course, we already know the asymptotic behaviour of these last two quantities. Working through this analysis, we reach the following conclusions:

Proposition 15.21 *For any* $|\,\theta\,| < \pi$ *we have*

$$\mathsf{Exp}\,\{\mathbf{\Delta}\,[\varphi]\,;\,\eta_s[\theta]\} \;=\; \theta + \mathrm{O}\Big(\frac{\log s}{(s+1)^{\frac{3}{4}}}\Big), \qquad s \to \infty\,, \tag{15.4.24}$$

with this identity being uniformly true if θ *is restricted to any compact subset of* $(-\pi,\pi)$, *while for any* $0 < |\,\theta\,| < \pi$ *we have*

$$\|\,\mathbf{\Delta}\,[\varphi]\,\eta_s[\theta]\,\|^2 \;=\; \theta^2 + \mathrm{O}\Big(\frac{\log s}{(s+1)^{\frac{1}{4}}}\Big), \tag{15.4.25.a}$$

$$\mathsf{Var}\,\{\mathbf{\Delta}\,[\varphi]\,;\,\eta_s[\theta]\} \;=\; \mathrm{O}\Big(\frac{\log s}{(s+1)^{\frac{1}{4}}}\Big), \tag{15.4.25.b}$$

as $s \to \infty$, *with these identities being uniformly true if* θ *is restricted to any compact subset of* $(-\pi,0)\cup(0,\pi)$. *Consequently* $\big(\eta_s[\theta]\big)_{s\geqslant 1}$ *is a sequence of approximating eigenvectors for the approximate eigenvalue* θ *of* $\mathbf{\Delta}\,[\varphi]$ *for any* $0 < |\,\theta\,| < \pi$.

In all likelihood, the above asymptotic formulæ are not sharp and can be improved. However, it is quite certain that the omission of the value $\theta = 0$ is no accident, since it can be shown that $\mathsf{Var}\,\{\mathbf{\Delta}\,[\varphi]\,;\,\eta_s[0]\}$ fails to converge to 0 as $s \to \infty$.

15.4.4 *Smeared LHW States*

In all the results presented in the preceding Subsections, we showed that the asymptotic behaviour of the expectations and variances of the various phase observables with respect to the transformed LHW states was either uniform or (in some sense) locally uniform in the angle parameter θ. Although the actual form of the uniform or local uniformity varies with the choice of phase observable, in all cases it is true that for any $\varepsilon > 0$ we can find a subset U_ε of $[-\pi, \pi]$ of Lebesgue measure less than ε such that all the asymptotic formulæ discussed are uniformly valid on $[-\pi, \pi] \setminus U_\varepsilon$. This observation permits us to smear the transformed LHW states, and consider mixed states of the form $\eta_s[F]$, where $F \in L^1[-\pi, \pi]$ is a positive function with integral equal to 1, where

$$\eta_s[F] = \int_{-\pi}^{\pi} F(\theta) \,|\, \eta_s[\theta] \,\rangle \,\langle\, \eta_s[\theta] \,|\, d\theta \,. \tag{15.4.26}$$

In practice, we would be interested in such states for functions F whose support is in a small interval around some particular value $\theta \in [-\pi, \pi]$. It is now clear that, no matter what phase observable we are considering, its expectation in the state $\eta_s[F]$ converges to the limit

$$\int_{-\pi}^{\pi} \theta \, F(\theta) \, d\theta \,, \tag{15.4.27}$$

as $s \to \infty$, while its variance converges to 0. Given that smeared LHW states are much more physically realistic than the idealized LHW states, this result gives a more practical confirmation of the asymptotic behaviour of these phase observables in the LHW states.

15.5 Asymptotics: Conclusions

Looking over the results recorded in this Chapter, we are reminded of the older classic monographs on plasmas, containing exhaustive surveys of different types of plasma waves. Since each type is found in nature, such a compendium has a purpose, at least for specialists. But what is the purpose of our list of asymptotic expressions and limits? The list arises from the unfortunate circumstance that no one has been able to pin down some single master quantum observable from which all phase phenomena will flow. Given this lack of certainty, compromise models are adopted. Certain states and certain operators seem to have some association (not always very clear, and typically only indirect) with phase. Clarifying the nature of that connection must be done through the classical limit, as made clear by Bohr. The results in this Chapter provide some details which enable this clarification to be done, at least to some extent.

What conclusions can we draw? Every operator in every state has the large parameter limit that would be expected on purely classical grounds. For large parameter values, before the limit is taken, things are quite different for each phase observable. Some authors assert that a better candidate for phase operator or pure phase state is one in which the classical limit is more nearly approached for smaller parameter values. We disagree. The finite parameter values keep us firmly in the realm of quantum mechanics, and the different behaviour that we find can be expected to be measurable, in principle, and reflect purely quantal phenomena. We should not be in too much of a hurry to remove these quantum mechanical effects — after all, if they were not there, there would be no quantum phase properties to discover, and we could all go home early!

Chapter 16

Measurements

It does not help to point out that we could *have measured B had we wished. The fact is that we did not.*

– R. P. Feynman

16.1 Introduction

When developing quantum theory in previous Chapters we observed that, in certain cases, it was necessary to approximate operators so as to make them fit within the formalism of the current model of quantum mechanics (whether smooth or bounded). We also noted (in a general way) that measurements of observables sometimes required further approximations, to reflect experimental error, and we discussed instruments and instrument observables as a theoretical method for accommodating such concerns.

However, the practical applications of such matters have not yet been considered, and we intend to rectify some of that omission in this Chapter. In particular, we shall discuss some of the practical problems inherent in measuring observables whose spectrum has a continuous component. Since all but one of the phase-related operators considered in this book are of this form, it stands to reason that it is important to understand the issues related to the measurement of such observables when interpreting the results of quantum optics experiments.

We shall not attempt to develop a complete consideration of these problems, for that is a major undertaking. However, there are two points that we wish to emphasize. The first is very general — we argue that a measurement device (apparatus) can only be associated with an observable with discrete spectrum, and hence any measurement of an observable with continuous spectrum is necessarily approximate (although in principle, if not in practice, that approximation can be made as good as we please). The second point is that a measurement device describes not only the measurement itself but also the output state of that measurement, and hence a good measurement device must describe the theoretical output states accurately, as well as giving a close representation of the theoretical spectrum.

The theory of Barnett & Pegg will be examined from this standpoint, and is found wanting in that the output states associated with it cannot be associated with a phase operator (or any operator, come to that).

16.1.1 *The Collapse Formulæ*

In Chapter 6, formulæ were given for the output state upon measurement of an observable and registration of a result. These merit a brief review.

As a prototype for the measurement process, consider the incidence of a beam of "particles" on a device (or apparatus) which measures the property represented by a self-adjoint observable A. The device has a mechanism which registers the values of the property possessed by the beam. The beam is represented by a state, and the registered values are restricted to the spectrum of the operator representing the observable. Accepting the standard interpretation, the measurement device is essentially macroscopic[1], and upon registration of the spectral value, the initial state collapses to the corresponding state, which is the new output state of the beam. When the observable A has a discrete spectrum $\{a_j : j \in \mathbb{N}\}$, possibly degenerate, it may be expressed as

$$A = \sum_{j\geqslant 1} a_j Q_j \,, \tag{16.1.1}$$

where Q_j is the orthogonal projection onto the eigenspace corresponding to the eigenvalue a_j and, after registration of some value a_k, an input state ρ_{in} collapses to

$$\rho_{\text{out}} = \frac{Q_k \rho_{\text{in}} Q_k}{\text{Tr}\left[Q_k \rho_{\text{in}} Q_k\right]} \,. \tag{16.1.2}$$

If the machine is set to register the occurrence of any eigenvalues in the subset Δ of the spectrum $\sigma(A)$ of A, but does not distinguish which amongst them occurred, the output state is

$$\rho_{\text{out}} = \frac{Q(\Delta) \rho_{\text{in}} Q(\Delta)}{\text{Tr}\left[Q(\Delta) \rho_{\text{in}} Q(\Delta)\right]} \,, \tag{16.1.3}$$

[1]This does not preclude that it can be described in a wholly quantum mechanical formalism, without invoking classical mechanics.

where

$$Q(\Delta) = \sum_{a_j \in \Delta} Q_j \tag{16.1.4}$$

is the orthogonal projection operator onto the subspace associated with the eigenvalues in Δ.

If, instead, the observable is bounded and has an absolutely continuous spectrum (we are working within the bounded model) then, given the spectral representation

$$A = \int_{\sigma(A)} \lambda \, dE_A(\lambda), \tag{16.1.5}$$

for A, if a measurement of A results in registration of some spectral value within the Borel set Δ of the spectrum $\sigma(A)$, the input state ρ_{in} collapses to

$$\rho_{\text{out}} = \frac{E_A(\Delta)\rho_{\text{in}}E_A(\Delta)}{\text{Tr}\,[E_A(\Delta)\rho_{\text{in}}E_A(\Delta)]}. \tag{16.1.6}$$

These are the appropriate von Neumann-Lüders formulæ. For an observable with non-overlapping discrete and absolutely continuous spectral components the collapse formula is simply a combination of the above two cases.

16.1.2 *Significant Figures*

These formulæ make an assumption of perfection, in that they rely on the existence of perfect measurements. Given inevitable experimental error, this situation is, in general, unreasonable.

However, in the discrete case, this objection can be avoided. If the spectrum possesses no limit points, then the eigenvalues in the spectrum are isolated and so, provided solely that the experimental apparatus is sufficiently accurate to distinguish between neighbouring eigenvalues, we can infer ideal spectral results even from imperfect measurements. We can also determine when our device is not sufficiently accurate, since an inaccurate device will yield output states with components in the wrong eigenspaces. Thus we can achieve ideal measurements in this case.

But this is no longer true in the continuous case, and this fact has significant consequences. The von Neumann-Lüders equation (16.1.6) implies that it is possible to distinguish one Borel subset of $\sigma(A)$ from another.

Since, for example, the intervals $[a, b]$ and $[a, b + \delta]$ are distinct Borel sets, *no matter how small* the positive value δ, this implication leaves no room for experimental error. Since it is not possible to measure any physical quantity with arbitrary precision, the assumption of the von Neumann-Lüders equation is untenable. To resolve this problem, we argue that any measurement apparatus actually measures an observable with a discrete spectrum. Let us examine how this comes about.

Suppose that an apparatus is designed which intends to measure a self-adjoint observable A with purely continuous spectrum, which has a spectral representation described as in equation (16.1.5). Any measurement recorded by this apparatus is subject to experimental error, and we quantify this by assigning an error bound to the readings obtained from the apparatus. For example, when measuring distances with an ordinary meter stick, measurements can only realistically be obtained to within an accuracy of a millimeter. More generally, we would state the number of significant figures for which are results are accurate.

Suppose now that the spectrum of A is the interval $[0, 1]$, and that the apparatus used to measure A can only be guaranteed to be accurate to 2 decimal places. Effectively, then, we are not measuring A but rather some other observable A_d whose spectrum consists of the 101 points 0, 0.01, ..., 0.99, 1. It is then appropriate to ask what the relationship of A_d to A is. It might be reasonable to assume that A_d is obtained from A through some application of the spectral calculus. For example, if $h : [0, 1] \to [0, 1]$ is the function such that $h(t)$ is equal to t rounded to 2 decimal places, we might choose to define $A_d = h(A)$. Crucially, we are making a distinction between the observable A that we wish to measure and the observable A_d which the experimental apparatus actually measures. The observable A_d is an example of an *instrument*, or *device*, *observable*[2]. The key point to note about A_d is that it is an example of an instrument observable whose form is motivated by practical issues of real-world measurement, rather than theoretical niceties. We emphasize that the above form of A_d is not the only one possible, and other operators might be considered.

Viewed in this light, the aim of an experimenter is to devise an apparatus, with corresponding device observable A_d, which yields an acceptably good approximation to the target observable A. Amongst other matters, a good approximation would require A_d to have a large number of eigenvalues

---

[2]Device observables will be discussed in detail in the next Subsection.

very densely packed[3] within the spectrum of A. In this sense, the analysis of observables with continuous spectrum can be subsumed by the analysis of observables with discrete spectrum, although in a rather special way.

Before considering what might constitute a good choice of A_d, we need to make one further refinement to the above observations. We have said that it is always theoretically possible to measure the eigenvalues of observables with discrete spectra accurately, given that measurements can be made which distinguish between neighbouring points of the spectrum. However this is not true if the discrete spectrum possesses a limit point, as does the Hamiltonian of the Hydrogen atom. If the spectrum of the observable possesses a limit point λ, and if the experimental apparatus measures spectral values with an error of δ, then the infinite number of eigenvalues within δ of λ will be indistinguishable. Thus no matter how small δ is, and hence no matter how accurate our apparatus is, we cannot distinguish all eigenvalues. Thus we add to the above observations the requirement that a device observable A_d must have discrete spectrum without any limit points. This is a very strong requirement, but an unavoidable one in operational terms.

It is worth emphasizing that the inaccuracies of measurement which led to our introducing the device observable A_d are macroscopic in nature, relating to the practical issues of registering and interpreting the results of experiment, and are therefore different from those inaccuracies which result from the Uncertainty Principle.

16.2 Good Device Observables

In the previous Section, we argued that it is impossible to measure observables with continuous spectra perfectly. Observables that can be measured perfectly have discrete spectra with no limit points. Any apparatus used to measure some physical quantity must describe such an observable, for what that device does is record spectral values, which have been constrained to lie in some discrete set, and emit output states appropriate to the recorded spectral value, and this information defines a unique observable of this type.

[3] There is, however, a limit on the density of this packing. For example, if the observable A represents some form of distance, then neighbouring eigenvalues of A_d cannot be closer together than the diameter of an electron, and in practice will need to be more widely spaced than that.

To emphasize this point, we are calling such observables *device observables*, and the measurement devices that they represent *ideal instruments*. Although we have discussed a more general notion of instrument observable earlier on in this book (and further details of this more general concept can be found in [41], [52]), the arguments of the previous Section indicate that, for practical purposes, only ideal instruments are physically realizable.

A well-constructed measurement apparatus, intended to measure some abstract quantum mechanical observable A, is one whose associated device observable A_d is, in some sense, a "good approximation" for the target observable A. In this Section we shall discuss criteria which, we believe, should be used to measure the adequacy of any device observable. For simplicity, we shall assume that our target observable A is bounded with absolutely continuous spectrum $[a, b]$.

Any device observable A_d will then be an operator of the form

$$A_d = \sum_{\lambda \in D} \lambda P_\lambda , \tag{16.2.1.a}$$

where D is a finite subset of $[a, b]$, and $\{P_\lambda : \lambda \in D\}$ is a family of orthogonal projections such that

$$\sum_{\lambda \in D} P_\lambda = I , \tag{16.2.1.b}$$

$$P_\lambda P_\mu = 0 , \qquad \lambda, \mu \in D , \lambda \neq \mu . \tag{16.2.1.c}$$

Thus the elements of D are the actual values that the ideal instrument represented by A_d can register and, if a value of $\lambda \in D$ is recorded when the system is in the state ρ_{in}, the resulting output state is

$$\rho_{\text{out}} = \frac{P_\lambda \rho_{\text{in}} P_\lambda}{\operatorname{Tr}(P_\lambda \rho_{\text{in}} P_\lambda)} .$$

Note that we have placed no restriction on the nature of the projections P_λ — they can have finite or infinite rank — this flexibility allows for a wide range of device observables.

What makes a good device observable? As we have already remarked, a good device observable A_d should have spectrum D which has a large number of points closely spaced throughout the entire spectrum $[a, b]$ of A, so that the ideal instrument can record spectral values with a uniformly fine degree of accuracy. We can quantify this requirement as follows. The function $d_D : [a,b] \to [0,\infty)$ is a continuous function, where

$$d_D(x) = \min_{\lambda \in D} |x - \lambda| , \qquad x \in [a,b] , \tag{16.2.2}$$

is the distance from the point $x \in [a,b]$ to the set D. We will regard D as being closely packed throughout the entire spectrum $[a,b]$ of A if the quantity[4]

$$\| D \| = \sup_{x \in [a,b]} d_D(x) \tag{16.2.3}$$

is small. That this criterion is appropriate is evident, for every element of $[a,b]$ is within $\| D \|$ of some element of D, and so the smaller the value of $\| D \|$, the greater the accuracy of the ideal instrument in recording spectral values. We call the requirement for $\| D \|$ to be small the *spectral accuracy condition.*

However we need more than spectral accuracy. As we have observed, a device observable is defined not solely by its spectral values, but also by its output states, and consequently a good device observable A_d must also yield output states which provide good approximations for the ideal output states that the target observable A would, in theory, yield. If the system is in the initial state ρ_{in}, and a measurement is made which yields a result lying in the Borel subset $\Delta \subseteq [a,b]$, then the observable A should, in theory, yield the output state

$$\rho_{\text{out}} = \frac{E(\Delta)\rho_{\text{in}}E(\Delta)}{\text{Tr}\,(\,E(\Delta)\rho_{\text{in}}E(\Delta)\,)} ,$$

where E is the spectral measure for A, while the instrument observable A_d will in fact yield the output state

$$\frac{E_d(\Delta)\rho_{\text{in}}E_d(\Delta)}{\text{Tr}\,(\,E_d(\Delta)\rho_{\text{in}}E_d(\Delta)\,)} ,$$

where E_d is the spectral measure for A_d, so that

$$E_d(\Delta) = \sum_{\lambda \in \Delta} P_\lambda .$$

[4]Topologically speaking, the quantity $\| D \|$ is the Euclidean distance between the closed set D and $[a,b]$.

Thus a good device observable is one for which the state $E_d(\Delta)\rho_{\rm in}E_d(\Delta)$ is a good approximation for $E(\Delta)\rho_{\rm in}E(\Delta)$ for any input state $\rho_{\rm in}$ and any subset $\Delta \subseteq [a,b]$.

It is still not easy to quantify this requirement. However, we can still gain some insight into this requirement in the following manner. In practice, we shall consider collections of device observables $A_d^{(n)}$, with associated spectra $D^{(n)}$ and spectral measures $E^{(n)}$, which are parametrized by some integer n. All of these device observables will be of a similar type, and the parameter n is to represent some measure of the precision of the device observable, in that the device observable $A_d^{(n)}$ is intended to be increasingly accurate as n increases. To this end, we shall assume that our sequence of device observables $A_d^{(n)}$ satisfies the spectral accuracy condition in the sense that $\|D^{(n)}\| \to 0$ as $n \to \infty$.

Increasing accuracy for output states (in the case of pure input states at least) will be obtained provided that the sequence $\left(E_d^{(n)}(\Delta)\phi\right)_{n\geqslant 1}$ converges to $E(\Delta)\phi$ in some sense as $n \to \infty$ for all input vectors ϕ and Borel subsets Δ of $[a,b]$.

The key question to consider is in what sense the above sequences should converge — for example, should they be required to converge weakly or strongly? Some thought shows that the appropriate requirement is that of strong convergence since, among other things, this implies the appropriate convergence for general mixed states as well as pure ones.

This requirement, that all sequences of spectral projections should converge strongly, is still a very general one, but fortunately it can be characterized by a simpler condition. Moreover, this latter condition is one which is much more susceptible of being tested, since it relates to the observables $A_d^{(n)}$ and A themselves, rather than to their spectral projections. It is a standard theorem of functional analysis[5] that the above condition concerning the spectral projections of these device observables is equivalent to the simple requirement that the sequence of device observables $\left(A_d^{(n)}\right)_{n\geqslant 1}$ converges strongly to A [55].

We summarize these observations in the following Definition.

Definition 16.1 *The sequence* $\left(A_d^{(n)}\right)_{n\geqslant 1}$ *of device observables is a* good approximating sequence *of observables for the target observable A if the norm* $\|D^{(n)}\| \to 0$ *as* $n \to \infty$ *and also if* $A_d^{(n)} \to A$ *strongly as* $n \to \infty$.

[5]The boundedness of the operator A is crucial here.

It may be possible to define sequences $\left(A_d^{(n)}\right)_{n\geqslant 1}$ of device observables which converge to A uniformly, and not just strongly, in that $\left\|A_d^{(n)} - A\right\| \to 0$ as $n \to \infty$, and this is highly desirable if true. However, in general, it is not often the case that we will be able to achieve uniform convergence, and so we shall frequently have to be content with strong convergence.

It is important to realize that weak convergence of the sequence of device observables $\left(A_d^{(n)}\right)_{n\geqslant 1}$ to A does not imply convergence of the sequences of spectral projections, and hence a sequence of device observables which only converges weakly to A does not yield output states which closely approximate the ideal states to be output by the target observable A. As we have already remarked (if in a different context) and shall discuss again below, the fact that the device observables considered by the theory of Barnett & Pegg only converge weakly to their target observable X is, we feel, a major impediment to drawing robust interpretations from that theory.

16.2.1 *Device Observables, Good And Bad*

We shall now consider two examples of sequences of device observables. The first is general in nature, and represents (arguably) a fairly realistic model for physical device observables. The second is the sequence of device observables introduced in the theory of Barnett & Pegg. We shall show that this sequence does not satisfy the above criteria for being a good approximating sequence of device observables.

16.2.1.1 *Using The Spectral Calculus*

The spectral calculus can be applied to the observable A to yield projections defining device observables which are closely conformed to A. This construction is not new, and was orginally introduced by von Neumann [230]. We have already introduced a particular example of such a device observable in Section 16.1.2.

As usual, a partition of $[a,b]$ is a finite subset P of $[a,b]$ containing the endpoints a and b. If we write $P = \{a = x_0 < x_1 < \cdots < x_N = b\}$, then the partition P defines the closed intervals $I_n^{(P)} = [x_{n-1}, x_n]$ for $1 \leqslant n \leqslant N$, and the norm $\|P\|$ of the partition is the maximum width of these intervals,

$$\|P\| = \max_{1\leqslant n\leqslant N} |x_n - x_{n-1}| \,. \qquad (16.2.4)$$

Given a partition P of $[a, b]$, choose a subset $D^{(P)} = \{\lambda_1, \ldots, \lambda_N\}$ of N distinct points of $[a, b]$ such that $\lambda_n \in I_n^{(P)}$ for each $1 \leqslant n \leqslant N$. It is then clear that the norm $\| P \|$ of the partition and the norm $\|D^{(P)}\|$ of the set D are related by the inequality

$$\|D^{(P)}\| \leqslant \| P \| \leqslant 2 \|D^{(P)}\| . \tag{16.2.5}$$

For any $1 \leqslant n \leqslant N$, let $\Pi_n^{(P)}$ be the orthogonal projection

$$\Pi_n^{(P)} = E(I_n^{(P)}) , \tag{16.2.6}$$

where E is the spectral measure of the observable A. With these choices, we define the device observable

$$A_d^{(P)} = \sum_{n=1}^{N} \lambda_n \, \Pi_n^{(P)} . \tag{16.2.7}$$

In view of inequality (16.2.5), the norm of the spectrum $D^{(P)}$ of $A_d^{(P)}$ is small if and only if the norm of the partition P is small, and so choosing a partition with a small norm yields a device observable which satisfies the spectral accuracy condition. However, we can do better than this, since the norm of the partition also controls the degree of accuracy of the output states of this device observable, and in a particularly strong manner.

Proposition 16.2 *The target observable A is uniformly approximated by $A_d^{(P)}$, in that*

$$\|A - A_d^{(P)}\| \leqslant \| P \| . \tag{16.2.8}$$

Proof: From the spectral resolution of A it is clear that

$$\|(A - A_d^{(P)})\phi\|^2 = \sum_{n=1}^{N} \int_{I_n^{(P)}} |\lambda - \lambda_n|^2 \, d\langle \phi, E(\lambda)\phi \rangle ,$$

for any vector ϕ, and this leads to the inequality

$$\begin{aligned} \|(A - A_d^{(P)})\phi\|^2 &\leqslant \sum_{n=1}^{N} |x_n - x_{n-1}|^2 \int_{I_n^{(P)}} d\langle \phi, E(\lambda)\phi \rangle \\ &\leqslant \| P \|^2 \, \| \phi \|^2 \end{aligned}$$

for any vector ϕ, which indeed implies that $\|A - A_d^{(P)}\| \leqslant \| P \|$, as required. ∎

Thus, if $(A_d^{(P_n)})_{n\geqslant 1}$ is a sequence of device observables obtained by the above construction from a sequence of partitions $(P_n)_{n\geqslant 1}$ of $[a,b]$ for which the sequence of norms $(\|P_n\|)_{n\geqslant 1}$ converges to 0 as $n\to\infty$, then it is clear that $(A_d^{(P_n)})_{n\geqslant 1}$ is a good approximating sequence of device observables for A, and is moreover a sequence of device observables which converges uniformly to A as $n\to\infty$.

Thus we have an extremely natural construction for device observables which are evidently very closely related to the target observable A (since they are obtained from A via the spectral calculus) and which do indeed provide good approximations to that observable. The good nature of this approximation can be further seen by the fact that the following convergence result can be obtained, showing how the generalized eigendistributions of A can also be derived from the above sequence of device observables. For clarity we shall assume that the observable A is cyclic and belongs to $\mathcal{L}^+(\mathcal{S}(\mathbb{R}))$ (although these requirements can be avoided if necessary).

Proposition 16.3 *Suppose $(A_d^{(P_n)})_{n\geqslant 1}$ is a sequence of device observables of the above form, where $\|P_n\|\to 0$ as $n\to\infty$, and let $\{T_\lambda : \lambda\in[a,b]\}$ be the generalized eigendistributions of A. If $\lambda\in[a,b]$, for any $n\geqslant 1$ choose the integer $j(n,\lambda)$ such that $\lambda\in I_{j(n,\lambda)}^{(P_n)}$. Then the identity*

$$\lim_{n\to\infty} \left|I_{j(n,\lambda)}^{(P_n)}\right|^{-1} \langle h\,,\,\Pi_{j(n,\lambda)}^{(P_n)} f\rangle \;=\; [\![T_\lambda f\,,\,\overline{h}]\!]\,, \qquad f,h\in\mathcal{S}(\mathbb{R})\,, \quad (16.2.9)$$

holds for almost all $\lambda\in[a,b]$ (with respect to the Lebesgue measure).

Proof: There exists a unitary operator $U : L^2(\mathbb{R})\to L^2([a,b],d\mu)$ which diagonalizes A, so that $(UAh)(\lambda) = \lambda(Uh)(\lambda)$ for any $h\in L^2(\mathbb{R})$. Consequently $(UE(\Delta)h)(\lambda) = \chi_\Delta(\lambda)(Uh)(\lambda)$ for any Borel set Δ and $h\in L^2(\mathbb{R})$. Since the spectrum of A is absolutely continuous, the Radon-Nikodym Theorem implies that $d\mu(\lambda) = w(\lambda)d\lambda$ for some measurable function w. Then

$$\langle h\,,\,\Pi_{j(n,\lambda)}^{(P_n)} f\rangle \;=\; \int_{I_{j(n,\lambda)}^{(P_n)}} \overline{(Uh)(\lambda)}\,(Uf)(\lambda)\,w(\lambda)\,d\lambda$$

for any $f,h\in\mathcal{S}(\mathbb{R})$. Since a theorem of Lebesgue states that the identity

$$\lim_{r\to 0}\frac{1}{2r}\int_{a-r}^{a+r} G(x)\,dx \;=\; G(a)$$

holds for almost all a whenever G is locally integrable on the interval $[a-r, a+r]$, it follows that

$$\lim_{n\to\infty} \left|I^{(P_n)}_{j(n,\lambda)}\right|^{-1} \int_{I^{(P_n)}_{j(n,\lambda)}} \overline{(Uh)(\lambda)}\,(Uf)(\lambda)\,w(\lambda)\,d\lambda$$

$$= \overline{(Uh)(\lambda)}(Uf)(\lambda)w(\lambda) = [\![T_\lambda f\,,\,\overline{h}]\!]$$

for $f, h \in \mathcal{S}(\mathbb{R})$ and almost all λ, as required. ■

In other words, the matrix elements of T_λ can be obtained by taking the limits of the weighted matrix coefficients of $\Pi^{(P_n)}_{j(n,\lambda)}$ as shown in equation (16.2.9). Given that the nature of the generalized eigendistributions of A, a stronger result than this cannot be expected.

All the self-adjoint phase and phase-related operators (other than the operators X_s of the theory of Barnett & Pegg) can be approximated by device observables in this manner. However, it is not as easy to write down concrete representations of these device observables. In many cases, the spectral measure of the target observable is not known explicitly, and hence we do not have specific formulæ for the relevant spectral projections needed in the construction. In other cases, even if the spectral measure for A is known explicitly, it may not be possible to write down a closed form expression for the device observables A_d — this is the case, so far as we know, for the Toeplitz phase operator X. However, device observables of this type could be written down for the exponentiated Toeplitz phase observables C and S, and doing so might well be useful in various numerical calculations.

16.2.1.2 *Barnett & Pegg Device Observables*

While we believe that the procedure described above provides a particularly good method for describing device observables for the target observable A, it has the drawback that it is, typically, not possible to obtain explicit expressions for these device observables. For calculational purposes, then, device observables derived from the spectral calculus are not particularly useful.

It is therefore tempting to attempt to construct device observables which are simple to derive and work with in calculations. As we have observed, defining a device observable A_d involves choosing not only the choice of the spectrum D but also the projections which define the output states.

One particularly simple approach would be to choose these projections to have rank one, so that each element of D is a nondegenerate eigenvalue of A_d and P_λ is the one-dimensional projection onto the eigenspace of λ for any $\lambda \in D$. Such a device observable is then constrained by the choice of eigenvectors for the elements of D.

The operators introduced by the theory of Barnett & Pegg are just such device observables, and their target observable is the Toeplitz phase operator X (since they converge to X weakly). For any integer $s \geqslant 1$, we consider the set $D^{(s)}$

$$D^{(s)} = \{\theta_{s,j} : 0 \leqslant j \leqslant s\} \cup \{0\}, \tag{16.2.10.a}$$

where, as usual,

$$\theta_{s,j} = -\pi + \frac{2\pi j}{s+1}, \qquad 0 \leqslant j \leqslant s. \tag{16.2.10.b}$$

For any $0 \leqslant j \leqslant s$, the element $\theta_{s,j}$ of $D^{(s)}$ is associated with the one-dimensional projection

$$P_{s,j} = |\,\eta_{s,j}\,\rangle\,\langle\,\eta_{s,j}\,| \tag{16.2.10.c}$$

associated with the transformed LHW state $\eta_{s,j} = \eta_s[\theta_{s,j}]$. Viewed in this light, the Barnett & Pegg observable X_s is then the device observable[6]

$$X_s = \sum_{j=0}^{s} \theta_{s,j}\, P_{s,j}\,. \tag{16.2.10.d}$$

It is clear that $\|D^{(s)}\| \leqslant 2\pi/(s+1)$, and hence the norm $\|D^{(s)}\|$ converges to 0 as $s \to \infty$. In this sense, then, the Barnett & Pegg device observables are good, in that they satisfy the spectral accuracy condition. However, as we have already observed, the sequence of operators $\big(X_s\big)_{s\geqslant 1}$ converges weakly to X, but does not do so strongly. Thus the spectral projections of X_s do not converge strongly to the spectral projections of X and so this sequence of device observables is not a good approximating sequence of device observables in our sense.

---

[6]The point 0 must be included explicitly in the definition of the spectrum $D^{(s)}$ of the operator X_s, since $X_s\, h_n = 0$ for all $n > s$, and so the eigenvalue 0 of X_s is associated with the projection $I - P^{(s)}$ (and possibly with a projection $P_{s,k}$, if k exists for which $\theta_{s,k} = 0$).

It is perhaps worth noting that the spectral projections for the device observables X_s also converge weakly, for since

$$\begin{aligned}\langle e_m\,,\,E^{(s)}(\Delta)e_n\rangle &= \sum_{\theta_{s,j}\in\Delta}\langle e_m\,,\,P_{s,j}e_n\rangle \\ &= \sum_{\theta_{s,j}\in\Delta}\langle e_m\,,\,\eta_{s,j}\rangle\langle\eta_{s,j}\,,\,e_n\rangle \\ &= \frac{i^{m-n}}{s+1}\sum_{\theta_{s,j}\in\Delta}e^{i(n-m)\theta_{s,j}}\,,\end{aligned}$$

for any $m,n\geqslant 0$, subinterval Δ of $[-\pi,\pi]$ and $s\geqslant 1$, it follows that

$$\lim_{s\to\infty}\langle e_m\,,\,E^{(s)}(\Delta)e_n\rangle = \frac{i^{m-n}}{2\pi}\int_\Delta e^{i(n-m)\beta}\,d\beta$$

for any $m,n\geqslant 0$ and subinterval Δ of $[-\pi,\pi]$. But this last quantity is not the expectation $\langle e_m\,,\,E(\Delta)e_n\rangle$ of the spectral projection $E(\Delta)$ of X — instead it is the expectation $\langle e_m\,,\,\overline{\mathcal{M}}(\chi_\Delta)e_n\rangle$. Thus we can show that the sequence of spectral projections $\big(E^{(s)}(\Delta)\big)_{s\geqslant 1}$ converges weakly to the operator $\overline{\mathcal{M}}(\chi_\Delta)$ for any subinterval Δ of $[-\pi,\pi]$, but the fact that this limit is not the spectral projection $E(\Delta)$ of X shows that, even weakly, the Barnett & Pegg device observables X_s do not provide a good approximating sequence for the Toeplitz phase operator (or for any other operator). Since the matrix coefficients of the spectral projections of an observable provide, in the sense of quantum logic, the "answers" to the quantum mechanical questions concerning that operator, we must conclude that the Barnett & Pegg device observables are asking the wrong questions!

16.2.2 *SAE Instruments*

While the device observables X_s from Barnett & Pegg theory do not form a good approximating sequence for the Toeplitz phase operator, they are nonetheless of interest. They are, of course, device observables which satisfy the spectral accuracy condition, and the spectral projections of the observables X_s, being defined in terms of the transformed LHW states, can be argued to have many properties often thought to be appropriate to phase observables.

Abstracting these properties leads us to the observation that there is a further manner in which a device observable can be seen as modelling the behaviour of a more general observable. If a measurement of the target observable is made of a system in the pure state defined by the unit vector ϕ, and the result of that measurement is that the value of that observable lies in the interval Δ, then the output state after measurement is the pure state defined by the vector $E(\Delta)\phi$. Since

$$\frac{1}{\|E(\Delta)\phi\|}\,\|(A-\lambda)E(\Delta)\phi\| \;\leqslant\; |\Delta|\,, \tag{16.2.11}$$

for any $\lambda \in \Delta$, in some sense the degree to which $E(\Delta)\phi$ is almost an eigenvector of A, with eigenvalue $\lambda \in \Delta$, is controlled by the size of the measurement interval Δ. Thus if $A_d^{(n)}$ is a sequence of device observables which satisfy the spectral accuracy condition, so that $\|D^{(n)}\| \to 0$ as $n \to \infty$, then this sequence could be seen as representing some of the properties of the target observable A if output states increasingly approximate eigenvectors of A.

This leads us to consider sequences of device observables which are based upon sequences of approximating sequences of eigenvectors for approximate eigenvalues of the target observable A. We have mentioned the concept of an approximate eigenvalue a number of times previously, particularly in Chapter 15, but it is now appropriate to present a formal definition of the concept, in order to clarify our later discussion.

Definition 16.4 *If A is a bounded operator on the Hilbert space $\mathcal{H}$, then the complex number λ is an* approximate eigenvalue *for A if there is a sequence $\big(\psi_n[\lambda]\big)_{n\geqslant 1}$ of unit vectors in $\mathcal{H}$ such that*

$$\lim_{n\to\infty}\,\|(A-\lambda)\psi_n[\lambda]\| \;=\; 0\,. \tag{16.2.12}$$

If λ is an approximate eigenvalue for A, any such sequence $\big(\psi_n[\lambda]\big)_{n\geqslant 1}$ is called a sequence of approximating eigenvectors, *or SAE*[7], *for λ and A.*

Clearly, every eigenvalue λ of A is an approximate eigenvalue of A, since we can choose a SAE for λ and A by setting $\psi_n[\lambda] \;=\; \psi$ for all $n \geqslant 1$, where ψ is any unit eigenvector of A for the eigenvalue λ. On the other hand, every approximate eigenvalue of A belongs to the spectrum $\sigma(A)$ of A. If A is a normal operator, then every element of the spectrum $\sigma(A)$

[7]The SAE in general does not converge — indeed it converges in $\mathcal{H}$ if and only if λ is an eigenvalue of A, in which case the limit of the SAE is an eigenvector of A.

of A is an approximate eigenvalue. It should be noted, however, that any approximate eigenvalue λ can be associated with many different SAEs.

The fact that the sequence $(X_s)_{s\geqslant 1}$ of device observables associated with the theory of Barnett & Pegg mirrors the properties of X reflected in equation (16.2.11) can be described through the fact that these device observables can be constructed using families of SAEs. We now discuss, in a general context, how this construction is achieved.

Working again with the target observable A, we note that every element of the spectrum $\sigma(A) = [a, b]$ of A is an approximate eigenvalue. For each $\lambda \in [a, b]$, we choose a SAE $(\psi_n[\lambda])_{n\geqslant 1}$ for λ and A. For any $\lambda \in [a, b]$ and $n \in \mathbb{N}$, let $P_n[\lambda]$ denote the one-dimensional projection

$$P_n[\lambda] = |\,\psi_n[\lambda]\,\rangle\,\langle\,\psi_n[\lambda]\,|\,. \tag{16.2.13}$$

Suppose that, for any $n \in \mathbb{N}$, it is possible to choose a subset $D^{(n)}$ of $[a, b]$ which contains n elements such that $\{\psi_n[\lambda] \,:\, \lambda \in D^{(n)}\}$ is an orthonormal collection of vectors in $\mathcal{H}$. Moreover, suppose that $\|D^{(n)}\| \to 0$ as $n \to \infty$. Then, for any $n \geqslant 1$, the device observable

$$A_d^{(n)} = \sum_{\lambda \in D^{(n)}} \lambda\, P_n[\lambda] \tag{16.2.14}$$

has spectrum $D^{(n)} \cup \{0\}$ and n-dimensional range, and the sequence of device observables $(A_d^{(n)})_{n\geqslant 1}$ satisfies the spectral accuracy condition. Any such sequence of device observables is called a sequence of SAE device observables.

It is clear that this is a highly intricate construction, but it is justified by noting that the operators X_s of Barnett & Pegg form a sequence of SAE device observables through choosing $\psi_s[\lambda] = \eta_s[\lambda]$ for any $s \geqslant 1$ and $\lambda \in [-\pi, \pi]$.

In view of the discussion surrounding equation (16.2.11), we are led to consider the quantities

$$\varepsilon_{n,K}(A) = \sup_{\lambda \in K} \|(A - \lambda)\psi_n[\lambda]\|\,, \tag{16.2.15}$$

where $n \geqslant 1$ and K is a compact subset of $\sigma(A)$, and we would like to require that $\varepsilon_{n,K}(A) \to 0$ as $n \to \infty$ for various compact subsets K of $[a, b]$. Ideally, we would like to choose K to be the complete spectrum $[a, b]$ of A, but this may not be practicable, and so we make the following Definition.

Definition 16.5 *A sequence of SAE device observables for the target observable A satisfies the* spectral uniformity condition *if*

$$\lim_{n\to\infty} \varepsilon_{n,K}(A) = 0 \tag{16.2.16}$$

for all compact subsets K of some open dense subset U of $\sigma(A) = [a,b]$.

Another condition which is felt to be of physical importance is that it should be possible to approximate vectors in $\mathcal{H}$ increasingly accurately by vectors in the range of the device observables $A_s^{(n)}$. Thus we need to consider the n-dimensional subspace $\mathcal{H}^{(n)}$ of $\mathcal{H}$ which is spanned by the vectors $\{\psi_n[\lambda] : \lambda \in D^{(n)}\}$.

Definition 16.6 *A sequence of SAE device observables for the target observables A satisfies the* ascending subspace condition *if $\mathcal{H}^{(n)} \subseteq \mathcal{H}^{(n+1)}$ for all $n \geqslant 1$ and if, moreover, the union $\bigcup_{n\geqslant 1} \mathcal{H}^{(n)}$ is dense in $\mathcal{H}$.*

We now see from the results of Chapter 15 that the device observables $(X_s)_{s\geqslant 1}$ of the theory of Barnett & Pegg form a sequence of SAE device observables for the Toeplitz phase operator X which satisfies the ascending subspace condition, as well as the spectral uniformity condition, by choosing U to be the open dense subspace $(-\pi,\pi)$ of the spectrum $[-\pi,\pi]$ of X. This is all well and good. The results in Chapter 15 which demonstrate that the device observables of Barnett & Pegg satisfy the spectral uniformity condition for X are of intrinsic interest, and it is therefore useful to see these results employed as an estimator of the effectiveness of the observable X_s as an approximant for X. However, a serious problem with this approach is the fact that the spectral uniformity condition does not uniquely identify the target observable A from the sequence of SAE device observables, since a given sequence of SAE device observables can satisfy the spectral uniformity condition for more than one target observable. For example, it is clear from Chapter 15 that the sequence $(X_s)_{s\geqslant 1}$ not only satisfies the spectral uniformity condition for X, but also does so for the Bargmann-Segal phase operator $\Xi(\varphi)$ (choosing U to be the open set $(-\pi,\pi)$). If it were the case (as we suspect) that the Weyl phase operator $\mathbf{\Delta}[\varphi]$ has spectrum $[-\pi,\pi]$, then the same sequence $(X_s)_{s\geqslant 1}$ satisfies the spectral uniformity condition for $\mathbf{\Delta}[\varphi]$ as well (choosing U to be the open set $(-\pi,0)\cup(0,\pi)$). Thus, while the spectral accuracy and uniformity conditions (and the ascending subspace condition) are interesting, they do not characterize the target observable that is being approximated — something else is needed.

Of course, the Toeplitz phase operator X is distinguished amongst the various observables for which $(X_s)_{s\geqslant 1}$ satisfies the spectral uniformity condition by the fact that it is the weak operator limit of that sequence. But, as we have already remarked, weak operator convergence is not sufficient for the purposes of analysis.

Thus, while the concept of sequences of SAE device observables is of interest, providing some heuristic justification for the construction of the operators of Barnett & Pegg, it is still of itself not sufficient to provide good device observable with which calculations can be performed with confidence.

We end by observing that, in principle, there are conditions under which a sequence of SAE device observables can provide a good approximating sequence of device observables for the target observable A. We simply need to strengthen the spectral uniformity condition.

Proposition 16.7 *If $(A_d^{(n)})_{n\geqslant 1}$ is a sequence of SAE device observables for the target observable A such that the ascending subspace condition is satisfied, and moreover such that*

$$\lim_{n\to\infty} \sqrt{n}\varepsilon_{n,[a,b]}(A) \;=\; 0\,, \tag{16.2.17}$$

then $(A_d^{(n)})_{n\geqslant 1}$ is a good approximating sequence of device observables for A.

Proof: If $\psi \in \mathcal{H}^{(N)}$, then $\psi \in \mathcal{H}^{(n)}$ for all $n \geqslant N$, so we can write

$$\psi \;=\; \sum_{\lambda\in D^{(n)}} f_n(\lambda)\,\psi_n[\lambda]$$

for any $n \geqslant N$. But then

$$\begin{aligned}
&\left\| (A - A_\lambda^{(n)})\psi \right\|^2 \\
&\quad\leqslant\; \left(\sum_{\lambda\in D^{(n)}} |\, f_n(\lambda)\,|^2\right)\left(\sum_{\lambda\in D^{(n)}} \left\| (A - A_d^{(n)})\psi_n[\lambda] \right\|^2\right) \\
&\quad=\; \|\,\psi\,\|^2 \left(\sum_{\lambda\in D^{(n)}} \left\| (A - A_d^{(n)})\psi_n[\lambda] \right\|^2\right) \\
&\quad\leqslant\; n\,\|\,\psi\,\|^2\, \varepsilon_{n,[a,b]}(A)^2
\end{aligned}$$

for any $n \geqslant N$, so that

$$\left\|(A - A_d^{(n)})\psi\right\| \leqslant \sqrt{n}\,\varepsilon_{n,[a,b]}(A)\ \|\psi\|$$

for any $n \geqslant N$, which implies that $\left\|(A - A_d^{(n)})\psi\right\| \to 0$ as $n \to \infty$. Since the sequence $\left(A - A_d^{(n)}\right)_{n\geqslant 1}$ is uniformly bounded and $\bigcup_{n\geqslant 1} \mathcal{H}^{(n)}$ is dense in $\mathcal{H}$, it follows that $\left(A_d^{(n)}\right)_{n\geqslant 1}$ converges strongly to A, and hence is a good approximating sequence of device observables for A. ∎

Thus, although current models (such as the model of Barnett & Pegg) do not provide sequences of SAE device observables which form good approximating sequences of device observables for a given target observable, the above result shows that it is possible, in principle, that some analogous (but stronger) construction might be able to do so. If a simple example of such a sequence of SAE device observables could be found, we would then have a framework within which it was simple to perform calculations and yet which (in the limit as $n \to \infty$) yields reliable approximate results.

16.2.3 *The Vorontsov-Rembovksy Rebuttal*

We have already mentioned the fact that any system of measurement is subject to unavoidable tolerance errors. Indeed, it was to deal with such errors that the concept of a device observable has been introduced. This, together with the fact that the sequence $\left(X_s\right)_{s\geqslant 1}$ of SAE device observables only converges weakly (and does not possess any more useful convergence properties) has significant consequences. These consequences, and their physical implications, have been considered by Vorontsov & Rembovksy [232], who have shown that an interpretation of the Barnett & Pegg theory can be said in some cases not to preserve probability. We set out, and extend, their ideas here.

The insight of Vorontsov & Rembovsky is to inquire about the behaviour of the Barnett & Pegg family of operators under successive measurements, first of the angle operator and then of the number operator. The difficulty then arises in the limit as $s \to \infty$, which is (as usual) to be taken at the end of all calculations.

While the operators X_s of Barnett & Pegg theory are certainly device observables, the process of taking the limit as $s \to \infty$ is inconsistent with the need to allow for tolerance errors. This is because the device observable X_s, if it is to be practicable, requires a measurement apparatus which can distinguish between spectral values as little as $2\pi/(s+1)$ (the difference between successive values of $\theta_{s,j}$) apart. So if an experimental apparatus is such that all measurements are subject to some tolerance error of size $\delta > 0$, then X_s is no longer a valid device observable for this apparatus once $s > 2\pi\delta^{-1}$.

Vorontsov & Rembovksy propose the following modification of Barnett & Pegg theory. They suppose that a measurement apparatus has been designed to provide information about the device observable X_s, but that the experimental apparatus has a tolerance error of $\delta > 0$. Consequently, should the device register a measurement of $\Theta \in [-\pi, \pi)$, then it is possible that any spectral value $\theta_{s,j}$ of X_s, for which $|\,\theta_{s,j} - \Theta\,| < \delta$, might have been recorded, and that it is impossible to determine which of these possible values was actually registered.

In their paper, Vorontsov & Rembovsky do not discuss the nature of the set D of possible values Θ that might be recorded by such an apparatus. Since the device has a tolerance error of δ, the elements of D must be presumed to be spaced over the interval $[-\pi, \pi)$ in such a manner that successive elements of D are at least δ apart (and hence can be distinguished). Additionally, since it must be possible to register every element of D, it must be the case that the interval $(\Theta - \delta, \Theta + \delta)$ contains at least one point $\theta_{s,j}$ for every $\Theta \in D$.

There are many ways in which these requirements can be met. What is clear, however, is that any choice of D which does so must define a device observable which describes the measurement apparatus. Let us make the following simple (and reasonable) choice. Suppose that $\sigma \in \mathbb{N}$, and that $D(\sigma) = \{\Theta_{\sigma,j} : 0 \leqslant j \leqslant \sigma\}$, where

$$\Theta_{\sigma,j} = -\pi + \frac{\pi(2j+1)}{\sigma+1} = \frac{1}{2}\left[\theta_{\sigma,j} + \theta_{\sigma,j+1}\right], \qquad (16.2.18)$$

for $0 \leqslant j \leqslant \sigma$ (where we are writing $\theta_{\sigma,\sigma+1} = \pi$ for convenience). For any $s \geqslant \sigma$, the device observable $X_s(\sigma)$ is defined by introducing the orthogonal projections

$$\Pi_{\sigma,j}^{(s)} = E^{(s)}\left[\theta_{\sigma,j}, \theta_{\sigma,j+1}\right), \qquad 0 \leqslant j \leqslant \sigma, \qquad (16.2.19.a)$$

(where, as usual, $E^{(s)}$ is the spectral measure for the Barnett & Pegg device observable X_s), and defining

$$X_s(\sigma) = \sum_{j=0}^{\sigma} \Theta_{\sigma,j}\, \Pi^{(s)}_{\sigma,j}\,. \tag{16.2.19.b}$$

In other words, the effect of the device observable $X_s(\sigma)$ is to take a measurement of X_s, and to register as output the midpoint $\Theta_{\sigma,j}$ of the subinterval $[\theta_{\sigma,j}, \theta_{\sigma,j+1})$ of $[-\pi, \pi)$ to which that measurement belongs. The nature of the projections $\Pi^{(s)}_{\sigma,j}$ shows how the device observable $X_s(\sigma)$ fails to distinguish between the various points $\theta_{s,j}$ which belong to the interval $[\theta_{\sigma,j}, \theta_{\sigma,j+1})$. Barring some technical concerns about the endpoints of these intervals, the device observable[8] $X_s(\sigma)$ implements the requirements stated above with tolerance error $\delta = \pi/(\sigma+1)$.

Thus, if the initial state of the system is ρ_{in}, then the output state (given a recorded value of $\Theta_{\sigma,j} \in D(\sigma)$ for $X_s(\sigma)$) is

$$\rho^{\sigma,s,j}_{\text{out}} = \frac{\Pi^{(s)}_{\sigma,j}\, \rho_{\text{in}}\, \Pi^{(s)}_{\sigma,j}}{\operatorname{Tr}\left(\Pi^{(s)}_{\sigma,j}\, \rho_{\text{in}}\, \Pi^{(s)}_{\sigma,j}\right)}\,. \tag{16.2.20}$$

Given the nature of the set $D(\sigma)$ outlined above, it follows that for any $\Theta_{\sigma,j} \in D(\sigma)$ we can find integers $0 \leqslant m^{(s)}_{\sigma,j} \leqslant M^{(s)}_{\sigma,j} \leqslant s$ such that

$$\left\{0 \leqslant k \leqslant s : \theta_{s,k} \in \left[\theta_{\sigma,j}, \theta_{\sigma,j+1}\right)\right\} = \left\{m^{(s)}_{\sigma,j}, m^{(s)}_{\sigma,j}+1, \ldots, M^{(s)}_{\sigma,j}\right\}, \tag{16.2.21}$$

in which case

$$\Pi^{(s)}_{\sigma,j} = \sum_{k=m^{(s)}_{\sigma,j}}^{M^{(s)}_{\sigma,j}} P_{s,k}. \tag{16.2.22}$$

---

[8] It is worth noting that the device observable $X_s(\sigma)$ can (almost) be obtained from X_s via the spectral calculus, in a manner similar to that discussed in Subsection 16.1.2, for $X_s(\sigma)$ is almost the same as $h_\sigma(X_s)$, where $h_\sigma : [-\pi, \pi] \to \mathbb{R}$ is the function

$$h_\sigma(t) = \begin{cases} \Theta_{\sigma,j}, & \theta_{\sigma,j} \leqslant t < \theta_{\sigma,j+1},\ 0 \leqslant j \leqslant \sigma, \\ \Theta_{\sigma,\sigma}, & t = \pi. \end{cases}$$

The only difference between $X_s(\sigma)$ and $h_\sigma(X_s)$ is to be found in the action of these operators on states h_n for $n > s$, and these differences will disappear in the limit as $s \to \infty$.

Suppose now that $\rho_{\text{in}} = |\, h_n \rangle \langle h_n \,|$ is the pure state determined by the Hermite-Gauss vector h_n, where $0 \leqslant n \leqslant s$. Then

$$\Pi^{(s)}_{\sigma,j} \rho_{\text{in}} \Pi^{(s)}_{\sigma,j} = |\, \Pi^{(s)}_{\sigma,j} h_n \rangle \langle \Pi^{(s)}_{\sigma,j} h_n \,| ,$$

and so

$$\operatorname{Tr}\left(\Pi^{(s)}_{\sigma,j} \rho_{\text{in}} \Pi^{(s)}_{\sigma,j} \right) = \left\| \Pi^{(s)}_{\sigma,j} h_n \right\|^2 = \frac{1}{s+1} \left(M^{(s)}_{\sigma,j} - m^{(s)}_{\sigma,j} + 1 \right) ,$$

and we observe that $\rho^{\sigma,s,j}_{\text{out}}$ is a pure state.

If, after such an (approximate) measurement of X_s by this apparatus, a subsequent measurement of the number operator N is made, then the probability of recording the integer value $m \geqslant 0$ is given by the formula $p^{(\sigma,s)}_m(j) = \langle h_m, \rho^{\sigma,s,j}_{\text{out}} h_m \rangle$, and this quantity can be shown to be equal to

$$\frac{\sin^2 \left[\dfrac{\pi(n-m)(M^{(s)}_{\sigma,j} - m^{(s)}_{\sigma,j} + 1)}{s+1} \right]}{(s+1)\left(M^{(s)}_{\sigma,j} - m^{(s)}_{\sigma,j} + 1\right) \sin^2 \left[\dfrac{\pi(n-m)}{s+1} \right]} , \tag{16.2.23.a}$$

if $0 \leqslant m \leqslant s$ and $m \neq n$, while it is equal to

$$\frac{1}{s+1} \left(M^{(s)}_{\sigma,j} - m^{(s)}_{\sigma,j} + 1 \right) \tag{16.2.23.b}$$

if $m = n$, and is zero for all $m > s$. From the definition of the integers $m^{(s)}_{\sigma,j}$ and $M^{(s)}_{\sigma,j}$ it is clear that

$$\left| \frac{1}{s+1} \left(M^{(s)}_{\sigma,j} - m^{(s)}_{\sigma,j} + 1 \right) - \frac{1}{\sigma+1} \right| \leqslant \frac{1}{s+1} ,$$

and so, if we define

$$p^{(\sigma,s)}_m = \begin{cases} \dfrac{\sigma+1}{(s+1)^2} \dfrac{\sin^2 \left[\dfrac{\pi(n-m)}{\sigma+1} \right]}{\sin^2 \left[\dfrac{\pi(n-m)}{s+1} \right]} , & 0 \leqslant m \leqslant s,\ m \neq n, \\ \dfrac{1}{\sigma+1} , & m = n, \\ 0, & m > s, \end{cases} \tag{16.2.24}$$

we can then show that

$$\left| p^{(\sigma,s)}_m(j) - p^{(\sigma,s)}_m \right| \leqslant \frac{Am+B}{s+1} , \qquad m \geqslant 0 , \tag{16.2.25}$$

where the constants A and B depend only upon σ and n.

It is often claimed of the Barnett & Pegg device observables that the number operator is uniformly distributed over their eigenstates. This is true when $s = \sigma$, for then[9] $m^{(\sigma)}_{\sigma,j} = M^{(\sigma)}_{\sigma,j} = j$ for any $0 \leqslant j \leqslant \sigma$, which implies that

$$p^{(\sigma,\sigma)}_m(j) \;=\; \begin{cases} \dfrac{1}{\sigma+1}, & 0 \leqslant m \leqslant \sigma, \\ 0, & m > \sigma, \end{cases} \tag{16.2.26}$$

for any $0 \leqslant j \leqslant \sigma$. However, this is not true when $s > \sigma$, and in general the probabilities $p^{(\sigma,s)}_m(j)$ are large when $m - n$ is nearly divisible by $s+1$ (and $0 \leqslant m \leqslant s$), but small for other values of m. This can be seen in the following figure, which shows the relative sizes of $p = p^{(\sigma,s)}_m$ as m ranges from 0 to s, in the particular case that $s = 20000$, $\sigma = 2500$ and $n = 200$. Inevitably, the individual probabilities $p^{(\sigma,s)}_m$ are small, with the largest (in this case) being approximately 4×10^{-4}.

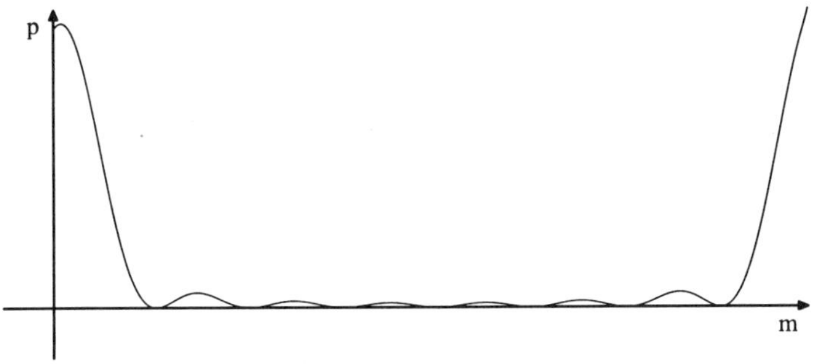

Fig. 16.1 Relative Values of the Probabilities $p^{(\sigma,s)}_m$.

Consequently, even if the initial state ρ_{in} is associated with a small eigenvalue n of the number operator, there is always a high probability of a measurement of N — at the end of this dual measurement process — yielding large values of the order of $s+1$, *no matter how large s is.* This is due to the fact that the (exponential of the) observable X_s links the ground eigenstate h_0 directly to the eigenstate h_s, and consequently the Barnett &

[9]This is not surprising, for

$$X_\sigma(\sigma) \;=\; X_\sigma + \frac{\pi}{\sigma+1}\,I\,,$$

and so measuring $X_\sigma(\sigma)$ is (effectively) making a perfect measurement of X_σ.

Pegg approximate device observables $X_s(\sigma)$ act to "pump information to infinity". One consequence of this can be a loss of probability, which would be nonphysical.

To see this last point, it is important to note that formulæ (16.2.24) and (16.2.25) are valid uniformly for all values of $\Theta_{\sigma,j} \in D(\sigma)$. Let us modify the previous experiment, and consider the situation where the input beam $\rho_{\text{in}} = |\, h_n \rangle \langle h_n \,|$ is passed through the apparatus which makes the approximate measurement of X_s by measuring $X_s(\sigma)$, and is then passed to the device measuring the number operator, *without the outcome of the first measurement being registered.* Since the intermediate value $\Theta_{\sigma,j}$ in $D(\sigma)$ is now not known, the probabilities for registering the various possible values for the number operator will have to be weighted averages of the probabilities $p_m^{(\sigma,s)}(j)$ considered above. Since we know nothing about the first measurement, we do not know the nature of the weighted average, but this does not matter, since the uniformity of (16.2.24) and (16.2.25) with respect to j tells us that, no matter what weighted sum over $D(\sigma)$ is used, the probability of recording a value of $m \geqslant 0$ for the number operator will be an expression of the form

$$p_m^{(\sigma,s)} + E_m^{(\sigma,s)} ,$$

where

$$\left| E_m^{(\sigma,s)} \right| \leqslant \frac{Am+B}{s+1}, \qquad m \geqslant 0 .$$

Having calculated these probabilities, Barnett & Pegg theory requires that the limit as $s \to \infty$ be taken. Then the "probability" of recording a value of $m \geqslant 0$ for the number operator after this dual measurement process would be

$$\begin{aligned} p_m^{(\sigma)} &= \lim_{s\to\infty} p_m^{(\sigma,s)} \\ &= \begin{cases} \dfrac{\sigma+1}{\pi^2(n-m)^2} \sin^2 \left[\dfrac{\pi(n-m)}{\sigma+1}\right], & m \neq n, \\ \dfrac{1}{\sigma+1}, & m = n. \end{cases} \end{aligned} \tag{16.2.27}$$

Since the results of the first experiment are not recorded, we might initially expect that the first measurement has no effect on the second, in which case we would have $p_m^{(\sigma)} = \delta_{mn}$, which is clearly not the case. However, taking into account the observations made in Section 6.3 concerning the lack of strict repeatability for approximate observables, it would be more reasonable to expect the numbers $p_m^{(\sigma)}$ to be large when m is close to n, and small for other values of m. This is indeed the case. However, and crucially, the above "probabilities" $p_m^{(\sigma)}$ do not define a probability distribution, since elementary Fourier analysis shows that

$$\begin{aligned}\sum_{m=0}^{\infty} p_m^{(\sigma)} &= \frac{1}{\sigma+1} + \frac{\sigma+1}{\pi^2}\sum_{m=1}^{n}\frac{1}{m^2}\sin^2\left[\frac{\pi m}{\sigma+1}\right] \\ &\quad + \frac{\sigma+1}{\pi^2}\sum_{m=1}^{\infty}\frac{1}{m^2}\sin^2\left[\frac{\pi m}{\sigma+1}\right] \\ &= \frac{\sigma+2}{2\sigma+2} + \frac{\sigma+1}{\pi^2}\sum_{m=1}^{n}\frac{1}{m^2}\sin^2\left[\frac{\pi m}{\sigma+1}\right],\end{aligned}$$

which is always less than 1 — indeed, this expression only tends to 1 as $n \to \infty$. This time, then, we see that the linking that X_s creates between low and high eigenstates of the number operator has resulted in a loss of probability.

A recent paper by Vaccaro, Pegg & Barnett [226] has called into question some of the premises of the argument of Vorontsov & Rembovsky, and suggests possible alternative measurement systems which would provide more satisfactory results under such successive measurement arrangements. In particular they suggest, in effect, that the device observable $X_s(\sigma)$ should be replaced by $P^{(R)}\, X_s(\sigma)\, P^{(R)}$ for some positive integer $R \leqslant s$, on the grounds that only a finite range measurements of the number operator N should be possible in a given experiment, and that the integer R indicates the maximum possible value that can be registered for the number operator R. Such a device observable, they claim, would produce physical reasonable results.

In this description, were a measurement of $\Theta_{\sigma,j}$ to be recorded, when the input state was ρ_{in}, then the new output state would be

$$\rho_{\text{out}}^{R,\sigma,s,j} = \frac{P^{(R)}\Pi_{\sigma,j}^{(s)}P^{(R)}\,\rho_{\text{in}}\,P^{(R)}\Pi_{\sigma,j}^{(s)}P^{(R)}}{\mathrm{Tr}\left(P^{(R)}\Pi_{\sigma,j}^{(s)}P^{(R)}\,\rho_{\text{in}}\,p^{(R)}\Pi_{\sigma,j}^{(s)}P^{(R)}\right)}, \tag{16.2.28}$$

and the probability $\left\langle h_m, \rho^{R,\sigma,s,j} h_m\right\rangle$ of recording a value of m for a subsequent measurement of the number operator would then be

$$p_m^{(R,\sigma,s)}(j) = \begin{cases} \dfrac{\left\|\Pi_{\sigma,j}^{(s)}h_n\right\|^2}{\left\|P^{(R)}\Pi_{\sigma,j}^{(s)}h_n\right\|^2} p_m^{(\sigma,s)}(j), & 0 \leqslant m \leqslant R, \\ 0, & m > R, \end{cases}$$

$$= \begin{cases} \left[\sum_{l=0}^{R} p_l^{(\sigma,s)}(j)\right]^{-1} p_m^{(\sigma,s)}(j), & 0 \leqslant m \leqslant r, \\ 0, & m > R, \end{cases} \quad (16.2.29)$$

and we can show that the difference between this expression and

$$p_m^{(R,\sigma,s)} = \begin{cases} \left[\sum_{l=0}^{R} p_l^{(\sigma,s)}\right]^{-1} p_m^{(\sigma,s)}, & 0 \leqslant m \leqslant r, \\ 0, & m > R, \end{cases} \quad (16.2.30)$$

is smaller than a term of the form $C(s+1)^{-1}$ uniformly in m and j, where C is a positive constant that depends only upon R, σ and n. If we now consider the experiment where the outcome of the first measurement is passed to a measurement of the number operator, without the result of the first measurement being recorded, then the "probabilities" of recording a value m for the operator N can be shown to be (in the limit as $s \to \infty$),

$$p_m^{(R,\sigma)} = \begin{cases} \left[\sum_{l=0}^{R} p_l^{(\sigma)}\right]^{-1} p_m^{(\sigma)}, & 0 \leqslant m \leqslant R, \\ 0, & m > R, \end{cases} \quad (16.2.31)$$

and in this case we see that, since the values $p_m^{(R,\sigma)}$ add up to 1, there is no loss of probability. Moreover, these probabilities are large for values of m close to n, and are small for other values. At first sight, these observations seem promising.

This paper has been replied to by Vorontsov & Rembovsky [231]. They claim that the operator $P^{(R)}\,X_s(\sigma)\,P^{(R)}$ is not a valid device observable, since the projections $P^{(R)}\,\Pi_{\sigma,j}^{(s)}\,P^{(R)}$, $0 \leqslant j \leqslant s$ do not form a partition of unity. This is true, but then neither do the projections $\Pi_{\sigma,j}^{(s)}$, $0 \leqslant j \leqslant s$. However the first collection of projections, together with $I - P^{(R)}$, do form a partition of unity, just as the second collection of projections form a partition of unity in conjunction with $I - P^{(s)}$, so this particular objection is spurious. From the point of view of the discussion in this Chapter, however, an important objection lies in the fact that the projections $P^{(R)}\Pi_{\sigma,j}^{(s)}P^{(R)}$

are no longer mutually orthogonal, and hence the operator $P^{(R)}\,X_s(\sigma)\,P^{(R)}$ is no longer a device observable.

However Vorontsov & Rembovsky make other observations which are important. Firstly, they cast doubt on to what extent the proposed observable $P^{(R)}\,X_s(\sigma)\,P^{(R)}$ is physical, since it is not at all clear how a single measurement apparatus can be designed which measures phase angles while at the same time making restrictions on the nature of the possible values that can be measured for the number operator. (It is important to note in this regard that the proposed restriction on the number operator indicated by the integer R is an intrinsic part of the device observable $X_s(\sigma)$, and is not a feature of the second measurement apparatus to determine the value of the number operator N.) Secondly they emphasize that the definition of the operator X_s creates a link between the states h_0 and h_s, and therefore the difficulties of the nature outlined above are intrinsic to the definition of X_s, and are not special cases caused by unlikely designs of measurement apparatus. They argue that if X_s represents a physical quantity, then the device observable $X_s(\sigma)$ is a physically reasonable model of a measurement apparatus related to it. Whether physical or not, and whether it provides results in accord with experimental data or not, the putative validity of the observable $P^{(R)}\,X_s(\sigma)\,P^{(R)}$ does not affect the fact that the theory of Barnett & Pegg does not handle the device observable $X_s(\sigma)$ correctly, and it should.

It is reasonable, however, to presume that there ought to be limits on the measurement capability of any device designed to measure the number operator N. Since N has discrete spectrum with no limit points, a perfect measurement apparatus for N would be able to distinguish between eigenvalues absolutely, but it is perhaps unreasonable to expect the apparatus to be able to detect and determine any eigenvalue of N, no matter how large. This is the reason why Vaccaro, Barnett & Pegg wish to introduce the parameter R into the calculations. However, the problem lies in where that parameter should be introduced. It is the view of Vorontsov & Rembovsky that this parameter should not be introduced during the process of measuring phase angle. Perhaps, instead, it should be introduced when measuring the number operator itself.

Suppose that a device observable is designed which can register accurately the eigenvalues $0, 1, \ldots, R$ of the number operator, but which can not distinguish larger eigenvalues. What, then, is this device to do with these larger eigenvalues? It is reasonable to suppose that such a device might be able to record such values as being beyond its range to measure (perhaps even by blowing a fuse), and we choose to indicate this situation by letting the device observable assign the value -1 as the output of all such measurements. In other words, we consider the device observable

$$N_R = P^{(R)} N P^{(R)} - (I - P^{(R)}). \tag{16.2.32}$$

Now consider the situation in which the input state ρ_{in} is passed through the apparatus measuring the device observable $X_s(\sigma)$, with the resultant output state of that measurement being passed to the device measuring N_R, again without the result of the first angular measurement being recorded. Since the eigenvalue -1 of N_R merely exists to record measurements which are out of range, only those measurements which record a value within the acceptable range $0, 1, \ldots, R$ of eigenvalues are of interest. Thus the probabilities of interest are those of recording a particular nonnegative eigenvalue m, *given that some nonnegative eigenvalue has been recorded.* It is easy to show that these conditional probabilities are identical with those given in equation (16.2.31). Thus it is possible to envisage a system of measurement in which tolerance errors relating to angular measurement only affect measurements of angle (in that the device observable $X_s(\sigma)$ is used), and tolerance errors relating to measurement of the number operator only appear when measuring that operator (through the use of the device observable N_R), and which yet reproduces the calculational results of the formulation of Vaccaro, Barnett & Pegg. This approach would, we believe, overcome the physicality objections that Vorontsov & Rembovsky have to the formalism suggested in [226], and yet retain the results of that formalism.

However, the question still remains — what is meant by saying that the results of this theory are physical? If working with the operators $P^{(R)} X_s(\sigma) P^{(R)}$ yields physically interesting results, what observable are they measuring, and to what extent does that observable describe quantum phase? Unlike the standard theory of Barnett & Pegg, where the equivalent question is open to some debate, there is no problem in identifying the unique self-adjoint operator that is being described by this procedure. For any $m, n \geqslant 0$ and any $0 \leqslant j \leqslant \sigma$ we calculate that

$$\lim_{s\to\infty} \langle h_m, \Pi^{(s)}_{\sigma,j} h_n \rangle = \begin{cases} \dfrac{i^{n-m}}{\pi(n-m)} e^{\frac{\pi i(n-m)(2j+1)}{\sigma+1}} \sin\left[\dfrac{\pi(n-m)}{\sigma+1}\right], & m \neq n, \\ \dfrac{1}{\sigma+1}, & m = n. \end{cases}$$

Since, for any $m, n \geqslant 0$,

$$\lim_{s\to\infty} \langle h_m, X_s(\sigma) h_n \rangle = \lim_{s\to\infty} \sum_{j=0}^{\sigma} \Theta_{\sigma,j} \langle h_m, \Pi^{(s)}_{\sigma,j} h_n \rangle,$$

we deduce that the uniformly bounded sequence of operators $\big(X_s(\sigma)\big)_{s\geqslant 1}$ converges weakly to the bounded operator $X^{(\sigma)}$, where

$$\langle h_m, X^{(\sigma)} h_n \rangle = \begin{cases} \dfrac{i^{n-m+1}}{m-n}, & \sigma + 1 \not| \, m - n, \\ 0, & \text{otherwise}, \end{cases} \tag{16.2.33}$$

and that the sequence of operators $\big(P^{(R)} X_s(\sigma) P^{(R)}\big)_{s\geqslant 1}$ converges weakly to $P^{(R)} X^{(\sigma)} P^{(R)}$. But since each of the operators $P^{(R)} X_s(\sigma) P^{(R)}$ is of finite rank, being essentially an operator on the finite dimensional subspace $\mathcal{H}^{(R)}$ of $L^2(\mathbb{R})$ spanned by the Hermite-Gauss vectors $h_0, \ldots, h_R$, it follows that the sequence $\big(P^{(R)} X_s(\sigma) P^{(R)}\big)_{s\geqslant 1}$ converges strongly, even uniformly, to $P^{(R)} X^{(\sigma)} P^{(R)}$. Given the good nature of this convergence, the calculational approach of Barnett & Pegg can be discarded, since all expectations and variances obtained for the operators $P^{(R)} X_s(\sigma) P^{(R)}$ (for finite s) yield, in the limit as $s \to \infty$, the corresponding expectations and variances for the single operator $P^{(R)} X^{(\sigma)} P^{(R)}$. When all is said and done, therefore, this formalism is simply asking questions about the operator $P^{(R)} X^{(\sigma)} P^{(R)}$, and the explicit s-dependence of the theory of Barnett & Pegg is unnecessary.

What, then, is the operator $P^{(R)} X^{(\sigma)} P^{(R)}$, and to what extent is it a good descriptor of phase angle? Interestingly, this question depends upon the relative values of σ and R. For example, if $\sigma \geqslant R$, then inspection of equation (16.2.33) shows that

$$P^{(R)} X^{(\sigma)} P^{(R)} = P^{(R)} X P^{(R)}$$

is just the truncation to $\mathcal{H}^{(R)}$ of the Toeplitz phase operator X. As we have argued previously, X is an interesting but (we believe) inappropriate operator with which to describe quantum phase phenomena, so we do not feel that even this modification to the theory of Barnett & Pegg explains quantum phase adequately.

It is interesting to note that, when $\sigma \geqslant R$, the resulting operator is independent of σ, and so the concerns over tolerances in angular measurements have no affect on the outcome. It might be argued, therefore, that σ should be less than R, in which case the operator $P^{(R)}\, X^{(\sigma)}\, P^{(R)}$ is explicitly dependent upon both σ and R. However, even in this case, the resulting operator is still only a comparatively small modification of $P^{(R)}\, X\, P^{(R)}$, so we do not expect to find a close connection between this operator and the Weyl quantization of any function of angle.

Since the parameters σ and R in the above discussion impose limits on our ability to measure phase quantities and the number operator exactly, we should expect that, in the limit as σ and R both tend to infinity, we retrieve the ideal position where perfect measurements are possible. It is clear from the formulæ given above that the operators $P^{(R)}\, X^{(\sigma)}\, P^{(R)}$ converge weakly to the Toeplitz phase operator X as σ and R tend to infinity, no matter the manner in which these two parameters do so. We see this observation as further evidence that all of the formalism considered in this Section provides a complicated system for measuring the Toeplitz phase operator X, and no more.

Although the problem of approximate measurements is still far from resolved, the current discussion has raised the important issue of the need to consider tolerance errors for the number operator as well as for angular measurements, and it is to be hoped that a satisfactory theory can be found which accommodates all of the above concerns.

Bibliography

ỉw=f pw ḥꜣt=f r pḥ(wy)=fy mỉ gmyt m sš

It [the book] is finished, (from) its start to its finish, as was found in writing.

– *The Shipwrecked Sailor*, Leningrad Papyrus 1115, 186-87.

[1] Abraham R, Marsden JE & Ratiu T. *Manifolds, Tensor Analysis, and Applications* (Springer-Verlag, New York, 1988), 2nd edition.

[2] Aizenman M, Goldstein S, Lebowitz JL & Martin P. *On the equivalence between KMS-states and equilibrium states for classical systems.* Commun Math Phys, **138**, 209–220 (1977).

[3] Akhiezer NI & Glazman M. *Theory of Linear Operators in Hilbert Space, I& II* (Pitman, Boston, 1981), 2nd edition.

[4] Alicki R & Lendi K. *Quantum Dynamical Semigroups and Applications.* Lecture Notes in Physics 286 (Springer, Berlin, 1987).

[5] Alli G & Sewell GL. *New methods and structures in the theory of the multi-mode Dicke laser model.* J Math Phys, **36**, 5598–5626 (1995).

[6] Antoine JP, Inoue A & Trapani C. *Partial *-algebras of closable operators: a review.* Rev Math Phys, **8**, 1–42 (1996).

[7] Arnold V. *Méthodes Mathématiques de la Méchanique Classique* (MIR, Moscow, 1976).

[8] Aspect A, Dalibard J & Roger G. *Experimental tests of Bell's inequalities using time-varying analysers.* Phys Rev Lett, **49**, 1804–1807 (1982).

[9] Aspect A, Grangier P & Roger G. *Experimental tests of realistic local theories via Bell's theorem.* Phys Rev Lett, **47**, 460–463 (1981).

[10] —. *Experimental realization of EPR Gedankenexperiment: a new violation of Bell's inequalities.* Phys Rev Lett, **49**, 91–94 (1982).

[11] Auslander L. *Differential Geometry* (Harper and Row, New York, 1976).

[12] Bargmann V. *On a Hilbert space of analytic functions and an associated integral transform.* Commun Pure Appl Math, **14**, 187–214 (1961).

[13] —. *Remarks on a Hilbert space of analytic functions*. Proc Natl Acad Sci, **48**, 199–204 (1962).

[14] Barnett SM & Pegg DT. *On the Hermitian optical phase operator*. J Mod Optics, **36**, 7–19 (1989).

[15] Basov NG & Prokhorov AM. *Application of molecular beams to the radio spectroscopic study of the rotation spectra of molecules*. Zh Exsp Teor Fiz, **27**, 431–438 (1954).

[16] Baumgärtel H & Wollenberg M. *Causal Nets of Operator Algebras* (Akademie Verlag, Berlin, 1992).

[17] Bell JS. *On the Einstein Podolsky Rosen paradox*. Physics, **1**, 195–200 (1964).

[18] —. *Speakable and Unspeakable in Quantum Mechanics* (Cambridge University Press, Cambridge, 1987).

[19] Berezin FA & Shubin MA. *The Schrödinger Equation* (Kluwer, Dordrecht, 1991).

[20] Birkhoff G. *Lattice Theory* (Amer. Math. Soc., Providence, 1967).

[21] Birkhoff G & von Neumann J. *The logic of quantum mechanics*. Ann Math, **37**, 823–843 (1936).

[22] Bogolubov NN, Logunov AA, Oksak AI & Todorov IT. *General Principles of Quantum Field Theory* (Kluwer, Dordrecht, 1990).

[23] Bohm D. *Quantum Theory* (Prentice Hall, Englewood Cliffs, NJ, 1951).

[24] —. *Hidden variables and the implicate order,*. In *Quantum Implications*, editors Hiley BJ & Peat FD, 33–45 (Routledge, London, 1987).

[25] Böttcher A & Silbermann B. *Analysis of Toeplitz Operators* (Springer-Verlag, Berlin, 1990).

[26] Bourbaki N. *Topological Vector Spaces* (Springer-Verlag, Berlin, 1987).

[27] Braginsky VB & Khalili FY. *Quantum Measurement* (Cambridge University Press, Cambridge, 1992).

[28] Bratteli O & Robinson DW. *Operator Algebras and Quantum Statistical Mechanics, I, II* (Springer Verlag, New York, 1979, 1981).

[29] Brown A & Halmos P. *Algebraic properties of Toeplitz operators*. J Reine Agnew Math, **231**, 1–2, 89–102 (1963).

[30] Busch P, Grabowski M & Lahti PJ. *Operational Quantum Physics* (Springer-Verlag, Berlin, 1995).

[31] Cartan H. *Differential Forms* (Hermann, Paris, 1970).

[32] —. *Differential Calculus* (Hermann, Paris, 1971).

[33] Chari V & Pressley AN. *A Guide to Quantum Groups* (Cambridge UP, Cambridge, 1994).

[34] Chernoff PR. *Mathematical obstructions to quantization*. Hadronic Journal, **4**, 879–898 (1981).

[35] Connes A. *Noncommutative Geometry* (Academic Press, San Diego, 1994).

[36] Copson ET. *Asymptotic Expansions* (Cambridge University Press, Cambridge, 1965).

[37] Daubechies I. *On the distributions corresponding to bounded operators in*

the Weyl quantization. Commun Math Phys, **75**, 229–238 (1980).

[38] —. *Continuity statements and counterintuitive examples in connection with Weyl quantization*. J Math Phys, **24**, 1453–1461 (1983).

[39] DAVEY BA & PRIESTLEY HA. *Introduction to Lattices and Order* (Cambridge UP, Cambridge, 1990).

[40] DAVIES EB. *Markovian master equations*. Commun Math Phys, **39**, 91–110 (1974).

[41] —. *Quantum Theory of Open Systems* (Academic Press, London, 1976).

[42] DAVIS CC. *Lasers and Electro-Optics* (Cambridge University Press, Cambridge, 1996).

[43] DE MOEN B, VANHEUVERWIJN P & VERBEURE A. *Completely positive quasi-free maps of the CCR-algbra*. Rep Math Phys, **15**, 27–39 (1979).

[44] DEGROOT SR & SUTTORP LG. *Foundations of Electrodynamics* (North Holland, Amsterdam, 1972).

[45] DELIGNE P, ETINGOF P, FREED DS, JEFFREY LC, KAZHDAN D, MORGAN JW, D RM & WITTEN E, editors. *Quantum Fields and Strings: A Course for Mathematicians, I, II* (AMS, IAS, Princeton, 1999).

[46] DICKE R. *Coherence in spontaneous radiation processes*. Phys Rev, **93**, 99–110 (1954).

[47] DIRAC P. *The quantum theory of emission and absorption*. Proc Roy Soc London, **A114**, 243–265 (1927).

[48] —. *Lectures on modern quantum mechanics*. Unpublished, Dirac Archive, Churchill College, Cambridge (1928). Michaelmas term.

[49] DIRAC PAM. *The Principles of Quantum Mechanics* (Oxford at the Clarendon Press, Oxford, 1930, 1935, 1947, 1958, 1967), 1st edition.

[50] —. *Lectures on Quantum Field Theory* (Belfer Graduate School of Science, Yeshiva U, New York, 1966).

[51] DUBIN DA. *Solvable Models in Algebraic Statistical Mechanics* (Clarendon, Oxford, 1974).

[52] DUBIN DA & HENNINGS MA. *Quantum Mechanics, Algebras and Distributions*. Pitman Research Notes in Mathematics 238 (Longman, Harlow, Essex, 1990).

[53] DUBIN DA, HENNINGS MA & SMITH TB. *Quantization in polar coordinates and the phase operator*. Proc RIMS Kyoto, **30**, 479–532 (1994).

[54] —. *Mathematical theories of phase*. Int J Mod Phys B, **9**, 2597–2687 (1995).

[55] DUNFORD N, SCHWARTZ JT, BADE WG & BARTLE RG. *Linear Operators, I-III*. Wiley Classics Library (Interscience Publishers, New York, 1988).

[56] DUREN PL. *Theory of H^p Spaces* (Academic Press, New York, 1970).

[57] EMCH GG. *Algebraic Methods in Statistical Mechanics and Quantum Field Theory* (Wiley Interscience, New York, 1972).

[58] FARIS W. *Review of Roland Omnès's "The Interpretation of Quantum Mechanics"*. Notices of the AMS, **43**, 1328–13397 (1996).

[59] FEYNMAN RP. *Space-time approach to non-relativistic quantum mechanics*.

Rev Mod Phys, **76**, 367–387 (1948).

[60] —. *The Character of Physical Law* (MIT Press, Cambridge, Mass., 1965).

[61] FEYNMAN RP & HIBBS RG. *Quantum Mechanics and Path Integrals* (McGraw-Hill, New York, 1965).

[62] FLANDERS H. *Differential Forms* (Academic Press, New York, 1963).

[63] FOLLAND GB. *Harmonic Analysis in Phase Space* (Princeton UP, Princeton, N. J., 1989).

[64] FREYBERGER M & SCHLEICH W. *Photon counting, quantum phase, and phase-space distributions*. Phys Rev A, **47**, 30–33 (1993).

[65] FUCHS J. *Affine Lie Algebras and Quantum Groups* (Cambridge UP, Cambridge, 1992).

[66] GALINDO A. *Phase and number*. Lett Math Phys, **8**, 495–500 (1984).

[67] GALLAVOTTI G. *The Elements of Mechanics* (Springer-Verlag, New York, 1983).

[68] GARRISON JC & WONG J. *Canonically conjugate pairs, uncertainty relations, and phase operators,*. J Math Phys, **11**, 53–60 (1970).

[69] GELFAND IM, GRAEV MI & VILENKIN NY. *Generalized Functions: Integral Geometry and Representation Theory* (Academic Press, New York, 1964).

[70] GELFAND IM & SHILOV GE. *Generalized Functions: Properties and Operations* (Academic Press, New York, 1964).

[71] —. *Generalized Functions: Theory of Differential Equations* (Academic Press, New York, 1967).

[72] —. *Generalized Functions: Spaces of Fundamental and Generalized Functions* (Academic Press, New York, 1968).

[73] GELFAND IM & VILENKIN NY. *Generalized Functions: Applications of Harmonic Analysis* (Academic Press, New York, 1964).

[74] GERSTENHABER M & SCHACK SD. *Algebras, bialgebras, quantum groups, and algebraic deformations*. Contemporary Mathematics, **134**, 51–92 (1992).

[75] GIBBS JW. *Elementary Principles of Statistical Mechanics* (Dover Publ. (orig. Yale University Press, 1902), New York, 1960).

[76] GLAUBER R. *The quantum theory of optical coherence*. Phys Rev, **130**, 2529–2539 (1963).

[77] GLAUBER RJ. *Coherent and incoherent states of the radiation field*. Phys Rev, **131**, 2766–2788 (1963).

[78] —. *Coherence and quantum detection*. In *Quantum Optics* (Academic Press, New York, 1969).

[79] GLEASON AM. *Measures on the closed subspaces of a Hilbert space*. J Math and Mechanics, **6**, 885–893 (1957).

[80] GLIMM J & JAFFE A. *Quantum Physics, A Functional Integral Point of View* (Springer Verlag, New York, 1987), 2nd edition.

[81] GOHBERG I, GOLDBERG S & KAASHOEK MA. *Classes of Linear Operaors, I* (Birkhauser Verlag, Basel, 1990).

[82] Goldstein H. *Classical Mechanics* (Addison Wesley, Reading, Mass, 1950).

[83] Gordon JP, Zeiger HJ & Townes CH. *Molecular microwave oscillator and new hyperfine structure in the microwave spectrum of* NH_3. Phys Rev, **95**, 282–284 (1954).

[84] —. *The maser - new type of microwave amplifier, frequency standard, and spectrometer*. Phys Rev, **99**, 1264–1274 (1955).

[85] Gorini V, Kossakowski A & Sudarshan ECG. *Completely positive dynamical semigroups on N-level systems*. J Math Phys, **17**, 821–825 (1976).

[86] Gotay MJ. *Functional geometric quantization and Van Hove's theorem.* Int J Theor Phys, **19**, 139–161 (1980).

[87] Grabowski M. *On the phase operator*. Rep Math Phys, **29**, 377–382 (1991).

[88] Graham R & Haaken H. *Laser light - first examples of a second order phase transition far away from equilibrium*. Zeits Phys, **237**, 31–46 (1970).

[89] Greub WH. *Multilinear Algebra* (Springer-Verlag, Berlin, 1967).

[90] Groenewald HJ. *On the principles of elementary quantum mechanics*. Physica, **12**, 405–460 (1946).

[91] Grothendieck A. *Topological Vector Spaces* (Gordon-Breach, New York, 1973).

[92] Gudder S. *Quantum Probability* (Academic Press, Princeton, NJ, 1988).

[93] Guillemin V & Sternberg S. *Symplectic Techniques in Physics* (Cambridge UP, Cambridge, 1984).

[94] Haag R. *Local Quantum Physics* (Springer-Verlag, Berlin, 1992).

[95] Haken H. *Analogy between higher instabilities in fluids and lasers*. Phys Rev Lett, **53A**, 77–78 (1954).

[96] —. *Cooperative phenomena in systems far from thermal equilibrium and in non-physical systems*. Rev Mod Phys, **47**, 67–121 (1975).

[97] Halmos PR. *A Hilbert Space Problem Book, 2nd ed.* (Springer-Verlag, New York, 1982).

[98] Halmos PR & Sunder VS. *Bounded Integral Operators on* L^2 *Spaces* (McGraw-Hill, New York, 1974).

[99] Hannabuss K. *An Introduction to Quantum Theory* (Clarendon Press, Oxford, 1997).

[100] Hansen F. *The Moyal product and spectral theory for a class of infinite dimensional matrices*. Publ RIMS Kyoto Univ, **26**, 885–933 (1990).

[101] Harbison JP & Nahory RE. *Lasers: Harnessing the Atom's light* (Scientific American Library (HPHLP), New York, 1998).

[102] Hardy GH. *A Mathematician's Apology* (Cambridge UP, Cambridge, 1969).

[103] Heisenberg W. *Über quantentheoretische Umdeutung kinematischer und mechanischer Beziehungen*. Z Phys, **33**, 879–893 (1925).

[104] —. *The Physical Principles of the Quantum Theory* (Dover, New York, 1949).

[105] HEITLER W. *The Quantum Theory of Radiation* (Oxford University Press, Fair Lawn, NJ, 1944), 2nd edition.

[106] HELEMSKII AY. *Banach and Locally Convex Algebras* (Clarendon, Oxford, 1993).

[107] HENNEAUX M & TEITELBOIM C. *Quantization of Gauge Systems* (Princeton UP, Princeton, NJ, 1992).

[108] HENNINGS MA. *Some ideas concerning Hansen's version of spectral theory and the Moyal product* (1993). Preprint *Sidney Sussex College, University of Cambridge.*

[109] HENNINGS MA, DUBIN DA & SMITH TB. *Dequantization techniques for Weyl quantization*. Proc RIMS Kyoto, **34**, 325–354 (1998).

[110] —. *Existence theorems for ordered variants of Weyl quantization*. Proc RIMS Kyoto, **35**, 1–29 (1999).

[111] —. *Extended dynamics for the laser model* (1999). In preparation.

[112] HENNINGS MA, SMITH TB & DUBIN DA. *Approximations to the phase operator*. J Phys A, **28**, 6809–6856 (1995).

[113] —. *Asymptotics for the phase operator*. J Phys A, **28**, 6779–6807 (1995).

[114] HEPP K. *Quantum theory of measurement and macroscopic observables*. Helv Physica Acta, **45**, 237–248 (1972).

[115] HEPP K & LIEB E. *Phase transitions in reservoir driven open systems with applications to lasers and super-conductors*. Helv Physica Acta, **46**, 573–603 (1973).

[116] H.HAKEN. *Laser Theory*. In *Handbuch der Physik, XXV/2C* (Springer-Verlag, Berlin, 1970).

[117] HILBERT D. *Grundzüge einer allegemeinen Theorie der linearen Integralgleichungen* (Teubner, Leipzig, 1912).

[118] HOFFMAN K. *Banach Spaces of Analytic Functions* (Dover, New York, 1988).

[119] HOLEVO AS. *Probabilistic and Statistical Aspects of Quantum Theory* (North-Holland, Amsterdam, 1982).

[120] HÖRMANDER L. *The Analysis of Linear Partial Differential Operators, I* (Springer-Verlag, Berlin, 1990), 2nd edition.

[121] HOVE LV. *Sur certaines représentations unitaires d'un groupe infini de transformations*. Mem Acad Roy de Belgique, **Classe des Sci. 26** (1951).

[122] —. *Quantum mechanical perturbations giving rise to a statistical transport equation*. Physica, **21**, 517–540 (1955).

[123] HUDSON RL. *When is the Wigner quasi-probability density non-negative?*. Rep Math Phys, **6**, 249–252 (1974).

[124] ISHAM CJ. *Lectures on Quantum Theory* (Imperial College Press, London, 1995).

[125] ISRAEL RB. *Convexity in the Theory of Lattice Gases* (Princeton U. P.,

Princeton, NJ, 1979).
[126] ITÔ K, editor. *Encyclopedic Dictionary of Mathematics,II ed.* (MIT Press *for* Mathematical Society of Japan, Cambridge, Mass., 1986).
[127] ITZYKSON C & ZUBER JB. *Quantum Field Theory* (McGraw-Hill, New York, 1985).
[128] JAMMER M. *The Conceptual Development of Quantum Mechanics* (McGraw-Hill, New York, 1966).
[129] —. *The Philosophy of Quantum Mechanics* (John Wiley & sons, New York, 1974).
[130] JARCHOW H. *Locally Convex Spaces* (Teubner, Stuttgart, 1981).
[131] JAUCH JM. *Foundations of Quantum Mechanics* (Addison Wesley, Reading, Mass, 1968).
[132] JAVAN A, W. R. BENNET J & HERROTT DR. *Population inversion and continuous optical maser oscillation in a gas discharge containing a He-Ne mixture.* Phys Rev Lett, **6**, 106–110 (1961).
[133] JOSEPH A. *Quantum Groups and their Primitive Ideals* (Springer, Berlin, 1995).
[134] JOST R. *The General Theory of Quantum Fields* (AMS, Providence, 1965).
[135] KAMMERER JB. *L'algèbre des opérateurs de multiplication du star-produit de* $\mathbb{R}^{2n}$. C R Acad Sci Paris Ser I Math, **298**, 59–62 (1984).
[136] KATZ A. *An infinite system of classical particles.* J Math Phys, **8**, 2451–2459 (1967).
[137] KEMBLE EC. *The Fundamental Principles of Quantum Mechanics* (Dover, New York, 1958).
[138] KHINCHIN AI. *Mathematical Foundations of Statistical Mechanics* (Dover, New York, 1949).
[139] KLAUDER JR. *The design of radar signals having both high range resolution and high velocity resolution.* Bell System Tech J, **39**, 809–820 (1960).
[140] KLIMYK A & SCHMUEDGEN K. *Quantum Groups and their Representations* (Springer, Berlin, 1997).
[141] KOSSAKOWSKI A, FRIGERIO A, GORINI V & VERRI M. *Quantum detailed balance and the KMS condition.* Commun Math Phys, **57**, 97–110 (1977).
[142] KÖTHE G. *Topological Vector Spaces I,II* (Springer-Verlag, Berlin, 1969, 1979).
[143] KUBO R. *Statistical-mechanical theory of irreversible processes, I.* J Phys Soc Japan, **12**, 570–586 (1957).
[144] KÜMMERER B. *Markoff dilations on* W^**-algebras.* J Functional Anal, **63**, 139–177 (1985).
[145] LAMB JR WE. *Theory of an optical Maser.* Phys Rev, **134a**, 1429–1450 (1964).
[146] LANDSMAN NP. *Mathematical Topics Between Classical and Quantum Mechanics* (Springer, New York, 1998).
[147] LASSNER GA. *Operator symbols in the description of observable-state sys-*

tems. Rep Math Phys, **16**, 279 (1979).

[148] LEONHARDT U. *Measuring the Quantum State of Light* (Cambridge UP, Cambridge, 1997).

[149] LERNER EC, HUANG HW & WALTERS GE. *Some mathematical properties of oscillator phase operators*. J Math Phys, **11**, 1679–1684 (1970).

[150] LINDBLAD G. *On the generators of quantum dynamical semigroups*. Commun Math Phys, **48**, 119–130 (1976).

[151] LONDON F. *Über die Jacobischen Transformationen der Quantenmechanik*. Z Phys, **37**, 915–925 (1926).

[152] —. *Winkelvariable und Kanonische Transformationen in der Undulationmechanik*. Z Phys, **40**, 193–210 (1927).

[153] LOUDON R. *The Quantum Theory of Light* (Clarendon, Oxford, 1973).

[154] LOUISELL WH. *Amplitude and phase uncertainty relations*. Phys Lett, **7**, 60–61 (1963).

[155] —. *Quantum Statistical Properties of Radiation* (John Wiley & Sons, New York, 1973).

[156] LÜDERS G. *Über die Zustandsänderung durch Messprozess*. Ann der Physik, **8**, 322–328 (1951).

[157] LUDWIG G. *An Axiomatic Basis for Quantum Mechanics* (Springer-Verlag, Heidelberg, 1987).

[158] LUMER G & PHILLIPS RS. *Dissipative operators in a Banach space*. Pacific J Math, **11**, 679–698 (1961).

[159] LUSTZIG G. *Introduction to Quantum Groups* (Birkhaüser, Boston, 1993).

[160] LYNCH R. *The quantum phase problem: a critical review*. Physics Reports, **256**, 367–436 (1995).

[161] MACKEY G. *The Mathematical Foundations of Quantum Mechanics* (Benjamin, New York, 1963).

[162] MAIMAN TH. *Stimulated optical radiation in ruby*. Nature, **187**, 493–494 (1960).

[163] MALLIOS A. *Topological Algebras* (North-Holland, Amsterdam, 1986).

[164] MANDEL L & WOLF E. *Optical Coherence and Quantum Optics* (Cambridge University Press, Cambridge, 1995).

[165] MANIN YI. *Quantum Groups and Non-commutative Geometry*. Technical report, CRM, Montreal (1988).

[166] MARTI JT. *Introduction to the Theory of Bases* (Springer-Verlag, New York, 1969).

[167] MARTIN DC & SCHWINGER J. *Theory of many-particle systems, I*. Phys Rev, **115**, 1342–1373 (1959).

[168] MITTELSTAEDT P. *The Interpretation of Quantum Mechanics and the Measurement Process* (Cambridge UP, Cambridge, 1998).

[169] MOYAL JE. *Quantum mechanics as a statistical theory*. Proc Cam Soc, **45**, 99–124 (1949).

[170] NAIMARK MA. *On spectral functions of a symmetric operator*. Annals of Math, **7**, 285–296 (1943).

[171] NIETO MM. *Quantum phase and quantum phase operators : some physics and some history*. Physica Scripta, **T48**, 5–12 (1993).

[172] OMNÈS R. *The Interpetation of Quantum Mechanics* (Princeton University Press, Princeton, NJ, 1994).

[173] PALMER PF. *The singular coupling and weak coupling limits*. J Math Phys, **18**, 527–529 (1977).

[174] PALMER TW. *Banach Algebras and the General Theory of *-Algebras, I* (Cambridge University Press, Cambridge, 1994).

[175] PAUL H. *Phase of a microscopic electromagnetic field and its measurement*. Fortschr Physik, **22**, 657–689 (1974).

[176] PEGG DT & BARNETT SM. *Unitary phase operator in quantum mechanics*. Europhys Lett, **6**, 483–487 (1988).

[177] PERESSINI AL. *Ordered Topological Vector Spaces* (Harper and Row, New York, 1967).

[178] PEŘINOVÁ V, LUKŠ A & PEŘINA J. *Phase in Optics* (World Scientific, Singapore, 1998).

[179] PIRON C. *Foundations of Quantum Physics* (Benjamin, New York, 1976).

[180] PITT HR. *Integration, Measure and Probability* (Oliver & Boyd, Edingburgh, 1963).

[181] POOL JCT. *Mathematical aspects of the Weyl correspondence*. J Math Phys, **7**, 66–76 (1966).

[182] POPOV VN & YARUNIN VS. *Quantum and quasi-classical states of the photon phase operator*. J Mod Phys, **39**, 1525–1531 (1992).

[183] PRUGOVECKI E. *Quantum Mechanics in Hilbert Space* (Academic Press, New York, 1981).

[184] PUTNAM CR. *Commutation Properties of Hilbert Space Operators and Related Topics* (Springer-Verlag, New York, 1967).

[185] RAYMER MG. *Measuring the quantum mechanical wave function*. Contemporary Physics, **38**, 343–355 (1997).

[186] REED M & SIMON B. *Methods of Modern Mathematical Physics, I-IV* (Academic Press, New York, 1972,1975,1979,1978).

[187] RIESZ F & SZ.-NAGY B. *Functional Analysis* (F. Ungar, New York, 1955).

[188] ROBERTS JE. *Rigged Hilbert spaces in quantum mechanics*. Commun Math Phys, **3**, 98–109 (1966). *

[189] —. *The Dirac bra and ket formalism*. J Math Phys, **7**, 1097–1104 (1966). *

[190] ROBERTSON AP & ROBERTSON WJ. *Topological Vector Spaces* (Cambridge U. P., Cambridge, 1964).

[191] ROBERTSON HP. *The uncertainty principle*. Phys Rev, **34**, 163–164 (1929).

[192] —. *A general formulation of the uncertainty principle and its classical interpretation*. Phys Rev, **35**, 667 (1930).

[193] —. *An indeterminacy relation for several observables and its classical interpretation*. Phys Rev, **46**, 794–801 (1934).

[194] ROEPSTORFF G. *Path Integral Approach to Quantum Physics* (Springer-Verlag, Berlin, 1994).

[195] ROSE ME. *Elementary Theory of Angular Momentum* (John Wiley, New York, 1957).

[196] ROSENBLOOM M & ROVNYAK J. *Hardy Classes and Operator Theory* (Clarendon, Oxford, 1985).

[197] ROYER A. *Hermitian phase operators for the quantum harmonic oscillator* (1993). Preprint *Ecole Polytechnique, Montréal.*

[198] —. *Phase states and phase operators for the quantum harmonic oscillator*. Phys Rev A, **53**, 70–108 (1996).

[199] RUELLE D. *Statistical Mechanics* (Benjamin, New York, 1969).

[200] SAKAI S. *C^*-Algebras and W^*-Algebras* (Springer Verlag, Berlin, 1971).

[201] SCHAEFER HH. *Topological Vector Spaces* (Springer-Verlag, New York, 1971).

[202] SCHMIDT E. *Entwicklung willkürlicher Funktionen nach Systemen vorgeschreibener*. Math Annalen, **63**, 433–476 (1907).

[203] SCHMUEDGEN K. *Unbounded Operator Algebras and Representation Theory* (Akademie-Verlag, Berlin, 1990).

[204] SCHRÖDINGER E. *Quantisierung als Eigenwertproblem*. Ann der Phys, **79**, 361–376 (1926).

[205] —. *Die gegenwërtide Situation in der Quantenmechanik*. Naturwissenschaften, **23**, 807–812; 823–828; 844–849 (1935).

[206] SCHULMAN L. *Techniques and Applications of Path Integration* (John Wiley & sons, New York, 1981).

[207] SCHWARTZ L. *Théorie des Distributions, I, II* (Hermann, Paris, 1957, 1959).

[208] SCHWEBER S. *Relativistic Quantum Field Theory* (Row, Peterson and Co, Evanston, 1961).

[209] SEGAL IE. *Mathematical characterization of the physical vacuum*. Ill J Math, **6**, 500–523 (1962).

[210] SEWELL GL. *Quantum Theory of Collective Phenomena* (Oxford at the Clarendon Press, New York, 1986).

[211] SIMON B. *Trace Ideals and their Applications* (Cambridge U. P., Cambridge, 1979).

[212] SMITH TB. *Correspondence limit in the Wigner-Weyl picture*. Physics Letters, **95A**, 219–222 (1983).

[213] SMITH TB, DUBIN DA & HENNINGS MA. *The Weyl quantization of phase angle*. J Mod Optics, **39**, 1603–1608 (1992).

[214] STERNBERG S. *Lectures on Differential Geometry* (Prentice Hall, Englewood Cliffs, N. J., 1964).

[215] STREATER RF & WIGHTMAN AS. *PCT, Spin & Statistics, and All That* (W. A. Benjamin, New York, 1964).

[216] STROCCHI F. *General Properties of Quantum Field Theory* (World Scientific, Singapore, 1993).

[217] STROCCHI F & WIGHTMAN AS. *Proof of the charge superselection rule in local relativistic field theory*. J Math Phys, **15**, 2198–2224 (1974).

[218] Susskind L & Glogower J. *Quantum mechanical phase and time operator*. J Phys, **1**, 49–61 (1964).

[219] Taylor AE. *Introduction to Functional Analysis* (Wiley, New York, 1957).

[220] Thangavelu S. *Lectures on Hermite and Laguerre Expansions* (Princeton U. P., Princeton, NJ, 1993).

[221] Thirring W. *A Course in Mathematical Physics, I-IV* (Springer-Verlag, New York, 1979).

[222] Tolman RC. *The Principles of Statistical Mechanics* (Oxford University press, Oxford, 1938), 1st edition.

[223] Tomonaga SI. *Quantum Mechanics, I: Old Quantum Theory* (North-Holland, Amsterdam, 1968).

[224] Trèves F. *Topological Vector Spaces, Distributions and Kernels* (Academic press, New York, 1967).

[225] Turski LA. *The velocity operator for many-boson systems*. Physica, **57**, 432–444 (1972).

[226] Vaccaro JA, Pegg DT & Barnett SM. *The problem with "The problem of the Barnett-Pegg phase operator"*. Phys Lett A, **262**, 483–485 (1999).

[227] Vanheuverzwijn P. *Generators for quasi-free completely positive semi-groups*. Ann Inst H Poincare, **A 29**, 1211–138 (1978).

[228] —. *Errata to: Generators for quasi-free completely positive semi-groups*. Ann Inst H Poincare, **A 30**, 83 (1979).

[229] Varadarajan VS. *Geometry of Quantum Theory I,II* (Van Nostrand, Princeton, NJ, 1968, 1970).

[230] von Neumann J. *The Mathematical Foundations of Quantum Mechanics* (Princeton University Press, Princeton, NJ, 1955).

[231] Vorontsov YI & Rembovsky YA. *One more lookback at the problem of Pegg-Barnett's phase operator*. Phys Lett A, **262**, 486–488 (1999).

[232] —. *The problem of the Pegg-Barnett phase operator*. Phys Lett A, **254**, 7–10 (1999).

[233] Wallace PR. *Paradox Lost* (Springer-Verlag, New York, 1996).

[234] Webster R. *Convexity* (Oxford UP, Oxford, 1994).

[235] Weyl H. *Quantenmechanik und Gruppentheorie*. Z Physik, **46**, 1–46 (1927).

[236] —. *The Theory of Groups and Quantum Mechanics* (Dover, New York, 1930).

[237] Wheeler JA & Zurek WH. *Quantum Theory and Measurement* (Princeton University Press, Princeton, NJ, 1983).

[238] Whittaker ET. *Analytical Dynamics of Particles and Rigid Bodies* (Cambridge UP, Cambridge, 1964).

[239] Whittaker ET & Watson GN. *A Course of Modern Analysis* (Cambridge UP, Cambridge, 1961), 4th edition.

[240] Wiener N. *The Fourier Integral and Certain of its Applications* (Dover, New York, 1933).

[241] WIGNER EP. *On the quantum correction for thermodynamic equilibrium.* Phys Rev, **40**, 749–759 (1932).

[242] WILANSKY A. *Modern Methods in Topological Vector Spaces* (McGraw-Hill, New York, 1978).

[243] WOODHOUSE N. *Geometric Quantization* (Clarendon, Oxford, 1980).

[244] ZWANZIG R. *Statistical mechanics of irreversibility.* In *Lectures in Theoretical Physics, III*, editors BRITTIN WE, DOWNS BW & J. DOWNS E, Boulder Summer School in Theoretical Physics, 106–141 (Interscience, New York, 1961).

INDEX